Progress in Systems and Control Theory

Volume 25

Series Editor
Christopher I. Byrnes, Washington University

Dynamical Systems, Control, Coding, Computer Vision

New Trends, Interfaces, and Interplay

Giorgio Picci
David S. Gilliam
Editors

Birkhäuser Verlag
Basel · Boston · Berlin

Editors:

Giorgio Picci
Dipartimento di Elettronica e Informatica
Università di Padova
vai Gradenigo, 6/A
35131 Padova
Italia

David S. Gilliam
Department of Mathematics and Statistics
Texas Tech University
Box 41042
Lubbock, TX 79409-1042
USA

1991 Mathematics Subject Classification 93-06, 93-02

A CIP catalogue record for this book is available from the Library of Congress, Washington D.C., USA

Deutsche Bibliothek Cataloging-in-Publication Data

Dynamical systems, control, coding, computer vision : new trends, interfaces and interplay / Giorgio Picci ; David S. Gilliam, ed. – Basel ; Boston ; Berlin : Birkhäuser, 1999
(Progress in systems and control theory ; Vol. 25)
ISBN-13:978-3-0348-9848-5 e-ISBN-13: 978-3-0348-8970-4
DOI: 10.1007/978-3-0348-8970-4

Softcover reprint of the hardcover Ist edition 1999

Printed on acid-free paper produced of chlorine-free pulp. TCF ∞
ISBN-13:978-3-0348-9848-5

9 8 7 6 5 4 3 2 1

Contents

Preface

This book is a collection of essays devoted in part to new research directions in systems, networks, and control theory, and in part to the growing interaction of these disciplines with new sectors of engineering and applied sciences like coding, computer vision, and hybrid systems. These are new areas of rapid growth and of increasing importance in modern technology.

The essays, written by world-leading experts in the field, reproduce and expand the plenary and minicourse/minisymposia invited lectures which were delivered at the Mathematical Theory of Networks and Systems Symposium (MTNS-98), held in Padova, Italy, on July 6-10, 1998.

Systems, control, and networks theory has permeated the development of much of present day technology. The impact has been visible in the past fifty years through the dramatic expansion and achievements of the aerospace and avionics industry, through process control and factory automation, robotics, communication signals analysis and synthesis, and, more recently, even finance, to name just the most visible applications. The theory has developed from the early phase of its history when the basic tools were elementary complex analysis, Laplace transform, and linear differential equations, to present day, where the mathematics ranges widely from functional analysis, PDE's, abstract algebra, stochastic processes and differential geometry. Irrespective of the particular tools, however, the basic unifying paradigms of feedback, stability, optimal control, and recursive filtering, have remained the bulk of the field and continue to be the basic motivation for the theory, coming from the real world.

A peculiar aspect of the disciplines represented in this book (and of the MTNS symposium in particular) is the symbiosis of high-level mathematics with engineering and applied research. This symbiosis is the result of a tradition of collaboration and cross-fertilization between mathematicians and engineers, motivated by the interest in the analysis and synthesis of dynamical systems. In the early years the motivations came from circuit, amplifiers, and servomechanism analysis and design. Now the applications of systems and control theory and statistical filtering are far more sophisticated, ranging from precision control of complicated flexible structures for space or robotics applications, to guidance and path following of unmanned vehicles and autonomous robots operating in an unknown environment.

Here the communication aspects, and the new sensors, actuators, and computer technology are naturally playing a major role. To mention one example, internet is recently being considered as a communication channel for carrying the feedback and the remote sensing signals in certain distributed control and dynamic decision making processes. On a different side, dynamic vision and vision-in-the-loop offer now a concrete possibility of building autonomous navigation and control systems capable of learning on-line the shape of the environment in which they operate and to accomplish tasks via path-planning and obstacle avoidance in an unknown environment ("intelligent" behaviour).

The theory and the underlying mathematics is adapting to the new problems and undergoing a visible change in emphasis and sophistication. The scientific community working in this area is accordingly showing a peculiar ability to embrace and lead paradigm shifts. This book is in part a reflection of these changes and a proposal of possible new directions and new paradigms to cope with the latest technological advances of this era. In this spirit, we wish to thank the authors and all the contributors to MTNS-98, for making this conference an outstanding scientific and intellectual event.

Besides the numerous people who helped organizing MTNS98, it is a pleasure to thank here, in particular, Francesca Bettini, for the dedication and effort spent in editing, debugging, and putting together the final version of the manuscript of this book.

Giorgio Picci
Padova, Italy

David Gilliam
Lubbock, TX USA

Contributors List

B.D.O. Anderson Research School of Information Sciences and Engineering, The Australian National University, Canberra ACT 0200 Australia

D. Z. Arov South–Ukrainian State Pedagogical University, Odessa, Ukraine

A. Balluchi PARADES G.E.I.E., Via San Pantaleo, 66, 00186 Roma, Italy

P. Benner Zentrum für Technomathematik, Fachbereich 3/Mathematik und Informatik, Universität Bremen, D-28334 Bremen, FRG

R. Byers Department of Mathematics, University of Kansas, Lawrence, KS 66045-2142, USA

R. Cipolla Department of Engineering, University of Cambridge, Cambridge, UK, CB2 1PZ

M. Clerc CERMICS, Ecole Nationale des Ponts et Chaussées, 77455 Marne-la-Vallée Cedex 2, France

F. Dabbene CENS-CNR, Politecnico di Torino, Corso Duca degli Abruzzi 24, 10129 Torino, Italy

M.D. Di Benedetto Dipartimento di Ingegneria Elettrica, Università dell'Aquila, Poggio di Roio, 67040 L'Aquila, Italy

M. Fliess Centre de Mathématiques et de Leurs Applications, École Normale Supérieure de Cachan, 61 avenue du Président Wilson, 94235 Cachan, France.

G.D. Forney, Jr. Laboratory of Information and Decision Systems, MIT, Cambridge, MA 02139, USA

P.A. Fuhrmann Department of Mathematics, Ben-Gurion University of the Negev, Beer Sheva, Israel

I. Gohberg School of Mathematical Sciences, Raymond and Beverly Sackler Faculty of Exact Sciences, Tel-Aviv University, 69978 Tel-Aviv, Israel

A. Isidori Department of System Science and Mathematics, Washington University, St. Louis, MO 63130, USA; Dipartimento di Informatica e Sistemistica, Universita di Roma "La Sapienza", 00184 Roma, Italy

M.A. Kaashoek Department of Mathematics, Faculty of Sciences, Vrije Universiteit, De Boelelaan 1081a, 1081 HV Amsterdam, The Netherlands

H. Kimura Department of Mathematical Engineering and Information Physics, Faculty of Engineering, University of Tokyo, 7-3-1 Hongo Bunkyo-ku, Tokio 113, Japan

M. Kuijper Department of Electrical and Electronic Engineering, University of Melbourne, Parkville, Victoria 3052, Australia

P. Lancaster Department of Mathematics and Statistics, University of Calgary, Calgary, AB T2N 1N4, Canada

J. Malik EECS, Computer Science Division, University of California at Berkeley, Berkeley, CA 94720-1776, USA

S. Mallat Centre de Mathématiques Appliquées, Ecole Polytechnique, 91128 Palaiseau Cedex, France

C.F. Martin Department of Mathematics, Texas Tech University, Lubbock, Tx 79409, USA

V. Mehrmann Fakultät für Mathematik, TU Chemnitz, D-09107 Chemnitz, FRG.

P.R.S. Mendonça Department of Engineering, University of Cambridge, Cambridge, UK, CB2 1PZ

H. Mounier Département AXIS, Institut d'Électronique Fondamentale, Bâtiment 220, Université Paris-Sud, 91405 Orsay, France

P. Perona Electrical Engineering Dept., California Institute of Technology Pasadena, California 91125, USA; Dip. di Elettronica ed Informatica, Università di Padova, 35131 Padova, Italy

C. Pinello PARADES G.E.I.E., Via San Pantaleo, 66, 00186 Roma, Italy

J. Rosenthal Department of Mathematics, University of Notre Dame, Notre Dame, Indiana 46556, USA

A. Sangiovanni–Vincentelli Department of Electrical Engineering and Computer Science, University of California at Berkeley, CA 94720, USA; PARADES G.E.I.E., Via San Pantaleo, 66, 00186 Roma, Italy

G. Sapiro Electrical and Computer Eng., University of Minnesota, Minneapolis, MN 55455, USA

L. Schovanec Department of Mathematics, Texas Tech University, Lubbock, Tx 79409, USA

S. Soatto Department of Electrical Engineering, Washington University, One Brookings dr., St. Louis - MO 63130, USA; Dipartimento di Matematica ed Informatica, Università di Udine, 33100 Udine, Italy

E.D. Sontag Department of Mathematics, Rutgers University, New Brunswick, NJ 08903, USA

R. Tempo CENS-CNR, Politecnico di Torino, Corso Duca degli Abruzzi 24, 10129 Torino, Italy

P.C. Teo Computer Science, Stanford University, Stanford, CA 94305, USA

J.G. Thistle Department of Electrical and Computer Engineering, École Polytechnique de Montréal, C.P. 6079, succ. Centre-ville, Montreal, Quebec, Canada H3C 3A7

M.E. Valcher Dip. di Elettronica ed Informatica, Univ. di Padova, Via Gradenigo 6a, 35131 Padova, Italy

B.A. Wandell Psychology and Neuroscience, Stanford University, Stanford, CA 94305, USA

J.C. Willems Research Institute for Mathematics and Computing Science, University of Groningen, P.O. Box 800, 9700 AV Groningen, The Netherlands

H. Xu Fakultät für Mathematik, TU Chemnitz, D-09107 Chemnitz, FRG

Progress in Systems and Control Theory, Vol. 25

Riccati Equations, Network Theory and Brune Synthesis: Old Solutions for Contemporary Problems

Brian D.O. Anderson[1]

1 Introduction

Riccati equations have a natural connection with network theory. Classical passive network synthesis procedures, employing often frequency domain spectral factorization, can be mirrored by state variable procedures which rely on knowledge of a steady state Riccati equation solution. In contrast to many occurrences of steady state Riccati equations, it is possible (especially in network applications) to encounter equations which have strong, but not stabilizing solutions. Such equations constitute a problem for much software. Classical network synthesis procedures actually dealt with a frequency domain version of this problem using tools such as the Brune synthesis. The Riccati equation equivalent involves a deflation technique which will be exposed.

The aim of this paper is to expose certain connections between network theory and Riccati equations. Our special focus is on Riccati equations where there exist strong but not stabilizing solutions. These concepts are defined later; suffice it to say here that the "closed—loop" system matrix has eigenvalues in the closed but not open left half plane. In terms of spectral factorizations, one obtains spectral factors which have zeros on the imaginary axis.

In [1], a number of connections between Riccati differential equations, steady state Riccati equations and passive networks are explored. Given a positive real transfer function matrix $Z(s)$, it is a result of classical network theory [4, 7, 8] that there exists a spectral factorization

$$Z(s) + Z^T(-s) = W^T(-s)W(s) \tag{1.1}$$

where $W(s)$ is stable and has all zeros in $\text{Re}[s] \leq 0$ and sometimes $\text{Re}[s] < 0$. This result has its parallel in the solution of certain Riccati equations defined using a state—variable realization of $Z(s)$. Moreover, just as the factorizations (1.1) plays a major role in defining synthesis procedures (i.e. procedures for defining an interconnection of resistors, capacitors, inducators etc whose impedance is a prescribed $Z(s)$), so the solution of a Riccati

[1]The authors acknowledge funding of activities of the Cooperative Research Centre for Robust and Adaptive Systems by the Australian Commonwealth Government under the Cooperative Research Centres Program, funding of this research by the US Army Research Office, Far East, the Office of Naval Research, Washington.

equation associated with a state variable realization of $Z(s)$ can be used as a basis for synthesis.

Exploring these ideas is not the focus of this paper (though some are reviewed in the next section). Rather, we focus on a contemporary problem in the use of Riccati equations, that of understanding how one might solve an equation where a strong but not stabilizing solution exists. Such equations typically cause problems for software, but can arise in, for example, characterizing a transfer function which just achieves, rather than falls arbitrarily short of, a prescribed gain.

We shall explain how a deflation process can assist in solving such equations. Moreover, we motivate the process by an old procedure of network synthesis—the Brune synthesis—where the equivalent spectral factorization problem (requiring a spectral factor with imaginary axis zeros) is deliberately contrived, in order to allow a deflation type of solution to the synthesis problem.

Section 2 of the paper reviews the connection between Riccati equations and linear passive multiport networks. Section 3 characterizes some situations leading to Riccati equations with strong but not stabilizing solutions, and reviews (in outline form) a state—variable version of the Bruce synthesis. This motivates the development in Section 4 of a deflation process for solving what we term "difficult" Riccati equations. There is of course other literature on such equations. We note especially [10, 11, 12]. Section 5 offers brief concluding remarks.

2 Riccati Equations and Linear Passive Multiport Networks

Consider a p-port network with input current vector $u \in R^p$, and port voltage vector $v \in R^p$, with the network consisting of an interconnection of a finite number of time—invariant passive resistors, capacitors, inductors, transformers and gyrators. Suppose that an impedance function $Z(s)$ is well defined with bounded u leading to bounded y. Under these circumstances

$$Z(s) = H^T(sI - F)^{-1}G + J \tag{2.2}$$

for some minimal quadruple $\{F, G, H, J\}$. In writing (2.2), we make no claim that the implied state vector necessarily has entries comprising inductor currents and/or capacitor voltages.

It is well known that $Z(s)$ satisfies several properties, collectively known as the *positive real property.* These are (with * denoting transpose complex conjugate).

$$Z(s) \text{ is analytic in } \mathrm{Re}[s] > 0 \tag{2.3}$$

$$Z(s) + Z^*(s) \geq 0 \text{ for } \mathrm{Re}[s] > 0 \tag{2.4}$$

Equivalently,

$$Z(s) \text{ is analytic in } \mathrm{Re}[s] > 0 \tag{2.5}$$

$$\begin{array}{c}\text{Any pole of } Z(s) \text{ on } \mathrm{Re}[s] = 0 \text{ is simple, with residue} \\ \text{matrix that is nonnegative hermitian}\end{array} \tag{2.6}$$

$$\begin{array}{l} Z(j\omega) + Z^T(-j\omega) \geq 0 \text{ for all real } \omega \\ (\text{provided } j\omega \text{ is not a pole of } Z(s)) \end{array} \tag{2.7}$$

There are implications of these conditions for the state—variable description of $Z(s)$. For example, (2.7) implies (letting $\omega \to \infty$) that

$$J + J^T \geq 0 \tag{2.8}$$

The passivity of the network means that if the network is initially unexcited, and an arbitrary current is injected starting at time 0, then the energy absorbed by the network over an interval $[0, t_1]$ for arbitrary t_1 is necessarily nonnegative. More precisely, with

$$\dot{x} = Fx + Gu \quad x(0) = 0 \tag{2.9}$$

there holds (with $R = J + J^T$)

$$\int_0^{t_1} u^T y dt = \frac{1}{2} \int_0^{t_1} (u^T R u + 2x^T H u) dt \tag{2.10}$$

for all $u(\cdot)$. Equivalently, if the network is initially storing energy ($x(0) = x_0 \neq 0$), and an arbitrary $u(\cdot)$ is applied starting at time 0, there is an upper bound to the energy which can be extracted from the network over any interval, i.e. with

$$\dot{x} = Fx + Gu \quad x(0) = x_0 \tag{2.11}$$

there holds for all u

$$-\int_0^{t_1} u^T y dt \leq \alpha(x_0) \tag{2.12}$$

Here, $\alpha(x_0)$ is a nonnegative function of x_0, representing the initially stored energy. Equivalently,

$$V(x_0, u(\cdot), t_1) = \int_0^{t_1} u^T y = \frac{1}{2} \int_0^{t_1} (u^T R u + 2x^T H u) dt \tag{2.13}$$

is bounded below, independently of $u(\cdot)$ and t_1, by some function of x_0.

2.1 A Variational Problem

For each x_0, one can ask the question: by manipulating $u(\cdot)$, how can we extract the maximum energy from the network? More precisely, we ask: for the system (2.11) with performance index (2.13), how do we choose $u(\cdot)$ and t_1 to minimize the index?

Before addressing the solution, let us make several observtions.

(a) the problem makes sense precisely because of passivity, i.e. because $V(x_0, u(\cdot), t_1)$ is known to be bounded below independently of $u(\cdot)$ and t_1;

(b) consider two indices, $V(x_0, u(\cdot), t_1)$ and $V(x_0, u(\cdot), t_2)$ where $t_2 > t_1$. By choosing $u(t) = 0$ over $[t_1, t_2]$, the second index can always be made to equal the first. Hence

$$\inf_{u(\cdot)} V(x_0, u(\cdot), t_1)$$

for fixed x_0 is monotone decreasing in t_1.

(c) in the light of (2.12), and since the least value of $V(x_0, u(\cdot), t_1)$ obtainable by choosing $u(\cdot)$ is no greater than the value obtained with $u(\cdot) \equiv 0$, there holds

$$-\alpha(x_0) \leq \inf_{u(\cdot)} V(x_0, u(\cdot), t_1) \leq 0 \tag{2.14}$$

(d) it proves useful to examine both the case of arbitrary but fixed finite t_1, and the case $t_1 = \infty$. Then one can consider optimizing over t_1. In the light of (b), it will be no surprise that the case $t_1 = \infty$ delivers the infimum over all $u(\cdot)$ and t_1 of $V(x_0, u(\cdot), t_1)$.

(e) If $R = \frac{1}{2}(J + J^T)$ is singular, there is in general no smooth $u(\cdot)$ giving the desired minimum; if R is nonsingular, there is a smooth $u(\cdot)$ (these facts are not apriori obvious).

2.2 Solution to the Finite Interval Variational Problem

Let us make the assumption

$$R \text{ is nonsingular} \tag{2.15}$$

Then for fixed t_1, we have

$$V^*(x_0, t_1) = \inf_{u(\cdot)} V(x_0, u(\cdot), t_1) = x_0^T \Pi(0, t_1) x_0 \tag{2.16}$$

where $\Pi(., t_1)$ is the symmetric matrix satifying the Riccati differential equation

$$-\dot{\Pi} = \Pi(F - GR^{-1}H^T) + (F - GR^{-1}H^T)^T \\ \times \Pi - \Pi GR^{-1}G^T\Pi - HR^{-1}H^T \tag{2.17}$$

with boundary condition $\Pi(t_1, t_1) = 0$. The associated optimum $u(\cdot)$ is given in feedback form as

$$u^*(t) = -R^{-1}[G^T\Pi(t, t_1) + H^T]x(t) \tag{2.18}$$

so that the state trajectories of the network evolve according to

$$\dot{x} = [F - GR^{-1}H^T - GR^{-1}G^T\Pi(t, t_1)]x \tag{2.19}$$

The proof of this result, see eg. [1], uses fairly standard ideas of linear—quadratic optimization. It can also be established by a first principles, completion-of-square argument.

Note that (2.17) could only give trouble if it had a finite escape time. A minor variation on the argument leading to (2.14) shows that

$$-\alpha(x(t)) \leq x^T(t)\Pi(t, t_1)x(t) \leq 0 \tag{2.20}$$

It is not difficult to show that because this holds for all $x(t), \Pi(t, t_1)$ must have entries which are bounded independently of t and t_1, thus negating the possibility of an escape time .

From (2.14) and (2.16), it is clear that $\Pi(0, t_1) \leq 0$. We can in fact establish strict negative definiteness. Linear—quadratic variational theory actually yields that the minimizing control is unique. If there were to hold $x_0^T\Pi(0, t_1)x_0 = 0$ for some $x_0, u(t) \equiv 0$ would give this value for the index, and then the uniqueness argument would imply that

$$[G^T\Pi(t, t_1) + H^T]x(t) = 0 \quad \forall t \in [0, t_1]$$

Then (2.17) would yield (on right multiplication by x)

$$-\dot{\Pi}x = \Pi Fx + F^T\Pi x$$

or, since now $\dot{x} = Fx$ (with $u \equiv 0$),

$$\frac{d}{dx}(\Pi x) = -F^T(\Pi x)$$

At $t = t_1$, $\Pi x = 0$. Hence $\Pi(t, t_1)x(t) = 0 \ \forall t \in [0, t_1]$. Then $H^Tx(t) = 0 \ \forall t \in [0, t_1]$. Together with $\dot{x} = Fx$ on $[0, t_1]$, observability implies $x_0 = 0$.

For later reference, we record the fact that because $\inf_{u(\cdot)} V(x_0, u(\cdot), t_1)$ is monotone decreasing in t_1, the matrix $\Pi(0, t_1)$ is monotone decreasing in t_1, i.e.
$\Pi(0, t_1) \geq \Pi(0, t_2)$ for $t_2 \geq t_1$.

2.3 The Limiting Properties of $\Pi(t, t_1)$

With $\Pi(t, t_1)$ as defined above, it is trivial to see that

$$\Pi(0, t_1 - t) = \Pi(t, t_1) \tag{2.21}$$

which is a kind of shift—invariance property. Also, as commented earlier,

$$\inf_{u(\cdot)} V(x_0, u(\cdot), t_1) = x_0^T \Pi(0, t_1) x_0 \geq -\alpha(x_0) \tag{2.22}$$

is monotone decreasing in t_1.

Suppose $x_0 \in R^n$. By choosing $x_0 = e_1, e_2, \ldots, e_n$, we see that

$$\lim_{t_1 \to \infty} \Pi_{ii}(0, t_1) = \bar{\Pi}_{ii} \tag{2.23}$$

exists, and then by choosing $x_0 = (e_i + e_j)$ for all $i \neq j$ we see that

$$\lim_{t_1 \to \infty} \Pi_{ij}(0, t_1) = \bar{\Pi}_{ij} \tag{2.24}$$

exists. So there exists

$$\begin{aligned} \bar{\Pi} &= \lim_{t_1 \to \infty} \Pi(0, t_1) & (2.25) \\ &= \lim_{t_1 \to \infty} \Pi(0, t_1 - t) \quad \text{for fixed } t \\ &= \lim_{t \to -\infty} \Pi(0, t_1 - t) \\ &= \lim_{t \to -\infty} \Pi(t, t_1) \text{ by (2.21)} & (2.26) \end{aligned}$$

Thus $\bar{\Pi}$ is the limit as $t \to -\infty$ of solutions of (2.17).

With a little more work, one can prove that $\bar{\Pi}$ is a steady state solution of (2.17) i.e.

$$\begin{aligned} 0 = \bar{\Pi}(F - GR^{-1}H^T) + (F - GR^{-1}H^T)^T \\ \times \bar{\Pi} - \bar{\Pi} G R^{-1} G^T \bar{\Pi} - H R^{-1} H^T \end{aligned} \tag{2.27}$$

Obviously, since $\Pi(0, t_1) < 0$ and $\bar{\Pi} \leq \Pi(0, t_1)$, there holds

$$\bar{\Pi} < 0 \tag{2.28}$$

This can also be obtained directly by using the observability of (F, H) and (2.27).

An obvious question now is whether one can simply take limits in the equation for $u^*(\cdot)$ and the associated closed—loop trajectory, to assert for the the infinite time problem that

$$u^*(t) = -R^{-1}[G^T \bar{\Pi} + H^T] x(t) \tag{2.29}$$

and

$$\dot{x} = [F - GR^{-1}(G^T\bar{\Pi} + H^T)]x \tag{2.30}$$

The answer is not always. What can go wrong is the following. The optimal index $x_0^T\Pi(0,t_1)x_0$ is made up of the sum of two terms, viz

$$\frac{1}{2}\int_0^{t_1} u^{*T}Ru^*\,dt$$

and

$$\int_0^{t_1} x^T Hu^*\,dt$$

with $x(\cdot)$ computed along (2.19). The first term is always positive [if $x_0 \neq 0$], and the second necessarily negative, since $x_0^T\Pi(0,t_1)x_0 < 0$. When $t_1 \to \infty$, it may happen that the first term diverges to $+\infty$ and the second term to $-\infty$, even though their sum remains well behaved. If this divergence occurs, (2.29) and (2.30) come into question.

2.4 Spectral Factorization

Suppose we start with a positive real $Z(s)$ and associated realizing quadruple $\{F, G, H, J\}$, where $J + J^T = R > 0$. Let X be *any* solution of the same equation as satisfied by $\bar{\Pi}$, viz

$$\begin{aligned} 0 = X(F - GR^{-1}H^T) + (F - GR^{-1}H^T) \\ \times X - XGR^{-1}G^TX - HR^{-1}H^T \end{aligned} \tag{2.31}$$

and define

$$W(s) = R^{\frac{1}{2}} + R^{-\frac{1}{2}}(XG + H)^T(sI - F)^{-1}G \tag{2.32}$$

Then via algebraic mainpulation, one can show that

$$Z(s) + Z^T(-s) = W^T(-s)W(s) \tag{2.33}$$

Thus steady state solutions of (2.29) correspond to *spectral factors* $W(s)$ of $Z(s) + Z^T(-s)$, with the same matrices F and G occurring in the state—variable realization of $W(s)$ as occur in the state—variable realization of $Z(s)$.

Classical network synthesis of a positive real $Z(s)$ typically exploited spectral factors $W(s)$. State-variable approaches to network synthesis define the synthesising network in terms of F, G, H, J and a solution X to (2.31). [In fact, solutions to an inequality replacing (2.31) also will yield syntheses, with a nonminimal number of resistors, and all the syntheses obtained this way have a minimal number of reactive elements.]

2.5 Isolating the Riccati Equation Solution and Spectral Factor

For those who are well—versed in linear quadratic theory, it will be no surprise that there is interest in characterizing that particular solution $\bar{\Pi}$ of the steady state Riccati equation (2.31) obtained as the limit of the transient Riccati differential equation (2.17), and in particular using a characterization involving the eigenvalues of the limiting "closed—loop" matrix [see (2.30)], which is $F - GR^{-1}(G^T\bar{\Pi} + H^T)$. A key result is the following:

Theorem 2.1. *Let Z(s) with minimal realization $\{F, G, H, J\}$ be positive real, with $J + J^T > 0$. Let $\Pi(t, t_1)$ solve the Riccati differential equation (2.17) and let $\bar{\Pi} = \lim_{t \to -\infty} \Pi(t, t_1)$. Define*

$$\bar{F} = F - GR^{-1}(G^T\bar{\Pi} + H^T) \tag{2.34}$$

and $\bar{W}(s)$ by (2.32) with $X = \bar{\Pi}$, so that (2.33) with $\bar{W}$ replacing W holds. Then the eigenvalues of $\bar{F}$, which are the zeros of $\bar{W}(s)$, satisfy

$$Re\lambda_i(\bar{F}) \leq 0 \tag{2.35}$$

We remark that there is a world of difference between the more common results for linear quadratic poblems of the type $\mathrm{Re}\lambda_i(\bar{F}) < 0$ and the result of this theorem. Different proof techniques are needed. That aside, the crucial conclusion is that $\bar{\Pi}$ defines that particular spectal factor in (2.33) which is minimum phase (but not necessarily strictly so), and is known to be unique to within constant orthogonal matrix multiplication on the left, [2].

Proof. Define $V(t) = \Pi(t, t_1) - \bar{\Pi}$. From (2.17) and (2.27), a little manipulation yields

$$\dot{V} = -V\bar{F} - \bar{F}^T V + VGR^{-1}G^T V \tag{2.36}$$

$$V(t_1) = -\bar{\Pi} \tag{2.37}$$

Further, $V(t) \to 0$ as $t \to -\infty$, and since $\Pi(t, t_1) = \Pi(0, t_1 - t)$ is monotone increasing in t, $V(t)$ is nonnegative definite in t and monotone increasing as $t \to t_1$. Set $Y(t) = V(t_1 - t)$ for $t \geq 0$. Then

$$\dot{Y} = Y\bar{F} + \bar{F}^T Y - YGR^{-1}G^T Y \tag{2.38}$$

$$Y(0) = -\bar{\Pi} \tag{2.39}$$

with $Y(t)$ nonnegative and monotone decreasing to 0 as $t \to \infty$. Now set $Z^{-1}(t) = Y(t)$. Obviously Z is defined near $t = 0$, since $Y(0) = -\bar{\Pi}$ is nonsingular. Also,

$$\dot{Z} = -\bar{F}Z - Z\bar{F}^T + GR^{-1}G^T \tag{2.40}$$

$$Z(0) = -(\bar{\Pi})^{-1} \tag{2.41}$$

□

The linearity of the equation for $Z(t)$ ensures $Z(t)$ is defined for all $t \geq 0$. Further, $Z(t)$ is monotone increasing, with $Z(t)$ tending to infinity in the sense that $\lambda_{\min}[Z(t)] \to \infty$ as $t \to \infty$. [Note that $\lambda_{\min}(Z) = [\lambda_{\max}(Y)]^{-1}$ and the convergence of Y to zero ensures $\lambda_{\max}(Y) \to 0$]

To establish a contradiction, suppose that $\bar{F}$ has an eigenvalue with positive real part. For convenience, suppose it is real and $\bar{F}^T m = \lambda m$ for $m \neq 0$, $\lambda > 0$. Let $z(t) = m^T Z(t) m$. Then, with $n = m^T G R^{-1} G^T m$

$$\dot{z} = -2\lambda z + n \tag{2.42}$$

The solution is bounded for all $t \geq 0$, which contradicts

$$z(t) \geq \lambda_{\min}[Z(t)] m^T m.$$

Hence Re $\lambda_i(\bar{F}) \leq 0$, as required.

In studying the convergence of solution of Riccati differential equations to the steady state solution, it is common to link the rate of convergence (which is usually exponential) to the eigenvalues of the steady state closed—loop matrix.

If indeed, Re$\lambda_i(\bar{F}) < 0$, then one can argue that $\Pi(t, t_1) - \bar{\Pi}$ as $t \to \infty$ converges to zero exponentially fast [the difference being expressible in terms of $\exp(\bar{F}t)$]. What happens if Re$\lambda_i(\bar{F}) = 0$? It is easiest to work with $Y(t) = V(t_1 - t) = \Pi(0, t) - \bar{\Pi}$.

Corollary 2.1. *Adopt the same hypothesis as for Theorem 2.1, and let*

$$\dot{Y} = Y\bar{F} + \bar{F}^T Y - YGR^{-1}G^T Y \tag{2.43}$$

$$Y(0) = -\bar{\Pi} \tag{2.44}$$

Suppose that for some eigenvalue of $\bar{F}$, there holds Re$\lambda_i(\bar{F}) = 0$. Then $Y(t) \to 0$ at rate $0(t^{-1})$.

Proof. For convenience, suppose that the eigenvalue with zero real part is real, i.e. zero. Thus suppose that for $m \neq 0$, there holds $\bar{F}m = 0$. Let

$Z^{-1}(t) = Y(t)$, and $z(t) = m^T Z(t)m$. Then the proof of the theorem (see (2.42)) yields that

$$z(t) = -m^T \bar{\Pi}^{-1} m + tn$$

Now

$$m^T Y(t)mm^T Z(t)m \geq (m^T m)^2$$

so that

$$m^T Y(t)m \geq \frac{(m^T m)^2}{nt - m^T \bar{\Pi}^{-1} m}$$

This shows that $Y(t)$ can converge no faster than at rate $O(t^{-1})$.

□

The fact that $Y(t)$ converges no slower than this rate relies on a lengthy calculation explicitly computing $Y(t)$, when $\bar{F}$ is in Jordan form and will be omitted here.

2.6 Summary to this point

1. Transient Riccati equations can be associated with a minimal realization of a positive real transfer function matrix $Z(s)$ for which $Z(\infty) + Z^T(\infty)$ is nonsingular.

2. The solution of the transient equation approaches a steady state solution of the equation.

3. Steady state solutions to the Riccati equation immediately yield a synthesis of $Z(s)$.

4. The rate of approach of transient equation solution to a steady state is exponential when

$$Z(j\omega) + Z^*(j\omega) > 0$$

for all real ω (apart from $j\omega$ which are poles of $Z(s)$. Otherwise it is $O(1/t)$.

5. The limiting solution of the Riccati equation defines a "closed—loop" system matrix with eigenvalues in the closed left half plane, and a spectral factor $W(s)$ of $Z(s)$ satisfying

$$Z(s) + Z^T(-s) = W^T(-s)W(s)$$

which is minimum phase (and so may have zeros on the imaginary axis).

3 Difficult Riccati Equations

What does the section title refer to? It refers to Riccati equations of the form

$$XA + A^T X - XBB^T X + Q = 0 \tag{3.45}$$

in which the associated Hamiltanian matrix

$$\mathcal{H} = \begin{bmatrix} A & -BB^T \\ -Q & -A^T \end{bmatrix} \tag{3.46}$$

has eigenvalues which are pure imaginary. As is well known, [3] if (3.45) has a stabilizing solution, call it $\bar{X}$, in the sense that

$$\text{Re}\lambda\ _i(A - BB^T\bar{X}) < 0 \tag{3.47}$$

the the eigenvalues of H must split into the disjoint sets $\{\text{Re}\lambda_i(A-BB^T\bar{X}\}$ and $\{-\text{Re}\lambda_i(A-BB^T\bar{X})\}$. and $\bar{X}$ is in fact unique. In these circumstances, much attention has been given to the construction of effective numerical procedures for finding $\bar{X}$, often using eigenspaces associated with these sets.

There are known situations where such an eigenvalue split is not possible.

(a) Recall the ideas of the last section. Suppose that $Z(s)$ defined by a minimal $\{F, G, H, J\}$ with
$J + J^T > 0$ is positive real with

$$Z(j\omega) + Z^T(-j\omega)$$

singular for some real ω. Then the Riccati equation

$$\begin{aligned} X(F - GR^{-1}H^T) + (F - GR^{-1}H^T)^T X - X \\ \times GR^{-1}G^T X - HR^{-1}H^T = 0 \end{aligned} \tag{3.48}$$

has a solution $\bar{\Pi}$ for which

$$\text{Re}\lambda_i(F - GR^{-1}H^T - GR^{-1}G^T\bar{\Pi}) \leq 0 \tag{3.49}$$

and strict inequality is not valid. The Hamiltonian matrix here is

$$\mathcal{H} = \begin{bmatrix} F - GR^{-1}H^T & -GR^{-1}G^T \\ -HR^{-1}H^T & -F^T + HR^{-1}G^T \end{bmatrix} \tag{3.50}$$

(b) Let $S(s)$ be a stable strictly proper transfer function matrix with minimal realization $\{F, G, H\}$ and $S(\infty) = 0$. When does $S(s)$ obey

$$\|S(j\omega)\| \leq \gamma$$

for all real ω? When $\gamma = 1$, this is simply asking the question, when is $S(s)$ a scattering matrix, an old network theory construct [4]. For arbitrary γ, the question has come into much prominence in connection with H_∞ theory. (The question can of course also be considered for $S(\infty) \neq 0$, and the ideas are effectively the same, but algebraically more complicated to explain.) The following is by now a well known result, see eg [5].

Theorem 3.1. *Let $S(s) = H^T(sI - F)^{-1}G$, with the $Re(F) < 0$. Then $\|S(j\omega)\| \leq \gamma$ for all ω if and only if the following equation is solvable:*

$$XF + F^T X + \gamma^{-2} XGG^T X + HH^T = 0 \tag{3.51}$$

This is a Riccati equation. When $\|S(j\omega\| < \gamma$, then and only then there exist a solution $\bar{X}$ which is stabilizing, i.e.

$$\text{Re}\lambda_i \left[F + \gamma^{-2} GG^T \bar{X}\right] < 0 \tag{3.52}$$

Else, we can simply require

$$\text{Re}\lambda_i \left[F + \gamma^{-2} GG^T \bar{X}\right] \leq 0 \tag{3.53}$$

In the first case, $\bar{X}$ is the limit approached exponentially fast by the solution of a Riccati differential equation; also, it can be obtained by any one of a number of numerical algorithms. In the second case, $\bar{X}$ is still the limit of a differential equation solution (but the approach rate is not exponential), and most numerical algorithms give trouble.

(c) In H_∞ theory in order to find a stabilizing controller yielding closed—loop gain less than some prescribed γ, two Riccati equations must have stabilizing solutions, call then $\bar{X}$ and $\bar{Y}$, and the product $\bar{X}\bar{Y}$ must also satisfy a strict inequality on its singular values in terms of γ. If the H_∞ problem is solvable for some γ, then there is an infimum of such values γ. At this infimum, it may be that one of the Riccati equations has a strong but not stabilizing solution.

3.1 Network Theory Approach to Difficult Riccati Equations

There is an old approach to network synthesis which provides one approach to dealing with difficult Riccati equations, the Brune synthesis procedure, [6, 8].

From a classical point of view, what happens is the following. A positive real $Z(s)$ is prescribed, and a synthesis is required, (i.e. one seeks a description of a network of passive elements whose impedance is $Z(s)$). In outline the procedure is as follows.

1. A positive scalar ρ is found so that
$$Z(j\omega) + Z^T(-j\omega) - \rho I \geq 0 \tag{3.54}$$
for all ω, with singularity of the object on the left for some ω_0

2. An orthogonal constant T is found so that with Z replaced by $\bar{Z} = T^T ZT$, there holds
$$\bar{Z}_{11}(j\omega_0) + \bar{Z}_{11}(-j\omega_0) - \rho = 0 \tag{3.55}$$

3. $\bar{Z}(s) - (\rho/2)I$ is now regarded as the entity to synthesise (and from this a synthesis of $Z(s)$ will follow). Note that at $s = j\omega_0$, the real part of the 1-1 term is zero.

4. $\bar{Z}(s) - (\rho/2)I$ is expressed as the impedance of a low complexity (degree 2 in general) $2n$-port network, whose elements are readily computable, cascaded with an n-port of complexity less than that of $\bar{Z}(s) - (\rho/2)I$. (Complexity here means McMillan degree, or order of a minimal state—variable realization).

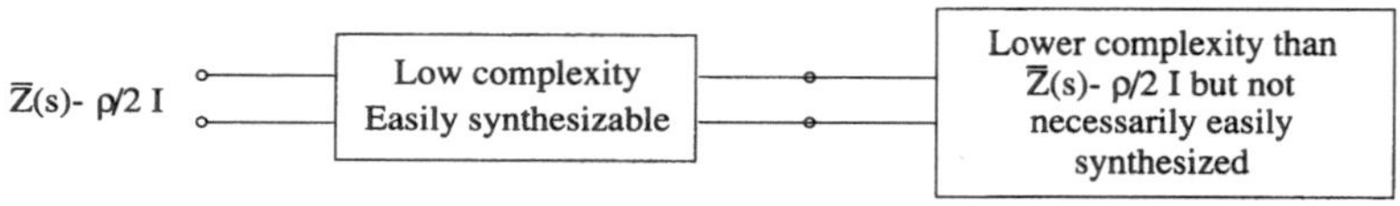

Figure 1: Cascade (Brune) synthesis

One now has the basis for a recursion—what remains to be synthesised after each step is always less complicated than what was there before. Repeating the steps eventually leads to termination of the procedure.

There is a state—variable view of this procedure, set out in [9]. The key to obtaining a synthesis is to solve a Riccati equation. Steps 1, 2 and 3 are the same. In step 4, a coordinate basis change is obtained and in the new coordinate basis, the state vector of $\bar{Z}(s) - (\rho/2)I$ can be partitioned into two subvectors, one corresponding to the easily synthesised, low complexity network, and one to the "lower" complexity but not-necessarily-easily-synthesised network. In the changed coordinate basis, the Riccati equation solution, call it P, is of the form
$$P = \begin{bmatrix} P_1 & 0 \\ 0 & P_2 \end{bmatrix} \tag{3.56}$$
with P_1 a readily computable matrix (in general 2×2,) and P_2 satisfying a lower dimmesion Riccati equation than P. The eigenvalues of the Hamiltonian matrix associated with the P equation are the union of $\pm j\omega_0$, perhaps

repeated, and the eigenvalues of the Hamiltonian matrix associated with the P_2 equation.

Thus in terms of the Riccati equation, *one step of the Brune synthesis procedure corresponds to changing the coordinate basis of the Riccati equation so that part of the solution can be explicitly calculated and pulled out, while the remaining part of the solution satisfies a reduced order Riccati equation, which will be "less" difficult, in that the associated Hamiltonian will have at least two fewer imaginary eigenvalues.*

For scalar positive real transfer functions, some more detail is as follows. Suppose that
$z(s) = \frac{1}{2}R + h_a^T(sI - F_a)^{-1}g_a$ with $\text{Re}z(j\omega_0) = 0$. Let T_a be a nonsingular matrix with last two columns
$(\omega_0^2 I + F_a^2)^{-1}g_a, -F_a(\omega_0^2 I + F_a^2)^{-1}g_a$. Set
$F_b = T_a F_a T_a^{-1}, g_b = T_a g_a, h_b^T = h_a^T T_a^{-1}$. Compute

$$\begin{bmatrix} h_b^T(\omega_0^2 I + F_b^2)^{-1} \\ h_b^T(\omega_0^2 I + F_b^2)^{-1} F_b \end{bmatrix} = \begin{bmatrix} K_{12} & K_{22} \end{bmatrix} \tag{3.57}$$

where K_{22} is 2×2. Set

$$T_b^{-1} = \begin{bmatrix} I & 0 \\ -K_{22}^{-1}K_{12} & I \end{bmatrix} \tag{3.58}$$

where the $2-2$ block entry of T_b^{-1} is 2×2. Also set

$$F_c = T_b F_b T_b^{-1},\ g_c = T_b g_b,\ h_c^T = h_b^T T_b^{-1}.$$

This ensures that for any $z(s)$, whether or not $\text{Re}z(j\omega_0) = 0$, that

$$\begin{bmatrix} (\omega_0^2 I + F_c^2)^{-1} g_c & -F_c(\omega_0^2 I + F_c^2)^{-1} g_c \end{bmatrix} = \begin{bmatrix} 0 & 0 \\ \vdots & \vdots \\ 0 & 0 \\ 1 & 0 \\ 0 & 1 \end{bmatrix} \tag{3.59}$$

and

$$\begin{bmatrix} h_c^T(\omega_0^2 I + F_c^2)^{-1} \\ h_c^T(\omega_0^2 I + F_c^2)^{-1} F_c \end{bmatrix} = \begin{bmatrix} 0 \cdots 0 & \\ & K_{22} \\ 0 \cdots 0 & \end{bmatrix} \tag{3.60}$$

and if Re $z(j\omega_0) = 0$ with $d/ds[\text{Re}Z(s)] = 0$ at $s = j\omega_0$, there also holds

$$K_{22} = \begin{bmatrix} \alpha^2 & 0 \\ 0 & \beta^2 \end{bmatrix} \tag{3.61}$$

for some $\alpha, \beta \neq 0$. All solutions of the Riccati equation

$$\begin{aligned} &P(F - gR^{-1}h^T) + (F - gR^{-1}h^T)^T P - PgR^{-1}g^T \\ &\quad \times P - hR^{-1}h^T = 0 \end{aligned} \tag{3.62}$$

are of the form

$$P = P_1 \dot{+} \begin{bmatrix} \alpha^2 & 0 \\ 0 & \beta^2 \end{bmatrix} \tag{3.63}$$

where P_1 satisfies an equation of smaller dimension:

$$\begin{aligned} &P_1(F_1 - g_1 R^{-1} h_1^T) + (F_1 - g_1 R^{-1} h_1^T)^T P_1 - P_1 g_1 \\ &\quad \times R^{-1} g_1^T P_1 - h_1 R^{-1} h_1^T = 0 \end{aligned}$$

and $z_1(s) = \frac{1}{2}R + h_1^T(sI - F_1)^{-1}g_1$ is a positive real transfer function, of lesser degree than $z(s)$.

4 Deflation and Difficult Riccati Equations

We return to (3.45) and (3.46) repeated for convenience with trivial change of notation as

$$PA + A^T P - PBB^T P + Q = 0 \tag{4.64}$$

$$\mathcal{H} = \begin{bmatrix} A & -BB^T \\ -Q & -A^T \end{bmatrix} \tag{4.65}$$

We shall also use the function

$$\Phi(j\omega) = I + B^T(-j\omega I - A^T)Q(j\omega I - A)^{-1}B \tag{4.66}$$

In the sequel, we shall have occasion to use some tranformations on (4.64) through (4.66). The transformations relate to replacing an optimization problem for $\dot{x} = Ax + Bu$ by one for $\dot{x} = (A - BB^T L)x + Bu$ for some symmetric L. Replacing (4.64) we have

$$\begin{aligned} &P_L(A - BB^T L) + (A - BB^T L)^T P_L - P_L BB^T P_L \\ &\qquad + (Q - LA - A^T L + LBB^T L) = 0 \end{aligned} \tag{4.67}$$

It is trivial to observe that P satisifies (4.64) if and only if $P_L = P - L$ satisfies (4.67). The "closed—loop" matrix in both cases is $A - BB^T P$ and so P is stabilizing or strong for (4.64) if and only if P_L is stabilizing or strong for (4.67). The Hamiltonian matrix associated with (4.67) is

$$\mathcal{H}_L = \begin{bmatrix} A - BB^T L & -BB^T \\ Q - LA - A^T L - LBB^T L & -A^T + BB^T L \end{bmatrix} \tag{4.68}$$

Observe that

$$\mathcal{H}_L = \begin{bmatrix} I & 0 \\ -L & I \end{bmatrix} \mathcal{H} \begin{bmatrix} I & 0 \\ L & I \end{bmatrix} \tag{4.69}$$

and so $\mathcal{H}$ and $\mathcal{H}_L$ have the same eigenvalues. Next, we have

$$\begin{aligned}\Phi_L(j\omega) &= I + B^T(-j\omega I - A^T - LBB^T)^{-1} \\ &\quad \times (Q - LA - AA^T L + LBB^T L) \\ &\quad \times (j\omega I - A - BB^T L)^{-1} B \qquad (4.70)\\ &= [I + B^T(-j\omega I - A^T)^{-1} LB]\Phi(j\omega) \\ &\quad \times [I + B^T L(j\omega I - A)^{-1} B] \qquad (4.71)\end{aligned}$$

Thus at points ω_0 which are not poles or zeros of Φ and Φ_L we conclude that $\Phi(j\omega_0)$ is positive (nonnegative) definite if and only if $\Phi_L(\omega_0)$ is positive (nonnegative)definite.

If (A, B) is stabilizable, it is clear that without loss of generality, we can assume there exists L ensuring that $\Phi_L(j\omega)$ has no poles on the imaginary axis.

In this section, we shall restrict attention to Riccati equation (4.64) *for which* (A, B) *is stabilizable.* Without loss of generality therefore, in the remainder of this section, we shall make the assumption:

Assumption 4.1. *In the Riccati equation* (4.64), *the pair* (A, B) *is stabilizable and* A *has no eigenvalues on the imaginary axis.*

In this case, the following result is well known, see [3].

Theorem 4.1. *Let* A, B, Q *be matrices of compatible dimensions such that* $Q = Q^T$, *and assumption 1 holds. Define* Φ *as in* (4.66). *Then*

1. *The following statements are equivalent*

   (a) $\Phi(j\omega) > 0$ *for all* $0 \leq \omega \leq \infty$

   (b) *There exists a unique stabilizing real solution* $P_s = P_s^T$ *to* (4.64) *i.e.* $Re\lambda_i(A - BB^T P_s) < 0$

   (c) *The Hamiltonian matrix* $\mathcal{H}$ *has no* $j\omega$ *-axis eigenvalue*

2. *The following statements are equivalent*

   (a) $\Phi(j\omega) \geq 0$ *for all* $0 \leq \omega \leq \infty$

   (b) *There exists a unique strong real solution* $P_s = P_s^T$ *to* (4.64), *i.e.* $Re\lambda_i[A - BB^T P_s] \leq 0$

Our principal concern in this section is of course with case 2 in the above theorem.

We now need to record several more results, which are an amalgam and development of ideas of [10] and [11].

Theorem 4.2. *Adopt the hypothesis of Theorem 4.1. Suppose that*

$$\begin{bmatrix} A & -BB^T \\ -Q & -A^T \end{bmatrix} \begin{bmatrix} X \\ Y \end{bmatrix} = \begin{bmatrix} X \\ Y \end{bmatrix} \Lambda \tag{4.72}$$

where the matrices are of compatible dimensions, Λ has dimension no greater than that of A and $\begin{bmatrix} X^T & Y^T \end{bmatrix}^T$ has full column rank. Suppose also that eigenvalues of $-\Lambda^*$ are controllable modes of (A, B), and that $X^*Y = Y^*X$. Then X has full column rank.

Proof. Let real ω_0 be chosen such that
$(j\omega_0 I - A), (j\omega_0 I - \Lambda)$ and $\Phi(j\omega_0)$ are all nonsingular and $\Phi(j\omega_0) > 0$. (This is possible since $\lim_{\omega \to \infty} \Phi(j\omega) = I$, and does not rely on an assumption that
$\Phi(j\omega) \geq 0$ for all ω.) Suppose in order to obtain a contradiction that X does not have full column rank. Then there exists a nonzero α such that $X(j\omega_0 I - \Lambda)\alpha = 0$. From (4.72), we derive

$$(j\omega_0 I - A)X\alpha + BB^T Y\alpha = 0 \tag{4.73}$$

or

$$X\alpha = -(j\omega_0 I - A)^{-1} BB^T Y\alpha \tag{4.74}$$

□

Then from (4.72) we obtain

$$Q(j\omega_0 I - A)^{-1} BB^T Y\alpha - (j\omega_0 I + A^T)Y\alpha = Y(\Lambda - j\omega_0 I)\alpha$$

whence

$$\Phi(j\omega_0)B^T Y\alpha = B^T(-j\omega_0 I - A^T)^{-1} Y(\Lambda - j\omega_0 I)\alpha$$

Premultiplying by $\alpha^* Y^* B^T$ and using (4.73) again gives

$$\begin{aligned} \alpha^* Y^* B^T \Phi(j\omega_0) B^T Y\alpha &= \alpha^* X^* Y(\Lambda - j\omega_0 I)\alpha \\ &= \alpha^* Y^* X(\Lambda - j\omega_0 I)\alpha = 0 \end{aligned}$$

By the positive definiteness of $\Phi(j\omega_0)$, this yields

$$B^T Y\alpha = 0$$

and by (4.74)

$$X\alpha = 0$$

Hence the null space of X is $(j\omega_0 I - \Lambda)^{-1}$ invariant, and so there exists λ and $\beta \neq 0$ such that $X\beta = 0$ and

$$(j\omega_0 I - \Lambda)^{-1}\beta = (j\omega_0 - \lambda)^{-1}\beta$$

whence $\Lambda\beta = \lambda\beta$ and $X(j\omega_0 I - \Lambda)\beta = 0$. Repeating the above argument starting not from $X(j\omega_0 I - \Lambda)\alpha = 0$ but $X(j\omega_0 I - \Lambda)\beta = 0$ yields $B^T Y\beta = 0$. From $-QX - A^T Y = Y\Lambda$ and $\Lambda\beta = \lambda\beta$ follows $-A^T(Y\beta) = \lambda(Y\beta)$. Since $X\beta = 0, \beta \neq 0$ and $\begin{bmatrix} X^T & Y^T \end{bmatrix}^T$ is full rank, $Y\beta \neq 0$. The controllability hypothesis is then contradicted.

In reference to Theorem 4.1, our principal concern in this section is with case 2. Notice that in contrast to case 1, there is no statement (c). This deficiency is in fact remedied in the following theorem, which uses Theorem 4.2 in its proof.

Theorem 4.3. *Adopt the same hypotheses as Theorem 4.1. Then condition 2a and 2b of Theorem 4.1 are equivalent to*

1. *(4.72) holds with square* Λ

2.
$$Re\lambda_i(\Lambda) \leq 0 \tag{4.75}$$

3.
$$\begin{bmatrix} X^T & Y^T \end{bmatrix}^T \text{ is of full column rank} \tag{4.76}$$

4.
$$X^*Y = Y^*X \text{ with } YX^{-1} \text{ real if } X \text{ is nonsingular} \tag{4.77}$$

Proof. Suppose a strong real solution $P_s = P_s^T$ exists to (4.64). Choose $X = I$ and $Y = P_s$ and $\Lambda = A - BB^T P_s$. Then properties (1)–(4) are immediately verified. Conversely, suppose properties (1)–(4) hold. By stabilizability, eigenvalues of $-\Lambda^*$ are controllable modes of (A, B). By Theorem 4.2, X is nonsingular. Define

$$P_s = YX^{-1} \tag{4.78}$$

Then the first row of (4.72) yields $A - BB^T P_s = X\Lambda X^{-1}$ and by (4.75) we obtain that $\mathrm{Re}\lambda_i(A - BB^T P_s) \leq 0$.
Since $A - BB^T P_s = X\Lambda X^{-1}, P_s A - P_s BB^T P_s = Y\Lambda X^{-1}$. The second row of (4.72) yields $-Q - A^T P_s = Y\Lambda X^{-1}$. The Riccati equation is immediate. From (4.77) and (4.78), we obtain that P_s is real and symmetric. □

The next result is more subtle. The motivation for it is as follows. One can regard solution procedures which obtain P_s as picking out eigenspaces of $\mathcal{H}$. If $\mathcal{H}$ has $j\omega$-axis eigenvalues and $\Phi(j\omega) \geq 0$ with singularity at $\omega = \omega_0$, such eigenvalues are always of even multiplicity. If one eigenvector corresponding to $j\omega_0$ is picked out, it is then not a priori clear what eigenvector of the (at least two dimensional eigenspace) corresponding to eigenvalue $-j\omega_0$ should be picked out. (Though it is not obvious, a particular selection does need to be made.)

Theorem 4.4. *Adopt the hypotheses of Theorem 4.1. Suppose that* $\Phi(j\omega) \geq 0$ *and that*

$$\begin{bmatrix} A & -BB^T \\ -Q & -A^T \end{bmatrix} \begin{bmatrix} x \\ y \end{bmatrix} = j\omega_0 \begin{bmatrix} x \\ y \end{bmatrix} \tag{4.79}$$

for some nonzero vector $\begin{bmatrix} x^T & y^T \end{bmatrix}^T$, ω_0 *real and nonzero. Then*

$$x^* y - y^* x = 0 \tag{4.80}$$

If (A, B) *is stabilizable,* $Re x$ *and* $Im x$ *are independent vectors.*

By hypothesis, A has no pure imaginary eigenvalue so $j\omega_0$ is not an eigenvalue of A. Then from (4.79)

$$x = -(j\omega_0 I - A)^{-1} BB^T y \tag{4.81}$$

and

$$y + (-j\omega_0 I - A^T)^{-1} Q (j\omega_0 I - A)^{-1} BB^T y = 0 \tag{4.82}$$

or

$$\Phi(j\omega_0) B^T y = 0$$

Since $\Phi(j\omega_0)$ is nonnegative, it follows that $f(j\omega) = y^* B\Phi(j\omega) B^T y$ has a minimum at $\omega = \omega_0$. Setting the derivative $f'(j\omega_0)$ to zero gives

$$\begin{aligned} & y^* BB^T (-j\omega_0 I - A^T)^{-1} Q (j\omega_0 I - A)^{-2} jBB^T y \\ & + y^* BB^T (-j\omega_0 I - A^T)^{-2} Q (j\omega_0 I - A)^{-1} \\ & \times (-j) BB^T y = 0 \end{aligned}$$

Simplifying with (4.82) gives

$$\begin{aligned} & - y^* (j\omega_0 I - A)^{-1} jBB^T y + y^* BB^T \\ & \times (-j\omega_0 I - A^T)^{-1} jy = 0 \end{aligned}$$

and from (4.81) there follows

$$x^*y - y^*x = 0$$

as required.

Obviously, (4.79) also implies

$$\begin{bmatrix} A & -BB^T \\ -Q & -A^T \end{bmatrix} \begin{bmatrix} \bar{x} \\ \bar{y} \end{bmatrix} = -j\omega_0 \begin{bmatrix} \bar{x} \\ \bar{y} \end{bmatrix}$$

and consequently with $x = u + jv, y = m + jn, (u, v, m, n \text{ real})$

$$\begin{bmatrix} A & -BB^T \\ -Q & -A^T \end{bmatrix} \begin{bmatrix} u & v \\ m & n \end{bmatrix} = \begin{bmatrix} u & v \\ m & n \end{bmatrix} \begin{bmatrix} 0 & -\omega_0 \\ \omega_0 & 0 \end{bmatrix} \tag{4.83}$$

Observe that since $\begin{bmatrix} x^T & y^T \end{bmatrix}^T$ and $\begin{bmatrix} \bar{x}^T & \bar{y}^T \end{bmatrix}^T$ correspond to different eigenvalues, they are independent vectors. Hence the matrix $\left[\begin{smallmatrix} u & v \\ m & n \end{smallmatrix}\right]$ has full column rank. Also, Theorem 4.4 implies that

$$(u^T - jv^T)(m + jn) - (m^T - jn^T)(u + jv) = 0$$

or

$$u^T n = v^T m$$

i.e. we have the following symmetry:

$$\begin{bmatrix} u & v \end{bmatrix}^T \begin{bmatrix} m & n \end{bmatrix} = \begin{bmatrix} m & n \end{bmatrix}^T \begin{bmatrix} u & v \end{bmatrix} \tag{4.84}$$

By Theorem 4.2, and using the stabilizability, it follows that $\left[\begin{smallmatrix} u & v \end{smallmatrix}\right]$ has full column rank.

We now have the basis for deflation of the problem of finding the strong solution P to (4.64), under a stabilizability assumption and nonnegativity of $\Phi(j\omega)$ assumption.

We are of course interested in deflating out the pure imaginary eigenvalues of $\mathcal{H}$.

Our starting point is the matrix of two columns:

$$\begin{bmatrix} X \\ \cdots \\ Y \end{bmatrix} = \begin{bmatrix} u & v \\ \cdots & \cdots \\ m & n \end{bmatrix} \tag{4.85}$$

Let S be a nonsingular matrix such that

$$SX = \begin{bmatrix} I \\ 0 \end{bmatrix} \tag{4.86}$$

Replace (4.64), (4.65) and (4.83) by

$$\begin{aligned}(S^{-T}PS^{-1})SAS^{-1} &+ (SAS^{-1})^T(S^{-T}PS^{-1})\\ &- (S^{-T}PS^{-1})SBB^TS^T(S^{-T}PS^{-1})\\ &+ S^{-T}QS^{-1} = 0\\ \mathcal{H}_s &= \begin{bmatrix} SAS^{-1} & -SBB^TS^T \\ -S^{-T}QS^{-1} & -(SAS^{-1})^T \end{bmatrix}\\ &= \begin{bmatrix} S & 0 \\ 0 & S^{-T} \end{bmatrix} \mathcal{H} \begin{bmatrix} S^{-1} & 0 \\ 0 & S^T \end{bmatrix}\\ H_s \begin{bmatrix} SX \\ S^{-T}Y \end{bmatrix} &= \begin{bmatrix} SX \\ S^{-T}Y \end{bmatrix} \Omega \end{aligned} \tag{4.87}$$

with obvious definition of Ω.

The expression for Φ, if computed for $\bar{A} = SAS^{-1}$, $\bar{B} = SB$ etc remains unchanged. Obviously, all that is involved here is a change of basis.

It is crucial to observe at this point that (4.84) is equivalent to

$$X^TY = Y^TX$$

and accordingly after transformation, where $\bar{X} = SX, \bar{Y} = S^{-T}Y$,

$$\bar{X}^T\bar{Y} = \bar{Y}^T\bar{X}$$

or, with $\bar{Y} = [\, \bar{Y}_1^T \; \bar{Y}_2^T \,]^T$,

$$\bar{Y}_1 = \bar{Y}_1^T$$

Hence there exists a *symmetric* L such that

$$L \begin{bmatrix} I \\ 0 \end{bmatrix} + \bar{Y} = 0 \tag{4.88}$$

with L computable from the original vectors u, v, m, n associated with the eigenvectors of $\mathcal{H}$. If $\bar{Y}_1$ is nonsingular, one possibility is

$$L = -\begin{bmatrix} I \\ \bar{Y}_2\bar{Y}_1^{-1} \end{bmatrix} \bar{Y}_1 \begin{bmatrix} I & \bar{Y}_1^{-1}\bar{Y}_2 \end{bmatrix}$$

In light of (4.86) through (4.88), we now have

$$\begin{bmatrix} \bar{A} - \bar{B}\bar{B}^TL & -\bar{B}\bar{B}^T \\ -\bar{Q} + L\bar{A} + \bar{A}^TL - L\bar{B}\bar{B}^TL & -\bar{A}^T - L\bar{B}\bar{B}^T \end{bmatrix} \begin{bmatrix} \begin{bmatrix} I \\ 0 \end{bmatrix} \\ 0 \end{bmatrix}$$

$$= \begin{bmatrix} \begin{bmatrix} I \\ 0 \end{bmatrix} \\ 0 \end{bmatrix} \begin{bmatrix} 0 & -\omega_0 \\ \omega_0 & 0 \end{bmatrix}$$

With $\tilde{A} = \bar{A} - \bar{B}\bar{B}^T L$ etc,

$$\begin{bmatrix} \tilde{A}_{11} & \tilde{A}_{12} & -\bar{B}_1\bar{B}_1^T & -\bar{B}_1\bar{B}_2^T \\ \tilde{A}_{21} & \tilde{A}_{22} & -\bar{B}_2\bar{B}_1^T & -\bar{B}_2\bar{B}_2^T \\ -\tilde{Q}_{11} & -\tilde{Q}_{21}^T & -\tilde{A}_{11}^T & -\tilde{A}_{21}^T \\ -\tilde{Q}_{21} & -\tilde{Q}_{22} & -\tilde{A}_{12}^T & -\tilde{A}_{22}^T \end{bmatrix} \begin{bmatrix} I \\ 0 \\ 0 \\ 0 \end{bmatrix} = \begin{bmatrix} \Omega \\ 0 \\ 0 \\ 0 \end{bmatrix} \tag{4.89}$$

At once, we see that $\tilde{A}_{21} = 0, \tilde{Q}_{11} = 0$ and $\tilde{Q}_{21} = 0$.

Accordingly,

$$\Phi_L(j\omega) = I + \bar{B}_2^T(-j\omega I - \tilde{A}_{22}^T)^{-1}\tilde{Q}_{22}(j\omega I - \tilde{A}_{22})^{-1}\bar{B}_2 \tag{4.90}$$

and $(\tilde{A}_{22}, \bar{B}_2)$ is stabilizable because $(\bar{A}, \bar{B})$ is stabilizable.

Evidently, we have achieved a degree reduction. It is trivial to check that the Hamiltonian matrix in (4.89) after row and column permutation is

$$\begin{bmatrix} \tilde{A}_{11} & * & * & * \\ 0 & -\tilde{A}_{11}^T & * & * \\ 0 & 0 & \tilde{A}_{22} & -\bar{B}_2\bar{B}_2^T \\ 0 & 0 & -\tilde{Q}_{22} & -\tilde{A}_{22}^T \end{bmatrix}$$

which shows that the Hamiltonian matrix associated with the new form of Φ i.e. with $\tilde{A}_{22}, \bar{B}_2$ and $\tilde{Q}_{22}$ has the same eigenvalues as the original one, *excluding* two eigenvalues at $j\omega_0$ and $-j\omega_0$.

We also have the following key result.

Theorem 4.5. *With notation as above, there exists a strong solution $\tilde{P}_{s22}$ for the Riccati equation of the reduced order problem defined by $\tilde{A}_{22}, \bar{B}_2, \tilde{Q}_{22}$ with $\tilde{A}_{22}, \bar{B}_2$ stabilizable if and only if there exists a strong solution P_s of the Riccati equation for the original problem defined by A, B, Q. The solutions are related by*

$$P_s = S\left(L + \begin{bmatrix} 0_{2\times 2} & 0 \\ 0 & \tilde{P}_{s22} \end{bmatrix}\right) S^T \tag{4.91}$$

Proof. Suppose first that P_s exists and is strong. Then $\bar{P}_s = S^{-1}P_sS^{-T}$ is a strong solution for the Riccati equation for $\bar{A}, \bar{B}, \bar{Q}$. Then

$$\tilde{P}_s = S^{-1}P_sS^{-T} + L \tag{4.92}$$

is the strong solution of the Riccati equation for $\tilde{A}, \bar{B}$ and $\tilde{Q}$. Write this as

$$\tilde{P}_s \begin{bmatrix} \tilde{A}_{11} & \tilde{A}_{12} \\ 0 & \tilde{A}_{22} \end{bmatrix} + \begin{bmatrix} \tilde{A}_{11}^T & 0 \\ \tilde{A}_{12}^T & \tilde{A}_{22}^T \end{bmatrix} \tilde{P}_s - \tilde{P}_s \begin{bmatrix} \bar{B}_1 \\ \bar{B}_2 \end{bmatrix} \times \begin{bmatrix} \bar{B}_1^T & \bar{B}_2^T \end{bmatrix} \tilde{P}_s + \begin{bmatrix} 0 & 0 \\ 0 & \tilde{Q}_{22} \end{bmatrix} = 0 \tag{4.93}$$

Because P_s exists and is strong, by Theorem 4.1
$\Phi(j\omega) \geq 0$ (with Φ computed as in (4.66)). Then
$\Phi_L(j\omega) \geq 0$ with $(\tilde{A}_{22}, \bar{B}_2)$ stabilizable. Consequently, again by Theorem 4.1 there exists a strong solution to

$$\tilde{P}s_{22}\tilde{A}_{22} + \tilde{A}_{22}^T\tilde{P}s_{22} - \tilde{P}s_{22}\bar{B}_2\bar{B}_2^T\tilde{P}s_{22} + \tilde{Q}_{22} = 0 \tag{4.94}$$

It is trivial to see then that

$$\tilde{P}_s = \begin{bmatrix} 0 & 0 \\ 0 & \tilde{P}s_{22} \end{bmatrix} \tag{4.95}$$

is a strong solution to (4.93). This solution is unique. Hence (4.91) follows for (4.92) and (4.95). The argument is clearly reversible. □

Let us now try to sum up how we deflate pure imaginary eigenvalues.

(a) We start with (4.64) and (4.65) in which (A, B) is stabilizable and $\Phi(j\omega)$ is nonnegative, but may have $j\omega$-axis poles

(b) to avoid possible confusion of $j\omega$ axis poles and zeros of $\Phi(j\omega)$, we find a symmetric L such that $A - BB^TL$ has no $j\omega$ - axis eigenvalues, and transform along the lines of (4.67) through (4.71). We now drop this L.

(c) Let $\begin{bmatrix} x^T & y^T \end{bmatrix}$ be an eigenvector of $\mathcal{H}$ corresponding to eigenvalue $j\omega_0, 0 < \omega_0 < \infty$. Let $x = u + jv$,
$y = m + jn$, with u, v, m, n real. With

$$\begin{bmatrix} X \\ Y \end{bmatrix} = \begin{bmatrix} u & v \\ m & n \end{bmatrix}$$

find S so that $SX = \begin{bmatrix} I \\ 0 \end{bmatrix}$ and L symmetric so that

$$L\begin{bmatrix} I \\ 0 \end{bmatrix} + S^{-T}Y = 0$$

(d) Replace A by $\bar{A} = SAS^{-1}$, B by $\bar{B} = SB, Q$ by $\bar{Q} = S^{-T}QS^{-1}$ and then $\bar{A}$ by $\tilde{A} = \bar{A} - \bar{B}\bar{B}^T L$, and $\bar{Q}$ by $\tilde{Q} = \bar{Q} - L\bar{A} - \bar{A}^T L + L\bar{B}\bar{B}^T L$

(e) The new Hamiltonian matrix can be replaced by a smaller one associated with $\tilde{A}_{22}, \bar{B}_2, \tilde{Q}_{22}$ and

$$P_s = S\left(L + \begin{bmatrix} 0_{2\times 2} & 0 \\ 0 & \tilde{P}_{s22} \end{bmatrix}\right) S^T$$

relates a strong solution for the associated smaller dimension Riccati equation to the strong solution of the original Riccati equation.

Above, we have described how to deal with a pure imaginary $j\omega$ axis eigenvalue. Dealing with an eigenvalue at the origin is easier.

Suppose that for real x and y,

$$\begin{bmatrix} A & -BB^T \\ -Q & -A^T \end{bmatrix} \begin{bmatrix} x \\ y \end{bmatrix} = 0$$

We change the coordinate basis so that
$\bar{x} = \begin{bmatrix} 1 & 0 & \cdots & 0 \end{bmatrix}^T$ and choose a symmetric L such that

$$L \begin{bmatrix} 1 \\ 0 \\ \vdots \\ 0 \end{bmatrix} + \bar{y} = 0$$

(where $\bar{y}$ is the result of transforming y through the coordinate basis change). Then as a replacement to (4.89), we have

$$\begin{bmatrix} \tilde{A}_{11} & \tilde{A}_{12} & -\bar{B}_1\bar{B}_1^T & -\bar{B}_1\bar{B}_2^T \\ \tilde{A}_{21} & \tilde{A}_{22} & -\bar{B}_2\bar{B}_1^T & -\bar{B}_2\bar{B}_2^T \\ -\tilde{Q}_{11} & -\tilde{Q}_{12} & -\tilde{A}_{11}^T & -\tilde{A}_{21}^T \\ -\tilde{Q}_{12}^T & -\tilde{Q}_{22} & -\tilde{A}_{12}^T & -\tilde{A}_{22}^T \end{bmatrix} \begin{bmatrix} 1 \\ 0 \\ 0 \\ 0 \end{bmatrix} = \begin{bmatrix} 0 \\ 0 \\ 0 \\ 0 \end{bmatrix}$$

Evidently the first column is zero, Φ_L is now given by (4.90), the effective Hamiltonian matrix has dimension 2 less than the original one, with the same eigenvalues apart from two eigenvalues at the origin.

5 Concluding Remark

Let us note several directions in which these ideas could be extended. First, suppose that the Hamiltonian matrix has an eigenvalue of $j\omega_0$, ω_0 real, of multiplicity $2r$ for $r > 1$. Is there some straightforward procedure for deflating not in r steps, but in one? Second, any $j\omega$-axis eigenvalue of the Hamiltonian matrix is multiple. This means that numerical problems can arise in finding the eigenvalue. How best may these be addressed? Finally, in H_∞ problems, it may be the case that the infimum closed—loop gain gives rise to a Riccati equation with indefinite quadratic term and Hamiltonian matrix with pure imaginary eigenvalues. Variation on the deflation procedure to deal with this situation would be welcome.

References

[1] B.D.O. Anderson and S. Vongpanitlerd, *Network Analysis and Synthesis,* Prentice Hall Inc, Englewood Cliffs, N.J., 1973.

[2] D. C. Youla, "On the factorization of rational matrices", *IRE Trans. on Information Theory*, vol IT-7, 1961, pp 179–189.

[3] K. Zhou, J.C. Doyle and K. Glover, *Robust and Optimal Control*, Prentice Hall, N.J., 1996.

[4] R. W. Newcomb, *Linear Multiport Synthesis*, McGraw Hill, New York, 1966.

[5] M. Green and D.J.N. Limebeer, *Linear Robust Control*, Prentice Hall, N.J., 1995.

[6] O. Brune, "Synthesis of a finite two—terminal network whose driving point impedance is a prescribed function of frequency", *J. Math Phys*, vol 10, 1931, pp 191–216.

[7] E.A. Guillemin, *Synthesis of Passive Netowrks*, McGraw Hill, New York, 1957.

[8] L. Weinberg, *Network Analysis and Synthesis*, McGraw Hill, New York, 1962.

[9] B.D.O. Anderson and P.J. Moylan, "The Brune synthesis in state—space terms", *Circuit Theory and Applications*, vol 3, 1975, pp 193–199.

[10] P. Wangham and T. Mita, "Spectral factorization of singular systems using generalized algebraic Riccati equations", Proc. 36th IEEE CDC, San Diego, 1997, pp 4824–4829.

[11] D.J. Clements, B.D.O. Anderson, A.J. Laub and J.B. Matson, "Spectral factorization with imaginary—axis zeros", *Linear Algebra and its Applications*, vol 250, 1997, pp 225–252

[12] K-C Goh and M. Safonov, "The extended $j\omega$-axis eigenstructure of a Hamiltonian matrix pencil", Proc. 31st IEEE CDC, 1992, pp 1897–1902.

Research School of Information Sciences and Engineering, The Australian National University, Canberra ACT 0200 Australia

Progress in Systems and Control Theory, Vol. 25

Passive Linear Systems and Scattering Theory

D. Z. Arov[1]

1 Introduction

Passive Linear Time–Invariant Systems (PLTIS's) theory has been developed in connection with quantum mechanics and mathematical physics (spectral, scattering and other problems), with networks, control and stochastic processes theories (synthesis, stability, prediction and other problems) and there is a considerable literature on these topics (see, for example, the books [25], [28], [12], [42], [35], [21], the papers [23], [16], [17], [11], [40], [22], [7], [8] and references in these books and papers). In the first half of the 20th century the impedance formalism was developed, in which the *transfer functions* (t.f.'s) of PLTIS's were the so called resistance or impedances matrices. For *Conservative Linear Time–Invariant Systems* (CLTIS's) this development was intimately connected with the Riesz–Herglotz integral representation of positive-real functions with scalar, matrix or operator values. Connected to this representation and to the resolvent and spectral theory of selfadjoint and unitary operators in Hilbert space is Cauer method of synthesis of lossless electrical n–ports, etc. At the second half of this century the scattering and transmission (or chain scattering) formalism was also developed in the PLTIS's theory. In this formalism the t.f.'s of PLTIS's, that are called scattering and transmission matrices, are contractive and J–contractive functions, respectively. Development of such CLTIS's theory was intimately connected with the Blaschke–Riesz–Herglotz–Potapov multiplicative representation of these functions with matrix or operator values. In connection to the cascade method, the theory of dissipative and nondissipative operators and of corresponding Livšic–Brodskii nodes, of contractive and noncontractive operators and of corresponding unitary and J–unitary nodes, respectively, was developed. Here the scattering and transmission matrices of CLTIS's are called characteristic functions of the basic operators of these systems or of the corresponding operator nodes (see [25], [28], [13]). This cascade method was used in the synthesis of lossless electrical n–ports ([12], [25], [19]). For scattering CLTIS's different functional models exist by which these systems or only their basic operators are constructed from the t.f.'s (the characteristic functions), see [16], [17], [31], [32], [33], [11], etc.. There exists a one-to-one connection between scattering PLTIS's (CLTIS's) and the Lax–Phillips dissipative (conservative, respectively) scattering scheme [26], [27] (see [1], [7], [39], [5]). The Heisenberg scattering

[1]This research was made possible in part by Grant No.UM1–298 from US CRDF and the Unkrainian Goverment.

matrix of the Lax–Phillips scheme coincides with the scattering matrix of the corresponding LTIS.

In networks and control theories more general classes of PLTIS's than the above mentioned resistance, scattering or transmission systems, are considered which correspond to more general so-called "supply rates". For these systems in particular Kalman posed the problem of minimal dissipative realizations of t.f.'s. The theory is closely connected to Lyapunov equations, Riccati equations and inequalities, the Kalman–Yakubovich–Popov lemma, etc. (see [23], [35]),[21]).

Simple connections exist between PLTIS's (resistance, scattering, transmission and others), with discrete and continuous time. They are based on transforms of Cayley, Potapov–Ginzburg and more general types. These transforms permit to obtain results for one type of PLTIS's by analogous results for the other type (see [7], [5], [30]).

In this work we formulate the results for passive scattering of linear, continuous, time–invariant systems (PSLCTIS's) and also for conservative (CSLCTIS's) systems.

Analogous results hold for other types of PLTIS's and CLTIS's, sometimes under some extra assumptions.

The material of § 2–4 was mentioned in a previous survey by the author [8]. The results of § 5–6 are new: the criteria of unitary equivalence of minimal PSLCTIS's realizations is taken from a joint work with M. Nudelman, submitted for publication [6].

2 Passive scattering linear systems

2.1 The general definition of LCTIS

We start with a definition of a linear continuous time–invariant system (LCTIS). This notion was proposed in connection with the Lax–Phillips scattering scheme (see [39]). Definitions of this kind also arose in control theory see [36].

A LCTIS $\Sigma = \left(A_0, B, N; X, \mathfrak{N}^-, \mathfrak{N}^+\right)$ is the colligation of three separable Hilbert spaces X, $\mathfrak{N}^-$ and $\mathfrak{N}^+$ of vectors x, φ^- and φ^+ (inner states, input and output data, respectively) and linear operators A_0, B and N (basic, input and output operators, respectively) with some restrictions which we formulate below. The basic operator A_0 acts in X. It is the generator of a C_0 - semigroup $T_\Sigma(t)$ of linear bounded operators in X, which describes the evolution of states when $\varphi^-(t) \equiv 0$: $x(t) = T_\Sigma(t)x(0)$. Let $X_+ \subset X \subset X_-$ be the rigged space, defined by the operator A_0^*, i.e. $X_- = X_-(A_0^*) = X_+^*$, $X_+ = X_+(A_0^*)$ be the Hilbert space of the vectors from the domain of definition $\mathcal{D}(A_0^*)$ of the operator A_0^* with the corresponding A_0^*–graph metric. The input operator B is a linear bounded operator acting between the Hilbert spaces $\mathfrak{N}^-$ and X_-, i.e. $B \in \mathcal{L}(\mathfrak{N}^-, X_-)$. For A_0 we consider its natural extension $A = \widehat{A}_0$ into X_-: $\widehat{A}_0$ is the adjoint operator of A_0^*, where A_0^* is considered as an operator in $\mathcal{L}(X_+, X)$ $\big(\|A_0^*\|_{+,0} \le 1\big)$, so that $\widehat{A}_0 \in \mathcal{L}(X, X_-)$, $\mathcal{D}(A_0) = \{x \in X : Ax \in X\}$, $A_0 = A|\mathcal{D}(A_0)$.

The Hilbert space $\mathfrak{D}_{A_0,B}$ is defined via the operators A_0 and B:

$$\mathfrak{D}_{A_0,B} = \big\{(x, \varphi^-) \in X \times \mathfrak{N}^- : Ax + B\varphi^- \in X\big\},$$

$$\big\|(x, \varphi^-)\big\|^2_{\mathfrak{D}_{A_0,B}} = \|x\|^2 + \|\varphi^-\|^2 + \|Ax + B\varphi^-\|^2, \qquad (x, \varphi^-) \in \mathfrak{D}_{A_0,B}.$$

The output operator N is a linear bounded operator acting between Hilbert spaces $\mathfrak{D}_{A_0,B}$ and $\mathfrak{N}^+$, i.e. $N \in \mathcal{L}(\mathfrak{D}_{A_0,B}, \mathfrak{N}^+)$.

The evolution of a LCTIS $\Sigma = \big(A_0, B, N; X, \mathfrak{N}^-, \mathfrak{N}^+\big)$ is described by the equations

$$\begin{cases} \dfrac{dx}{dt} &= Ax(t) + B\varphi^-(t), \\ \varphi^+(t) &= N\big(x(t), \varphi^-(t)\big), \qquad t \ge 0, \end{cases} \tag{2.1}$$

considered for so-called admissible pairs $\big(x(0), \varphi^-(\cdot)\big)$ of initial states $x(0)$ and input data $\varphi^-(\cdot)$, such that:

(i) $\big(x(0), \varphi^-(0)\big) \in \mathfrak{D}_{A_0,B}$,

(ii) $\varphi^-(\cdot) \in W_2^1(\mathfrak{N}^-)$, i.e. $\varphi^-(t) = \varphi^-(0) + \int_0^t \psi(s)\, d(s)$,

where $\psi(s)$ is a measurable (on $[0,\infty)$) function with values in $\mathfrak{N}^-$, and $\|\varphi^-(t)\|^2 + \|\psi(t)\|^2 \in L_1(0,\infty)$. Under some extra assumption on Σ (see [39]) for each admissible pair $\big(x(0), \varphi^-(\cdot)\big)$ the solution $x(t) \in X_-$ of the first equation in (2.1) has property that $x(t) \in X$ and, moreover, $\big(x(t), \varphi^-(t)\big) \in \mathfrak{D}_{A_0,B}$ for all $t \ge 0$.

The transfer function (t. f.), $S_\Sigma(\lambda)$, of a LCTIS $\Sigma = \big(A_0, B, N; X, \mathfrak{N}^-, \mathfrak{N}^+\big)$ is defined by formula

$$S_\Sigma(\lambda)\varphi^- = N\big((\lambda I - \widehat{A}_0)^{-1}B\varphi^-, \varphi^-\big), \qquad \varphi^- \in \mathfrak{N}^-. \tag{2.2}$$

We can consider $X_+(A_0)$ as a subspace $X_+(A_0) \bigoplus \{0\} \subset \mathfrak{D}_{A_0,B}$. The inner-output operator C is defined by formula

$$C = N \,|\, X_+(A_0) \in \mathcal{L}\big(X_+(A_0), \mathfrak{N}^+\big). \tag{2.3}$$

If $B\mathfrak{N}^- \subset X$ then the input–output operator

$$D = N \,|\, \mathfrak{N}^- \in \mathcal{L}(\mathfrak{N}^-, \mathfrak{N}^+) \tag{2.4}$$

is defined, too. In this case $N = [C, D]$ and

$$S_\Sigma(\lambda) = D + C(\lambda I - \widehat{A}_0)^{-1} B. \tag{2.5}$$

In the general case we have, for $(x, \varphi^-) \in \mathfrak{D}_{A_0,B}$:

$$N(x, \varphi^-) = C\bigl(x - (\lambda I - \widehat{A}_0)^{-1} B\varphi^-\bigr) + S_\Sigma(\lambda)\varphi^-. \tag{2.6}$$

For a LCTIS $\Sigma = \bigl(A_0, B, N; X, \mathfrak{N}^-, \mathfrak{N}^+\bigr)$ the adjoint system is defined as the system $\Sigma^* = \bigl(A_0^*, C^*, N_*; X, \mathfrak{N}^+, \mathfrak{N}^-\bigr)$, in which N_* is defined by formula

$$N_*(x, \varphi^+) = B^*\bigl(x - (\lambda I - \widehat{A}_0^*)^{-1} C^* \varphi^+\bigr) + S_\Sigma^*(\overline{\lambda})\varphi^+, \tag{2.7}$$

where $(x, \varphi^+) \in \mathfrak{D}_{A_0^*, C^*}$. This definition of N_* doesn't depend on the choice of the point λ.

2.2 Passive scattering LCTIS's (PLCTIS's)

A LCTIS Σ is called passive scattering if for all admissible pairs $\bigl(x(0), \varphi^-(\cdot)\bigr)$

$$\|\varphi^-(t)\|^2 - \|\varphi^+(t)\|^2 \geq \frac{d}{dt}\|x(t)\|^2 \qquad (\forall t \geq 0), \tag{2.8}$$

This property is equivalent to the following:

$$\|\varphi^-\|^2 - \|N(x, \varphi^-)\|^2 \geq 2\mathrm{Re}\,(\widehat{A}_0 x + B\varphi^-, x), \qquad (x, \varphi^-) \in \mathfrak{D}_{A_0,B}. \tag{2.9}$$

Theorem 1 *Let $\Sigma = \bigl(A_0, B, N; X, \mathfrak{N}^-, \mathfrak{N}^+\bigr)$ be a PSLCTIS. Then the adjoint system $\Sigma^* = \bigl(A_0^*, C^*, N_*; X, \mathfrak{N}^+, \mathfrak{N}^-\bigr)$ is a PSLCTIS, too.*
Here C and N_* are defined by formulas (2.3) *and* (2.7).

Let $S(\mathfrak{N}^-, \mathfrak{N}^+)$ be the Schur class

$$S(\mathfrak{N}^-, \mathfrak{N}^+) = \Bigl\{S(\lambda) : S \in H_\infty(\mathfrak{N}^-, \mathfrak{N}^+), \quad \|S\|_\infty \leq 1\Bigr\}$$

of holomorphic functions on the right half-plane $\pi_+ = \{\lambda : \mathrm{Re}\,\lambda > 0\}$ whose values are linear contractive operators acting between the Hilbert spaces $\mathfrak{N}^-$ and $\mathfrak{N}^+$.

Theorem 2 *Let $\Sigma = \bigl(A_0, B, N; X, \mathfrak{N}^-, \mathfrak{N}^+\bigr)$ be a PSLCTIS. Then $S_\Sigma \in S(\mathfrak{N}^-, \mathfrak{N}^+)$.*

The t. f. S_Σ of a PSLCTIS Σ is called the scattering matrix of Σ. It is precisely the so-called Heisenberg scattering matrix in the Lax–Phillips dissipative scattering scheme. This scheme corresponds to Σ in a natural way (see [1], [39], [5]).

3 Conservative scattering linear systems

3.1 The conservative scattering realization theorem

A LCTIS $\Sigma = \left(A_0^{\prime}, B, N; X, \mathfrak{N}^-, \mathfrak{N}^+\right)$ is called conservative scattering (CSLCTIS) if for each admissible pair $\left(x(0), \varphi^-(\cdot)\right)$ we have

$$\|\varphi^-\|^2 - \|\varphi^+\|^2 = \frac{d}{dt}\,\|x(t)\|^2 \qquad (\forall\, t \geq 0) \tag{3.1}$$

and if the adjoint system Σ^* has the same kind of property, i. e. if

$$\begin{aligned} &\|\varphi^-\|^2 - \|N(x,\varphi^-)\|^2 = 2\mathrm{Re}\left(\widehat{A}_0\, x + B\,\varphi^-, x\right), \quad (x,\varphi^-) \in \mathfrak{D}_{A_0,B}, \\ &\|\varphi^+\|^2 - \|N_*(x,\varphi^+)\|^2 = 2\mathrm{Re}\left(\widehat{A}_0^*\, x + C^*\varphi^+, x\right), \quad (x,\varphi^+) \in \mathfrak{D}_{A_0^*,C^*}. \end{aligned} \tag{3.2}$$

From these equalities we have, in particular

$$\begin{aligned} 2\mathrm{Re}\left(A_0\, x, x\right) &= -\|Cx\|^2, \qquad x \in \mathcal{D}(A_0), \\ 2\mathrm{Re}\left(A_0^*\, x, x\right) &= -\|B^* x\|^2, \qquad x \in \mathcal{D}(A_0^*). \end{aligned} \tag{3.3}$$

Let, for simplicity, $A = A_0 \in \mathcal{L}(X)$ $(= \mathcal{L}(X,X))$. Then Σ is a CSLCTIS iff $B \in \mathcal{L}(\mathfrak{N}^-, X)$, $C \in \mathcal{L}(X, \mathfrak{N}^+)$, $\mathcal{D} \in \mathcal{L}(\mathfrak{N}^-, \mathfrak{N}^+)$ and the following relations hold:

$$D^* D = I_{\mathfrak{N}^-}, \; DD^* = I_{\mathfrak{N}^+}, \; C = -DB^*, \; 2\mathrm{Re}\, A = -C^*C = -BB^*. \tag{3.4}$$

For such CSLTIS Σ we have

$$S_\Sigma = D\left[I - B^*(\lambda I - A)^{-1} B\right] = \left[I - C^*(\lambda I - A)^{-1} C^*\right] D. \tag{3.5}$$

Assume we have more, namely $\mathfrak{N}^- = \mathfrak{N}^+ = \mathfrak{N}$ and $D = I_{\mathfrak{N}}$. Let

$$A_1 = -i\, A, \qquad B_1 = \frac{1}{\sqrt{2}}\, B,$$

then we obtain the colligation $\left(A_1, B_1, X, \mathfrak{N}\right)$ of Hilbert spaces X, $\mathfrak{N}$ and operators $A_1 \in \mathcal{L}(X)$, $B_1 \in \mathcal{L}(\mathfrak{N}, X)$ for which the following relation holds

$$\mathrm{Im}\, A_1 = B_1 B_1^*. \tag{3.6}$$

Such a colligation is called a Livšic–Brodskii dissipative node and the function

$$S_\Sigma(iz) = I + 2iB_1^*\left(zI - A_1\right)^{-1} B_1 \tag{3.7}$$

is called the characteristic function of this node [13]. Any bounded dissipative operator A_1 in the Hilbert space X can be included in a dissipative node as its basic operator: $\mathfrak{N} = \overline{(\operatorname{Im} A_1)X}$, $B_1 = (\operatorname{Im} A_1)^{1/2} | \mathfrak{N}$. Precisely this kind of node was considered by M. Livšic and its characteristic function was called by him the characteristic function of the dissipative operator A_1. Thus the notions of CSLCTIS and its scattering matrix are essentially generalizations of the notions of Livšic–Brodskii dissipative node and its characterinstic function.

The Cayley transform of operators and systems gives a connection between dissipative operators $-i\,A_0$ and contractions $T = (I + A_0)(I - A_0)^{-1}$, between CSLCTIS's Σ and corresponding conservative scattering linear discrete time–invariant systems $\Sigma_d = (T, F, G, H; X, \mathfrak{N}^-, \mathfrak{N}^+)$, in which

$$\begin{bmatrix} T & F \\ G & H \end{bmatrix} : X \oplus \mathfrak{N}^- \to X \oplus \mathfrak{N}^+$$

are unitary maps (see [5]). Such a collegation Σ_d is called a unitary node and the t. f. of system Σ_d is called the characteristic function of the unitary node. Each contractive operator $T \in \mathcal{L}(X)$ can be considered as the basic operator of a unitary node with

$$\mathfrak{N}^- = (I - TT^*)X, \qquad \mathfrak{N}^+ = \overline{(I - T^*T)X}, \qquad H = -T^* \,|\, \overline{(I - TT^*)X},$$
$$F = (I - TT^*)^{1/2} \,|\, \overline{(I - TT^*)\,X}, \qquad G = (I - T^*T)^{1/2}.$$

In the investigations of B. Sz.–Nagy and C. Foias and others (see [31]), essentially such unitary node was considered, and its characteristic function was called the characteristic function of the contraction T.

The development of the theory of characteristic functions of dissipative and contractive operators and of the corresponding dissipative and unitary nodes led to the following result in CSLCTIS's theory.

A LCTIS $\Sigma = \left(A_0, B, N; X, \mathfrak{N}^-, \mathfrak{N}^+\right)$ is called simple if for it we have $X = X^c_\Sigma \vee X^o_\Sigma$, where

$$X^c_\Sigma = \bigvee_{\operatorname{Re}\lambda > \gamma} (\lambda I - \widehat{A}_0)^{-1} B \mathfrak{N}^-, \qquad X^o_\Sigma = X^c_{\Sigma^*},$$

($\bigvee\limits_{\alpha} \mathcal{L}_\alpha$ denotes the closure of the linear span of all $\mathcal{L}_\alpha (\subset X)$).

A CSLCTIS Σ is simple iff iA_0 has no nontrivial selfajoint part.

Two LCTIS's $\Sigma_i = (A^{(i)}_0, B_i, N_i; X_i, \mathfrak{N}^-, \mathfrak{N}^+)$ $(i = 1, 2)$ are called unitary equivalent (resp., similar), if there exists a unitary (resp., bounded with bounded inverse) operator $R \in \mathcal{L}(X_1, X_2)$ such that

$$R\mathcal{D}(A^{(1)}_0) = \mathcal{D}(A^{(2)}_0), \qquad RA^{(1)}_0 = A^{(2)}_0 R \,|\, \mathcal{D}(A^{(1)}_0),$$
$$B^*_1 = R^* B^*_2, \qquad N_2(Rx, \varphi^-) = N_1(x, \varphi^-) \qquad ((x, \varphi^-) \in \mathfrak{D}_{A^{(1)}_0, B_1}).$$

Theorem 3 *Let $S(\lambda) \in S(\mathfrak{N}^-, \mathfrak{N}^+)$. Then a unique (up to unitary equivalence) simple CSLCTIS Σ exists with the scattering matrix $S_\Sigma(\lambda) = S(\lambda)$ (in π_+).*

3.2 Conservative resistance and transmission realizations

If in the left side of the passivity condition (2.8) we exchange the so called "scattering supply rate"

$$\|\varphi^-\|^2 - \|\varphi^+\|$$

with the "resistance supply rate" $\mathrm{Re}\,(\varphi_{\mathrm{in}}, \varphi_{\mathrm{out}})$, where φ_{in} and φ_{out} denote the input and output data from the Hilbert space $\mathfrak{N}_{\mathrm{in}} = \mathfrak{N}_{\mathrm{out}} = \mathfrak{N}$ and $(\cdot, \cdot)$ denotes the inner product, then a LCTIS with such passivity property is called passive-resistance LCTIS (PRLCTIS). For this kind of system, instead of the Schur class $S(\mathfrak{N}^-, \mathfrak{N}^+)$, one considers the Caratheodory–Nevanlinna class $\ell(\mathfrak{N})$ of functions $c(\lambda)$, holomorphic in π_+, whose values are linear bounded positive-real ($\mathrm{Re}\, c(\lambda) \geq 0$) operators acting in the Hilbert space $\mathfrak{N}$.

The Riesz–Herglotz–Nevanlinna formula for $c(\lambda) \in \ell(\mathfrak{N})$ gives a representation $c(\lambda)$ as t. f. (resistance matrix or impedance) of a unique (up to unitary equivalence) simple CRLCTIS. But such conservative resistance realization of $c(\lambda)$ was done under some extra assumption on $c(\lambda)$ (for, example, if $\lim\limits_{x \to +\infty} x^{-1}\, \mathrm{Sp}\, \mathrm{Re}\, c(x) = 0$, see [7], for more general case see [29]). The problem of general definition of PRLCTIS and CRLCTIS and the problem of CRLCTIS realization of a $c(\lambda) \in \ell(\mathfrak{N})$ without any extra assumption on $c(\lambda)$ is interesting. The situation is analogous for passive transmission LCTIS's. The transmission supply rate is the form $[\varphi_{\mathrm{in}}, \varphi_{\mathrm{in}}] - [\varphi_{\mathrm{out}}, \varphi_{\mathrm{out}}]$, where input and output data are vectors from the Hilbert spaces $\mathfrak{N}_{\mathrm{in}}$ and $\mathfrak{N}_{\mathrm{out}}$ with indefinite inner products $[\cdot, \cdot]$ defined by signature operators J_{in} and J_{out} in $\mathfrak{N}_{\mathrm{in}}$ and $\mathfrak{N}_{\mathrm{out}}$, resp. ($J^* = J^{-1} = J$). The t. f.'s of such systems have $(J_{\mathrm{in}}, J_{\mathrm{out}})$ – contractive values. In general, they can only be defined on a subset of π_+ and the poles of the t. f. may be in this subset.

If $\mathfrak{N}_{\mathrm{in}} = \mathfrak{N}_{\mathrm{out}} = \mathfrak{N}$, $J_{\mathrm{in}} = J_{\mathrm{out}} = J$, $A = A_0 \in \mathcal{L}(X)$, $D = I_{\mathfrak{N}}$ for a conservative transmission LCTIS Σ we can consider $A_1 = -iA$, $B_1 = \dfrac{1}{\sqrt{2}} B$ as it was done before for CSLCTIS's. Then the conservativity condition can be rewritten in the form: $\mathrm{Im}\, A_1 = B_1 J B^*$. The collegation $(A_1, B_1; J; X, \mathfrak{N})$ with this relation for A_1 and B_1 and a signature operator J is called a Livšic–Brodskii operator node with basic operator A_1 ($\in \mathcal{L}(X)$), and $S_\Sigma(iz)$ is called the characteristic function of this node. The operator $A_1 \in \mathcal{L}(X)$ is the basic operator of the operator node with $\mathfrak{N} = \overline{(\mathrm{Im}\, A_1)X}$, $B_1 = |\,\mathrm{Im}\, A_1\,|^{1/2} \,|\, \mathfrak{N}$ and $J = \mathrm{sign}\,(\mathrm{Im}\, A_1) \,|\, \mathfrak{N}$.

The development of the theory of characteristic functions of linear operators and operator nodes led to the result on conservative realization of J–contractive operator functions meromorphic in π_+ with some extra assumptions (see [13]). Similar results exist for discrete time–invariant systems as a result on the characteristic functions of J–unitary nodes (for example, see [7]).

3.3 Conservative realizations by n–ports with distributed parameters

It is a known result due to Krein that a scalar function $c(\lambda)$ of class ℓ which is real $(\overline{c(\overline{\lambda})} = c(\lambda))$ can be realized as the resistence function of a string, fixed at point L, with some distribution of mass on $[0, L]$, which may be singular $(L + \mathcal{M}(L) = +\infty)$, or regular, [24]. Instead of a string one can consider an electric ideal line on $[0, +\infty)$, the evolution of which is described by the equation

$$\frac{\partial x}{\partial s} = \frac{\partial x}{\partial t} H(s) J \qquad (0 \le s < +\infty); \quad J = \begin{bmatrix} 0 & -1 \\ -1 & 0 \end{bmatrix} \tag{3.8}$$

where $H(s) \ge 0$ is a measurable matrix function of size 2×2, real and diagonal, normalized by the condition $\operatorname{sp} H(s) = 1$; $x = \big(x_1(s,t), x_2(s,t)\big)$ is the vector of the values of current and voltage on the line at point s and at time t. The energy of the line is equal to

$$\|x\|^2 = \frac{1}{2} \int_0^\infty x H(s) x^* \, ds. \tag{3.9}$$

The power of state x at time t is

$$\begin{aligned} \frac{d}{dt} \|x\|^2 &= \frac{1}{2} \int_0^\infty \left(\frac{\partial x}{\partial t} H(s)\, x^* + x H(s) \frac{\partial x^*}{\partial t} \right) ds \\ &= \frac{1}{2} \int_0^\infty \left(\frac{\partial x}{\partial s} J\, x^* + x J \frac{\partial x^*}{\partial s} \right) ds \\ &= \operatorname{Re} x_1(0,t) x_2^*(0,t), \end{aligned} \tag{3.10}$$

assuming x has its support on a finite interval.

Thus, the considered line is a physical model of CRLCTIS with impedance $c(\lambda)$. L. de Branges considered not only diagonal real $H(s)$ with $h_{ii} \ge 0$ but matrices such that $h_{11}(s)$, $h_{22}(s)$, $ih_{12}(s)(= -ih_{21}(s))$ are real. For such $H(s)$ the corresponding resistence function $c(\lambda) \in \ell$ can be arbitrary, but not necessary real. On the basis of the fundamental result by de Branges [15] on the inverse spectral problem for canonical differential systems,

$$\frac{dx}{ds} = -\lambda x H(s) J \qquad (0 \le s < +\infty) \tag{3.11}$$

the following result was obtained: a function $c(\lambda) \in \ell$ is the impedance of a transmission line with distributed parameters such that the matrix $H(s) \geq 0$ has the property mentioned above. This line is uniquely determined by $c(\lambda)$ [41].

Instead of $c(\lambda)$ one can consider the scattering matrix $S(\lambda)$, which can be an arbitrary function from the Schur class S.

A general result of this kind for n–port lines with distributed parameters, when $H(s)$ is a matrix function of size $2n \times 2n$ with $n > 1$ is not known. The investigations of the inverse spectral and input scattering problem for canonical differential systems (3.10) with $J = \begin{bmatrix} 0 & -I_n \\ -I_n & 0 \end{bmatrix}$ provide, with some extra assumptions, results on the realization of matrix functions $c(\lambda) \in \ell^{n\times n}$ and $S(\lambda) \in S^{n\times n}$. Similarly, t. f.'s of conservative LCTIS's can be realized by n–lines with distributed parameters (see, for example [37], [38], [18]).

4 Optimal and star–optimal minimal passive scattering systems

4.1 Optimal PSLCTIS's

We denote by $x^{\Sigma}_{\varphi^-}(t)$ the state of a PSLCTIS Σ at time t for $x(0) = 0$, $\varphi^-(\cdot) \in W^1_2(\mathfrak{N}^-)$ and $\varphi^-(0) = 0$ ($\varphi^-(\cdot) \in \overset{\circ}{W}{}^1_2(\mathfrak{N}^-)$).

A PSLCTIS Σ_0 is called *optimal* if for all PSLCTIS's Σ with $S_\Sigma = S_{\Sigma_0}$(in π_+) we have

$$\|x^{\Sigma_0}_{\varphi^-}(t)\| \leq \|x^{\Sigma}_{\varphi^-}(t)\| \qquad (\forall t, \varphi^-(\cdot)).$$

For such a system $X^c_{\Sigma_0} \subset X^o_{\Sigma_0}$.

A LCTIS Σ is minimal if $X^c_\Sigma = X$ and $X^o_\Sigma = X$. Thus an optimal PSLCTIS Σ_0 is minimal iff it is controllable, i. e. if $X^c_{\Sigma_0} = X$ (the state space of Σ_0).

Theorem 4 *Let $S(\lambda) \in S(\mathfrak{N}^-, \mathfrak{N}^+)$. Then a unique (up to unitary equivalence) optimal minimal PSLCTIS Σ_0 exists with scattering matrix $S(\lambda)$ ($\in \pi_+$). One can obtain this system as the restriction (in a natural sense) of a CSLCTIS Σ with $S_\Sigma = S$ to the subspace $X_0 = \overline{P_{X^o_\Sigma} X^c_\Sigma}$.*

4.2 Star–optimal PSLCTIS's

A PSLCTIS Σ_{*0} is called star optimal if it is observable, i. e. $X^o_{\Sigma_{*0}} = X$, and if for all observable PSLCTIS's Σ with $S_\Sigma = S_{\Sigma_{*0}}$ we have

$$\|x^{\Sigma_{*0}}_{\varphi^-}(t)\| \geq \|x^{\Sigma}_{\varphi^-}(t)\| \qquad (\forall t, \varphi^-(\cdot)).$$

Theorem 5 *A minimal PSLCTIS Σ is optimal if and only if Σ^* is a minimal star–optimal PSLCTIS.*

Theorem 6 *Let $S(\lambda) \in S(\mathfrak{N}^-, \mathfrak{N}^+)$. Then a unique (up to unitary equivalence) star–optimal minimal PSLCTIS Σ_{*0} exists with $S_{\Sigma_{*0}}(\lambda) = S(\lambda)$ (in π_+). One can obtain this system as the restriction of a CSLCTIS Σ with $S_\Sigma = S$ on the subspace $X_{*0} = \overline{P_{X_\Sigma^c} X_\Sigma^o}$.*

4.3 Stable and bistable minimal optimal and star–optimal PSLCTIS's

A PSLCTIS Σ is called stable, if $T_\Sigma(t) \to 0$ (in strong sense) when $t \to +\infty$. It is called *bistable* if $T_\Sigma^*(t) \to 0$, as well. In these cases we will write that $\Sigma \in C_{0\bullet}$ and $\Sigma \in C_{00}$, resp. If for a PSLCTIS Σ we have that $\Sigma^* \in C_{0\bullet}$, i. e. if $T_\Sigma^*(t) \to 0$ $(t \to +\infty)$ then we will write that $\Sigma \in C_{\bullet 0}$.

Theorem 7 *Let Σ be a PSLCTIS of the class $C_{0\bullet}$. Then for $S(\lambda) = S_\Sigma(\lambda)$ $(\in S(\mathfrak{N}^-, \mathfrak{N}^+)$ the factorization problem*

$$I - S^*(i\mu)S(i\mu) = \varphi^*(i\mu)\,\varphi(i\mu) \qquad a.\ e. \tag{4.1}$$

has a solution $\varphi \in S(\mathfrak{N}^-, \overset{\circ}{\mathfrak{N}}{}^+)$ for some $\overset{\circ}{\mathfrak{N}}{}^+$.

2. *If, for $S(\lambda) \in S(\mathfrak{N}^-, \mathfrak{N}^+)$ the factorization problem (4.1) has a solution $\varphi \in S(\mathfrak{N}^-, \overset{\circ}{\mathfrak{N}}{}^+)$ for some $\overset{\circ}{\mathfrak{N}}{}^+$, then an optimal minimal PSLCTIS Σ_0 with $S_{\Sigma_0} = S$ belongs to the class $C_{0\bullet}$.*

Theorem 8 *Let Σ be a PSLCTIS of class $C_{\bullet 0}$. Then for $S(\lambda) = S_\Sigma(\lambda)$ $(\in S(\mathfrak{N}^-, \mathfrak{N}^+))$ the factorization problem*

$$I - S(i\mu)S^*(i\mu) = \psi(i\mu)\psi^*(i\mu) \qquad a.\ e. \tag{4.2}$$

has a solution $\psi \in S(\overset{\circ}{\mathfrak{N}}{}^-, \mathfrak{N}^+)$ for some $\overset{\circ}{\mathfrak{N}}{}^-$.

2. *If, for $S(\lambda) \in S(\mathfrak{N}^-, \mathfrak{N}^+)$ the factorization problem (4.2) has a solution $\psi \in S(\overset{\circ}{\mathfrak{N}}{}^-, \mathfrak{N}^+)$ for some $\overset{\circ}{\mathfrak{N}}{}^-$, then a star–optimal minimal PSLCTIS Σ_{*0} with $S_{\Sigma_{*0}} = S$ belongs to the class $C_{\bullet 0}$.*

Let $S(\lambda) \in S(\mathfrak{N}^-, \mathfrak{N}^+)$ and both factorization problems for $S(\lambda)$ be solvable. We can consider special solutions $\varphi = \varphi_r$ and $\psi = \varphi_l$ of these problems such that φ_r is an outher function, φ_l is a star–outher function, i. e. $\overline{\varphi_r(i\mu)H_2(\mathfrak{N}^-)} = H_2(\overset{\circ}{\mathfrak{N}}{}^+)$, $\overline{\varphi_l^*(i\mu)H_2(\mathfrak{N}^+)} = H_2(\overset{\circ}{\mathfrak{N}}{}^-)$.

Using these special solutions we can define an essentially unique function $h_s(i\mu) \in L_\infty(\overset{\circ}{\mathfrak{N}}{}^-, \overset{\circ}{\mathfrak{N}}{}^+)$ with $\|h_s\|_\infty \leq 1$ such that

$$h_s(i\mu)\varphi_l^*(i\mu) = -\varphi_r(i\mu)S^*(i\mu) \qquad \text{a. e.} \tag{4.3}$$

Moreover,

$$\widetilde{S}_{\mathrm{em}}(i\mu) \, : \, = \begin{bmatrix} \varphi_l(i\mu) & S(i\mu) \\ h_s(i\mu) & \varphi_r(i\mu) \end{bmatrix} \in L_\infty(\overset{\circ}{\mathfrak{N}}{}^- \oplus \mathfrak{N}^-, \mathfrak{N}^+ \oplus \overset{\circ}{\mathfrak{N}}{}^+) \tag{4.4}$$

has unitary values a. e. (see [9]).

Suppose that, for a given h_s, a biinner function with unitary boundary values $b \in S(\overset{\circ}{\mathfrak{N}}{}^-, \overset{\circ}{\mathfrak{N}}{}^-)$ exists, such that $h_s(i\mu)b(i\mu)$ is the boundary value of a function $S_{21} \in S(\overset{\circ}{\mathfrak{N}}{}^-, \overset{\circ}{\mathfrak{N}}{}^+)$. In this case we will write $h_s \in N_r(\overset{\circ}{\mathfrak{N}}{}^-, \overset{\circ}{\mathfrak{N}}{}^+)$. For such h_s there exists an essentialy unique biinner function $b_r \in S(\overset{\circ}{\mathfrak{N}}{}^-, \overset{\circ}{\mathfrak{N}}{}^-)$ which has the formulated property of b and which is a left divisor of all these b, i.e. $b_r^*(i\mu)b(i\mu)$ is the boundary value of some biinner function. The function b_r is called a minimal right denominator of $h_s \in N_r(\overset{\circ}{\mathfrak{N}}{}^-, \overset{\circ}{\mathfrak{N}}{}^+)$. In an analogous way we can introduce the class $N_l(\overset{\circ}{\mathfrak{N}}{}^-, \overset{\circ}{\mathfrak{N}}{}^+)$ and the minimal left denominator b_l for $h_s \in N_l(\overset{\circ}{\mathfrak{N}}{}^-, \overset{\circ}{\mathfrak{N}}{}^+)$.

Theorem 9 *An optimal minimal PSLCTIS Σ_0 belongs to the class C_{oo} if and only if for $S(\lambda) = S_{\Sigma_0}(\lambda)$ both factorization problems* (4.1) *and* (4.2) *are solvable and $h_s \in N_r(\overset{\circ}{\mathfrak{N}}{}^-, \overset{\circ}{\mathfrak{N}}{}^+)$.*

Theorem 10 *A star–optimal minimal PSLCTIS Σ_{*0} belongs to the class C_{oo} if and only if for $S(\lambda) = S_{\Sigma_{*0}}(\lambda)$ both factorization problems* (4.1) *and* (4.2) *are solvable and $h_s \in N_l(\overset{\circ}{\mathfrak{N}}{}^-, \overset{\circ}{\mathfrak{N}}{}^+)$.*

It is clear that for $S \in S(\mathfrak{N}^-, \mathfrak{N}^+)$ all minimal PSLCTIS's with scattering matrix $S(\lambda)$ belong to the class C_{oo} if and only if the optimal and star–optimal minimal PSLCTIS's with this scattering matrix belong to the same class.

This gives the following result.

Theorem 11 *Let $S(\lambda) \in S(\mathfrak{N}^-, \mathfrak{N}^+)$. Then all minimal PSLCTIS's with scattering matrix $S(\lambda)$ belong to the class C_{oo} if and only if both factorization problems* (4.1) *and* (4.2) *are solvable for $S(\lambda)$ and*

$$h_s \in N_l(\overset{\circ}{\mathfrak{N}}{}^-, \overset{\circ}{\mathfrak{N}}{}^+) \bigcap N_r(\overset{\circ}{\mathfrak{N}}{}^-, \overset{\circ}{\mathfrak{N}}{}^+).$$

In the case when $\dim \overset{\circ}{\mathfrak{N}}{}^\pm < \infty$ we have $N_l(\overset{\circ}{\mathfrak{N}}{}^-, \overset{\circ}{\mathfrak{N}}{}^+) = N_r(\overset{\circ}{\mathfrak{N}}{}^-, \overset{\circ}{\mathfrak{N}}{}^+)$.

If $\dim \overset{\circ}{\mathfrak{N}}{}^\pm < \infty$, then the conditions on $S(\lambda)$ formulated in Theorems 9, 10 and 11 are equivalent to the condition $S \in S\Pi(\mathfrak{N}^-, \mathfrak{N}^+)$, where

$S\Pi$ is the subclass of functions $S \in S(\mathfrak{N}^-, \mathfrak{N}^+)$ which have a pseudocontinuation into the left half–plane, i.e. such that the boundary values $S(i\mu)$ are boundary values of a meromorphic function in the left half–plane with bounded Nevanlinna characteristics (see [9]).

Let $S(\lambda)$ satisfy the conditions of Theorem 9. Then we can consider the so-called optimal minimal $\mathcal{D}$–representation of $S(\lambda)$ (in author's terminology, see [9]), as a matrix biinner function $\widetilde{S}_0$,

$$\widetilde{S}_0(\lambda) = \begin{bmatrix} \varphi_l(\lambda) b_r(\lambda) & S(\lambda) \\ S_{12}^{(r)}(\lambda) & \varphi_r(\lambda) \end{bmatrix}, \qquad \text{where } S_{12}^{(r)}(i\mu) = h_s(i\mu) b_r(i\mu).$$

Consider a lossless simple CSLCTIS $\widetilde{\Sigma}$ with $S_{\widetilde{\Sigma}} = \widetilde{S}_0$. Let $\widetilde{\Sigma} = (\widetilde{A}_0, \widetilde{B}, \widetilde{N}; \widetilde{X}; \overset{\circ}{\mathfrak{N}}^- \oplus \mathfrak{N}^-, \mathfrak{N}^+ \oplus \overset{\circ}{\mathfrak{N}}^+)$. Then $\widetilde{\Sigma}$ is minimal and belongs to the class C_{oo}.

Let now $\Sigma_0 = (A_0, B_0, N_0; X_0, \mathfrak{N}^-, \mathfrak{N}^+)$, where

$$X_0 = \widetilde{X}, \qquad A_0 = \widetilde{A}_0, \qquad B_0 = \widetilde{B} \mid \mathfrak{N}^-,$$
$$N_0 = P_{\mathfrak{N}^+} N \mid \mathfrak{D}_{A_0, B_0}.$$

Then Σ_0 is an optimal minimal PSLCTIS with scattering matrix $S(\lambda)$. It is a synthesis of $S(\lambda)$ by the so-called Darlington method (see [9]).

If the conditions of Theorem 10 hold for $S(\lambda)$, then in an analogous way we can obtain a star–optimal minimal PSLCTIS Σ_{*0} with $S_{\Sigma_{*0}} = S$, using the minimal star–optimal $\mathcal{D}$–representation of $S(\lambda)$, as a matrix biinner function $\widetilde{S}_{*0}$,

$$\widetilde{S}_{*0} = \begin{bmatrix} \varphi_l(\lambda) & S(\lambda) \\ S_{12}^{(l)}(\lambda) & b_l(\lambda) \varphi_r(\lambda) \end{bmatrix}.$$

In [3],[4] optimal minimal and star–optimal minimal passive scattering linear discrete time systems are considered. In [10] optimal minimal PRLDTIS's are considered ,too.

5 Unitary equivalence of minimal passive scattering realizations

5.1 The bistable case

Let $S(\lambda) \in S(\mathfrak{N}^-, \mathfrak{N}^+)$ and suppose that both factorization problems (4.1) and (4.2) are solvable for $S(\lambda)$, and that $h_s \in N_r(\overset{\circ}{\mathfrak{N}}^-, \overset{\circ}{\mathfrak{N}}^+) \bigcap N_l(\overset{\circ}{\mathfrak{N}}^-, \overset{\circ}{\mathfrak{N}}^+)$. Thus, by Theorem 11 we suppose that all minimal PSLCTIS's with scattering matrix $S(\lambda)$ are bistable. Then all such systems are unitary equivalent if and only if $h_s(i\mu)$ is the boundary value of a function $h_s \in S(\overset{\circ}{\mathfrak{N}}^-, \overset{\circ}{\mathfrak{N}}^+)$, i. e. if we can choose $b_l = I_{\overset{\circ}{\mathfrak{N}}^-}$, $b_r = I_{\overset{\circ}{\mathfrak{N}}^+}$. In this case

$$\widetilde{S}_{\text{em}} = \begin{bmatrix} \varphi_l & S \\ h_s & \varphi_r \end{bmatrix}$$

is a biinner function, minimal optimal and minimal star–optimal $\mathcal{D}$–representation of S, simultaneously.

5.2 The general case

Let now $S \in S(\mathfrak{N}^-, \mathfrak{N}^+)$ be an arbitrary Schur class function. For such a function essentially unique extremal solutions of the factorization inequalities

$$\begin{aligned} I - S^*(i\mu)S(i\mu) &\geq \varphi_r^*(i\mu)\,\varphi_r(i\mu) && \text{a. e.} \\ I - S(i\mu)S^*(i\mu) &\geq \varphi_l(i\mu)\,\varphi_l^*(i\mu) && \text{a. e.} \end{aligned} \tag{5.1}$$

exist such that $\varphi_r \in S(\mathfrak{N}^-, \overset{\circ}{\mathfrak{N}}^+)$, $\varphi_l \in S(\overset{\circ}{\mathfrak{N}}^-, \mathfrak{N}^+)$, φ_r is outer and φ_l is a star–outer function (see [31]). Moreover, there exists an essentially unique function $h_s(i\mu) \in L_\infty(\overset{\circ}{\mathfrak{N}}^-, \overset{\circ}{\mathfrak{N}}^+)$ such that $\widetilde{S}_{\text{em}}(i\mu)$ defined by formula (4.4) satisfies $\|\widetilde{S}_{\text{em}}\|_\infty \leq 1$.

Theorem 12 *Let $S \in S(\mathfrak{N}^-, \mathfrak{N}^+)$. Then all minimal PSLCTIS's with scattering matris $S(\lambda)$ are unitary equivalence if and only if $h_s(i\mu)$ is the boundary value of a function $h_s \in S(\overset{\circ}{\mathfrak{N}}^-, \overset{\circ}{\mathfrak{N}}^+)$.*

6 Similarity of minimal passive scattering realizations

6.1 The scalar case

Let $S(\lambda) \in S$ be a scalar function of Schur class. Let us assume it is an inner function. Then a simple CSLCTIS Σ with $S_\Sigma = S$ is minimal of class C_{oo}, and a simple PSLCTIS Σ with $S_\Sigma = S$ is CSLCTIS. Consenquently, in this case all minimal PSLCTIS's Σ with $S_\Sigma = S$ are unitarily equivalent. If $S(\lambda)$ isn't inner, but $\big(1 - |s(i\mu)|\big)/(1+\mu^2) \notin L_1$ then a simple CSLCTIS Σ with $S_\Sigma = S$ is not of class C_{oo} but it is minimal and all minimal PSLCTIS's Σ with $S_\Sigma = S$ are conservative, as it was for inner S. Thus, in this case all minimal PSLCTIS's with scattering matrix $S(\lambda)$ are also unitary equivalent. Let now

$$\big(1 - |S(i\mu)|\big)/(1+\mu^2) \in L_1. \tag{6.1}$$

Then we will consider two cases: $\alpha)$ $S \in S\Pi$ and $\beta)$ $S \notin S\Pi$. Under our assumptions the factorization problems (4.1) and (4.2) for S are solvable, moreover $\varphi_r = \varphi_l (= \varphi)$.

By Theorem 7 and 8 we have that a minimal optimal PSLCTIS Σ_0 with $S_{\Sigma_0} = S$ belongs to the class $C_{0\bullet}$ and a minimal star–optimal PSLCTIS Σ_{*0} belongs to the class $C_{\bullet 0}$. But if $S \notin S\Pi$ then by Theorems 9 and 10 we have that $\Sigma_0 \notin C_{00}$ and $\Sigma_{*0} \notin C_{00}$. Thus, in the case $\beta)$ we have that

$$\Sigma_0 \in C_{0\bullet}, \qquad \Sigma_0 \notin C_{\bullet 0} \qquad \text{and} \qquad \Sigma_{*0} \in C_{\bullet 0}, \qquad \Sigma_{*0} \notin C_{0\bullet}.$$

As we see, in this case the minimal systems Σ_0 and Σ_{*0} with the same scattering matrix $S(\lambda)$ are not similar.

We obtain that for similarity of all minimal PSLCTIS's with the same scalar scattering matrix $S(\lambda)$ it is necessary that $S(\lambda) \in S\Pi$, i. e. that $S(\lambda)$ has a pseudocontinuation. This condition is equivalent to saying that all minimal PSLCTIS's Σ with $S_\Sigma = S$ are bistable.

6.2 Similarity in the bistable case

Let now $S(\lambda) \in S(\mathfrak{N}^-, \mathfrak{N}^+)$ and $S(\lambda)$ satisfy to the conditions of Theorem 11. Thus, we consider the case when all minimal PSLCTIS's Σ with $S_\Sigma = S$ are bistable, i. e. they belong to the class C_{00}. In this case we can consider for S the functions b_l, b_r, $\widetilde{S}_0$ and $\widetilde{S}_{*0}$ which were defined in § 5. In terms of the functions b_l and $\widetilde{S}_{*0}$ we can formulate a condition which is necessary and sufficient to have similarity of all minimal PSLCTIS's with scattering matrix $S(\lambda)$. For this we consider the spaces:

$$\begin{aligned} \mathcal{H}(\widetilde{S}_{*0}) &= H_2(\mathfrak{N}^+ \oplus \mathring{\mathfrak{N}}^+) \ominus \widetilde{S}_{*0}\, H_2(\mathring{\mathfrak{N}}^- \oplus \mathfrak{N}^-), \\ \mathcal{H}(b_l) &= H_2(\mathring{\mathfrak{N}}^+) \ominus b_l H_2(\mathring{\mathfrak{N}}^-). \end{aligned}$$

Theorem 13 *Let $S \in S(\mathfrak{N}^-, \mathfrak{N}^+)$ be such that the conditions formulated in Theorem 11 hold. Then all minimal PSLCTIS's with scattering matrix $S(\lambda)$ are similar if and only if we have the following property*

$$\|P_{\{0\}\oplus\mathcal{H}(b_l)} | \mathcal{H}(\widetilde{S}_{*0})\| < 1. \tag{6.2}$$

This theorem gives the following sufficient condition for similarity of all minimal PSLCTIS's with scattering matrix $S(\lambda)$.

Theorem 14 *Let $S \in S(\mathfrak{N}^-, \mathfrak{N}^+)$ and assume the conditions of Theorem 11 hold. Let $\dim \mathcal{H}(b_l) < \infty$. Then all minimal PSLCTIS's with scattering matrix $S(\lambda)$ are similar.*

In the case, when $m = \dim \mathfrak{N}^- < \infty$ and $n = \dim \mathfrak{N}^+ < \infty$ we can consider, instead of $S(\mathfrak{N}^-, \mathfrak{N}^+)$, the Schur class $S^{n\times m}$ of matrix functions of size $n \times m$ holomorphic and contractive in π_+ . Then the conditions of Theorem 11 are equivalent to the condition that $S(\lambda) \in S^{n\times m}$ has a pseudocontinuation into the left half-plane, i. e. that $S \in S^{n\times m}\Pi$. Then Theorem 14 can be reformulated in the following form.

Theorem 15 *Let $S \in S^{n\times m}\Pi$ and suppose that the function b_l corresponding to S is a finite Blaschke–Potapov product, i. e. b_l is a rational matrix function. Then all minimal PSLCTIS's with scattering matrix $S(\lambda)$ are similar.*

Consider the following simple example to demonstrate an application of Theorem 13 in the case when dim $\mathcal{H}(b_l) \leq \infty$.

Let $S(\lambda) = kb(\lambda)$, where $b \in S(\mathfrak{N}, \mathfrak{N})$ is a biinner function, k is a scalar constant with $|k| < 1$.

From the formula for $\widetilde{S}_{*0}$ we obtain that

$$\mathcal{H}(\widetilde{S}_{*0}) = \{kg \oplus \varphi g \,:\, g \in \mathcal{H}(b)\}, \qquad \text{where} \qquad \varphi = (1-|k|^2)^{1/2}.$$

For $h = kg \oplus \varphi g \in \mathcal{H}(\widetilde{S}_{*0})$ we have

$$\|h\|^2 = \big(|k|^2 + \varphi^2\big)\|g\|^2 = \|g\|^2, \qquad \|h\| = \|g\|;$$

$$\begin{aligned} P_{\{0\}\oplus\mathcal{H}(b))}\, h &= 0 \oplus \varphi g, \\ \|P_{\{0\}\oplus\mathcal{H}(b))}\, h\| &= \varphi\|g\|. \end{aligned}$$

Consequently,

$$\|P_{\{0\}\oplus\mathcal{H}(b))} \,|\, (\widetilde{S}_{*0})\| = \varphi = \big(1-|k|^2\big)^{1/2} < 1.$$

Thus, if $S = kb$, where $b \in S(\mathfrak{N}, \mathfrak{N})$ is a biinner function, all minimal PSLCTIS's with scattering matrix S are similar.

Acknowledgement. The author thanks M. Nudelman for useful discussions which stimulated the investigation of the similarity problem considered in the last section.

REFERENCES

[1] V. M. Adamjan, D. Z. Arov, *On unitary couplings of semiunitary operators*, Mat. issled., Kishinev **1** (2) (1966), 3–66. (Russian).

[2] D. Z. Arov, M. A. Kaashoek, D. R. Pik, *Minimal and optimal linear discrete time-invariant dissipative scattering systems*, Integral Equations Operator Theory, **29** (1997), 127–154.

[3] D. Z. Arov, M. A. Kaashoek, D. R. Pik, *Optimal time–variant systems and factorization of operators*, I; *Minimal and optimal systems*, Integral Equations Operator Theory, **31**, No 4 (1998), 389-420.

[4] D. Z. Arov, M. A. Kaashoek, D. R. Pik, *Optimal time–varing systems and factorization of operators*, II; *factorization of operators*, J. Oper. Theory, (to appear).

[5] D. Z. Arov, M. A. Nudelman, *Passive linear stationary dynamical scattering systems with continuous time.* Integral Equations Operator Theory, **24** (1996), 1–45.

[6] D. Z. Arov, M. A. Nudelman, *Criterion of unitary similarity of minimal passive scattering systems with given transfer function*, Ukr. Math. J., (to appear).

[7] D .Z. Arov, *Passive linear stationary dynamical systems*, Sibirsk. Math. Journal, **20** (1979), 211–228.

[8] D. Z. Arov, *A survey on passive networks and scattering systems which are lossless or have minimal losses.* Archive für electronik und übertrgagungstechnik. International Journal of electronics and communications, **49**, N 5/6 (1995), 252–265.

[9] D. Z. Arov, *Stable dissipative linear stationary dynamical scattering systems*, J. Operator Theory, **2** (1979), 95–126.

[10] D. Z. Arov, *Optimal and stable passive systems.* Dokl. Akad. Nauk SSSR, **247** (1979), 265–268.

[11] J. A. Ball, N. Cohen, *De Branges–Rovnyak operator models and systems theory; a survey.* Oper. Theory: Advances and Appl., **50** (1990), 93-136.

[12] V. Belevich. *Classical network theory.* San Francisco: Holden Day, 1968.

[13] M. S. Brodsky, *Triangular and Jordan representations of linear operators.* Moscow: Nauka, 1969.

[14] T. Constantinescu, *Schur Parameters, Factorizations and Dilation Problems*, Birkhäuser Verlag, Basel, Operator Theory: Adv. and Appl., **82**, 1995.

[15] L. de Branges, *Some Hilbert spaces of entire functions.* Trans. Amer. Math. Soc., **96** (1960), 259–295; **99** (1961), 118–152; **100** (1960), 73–115; **105** (1962), 43–62.

[16] L. de Branges, J. Rovnyak, *Square Summable Power Series*, Holt, Rinehart and Winston, New York, 1966.

[17] L. de Branges, J. Rovnyak, *Appendix on square summable power series, Canonical models in quantum scattering theory, Perturbation Theory and its Applications in Quantum Mechanics*, C. H. Wilcox, New York, 1966.

[18] H. Dym, A. Iakob, *Positive definite extensions, canonical equations and inverse problems*, Operator theory: Advances and Applications, **12** (1984), 141–240.

[19] A. V. Efimov, V. P. Potapov, *J-expansive matrix-valued functions and their role in the analytic theory of electrical circuits.* Uspekhi Mat. Nauk, **28** (1973), 65–130. Russian Math. Surveys **28**, No 1 (1973), 69–140.

[20] J. Helton, *Systems with infinite-dimensional state space: the Hilbert space approach. Proc. IEEE*, **64** (1976), 145–160.

[21] A. Halany,V. Ionescu, *Time-Varying Discrete Linear Systems*, Birkhauser Verlag, Operator Theory: Adv.and Appl., **68**, 1994.

[22] D. J. Hill, P. J. Moylan, *Dissipative dynamical systems: basic input-output and state properties*, J. Franklin Inst. **309** (1980), 327–357.

[23] T. Kailath, *Norbert Wiener and the development of mathematical engineering*, in: Communications, computation, control and signal processing (eds. Paulraj, V. Roychowdhury, C. D. Schaper), 1997, 35–64.

[24] M. G. Krein, *On a generalization of investigations of Stiltjes*, Dokl. Akad. Nauk SSSR **93** (1953), 617–620.

[25] M. S. Livšic, *Operators, oscillations, waves (open systems)*. Moscow: Nauka, 1966.

[26] P. Lax, R. Phillips, *Scattering theory,* New York, 1967.

[27] P. Lax, R. Phillips R., *Scattering theory for dissipative hyperbolic systems*, J. Funct. Anal. **14** (2) (1975), 172–235.

[28] M. S. Livšic, A. A. Yantsevich, *The Theory of Operator Nodes in the Hilbert spaces.* Kharkov, 1977.

[29] M. M. Malamud, V. I. Mogilevskii, *On extensions of dual pairs of operators*, Dopov. Nats. Akad. Nauk Ukr. Mat. Prirodozn. Tekhn. Nauki, No 1 (1997), 30-37.

[30] M. A. Nudelman, *Optimal passive systems and semiboundedness of quadratic functionals*, (Russian) Sibirsk. Mat. Zh. **33**, No 1 (1992),78-86; transl. in Siberian Math. J. **33**, No 1 (1992), 62-69.

[31] B. Sz.–Nagy, C. Foias, *Analyse harmonique des operateurs de l'espace de Hilbert.* Academiai Kiado: Masson et Cie, 1967.

[32] N. K. Nikolskii, V. I. Vasyunin, *A unified approach to function models and the transcription problem*, in: The Gohberg Aniversary Collection, Operator Theory: Adv. and Appl. **41**, 1990, 405–434.

[33] N. Nikolski, V. Vasiunin, *Elements of spectral theory in terms of the free function model.* Part 1: Basic construction. Laboratorie de Mathematiques Pures de Bordeaux, E.R.S., 0127 C.N.R.S. Dec. 1996 96/29, 79 p.

[34] D. R. Pik, *Time varying dissipative systems*, Doctoral scriptie, Vrije Universiteit, Amsterdam, 1994.

[35] V. M. Popov, *Hyperstability of control systems.* Editura Academici, Bucuresti and Springer Verlag, Berlin, 1973.

[36] D. Salamon, *Realization theory in Hilbert space*, Math. Systems Theory, **21** (1989), 147–164.

[37] A. Sakhnovich, *Spectral functions of canonical system of order* $2n$, Math. USSR Sbornik, **71** (1992), 355–369.

[38] L. Sakhnovich, *The method of operator identities and problems of analysis*, Algebra and Analysis, **5** (1993), 4–80.

[39] Yu. L. Shmuljan, *Invariant subspaces of semigroups and the Lax–Phillips scheme*, Dep. in VINITI, N 8009–1386, Odessa, 1986, 49 p.

[40] J. C. Willems, *Dissipative dynamical systems,* Part I: General theory, Part II: Linar systems with quadratic supply rates. Archive for Rational Mechanics and Analysis **45** (1972), 321–393.

[41] H. Winkler, *The inverse spectral problem for canonical systems*, Integral Equations Operator Theory, **22** (1995), 360–374.

[42] M. R. Wohlers, *Lumped and Distributed Passive Networks*, Academic Press, 1969.

South–Ukrainian State Pedagogical University, Odessa

Progress in Systems and Control Theory, Vol. 25

Tracking Control and π-Freeness of Infinite Dimensional Linear Systems[1]

Michel Fliess, Hugues Mounier

Introduction

This is a report on recent works [1, 17, 19, 20, 21, 39, 40, 41, 43] on the control tracking of some infinite dimensional linear systems. It is based on a new property, called *π-freeness*, which allows the tracking of a reference trajectory in a way which bears some analogy with flat finite dimensional nonlinear systems (*see* [15, 16] and the references therein). Several examples are examined and simulations are provided.

π-freeness is an extension of the classic Kalman linear controllability, when viewed in the module-theoretic language of [10, 11, 17, 39]. Then a linear system is a finitely generated module over the principal ideal ring of linear differential operators. Kalman's controllability is equivalent to the freeness of this module, or what amounts to the same, to its torsion freeness. Infinite dimensional systems, like delay ones, yield modules over more general rings, where freeness and torsion freeness are no longer equivalent. Concrete case-studies (*see* [39, 41]) show that the corresponding modules are mostly not free but only torsion free. Freeness may nevertheless be recovered via a suitable localization, *i. e.*, by taking the inverse of an element π of the ring. Any basis of this free module is called a *flat*, or *basic*, *output*; it plays the same role as a *flat*, or *linearizing*, *output* of a flat finite dimensional system (*see also* [35]).

More ingredients should be added when studying partial differential equations such as the heat or the Euler-Bernoulli equations (*see* [19]). Operational calculus is first utilized in its version due Mikusiński [36, 37] which permits avoiding many of the severe analytical difficulties encountered with the classic Laplace transform (*see, e. g.,* [7]). Then a special type of C^∞-functions, called Gevrey functions [22], ensures the convergence of some infinite series.

The paper is organized as follows. We first introduce abstract linear systems over arbitrary commutative rings. Finite dimensional linear systems are then presented within this framework. We proceed to delay systems and relate them to the boundary control of the wave equation. The final sections are devoted to the heat and the Euler-Bernoulli equations.

[1]This work was partially supported by the European Commission's Training and Mobility of Researchers (TMR) Contract # ERBFMRX-CT970137, by the G.D.R. *Medicis* and by the G.D.R.-P.R.C. *Automatique.*

Acknowledgments *Several parts of the work presented here were done in collaboration with our colleagues and friends P. Rouchon (École des Mines de Paris, France) and J. Rudolph (Technische Universität Dresden, Germany).*

1 Abstract linear system theory

1.1 Basic definitions

Any ring R is commutative, with 1 and without zero divisors.
Notation The submodule spanned by a subset S of an R-module M is written $[P]$.

An *R-system* Λ, or a *system over R*, is an R-module. Two R-systems Λ_1 and Λ_2 are said to be *R-equivalent*, or *equivalent over R*, if the R-modules Λ_1 and Λ_2 are isomorphic.

An *R-dynamics*, or a *dynamics over R*, is an R-system Λ equipped with an *input*, *i. e.*, a subset $\boldsymbol{u}$ of Λ which may be empty, such that the quotient R-module $\Lambda/[\boldsymbol{u}]$ is torsion. The input $\boldsymbol{u}$ is *independent* if the R-module $[\boldsymbol{u}]$ is free, with basis $\boldsymbol{u}$.

An *output* $\boldsymbol{y}$ is a subset, which may be empty, of Λ.

An *input-output R-system*, or an *input-output system over R*, is an R-dynamics equipped with an output.

Remark 1.1.1 *Kalman's module-theoretic setting [26] is related to the state variable description, whereas our module description encompasses all system variables without any distinction.*

Let A be an R-algebra et and Λ be an R-system. The A-module $A \otimes_R \Lambda$ is an A-system, which *extends* Λ.

1.2 Relations

Let Λ be an R-system. There exists an exact sequence of R-modules [50]

$$0 \to N \to F \to \Lambda \to 0 \tag{1.2.1}$$

where F is free. The R-module N, which is sometimes called the *module of relations*, should be viewed as a *system of equations* defining Λ.

Associate to Λ a *free presentation* [50], *i. e.*, the short exact sequence of R-modules

$$F_1 \to F_0 \to \Lambda \to 0$$

where F_0 and F_1 are free. The R-module Λ is said to be *finitely generated*, or of *finite type*, if there exists a free presentation where any basis of F_0 is finite. It is said to be *finitely presented* if there exists a free presentation where any basis of F_0 and F_1 is finite. The matrix corresponding, for some given bases, to the mapping $F_1 \to F_0$ is called a *presentation matrix* of Λ. If the ring R is Noetherian, it is known [50] that the conditions of being of finite type and of being finitely presented coincide. Then (1.2.1) may be chosen such that both F and N are of finite type. This latter case will always be verified in the sequel.

Example 1.2.1 *Let us determine the R-module Λ corresponding to a system of R-linear equations*

$$\sum_{\kappa=1}^{\mu} a_{\iota\kappa}\xi_\kappa = 0, \quad a_{\iota\kappa} \in A, \iota = 1, \ldots, \nu$$

where $\xi_1, \ldots, \xi_\mu$ are the unknowns. Let F be the free R-module spanned by $f_1, \ldots, f_\mu$. Let $N \subseteq F$ be the module of relations, i.e., *the submodule spanned by $\sum_{\kappa=1}^{\mu} a_{\iota\kappa} f_\kappa$, $\iota = 1, \ldots, \nu$. Then, $\Lambda = F/N$. The ξ_κ's are the* residues *of the f_κ's,* i.e., *the canonical images of the f_κ's.*

1.3 Laplace functor

1.3.1 Mathematical preliminaries

Let K the quotient field of R. The K-vector space $\hat{\Lambda} = K \otimes_R \Lambda$ is called the *transfer K-vector space* of the R-system Λ. The functor $K \otimes_R \bullet$, from the category of R-modules to the category of K-vector spaces, is called the *Laplace functor* [12]. The *(formal) Laplace transform* of any $\lambda \in \Lambda$ is $\hat{\lambda} = 1 \otimes \lambda \in \hat{\Lambda}$. Let us briefly review some basic properties.

1. The set $\{\lambda_i \in \Lambda \mid i \in I\}$ is R-linearly independent if, and only if, the set $\{\hat{\lambda}_i \in \hat{\Lambda} \mid i \in I\}$ is K-linearly independent. The *rank* of Λ, which is written rk (Λ), is, by definition, the dimension of the K-vector space $\hat{\Lambda}$.

2. The kernel of the R-linear morphism $\Lambda \to \hat{\Lambda}$, $\lambda \mapsto \hat{\lambda}$, is the torsion submodule of Λ. The rank of an R-module is therefore zero if, and only if, it is torsion.

Consider an R-module Λ spanned by $\xi_1, \ldots, \xi_n$ which satisfy

$$M \begin{pmatrix} \xi_1 \\ \vdots \\ \xi_n \end{pmatrix} = 0$$

where $M \in A^{n \times n}$. The square matrix M is said to be of *maximum rank* if $\det(M) \neq 0$. The matrix M is therefore of maximum rank if, and only if, it is invertible as a matrix in $K^{n \times n}$.

Theorem 1.3.1 *The R-module Λ is torsion if, and only if, the matrix M is of maximum rank.*

Proof. Λ is torsion if, and only if, $\hat{\Lambda} = \{0\}$. Note that $\hat{\Lambda}$ is spanned by $\hat{\xi}_1, \ldots, \hat{\xi}_n$ which satisfy

$$M \begin{pmatrix} \hat{\xi}_1 \\ \vdots \\ \hat{\xi}_n \end{pmatrix} = 0$$

The zero solution is unique if, and only if, $M \in K^{n \times n}$ is invertible.

1.3.2 Transfer matrices

Consider an R-system, where the input $\boldsymbol{u} = (u_1, \ldots, u_m)$ and the output $\boldsymbol{y} = (y_1, \ldots, y_p)$ are, for simplicity's sake, finite. Since $\Lambda/[\boldsymbol{u}]$ is torsion, $\mathrm{span}_K(\hat{u}_1, \ldots, \hat{u}_m) = \hat{\Lambda}$. There exists a matrix $\mathcal{T} \in K^{p \times m}$, called the *transfer matrix* of the input-output system, such that

$$\begin{pmatrix} \hat{y}_1 \\ \vdots \\ \hat{y}_p \end{pmatrix} = \mathcal{T} \begin{pmatrix} \hat{u}_1 \\ \vdots \\ \hat{u}_m \end{pmatrix}$$

If $\boldsymbol{u}$ is independent, $\hat{u}_1, \ldots, \hat{u}_m$ is a basis of $\hat{\Lambda}$. Then, $\mathcal{T}$ is unique.

1.4 Input-output inversion

1.4.1 Generalities

Consider again the R-system Λ with finite input $\boldsymbol{u} = (u_1, \ldots, u_m)$ and finite output $\boldsymbol{y} = (y_1, \ldots, y_p)$. Its *output rank* is, by definition, $\rho = \mathrm{rk}\,([\boldsymbol{y}])$. The system is said to be *left invertible* (resp. *right invertible*) if $\rho = \mathrm{rk}\,([\boldsymbol{u}])$ (resp. $\rho = p$).

The quotient R-module $\Lambda/[\boldsymbol{y}]$ is the *residual R-dynamics*, or the *zero R-dynamics*. The residual dynamics is said to be *trivial* if $\Lambda/[\boldsymbol{y}] = \{0\}$.

Proposition 1.4.1 *The R-system Λ is left invertible if, and only if, its residual R-dynamics $\Lambda/[\boldsymbol{y}]$ is torsion.*

Proof. From $\mathrm{rk}(\Lambda/[\boldsymbol{y}]) = \mathrm{rk}\,(\Lambda) - \mathrm{rk}\,([\boldsymbol{y}])$, we obtain that $\mathrm{rk}\,(\Lambda/[\boldsymbol{y}]) = 0$ is equivalent to $\mathrm{rk}\,([\boldsymbol{y}]) = \mathrm{rk}\,(\Lambda) = \mathrm{rk}\,([\boldsymbol{u}])$.

Remark 1.4.1 *Left invertibility implies, since $\Lambda/[\boldsymbol{y}]$ is torsion, that $\boldsymbol{y}$ plays the role of an input. Any other system variable,* i.e., *any element of Λ, may be* determined *from the output. Right invertibility means that the components $y_1, \ldots, y_p$ of the output are R-linearly independent.*

1.4.2 Some illustrations

Assume that the input $\boldsymbol{u} = (u_1, \ldots, u_m)$ and the output $\boldsymbol{y} = (y_1, \ldots, y_p)$ are both finite and that $\boldsymbol{u}$ is independent, *i. e.*, that $\text{rk}([\boldsymbol{u}]) = m$. The next two propositions are clear.

Proposition 1.4.2 *An R-system with finite and independent input and finite output is left invertible (resp. right invertible) if, and only if, its transfer matrix is left invertible (resp. right invertible).*

The above system is said to be *square* if $m = p$.

Proposition 1.4.3 *Consider a square R-system with an independent input. Then, the two following properties are equivalent:*

- *the system is left invertible;*
- *the system is right invertible.*

An R-system with an independent input, which is both left and right invertible, is necessarily square.

Such a left and right invertible system is called *invertible.*

1.5 Different notions of controllability

An R-system Λ is said to be *R-torsion free controllable* (resp. *R-projective controllable*, *R-free controllable*) if the R-module Λ is torsion free (resp. projective, free). Elementary homological algebra (*see, e. g.*, [50]) yields the

Proposition 1.5.1 *R-free (resp. R-projective) controllability implies R-projective (resp. R-torsion free) controllability.*

Take an R-free controllable system Λ with a finite output y. This output is said to be *flat*, or *basic*, if y is a basis of Λ. The next proposition is clear.

Proposition 1.5.2 *Take an R-system Λ with a finite output. The output is flat if, and only if, the following two properties hold*

- *the system is right invertible,*
- *the residual dynamics is trivial.*

Then, Λ is R-free controllable. Moreover, if there is an independent input, the system is square.

1.6 π–freeness

The next result [17] follows at once from [51, Proposition 2.12.17, p. 233]:

Theorem and Definition 1.6.1 *Let Λ be an R–system, A an R–algebra, and $\mathcal{S}$ a multiplicative part of A such that Λ is $\mathcal{S}^{-1}R$-free controllable. Then, there exists an element π in $\mathcal{S}$ such that Λ is $R[\pi^{-1}]$–free controllable. The preceding system will then be called* π–free. *An output being a basis of $R[\pi^{-1}] \otimes_R \Lambda$ is called* π-flat *or* π-basic.

Corollary 1.6.1 *Let Λ be an R–torsion free controllable R–system and $\mathcal{S}$ a multiplicative part of R such that $\mathcal{S}^{-1}R$ is a principal ideal ring. Then, there exists $\pi \in \mathcal{S}$ such that Λ is $R[\pi^{-1}]$–free controllable and Λ is π–free.*

2 Finite dimensional linear systems

2.1 Modules over principal ideal rings

In this section R is the principal ideal ring $k[\frac{d}{dt}]$, whose elements are of the form $\sum_{finite} a_\alpha \frac{d^\alpha}{dt^\alpha}$, $a_\alpha \in k$, where k is a field. All $k[\frac{d}{dt}]$-modules are finitely generated and, therefore, finitely presented. The following two theorems are classic (*see, e. g.*, [30])

Theorem 2.1.1 *For a $k[\frac{d}{dt}]$-module M, the next two conditions are equivalent:*

1. *M is torsion,*
2. *the dimension of M as a k-vector space is finite,* i. e., $\dim_k M < \infty$.

Theorem 2.1.2 *Any $k[\frac{d}{dt}]$-module M may be written*

$$M \simeq \mathcal{F} \oplus t(M)$$

where $t(M)$ is the torsion submodule and $\mathcal{F}$ a free module.

Corollary 2.1.1 *Any $k[\frac{d}{dt}]$-module is free if, and only if, it is torsion free.*

2.2 State variable representation

Let M be a torsion $k[\frac{d}{dt}]$-module. The derivation $\frac{d}{dt}$ induces a k-linear endomorphism $\sigma : M \to M$. The next property is clear.

Proposition 2.2.1 *Let M be a torsion $k[\frac{d}{dt}]$-module. It corresponds to*

$$\frac{d}{dt}\begin{pmatrix} x_1 \\ \vdots \\ x_n \end{pmatrix} = F \begin{pmatrix} x_1 \\ \vdots \\ x_n \end{pmatrix} \tag{2.2.2}$$

where $F \in k^{n\times n}$ is the matrix of σ with respect to a basis $x_1, \ldots, x_n$ of the finite dimensional k-vector space M.

Consider a $k[\frac{d}{dt}]$-dynamics Λ with input $\boldsymbol{u} = (u_1, \ldots, u_m)$. Set $n = \dim_k(\Lambda/[\boldsymbol{u}])$. Take in Λ a set $\boldsymbol{\eta} = (\eta_1, \ldots, \eta_n)$ whose residue in $\Lambda/[\boldsymbol{u}]$ is a basis. From (2.2.2) we obtain

$$\frac{d}{dt}\begin{pmatrix} \eta_1 \\ \vdots \\ \eta_n \end{pmatrix} = F\begin{pmatrix} \eta_1 \\ \vdots \\ \eta_n \end{pmatrix} + \sum_{\alpha=1}^{\nu} G_\alpha \frac{d^\alpha}{dt^\alpha}\begin{pmatrix} u_1 \\ \vdots \\ u_m \end{pmatrix} \tag{2.2.3}$$

where $F \in k^{n\times n}$, $G_\alpha \in k^{n\times m}$. We say that $\boldsymbol{\eta}$ is a *generalized state*, and (2.2.3) a *generalized state variable representation.*

Let $\tilde{\boldsymbol{\eta}} = (\tilde{\eta}_1, \ldots, \tilde{\eta}_n)$ be another generalized state. As the residues of $\boldsymbol{\eta}$ and $\tilde{\boldsymbol{\eta}}$ in $\Lambda/[\boldsymbol{u}]$ are bases, we obtain

$$\begin{pmatrix} \tilde{\eta}_1 \\ \vdots \\ \tilde{\eta}_n \end{pmatrix} = P\begin{pmatrix} \eta_1 \\ \vdots \\ \eta_n \end{pmatrix} + \sum_{finite} Q_\gamma \frac{d^\gamma}{dt^\gamma}\begin{pmatrix} u_1 \\ \vdots \\ u_m \end{pmatrix} \tag{2.2.4}$$

where $P \in \mathrm{GL}_k(n)$, $Q_\gamma \in k^{n\times p}$. Note that the *generalized state variable transformation* (2.2.4) depends in general from the input and a finite number of its derivatives.

Assume that $\nu \geqslant 1$ and $J_\nu \neq 0$ in (2.2.3). Following (2.2.4) set

$$\begin{pmatrix} \eta_1 \\ \vdots \\ \eta_n \end{pmatrix} = \begin{pmatrix} \overline{\eta}_1 \\ \vdots \\ \overline{\eta}_n \end{pmatrix} + G_\nu \frac{d^{\nu-1}}{dt^{\nu-1}}\begin{pmatrix} u_1 \\ \vdots \\ u_m \end{pmatrix}$$

It yields

$$\frac{d}{dt}\begin{pmatrix} \overline{\eta}_1 \\ \vdots \\ \overline{\eta}_n \end{pmatrix} = F\begin{pmatrix} \overline{\eta}_1 \\ \vdots \\ \overline{\eta}_n \end{pmatrix} + \sum_{\alpha=1}^{\nu-1} \overline{G}_\alpha \frac{d^\alpha}{dt^\alpha}\begin{pmatrix} u_1 \\ \vdots \\ u_m \end{pmatrix}$$

The maximal order of derivation of $\boldsymbol{u}$ is at most $\nu - 1$. By induction we get

$$\frac{d}{dt}\begin{pmatrix} x_1 \\ \vdots \\ x_n \end{pmatrix} = A\begin{pmatrix} x_1 \\ \vdots \\ x_n \end{pmatrix} + B\begin{pmatrix} u_1 \\ \vdots \\ u_m \end{pmatrix} \tag{2.2.5}$$

where $A \in k^{n\times n}$, $B \in k^{n\times m}$. We say that (2.2.5) is a *Kalman state variable representation.* The n-tuple $\boldsymbol{x} = (x_1, \ldots, x_n)$ is a *Kalman state.* Two

Kalman states $\boldsymbol{x}$ and $\tilde{\boldsymbol{x}} = (\tilde{x}_1, \ldots, \tilde{x}_n)$ are related by

$$\begin{pmatrix} \tilde{x}_1 \\ \vdots \\ \tilde{x}_n \end{pmatrix} = P \begin{pmatrix} x_1 \\ \vdots \\ x_n \end{pmatrix} \tag{2.2.6}$$

where $P \in \mathrm{GL}_k(n)$. As a matter of fact the presence of $\boldsymbol{u}$ and of its derivatives as in (2.2.4) would yield in (2.2.5) derivatives of $\boldsymbol{u}$ of order $\geqslant 1$. We have proved [10] the

Theorem 2.2.1 *Any $k[\frac{d}{dt}]$-dynamics admits a Kalman state variable representation (2.2.5). Two Kalman states are related by (2.2.6).*

2.3 Controllability

2.3.1 Definition

The three notions of free, projective and torsion free controllability over $k[\frac{d}{dt}]$ coincide. A $k[\frac{d}{dt}]$-system Λ is therefore said to be *controllable* [10], if the $k[\frac{d}{dt}]$-module Λ is free.

2.3.2 Comparison

The dynamics (2.2.5), where u is assumed to be independent, is said to be *controllable à la Kalman* if

$$\mathrm{rk}(B, AB, \ldots, A^{n-1}B) = n$$

Theorem 2.3.1 *The Kalman state variable representation (2.2.5) is controllable if, and only if, it is controllable à la Kalman.*

Proof. First assume that (2.2.5) is uncontrollable in the Kalman sense. Write the uncontrollable part in the Kalman decomposition:

$$\frac{d}{dt}\begin{pmatrix} \xi_1 \\ \vdots \\ \xi_{n_0} \end{pmatrix} = F \begin{pmatrix} \xi_1 \\ \vdots \\ \xi_{n_0} \end{pmatrix}$$

where $F \in k^{n_0 \times n_0}$. We know that the corresponding module is torsion. Therefore the module Λ corresponding to (2.2.5) cannot be free. A similar argument shows that Λ free corresponds to a realization (2.2.5) which is controllable in the sense of Kalman.

Remark 2.3.1 *See [11] for the comparison with Willems' behavioral approach [56].*

3 Delay systems and controllability

Let R be the ring $k[\frac{d}{dt}, \delta_1, \ldots, \delta_r] = k[\frac{d}{dt}, \boldsymbol{\delta}]$ of polynomials in $r+1$ indeterminates over a commutative field k, where the δ_i's are (*localized*) *delay* operators of non commensurate amplitudes.

3.1 Controllability

The resolution of Serre's conjecture [52] due to Quillen [45] and Suslin [55] (*see also* [29, 59] for a detailed exposition) states that, on a polynomial ring, a projective module is free. Thus, in the present context, Quillen-Suslin's theorem may be stated as [17]:

Proposition 3.1.1 *A delay system is $k[\frac{d}{dt}, \boldsymbol{\delta}]$-free controllable if, and only if, it is $k[\frac{d}{dt}, \boldsymbol{\delta}]$-projective controllable.*

Very many notions can then be considered (through torsion freeness and freeness on the one hand, and through the variation of the ground ring on the other hand). Among these, the $k[\frac{d}{dt}, \boldsymbol{\delta}]$-free controllability is certainly the most appealing from an algebraic viewpoint. The existence of a basis is an extremely useful feature; but this notion seems quite rare in practice (*see, e. g.,* [39, 41]). The π-freeness retains the main advantage of freeness (existence of a basis) while being almost always satisfied in applications. Indeed, we have [17]

Proposition 3.1.2 *A $k[\frac{d}{dt}, \boldsymbol{\delta}]$-torsion free controllable $k[\frac{d}{dt}, \boldsymbol{\delta}]$-system Λ is π-free, where π may be chosen in $k[\boldsymbol{\delta}]$.*

3.1.1 Criteria for $k[\frac{d}{dt}, \boldsymbol{\delta}]$-free and $k[\frac{d}{dt}, \boldsymbol{\delta}]$-torsion free controllability

We establish [17] two criteria for $k[\frac{d}{dt}, \boldsymbol{\delta}]$-free and $k[\frac{d}{dt}, \boldsymbol{\delta}]$-torsion free controllability. The first one uses the resolution of Serre's conjecture [45, 55], and the second one[2] uses [58].

Theorem 3.1.1 *A delay system Λ with presentation matrix P_Λ of full generic rank β is $k[\frac{d}{dt}, \boldsymbol{\delta}]$-free controllable if, and only if,*

$$\forall (s, z_1, \ldots, z_r) \in \bar{k}^{r+1}, \quad rk_{\bar{k}}\, P_\Lambda(s, z_1, \ldots, z_r) = \beta$$

where $\bar{k}$ is the algebraic closure of k. This rank criterion is equivalent to the common minors of P_Λ of order β having no common zero in $\bar{k}^{r+1}$.

Theorem 3.1.2 *A delay system Λ is $k[\frac{d}{dt}, \boldsymbol{\delta}]$-torsion free controllable if, and only if, the gcd of the $\beta \times \beta$ minors of P_Λ belongs to k.*

[2] *See* [60] for related results.

Examples 3.1.1 1) *The system* $\dot{y} + \delta y = u$ *is* $k[\frac{d}{dt}, \boldsymbol{\delta}]$*-free controllable, with basis* y.
2) *The system* $\dot{y} = \delta u$ *is* $k[\frac{d}{dt}, \boldsymbol{\delta}]$*-torsion free controllable, but not* $k[\frac{d}{dt}, \boldsymbol{\delta}]$*-free controllable.*

3.1.2 π–freeness and transmission delay

Suppose we have two $k[\frac{d}{dt}]$–systems Λ_1 and Λ_2 in cascade with a transmission delay operator $\pi \in k[\boldsymbol{\delta}]$, $\pi \notin k$, between them. Denote this cascade by $\Lambda_1 \star \pi \star \Lambda_2$ and by $\Lambda_1 \star \Lambda_2$ the two systems in series without any transmission delay between them ($\Lambda_1 \star \Lambda_2$ is thus a $k[\frac{d}{dt}]$–system). For the sake of simplicity, take two $k[\frac{d}{dt}]$–systems with equations

$$A_i(\tfrac{d}{dt})\, \boldsymbol{y}_i = B_i(\tfrac{d}{dt})\, \boldsymbol{u}_i \tag{3.1.7}$$

where $\boldsymbol{u}_i$ and $\boldsymbol{y}_i$ have m elements, A_i, $B_i \in k[\frac{d}{dt}]^{m\times m}$, $i = 1, 2$, and

$$\boldsymbol{u}_2 = \pi \boldsymbol{y}_1$$

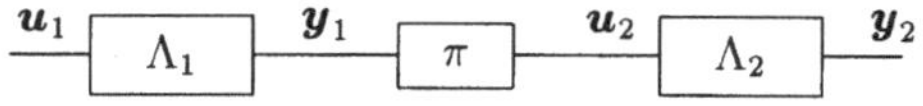

Figure 1 : Cascade connection.

Set $\boldsymbol{u} = \boldsymbol{u}_1$, $\boldsymbol{y} = \boldsymbol{y}_2$ and $\Lambda_1 \star \pi \star \Lambda_2 = [\boldsymbol{u}, \boldsymbol{y}_1, \boldsymbol{y}]$ with equations (3.1.7). We have the following result [17], which can be seen as a complement to [13]:

Proposition 3.1.3 *The* $k[\frac{d}{dt}]$*–system* $\Lambda_1 \star \Lambda_2$ *is controllable if, and only if, the delay system* $\Lambda_1 \star \pi \star \Lambda_2$ *is* π*–free.*

The following particular case of Proposition 3.1.3 is stated in a more classical fashion.

Corollary 3.1.1 *Consider the* $k[\frac{d}{dt}, \delta]$*-dynamics* $\Gamma = [\boldsymbol{x}, \boldsymbol{u}]$ *with equations*

$$\dot{\boldsymbol{x}} = F\boldsymbol{x} + G\pi\, \boldsymbol{u} \tag{3.1.8}$$

where $F \in k^{n\times n}$, $G \in k^{n\times m}$, $\pi \in k[\delta]$, $\pi \notin k$, *and the* $k[\frac{d}{dt}]$*-dynamics* $\widetilde{\Gamma} = [\tilde{\boldsymbol{x}}, \tilde{\boldsymbol{u}}]$ *with equations*

$$\dot{\tilde{\boldsymbol{x}}} = F\tilde{\boldsymbol{x}} + G\tilde{\boldsymbol{u}}$$

The following two properties are equivalent:

(i) The $k[\frac{d}{dt}, \delta]$*-dynamics* Γ *is* π*-free;*

(ii) The $k[\frac{d}{dt}]$*-dynamics* $\widetilde{\Gamma}$ *is controllable à la Kalman.*

Remark 3.1.1 *Arstein [2] has proved that dynamics of type* (3.1.8), *wich are most often encountered in practice (*see, e. g., *[39, 41]), may be reduced to the classic Kalman dynamics* (2.2.5). See *[18] for preliminary structural results on* (3.1.8).

3.1.3 Reachability, weak controllability

Consider [17] the $k[\frac{d}{dt}, \boldsymbol{\delta}]$–dynamics $\Gamma = [\boldsymbol{x}, \boldsymbol{u}]$ with equations

$$\dot{\boldsymbol{x}} = F(\boldsymbol{\delta})\,\boldsymbol{x} + G(\boldsymbol{\delta})\,\boldsymbol{u}$$

where $\boldsymbol{x} = (x_1, \ldots, x_n)$, $\boldsymbol{u} = (u_1, \ldots, u_m)$, and the matrices $F(\boldsymbol{\delta}) \in k[\boldsymbol{\delta}]^{n\times n}$ and $G(\boldsymbol{\delta}) \in k[\boldsymbol{\delta}]^{n\times m}$. The classic notions of *reachability* and *weak controllability* may be found in [38, 53].

Proposition 3.1.4 *The dynamics Γ is reachable if, and only if, Γ is free controllable over $k[\frac{d}{dt}, \boldsymbol{\delta}]$.*

Proposition 3.1.5 *The dynamics Γ is weakly controllable if, and only if, the $k(\boldsymbol{\delta})[\frac{d}{dt}]$-module $k(\boldsymbol{\delta})[\frac{d}{dt}] \otimes_{k[\frac{d}{dt}, \boldsymbol{\delta}]} \Gamma$ is free, where $k(\boldsymbol{\delta})$ denotes the quotient field of $k[\boldsymbol{\delta}]$.*

3.1.4 Spectral controllability

We use the ring $k[s, e^{-\boldsymbol{h}s}]$, viewed as a subring of the convergent power series ring $k\{\{s\}\}$ (where s plays the role of $\frac{d}{dt}$, and $e^{-\boldsymbol{h}s} = (e^{-h_1 s}, \ldots, e^{-h_r s})$, the h_i's ($h_i \in \mathbb{R}, h_i > 0$) being the *amplitudes* of the corresponding delays). The mapping $\frac{d}{dt} \mapsto s$, $\delta_i \mapsto e^{-h_i s}$ yields an isomorphism between the rings $k[\frac{d}{dt}, \boldsymbol{\delta}]$ and $k[s, e^{-\boldsymbol{h}s}]$. Thus, by a slight abuse of language, a finitely generated $k[s, e^{-\boldsymbol{h}s}]$–module will still be called a delay system [17, 40].

The following definition of *spectral controllability* extends previous ones (*see, e. g.*, [4, 48]) in our context.

Definition 3.1.1 *Let Λ be a delay system defined over the ring $k[s, e^{-\boldsymbol{h}s}]$, with presentation matrix P_Λ of full generic rank β. It is called* spectrally controllable *if*

$$\forall s \in \mathbb{C}, \quad rk_{\mathbb{C}}\, P_\Lambda(s, e^{-\boldsymbol{h}s}) = \beta$$

Set $\mathfrak{S}_r = k(s)[e^{-\boldsymbol{h}s}, e^{\boldsymbol{h}s}] \cap \mathfrak{E}$, where $\mathfrak{E}$ denotes the ring of entire functions. We have the following interpretation of spectral controllability [40]:

Proposition 3.1.6 *Let Λ be a delay system over $k[s, e^{-\boldsymbol{h}s}]$, such that Λ is $k[s, e^{-\boldsymbol{h}s}, e^{\boldsymbol{h}s}]$–torsion free controllable. Then Λ is spectrally controllable if, and only if, it is $\mathfrak{S}_r$–torsion free controllable.*

The following result [17] gives implication relationships between the notion of $\delta_1^{\alpha_1} \ldots \delta_r^{\alpha_r}$-freeness, $\alpha_1 + \ldots + \alpha_r > 0$, and the above quoted ones.

Proposition 3.1.7 *Let Λ be a $k[\frac{d}{dt}, \boldsymbol{\delta}]$-system. The following chain of implications is true*

$$\Lambda \text{ spectrally controllable} \Longrightarrow \Lambda \;\; \delta_1^{\alpha_1} \ldots \delta_r^{\alpha_r}\text{–free}$$
$$\Longrightarrow \Lambda \;\; k[\tfrac{d}{dt}, \boldsymbol{\delta}]\text{–torsion free.}$$

Proof. The proof follows directly from the inclusion chain

$$k[\tfrac{d}{dt}, \boldsymbol{\delta}] \subset k[\tfrac{d}{dt}, \boldsymbol{\delta}, \boldsymbol{\delta}^{-1}] \subset \mathfrak{S}_r.$$

3.2 The wave equation

3.2.1 The torsional behavior of a flexible rod

Consider [41] the torsional behavior of a flexible rod with a torque applied to one end. A mass is attached to the other end. The system is described by the one dimensional wave equation.

$$\begin{aligned} \sigma^2 \frac{\partial^2 q}{\partial \tau^2}(\tau, z) &= \frac{\partial^2 q}{\partial z^2}(\tau, z) \\ \frac{\partial q}{\partial z}(\tau, 0) = -u(\tau), &\quad \frac{\partial q}{\partial z}(\tau, L) = -J\frac{\partial^2 q}{\partial \tau^2}(\tau, L) \\ q(0, z) = q_0(z), &\quad \frac{\partial q}{\partial \tau}(0, z) = q_1(z) \end{aligned} \tag{3.2.9}$$

Here $q(\tau, z)$ denotes the angular displacement from the unexcited position at a point $z \in [0, L]$ at time $\tau \geqslant 0$, as shown in Figure 2; L is the length of the rod, σ the inverse of the wave propagation speed, J the inertial momentum of the mass, $u(\tau)$ the control torque and q_0, q_1 describe the initial angular displacement and velocity, respectively.

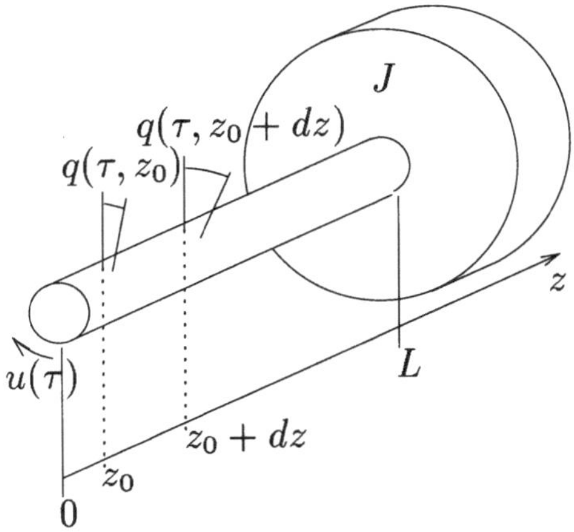

Figure 2 : The flexible rod

3.2.2 Delay system model

As well known, the general solution of (3.2.9) may be written

$$q(\tau, z) = \phi(\tau + \sigma z) + \psi(\tau - \sigma z)$$

where ϕ and ψ are one variable functions. The control objective will be to assign a trajectory to the angular position of the mass; the output is thus

$$y(\tau) = q(\tau, L)$$

Set $t = (\sigma/J)\tau$, $v(t) = (2J/\sigma^2)u(t)$ and $T = \sigma L$. Easy calculations (*see* [41] for details) yield the following delay system (compare with [8]):

$$\ddot{y}(t) + \ddot{y}(t - 2T) + \dot{y}(t) - \dot{y}(t - 2T) = v(t - T) \tag{3.2.10}$$

One readily has

$$v = (\delta^{-1} + \delta)\ddot{y} + (\delta^{-1} - \delta)\dot{y} \tag{3.2.11}$$

which implies

Proposition 3.2.1 *System* (3.2.10) *is δ-free, with basis y.*

3.2.3 Tracking

Equation (3.2.11), yields the open loop control (*see* figure 4)

$$v_d(t) = \ddot{y}_d(t + T) + \ddot{y}_d(t - T) + \dot{y}_d(t + T) - \dot{y}_d(t - T)$$

The displacements of the other points of the rod (*see* figure 5) can be obtained as (*see* [41])

$$q_d(z, t) = \frac{1}{2}\Big[y_d(t - z + T) + \dot{y}_d(t - z + T) + y_d(t - T + z) - \dot{y}_d(t - T + z)\Big]$$

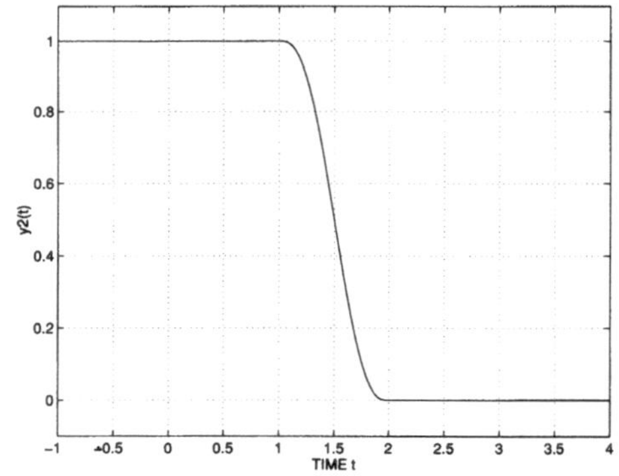

Figure 3 : The desired output y_2.

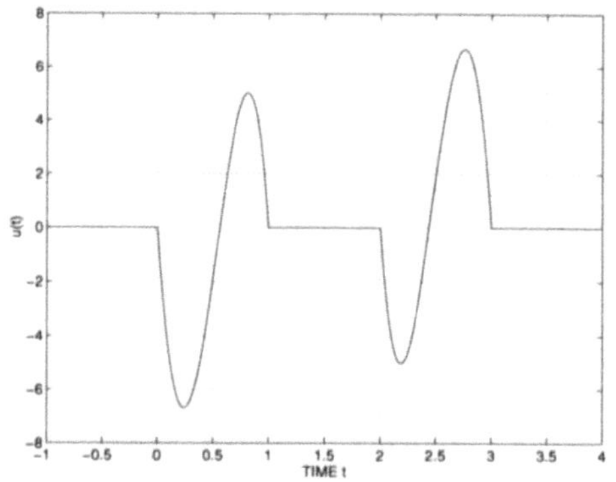

Figure 4 : The control u

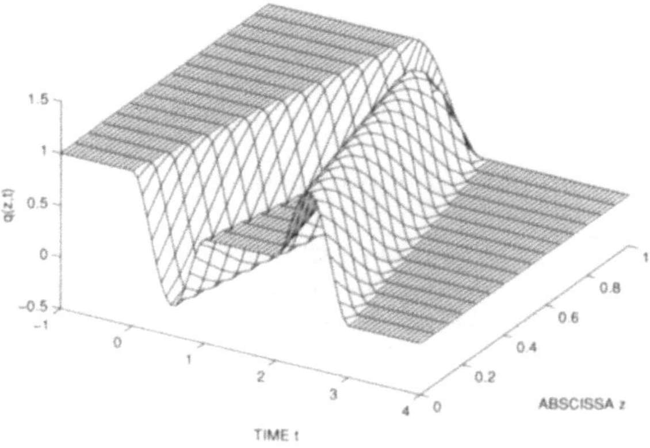

Figure 5 : Angular displacements $q(z,t)$

Remark 3.2.1 *Stabilization around the reference trajectory may be achieved by standard passivity methods or by the following feedback loop [41]*

$$v(t) = \lambda_0 \ddot{y}(t-T) - 2\dot{y}(t-T) + \lambda_1 y(t-T)$$

with $\lambda_0 \in]0,2[$, $\lambda_0 \neq 1$ *and* $\lambda_1 < 1/(\lambda_0 - 2)$ *(*see *[34]). Note that such type of feedback involving past derivatives of the state has already been used for stabilization purposes (*see, e. g., *[5]).*

4 Heat and Euler-Bernoulli equations

4.1 Mikusiński's operational calculus

The set of continuous functions $[0, +\infty) \to \mathbb{C}$ is a commutative ring $\mathcal{C}$ with respect to the pointwise addition $+$ where

$$(f+g)(t) = f(t) + g(t)$$

and the convolution product $\star$ where

$$(f \star g)(t) = (g \star f)(t) = \int_0^t f(\tau)g(t-\tau)d\tau = \int_0^t g(\tau)f(t-\tau)d\tau$$

According to a famous theorem due to Titchmarsh (*see* [36, 37, 57]), $\mathcal{C}$ does not possess zero divisors, *i. e.*,

$$f \star g = 0 \quad \Leftrightarrow \quad f = 0 \text{ or } g = 0$$

The quotient field $\mathcal{M}$ of $\mathcal{C}$ is called the *Mikusiński field.* Any element of $\mathcal{M}$ is called an *operator.*
Notations 1) A function $f(t)$ in $\mathcal{C}$ is sometimes written $\{f(t)\}$ when viewed as an operator in $\mathcal{M}$.
2) The (convolution) product of two operators $a, b \in \mathcal{M}$ is written ab.

Examples 4.1.1 1) *The neutral element* 1 *for the (convolution) product is the* Dirac operator. *It is the analogue of the Dirac distribution in Schwartz's distribution theory. The Dirac operator* 1 *should not be confused with the Heaviside function* $\{1\} \in \mathcal{C}$.
2) *The inverse in* $\mathcal{M}$ *of the Heaviside function* $\{1\}$ *is the differential operator* s. *It obeys to the classic rules,* i. e., *if* $f \in \mathcal{C}$ *is* C^1, *then* $\{sf\} = \{\dot{f}\} + \{f(0)\}$. *The meaning of operators in the subfield* $\mathbb{C}(s)$ *of rational functions in the variable* s *with complex coefficients is clear. The* fractional derivative $\sqrt{s}$ *appears as the inverse of* $\frac{1}{\sqrt{s}} = \{\frac{1}{\sqrt{2\pi t}}\}$.
3) *The field* $\mathcal{M}$ *contains the subring* S *of piecewise continuous functions* $\mathbb{R} \to \mathbb{R}$ *with left bounded supports,* i. e., *for any* $f \in S$, *there exists a constant* $\beta \in \mathbb{R}$, *such that, for* $t < \beta$, $f(t) = 0$. *The* translation operator e^{-hs}, $h \in \mathbb{R}$ *acts on* $f \in S$ *by* $e^{-hs}\{f(t)\} = \{f(t-h)\}$. *The inverse of* e^{-hs} *is* e^{hs}. *Notice that* $e^{s\sqrt{-1}}$ *is not an operator,* i. e., *it does not belong to* $\mathcal{M}$.

A sequence a_n, $n \geqslant 0$, of operators is said to be *operationally convergent* [36, 37] if there exists an operator p such that the $a_n p$'s belong to $\mathcal{C}$ and converge almost uniformly, *i. e.*, uniformly on any finite interval, to a function in $\mathcal{C}$. A series of operators $\sum_{\nu \geqslant 0} b_\nu$ is said to be *operationally convergent* if the sequence $\sum_{\nu=0}^{n} b_n$ is operationally convergent.

Example 4.1.1 *The operator $e^{\lambda\sqrt{s}}$, $\lambda \in \mathbb{C}$, may be defined by its Taylor expansion*

$$e^{\lambda\sqrt{s}} = \sum_{n\geqslant 0} \frac{\lambda^n s^{\frac{n}{2}}}{n!} \tag{4.1.12}$$

which is operationally convergent [37]. This property does not hold for the translation operator e^{-hs}, $h \in \mathbb{R}$.

An *operational function* [36] is a mapping $\mathbb{R} \to \mathcal{M}$. One can define the continuity, differentiability and integrability of operational functions.

4.2 Gevrey functions

A C^∞-function $\zeta(t)$ of the real variable t is said to be *Gevrey* [22][3] of *order*, or *class*, $\mu \geqslant 1$ if for any compact subset $K \subset \mathbb{R}$ and for any integer $n \geqslant 0$

$$|\zeta^{(n)}(t)| \leqslant C^{n+1}(n!)^\mu \qquad t \in K$$

where $C > 0$ is a constant depending on ζ and K.

The function

$$\phi(t) = \begin{cases} 0 & \text{if } t \leqslant 0 \\ e^{-1/t^d} & \text{if } t > 0 \end{cases}$$

where $d > 0$, is *flat*[4] at $t = 0$, *i. e.*, $\phi^{(\nu)}(0) = 0$ for $\nu \geqslant 1$. It is not analytic but Gevrey of class $(1 + d)/d$ (*see, e. g.*, [49]). The function

$$\eta_d(t) = \begin{cases} 0 & \text{if } t < 0 \\ \dfrac{\displaystyle\int_0^{t/T} \exp(-1/(\tau(1-\tau))^d)d\tau}{\displaystyle\int_0^1 \exp(-1/(\tau(1-\tau))^d)d\tau} & \text{if } t \in [0, T] \\ 1 & \text{if } t > T \end{cases} \tag{4.2.13}$$

is also Gevrey of class $(1 + d)/d$. It is flat at $t = 0$ and $t = 1$.

4.3 The heat equation

In order to illustrate our tools consider a boundary control problem of the heat equation

$$\frac{\partial^2 w(x,t)}{\partial x^2} = \frac{\partial w(x,t)}{\partial t} \qquad 0 < x < 1,\ t > 0 \tag{4.3.14}$$

---

[3] This notion has now become classic among analysts (*see, e. g.*, [24, 27, 33, 46, 49]). *See*, also, [23] and [37] for a slightly different setting.

[4] The word *flat* possesses a large variety of mathematical meanings. Think of flat coordinates, flat modules, etc.

The initial condition is $w(x,0) = 0$. Its boundary conditions are $\frac{\partial w(0,t)}{\partial x} = 0$ and $w(1,t) = u(t)$ where $u(t)$ designates the control variable. Operational calculus is replacing (4.3.14) with the ordinary differential equation, where x is the independent variable,

$$\hat{w}_{xx}(x,s) - s\hat{w}(x,s) = 0 \tag{4.3.15}$$

subject to the boundary conditions $\hat{w}_x(0,s) = 0$, $\hat{w}(1,s) = \hat{u}$ ($\hat{u}$ is an operator and $\hat{w}$ an operator function). The solution of (4.3.15) reads

$$\hat{w}(x,s) = \frac{\cosh(x\sqrt{s})}{\cosh(\sqrt{s})}\ \hat{u}(s) \tag{4.3.16}$$

Replace (4.3.16) by

$$\cosh(\sqrt{s})\ \hat{w} \ = \ \cosh(x\sqrt{s})\ \hat{u}$$

which defines a $\mathbb{C}[P,Q]$-module M, where $P = \cosh(x\sqrt{s})$, $Q = \cosh(\sqrt{s})$, which is torsion free, but not free. The localized $\mathbb{C}[P,Q,(PQ)^{-1}]$-module

$$\mathbb{C}[P,Q,(PQ)^{-1}] \otimes_{\mathbb{C}[P,Q]} M$$

is free, of rank 1, with basis $\hat{\zeta} = \hat{w}(0,s)$:

$$\hat{w} = \cosh(x\sqrt{s})\ \hat{\zeta}$$
$$\hat{u} = \cosh(\sqrt{s})\ \hat{\zeta}$$

When setting

$$w(x,t) = \Big(\sum_{n\geqslant 0} \frac{x^{2n}}{(2n)!}\frac{d^n}{dt^n}\Big)\,\zeta(t) \tag{4.3.17}$$

$$u(t) = \Big(\sum_{n\geqslant 0} \frac{1}{(2n)!}\frac{d^n}{dt^n}\Big)\,\zeta(t) \tag{4.3.18}$$

we assume that the function $\zeta(t)$, which corresponds to the operator $\hat{\zeta}$, is *flat* at $t = 0$.

Theorem 4.3.1 *The series* $\sum_{n\geqslant 0} \frac{x^{2n}}{(2n)!}\ \zeta^{(n)}(t)$ *and* $\sum_{n\geqslant 0} \frac{1}{(2n)!}\ \zeta^{(n)}(t)$ *are absolutely convergent if, and only if,* ζ *is a Gevrey function of class strictly less than* 2*. Moreover, if* ζ *is flat at* $t = 0$*, then their sums are respectively equal to* $\cosh(x\sqrt{s})\ \{\zeta\}$ *and* $\cosh(\sqrt{s})\ \{\zeta\}$.

Proof. The absolute convergence of the series follows at once from Stirling's formula

$$n! \sim e^{-n}n^{n-\frac{1}{2}}\sqrt{2\pi}, \qquad n \to +\infty$$

The second part follows from the operational convergence of the expansion (4.1.12).

4.4 The Euler-Bernoulli equation

4.4.1 The flexural behavior of a flexible rod

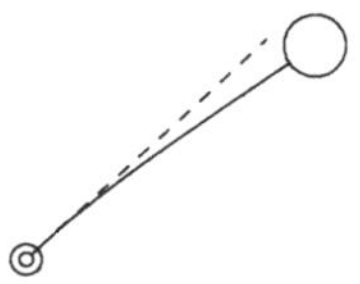

Figure 6 : A rotating Euler-Bernoulli beam with an end mass.

Take a flexible beam of length R, the end $x = 0$ of which is clamped into a motor's axle with angle θ, the other end $x = R$ being a mass m [19]. Suppose that its motion obeys to the Euler–Bernoulli equation (linear elasticity, weak flexion, inertia of the beam negligible with respect to the mass m, Coriolis forces negligible, *i. e.*, $\dot{\theta}$ small) with the boundary conditions:

$$\begin{aligned} &\frac{\partial^2 w}{\partial \tau^2} = -\frac{\partial^4 w}{\partial^4 x} && (4.4.19)\\ &w(0,\tau) = 0, \qquad \frac{\partial w(0,\tau)}{\partial x} = 0 \\ &\frac{\partial^2 w(1,\tau)}{\partial x^2} = 0, \qquad \frac{\partial^3 w(1,\tau)}{\partial x^3} = k_1 \frac{d^2}{d\tau^2}\theta(\tau) + k_2 \frac{\partial^2 w(1,\tau)}{\partial \tau^2} \end{aligned}$$

where $w(x,t)$ is the deformations field with respect to the rotating axis of angle θ and where $\frac{d^2\theta}{d\tau^2}$ is the control. For notational simplicity, the quantity $u(\tau) = k_1\theta(\tau) + k_2 w(1,\tau)$, will play the role of an input variable. The following parameters are the time $t = R^2\sqrt{\rho S/(EI)}\ \tau$, the length $r = Rx$ where E, I, R, ρ and S are the usual physical quantities.

4.4.2 Operational calculus and controllability

With initial conditions $w(x,0) = 0$, $\frac{\partial w(x,0)}{\partial \tau} = 0$, operational calculus associates to (4.4.19) the ordinary differential equation $s^2\hat{w} = -\hat{w}^{(4)}$. Its general solution reads

$$\hat{w}(x,s) = ae^{x\xi\sqrt{s}} + be^{-x\xi\sqrt{s}} + ce^{x\bar{\xi}\sqrt{s}} + de^{-x\bar{\xi}\sqrt{s}}$$

where $\xi = \exp(i\pi/4)$ and a, b, c, d are determined from the boundary conditions. After some calculations [19], one has

$$\hat{u} = -2\Big[2 + \cosh(\sqrt{2s}) + \cosh(i\sqrt{2s})\Big]\,\hat{y}$$

$$\begin{aligned}\hat{w}(x,s) = \Big[&(\cosh(\xi\sqrt{s}) + \cosh(\bar{\xi}\sqrt{s}))(i\sinh(x\xi\sqrt{s}) + \sinh(x\bar{\xi}\sqrt{s})) \\ &+(i\sinh(\xi\sqrt{s}) - \sinh(\bar{\xi}\sqrt{s}))(-\cosh(x\xi\sqrt{s}) + \cosh(x\bar{\xi}\sqrt{s})\Big]\,\frac{2\sqrt{s}}{\xi}\,\hat{y}\end{aligned}$$

where

$$\hat{y} = \pi^{-1}\frac{(c+d)}{2} \quad \text{and} \quad \pi = \frac{i\sinh(\xi\sqrt{s}) - \sinh(\bar{\xi}\sqrt{s})}{\xi s\sqrt{s}}$$

Set $\hat{u} = \gamma\hat{y}$ and $\hat{w} = \delta\hat{y}$. The next result [19] is obtained in the same manner as in Subsection 4.3.

Theorem 4.4.1 *The $\mathbb{C}[\gamma,\delta]$-module Λ spanned by $\hat{u}$ and $\hat{w}$ is torsion free, of rank 1, but not free. The localized $\mathbb{C}[\gamma,\delta,(\gamma\delta)^{-1}]$-module $\Lambda_{\gamma,\delta}$ is free, with basis $\hat{y}$.*

4.4.3 Tracking

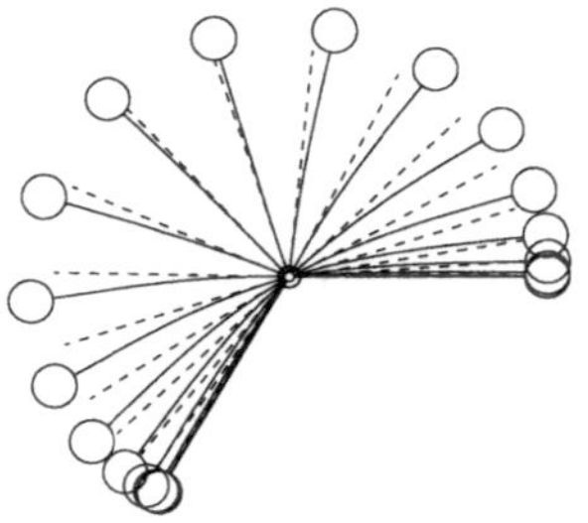

Figure 7 : The various positions of the beam through the time.

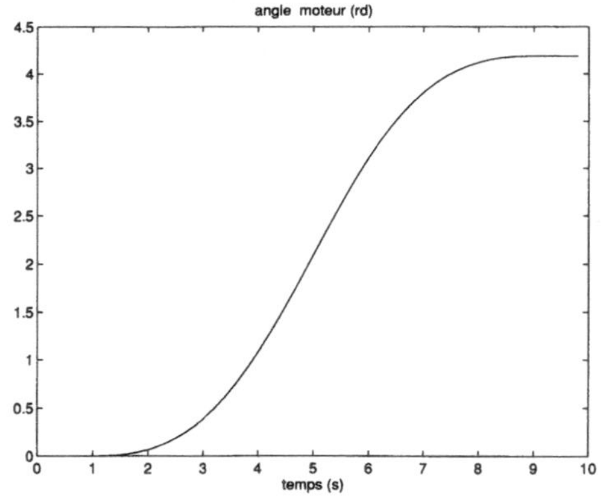

Figure 8 : Trajectory of $\theta(t)$.

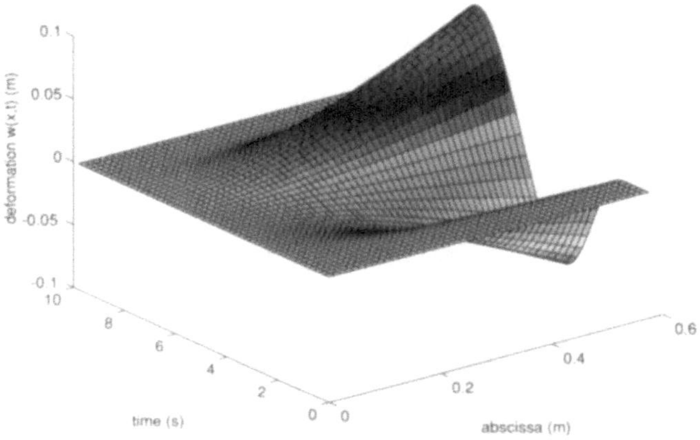

Figure 9 : The deformation $w(r,t)$.

The calculations [19] of Figures 7, 8 and 9 corresponds to an aluminium rod of 24 g, with a length of 0.575 m and a mass of $m = 250$ g ($EI = 0.0603$ N m^2, $\rho S = 0.0423$ kg m^{-1}). A motion of $4\pi/3$ rad is completed in 10 s with

$$y(\tau) \propto \eta_{10/9}(\tau)$$

where η is of the type (4.2.13). It guarantees the absence of vibrations at the end of the motion. The series, which are obtained as in (4.3.17) and (4.3.18), yield enough precision with only thirty terms.

Remark 4.4.1 *The stabilizing feedback around the reference trajectory [19] is obtained by a passivity based argument (*see *[47] for a complete solution of the stabilization of the Euler-Bernoulli equation).*

Remark 4.4.2 *See [1] for a practical implementation of our control synthesis.*

Conclusion

The comparison with other approaches [3, 6, 9, 28, 31, 32] on the control of partial differential equations will be made in forthcoming papers.

Related methods have been recently employed for linear systems with fractional derivatives [14, 25]. Other works in progress [42, 44] are also exploring the same circle of ideas for some types of infinite dimensional nonlinear systems.

References

[1] Y. AOUSTIN, M. FLIESS, H. MOUNIER, P. ROUCHON and J. RUDOLPH, Theory and practice in the motion planning control of a flexible robot arm using Mikusiński operators, *Proc. 4th Symp. Robotics Control*, Nantes, 1997, pp. 287–293.

[2] Z. ARSTEIN, Linear systems with delayed control: a reduction, *IEEE Trans. Automat. Contr.*, **27**, 1982, pp. 869–879.

[3] A. BENSOUSSAN, G. DA PRATO, M.C. DELFOUR and S.K. MITTER, *Representation and Control of Infinite Dimensional Systems*, vol. 1 & 2, Birkhäuser, Boston, 1992 & 1993.

[4] K. BHAT and H. KOIVO, Modal characterizations of controllability and observability for time-delay systems, *IEEE Trans. Automat. Contr.*, **21**, 1976, pp. 292–293.

[5] C.I. BYRNES, M. SPONG and T.J. TARN, A several complex variables approach to feedback stabilization of linear neutral delay-differential systems, *Math. Systems Theory*, **17**, 1984, pp. 97–133.

[6] R.F. CURTAIN and H.J. ZWART, *An Introduction to Infinite Dimensional Linear Systems Theory*, Springer, New York, 1995.

[7] G. DOETSCH, *Theorie und Anwendung der Laplace-Transformation*, Springer, Berlin, 1937.

[8] S. DRAKUNOV and U. ÖZGÜNER, Generalized sliding modes for manifold control of distributed parameter systems, in *Variable Structure and Lyapounov Control*, A. S. Zinober, Ed., Lect. Notes Control Inform. Sci. **193**, pp. 109–129. Springer, London, 1994.

[9] A. EL JAI and A.J. PRITCHARD, *Capteurs et actionneurs dans l'analyse des systèmes distribués*, Masson, Paris, 1986.

[10] M. FLIESS, Some basic structural properties of generalized linear systems, *Systems Control. Lett.*, **15**, 1990, pp. 391-396.

[11] M. FLIESS, A remark on Willems' trajectory characterization of linear controllability, *Systems Control Lett.*, **19**, 1992, pp. 43-45.

[12] M. FLIESS, Une interprétation algébrique de la transformation de Laplace et des matrices de transfert, *Linear Algebra Appl.*, **203-204**, 1994, pp. 429-442.

[13] M. FLIESS and H. BOURLÈS, Discussing some examples of linear system interconnections, *System Control Lett.*, **27**, 1996, pp. 1-7.

[14] M. FLIESS and R. HOTZEL, Sur les systèmes linéaires à dérivation non entière, *C.R. Acad. Sci. Paris*, **IIb-324**, 1997, pp. 99-105.

[15] M. FLIESS, J. LÉVINE, P. MARTIN and P. ROUCHON, Flatness and defect of non-linear systems: introductory theory and applications, *Internat. J. Control*, **61**, 1995, pp. 1327-1361.

[16] M. FLIESS, J. LÉVINE, P. MARTIN and P. ROUCHON, A Lie-Bäcklund approach to equivalence and flatness of nonlinear systems, *IEEE Trans. Automat. Contr.*, to appear.

[17] M. FLIESS and H. MOUNIER, Controllability and observability of linear delay systems: an algebraic approach, *ESAIM COCV*, **3**, 1998, pp. 301–314.
URL: `http://www.emath.fr/Maths/Cocv/cocv.html`

[18] M. FLIESS and H. MOUNIER, Quasi-finite linear delay systems: theory and applications, *Proc. IFAC Workshop Linear Time Delay Systems*, pp. 211–215, Grenoble, 1998.

[19] M. FLIESS, H. MOUNIER, P. ROUCHON and J. RUDOLPH, Systèmes linéaires sur les opérateurs de Mikusiński et commande d'une poutre flexible, *ESAIM Proc.*, **2**, 1997, pp. 183–193.
URL: `http://www.emath.fr/Maths/Proc/Vol.2/fliess/fliess.html`.

[20] M. FLIESS, H. MOUNIER, P. ROUCHON and J. RUDOLPH, Controlling the transient of a chemical reactor: a distributed parameter approach, *Proc. CESA 98*, Hammamet, Tunisia, 1998, to appear.

[21] M. FLIESS, H. MOUNIER, P. ROUCHON and J. RUDOLPH, A distributed parameter approach to the control of a tubular reactor: a multi-variable case, *Proc. CDC*, 1998, to appear.

[22] M. GEVREY, La nature analytique des solutions des équations aux dérivées partielles, *Ann. Sci. Éc. Norm. Sup.*, **25**, 1918, pp. 125–190.

[23] I.M. GUELFAND and G.E. CHILOV, *Les distributions*, t. 2 & 3, Dunod, Paris, 1964 & 1965.

[24] L. HÖRMANDER, *The Analysis of Partial Differential Operators*, vol. 1, Springer, Berlin, 1983.

[25] R. HOTZEL, Contributions à la théorie structurelle et la commande des systèmes linéaires fractionnaires, *Thèse*, Université Paris-Sud, Orsay, 1998.

[26] R.E. KALMAN, L. FALB and M.A. ARBIB, *Topics in Mathematical System Theory*, McGraw-Hill, New York, 1969.

[27] H. KOMATSU, Microlocal Analysis in Gevrey classes and in complex domains, in *Microlocal Analysis and Applications*, L. Cattabriga and L. Rodino, Eds., *Lect. Notes Math.* **1495**, pp. 161–236. Springer, Berlin, 1991.

[28] V. KOMORNIK, *Exact Controllability and Stabilization*, Wiley, Chichester & Masson, Paris, 1994.

[29] T.Y. LAM, *Serre's Conjecture*. Springer, Berlin, 1978.

[30] S. LANG, *Algebra*, 3rd ed., Addison-Wesley, Reading, MA, 1993.

[31] J.–L. LIONS, *Contrôle optimal des systèmes gouvernés par des équations aux dérivées partielles*, Dunod, Paris, 1968.

[32] J.–L. LIONS, *Contrôlabilité exacte, perturbations et stabilisation des systèmes distribués*, vol. 1 & 2, Masson, Paris, 1988.

[33] J.–L. LIONS and E. MAGENES, *Problèmes aux limites non homogènes et applications*, vol. 3, Dunod, Paris, 1970.

[34] J. MARSHALL, H. GÓRECKI, A. KORYTOWSKI and K. WALTON *Time delay systems stability and performance criteria with applications*, Ellis Horwood, New York, 1992.

[35] P. MARTIN, R.M. MURRAY and P. ROUCHON, Flat Systems, in *Plenary Lectures and Mini-Courses, ECC 97*, G. Bastin and M. Gevers, Eds., pp. 211-264, Brussels, 1997.

[36] J. MIKUSIŃSKI, *Operational Calculus*, vol. 1, Pergamon, Oxford & PWN, Warsaw, 1983.

[37] J. MIKUSIŃSKI and T.K. BOEHME, *Operational Calculus*, vol. 2, Pergamon, Oxford & PWN, Warsaw, 1987.

[38] A.S. MORSE, Ring models for delay-differential systems, *Automatica*, **12**, 1976, pp. 529–531.

[39] H. MOUNIER, Propriétés structurelles des systèmes linéaires à retards: aspects théoriques et pratiques, *Thèse*, Université Paris-Sud, Orsay, 1995.

[40] H. MOUNIER, Algebraic interpretations of the spectral controllability of a linear delay system, *Forum Mathematicum*, **10**, 1998, pp. 39–58.

[41] H. MOUNIER, P. ROUCHON and J. RUDOLPH, Some examples of linear systems with delays, *J. Europ. Syst. Autom.*, **31**, 1997, pp. 911–925.

[42] H. MOUNIER and J. RUDOLPH, Flatness based control of nonlinear delay systems: Example of a class of chemical reactors, *Internat. J. Contr.*, special issue *"Recent Advances in the Control of Non-linear Systems"*, to appear.

[43] H. MOUNIER, J. RUDOLPH, M. FLIESS and P. ROUCHON, Tracking control of a vibrating string with an interior mass viewed as delay system, *ESAIM COCV*, **3**, 1998, pp. 315–321.
URL: `http://www.emath.fr/Maths/Cocv/cocv.html`

[44] N. PETIT, Y. CREFF and P. ROUCHON, Motion planning for two classes of nonlinear systems with delays depending on the control, *Proc. CDC*, 1998, to appear.

[45] D. QUILLEN, Projective modules over polynomial rings, *Inventiones Math.*, **36**, 1976, pp. 167-171.

[46] J.–P. RAMIS, *Séries divergentes et théories asymptotiques*, Soc. Math. France, Marseille, 1993.

[47] B. Rao, Uniform stabilization of a hybrid system of elasticity, *SIAM J. Contr. Optim.*, **33**, 1995, pp. 440–454.

[48] P. Rocha and J.C. Willems, Behavioral controllability of D-D systems. *SIAM J. Contr. Opt.*, **35**, 1987, pp. 254–264.

[49] L. Rodino, *Linear Partial Differential Operators in Gevrey Spaces*, World Scientific, Singapore, 1993.

[50] J. Rotman, *An Introduction to Homological Algebra*, Academic Press, Orlando, 1979.

[51] L.H. Rowen, *Ring Theory*, Academic Press, Boston, 1991.

[52] J.–P. Serre, Faisceaux algébriques cohérents. *Annals. of Math.*, **61**, 1955, pp. 197–278.

[53] E. D. Sontag, Linear systems over commutative rings: a survey, *Richerche di Automatica*, **7**, 1976, pp. 1–34.

[54] M.W. Spong and T.J. Tarn, On the spectral controllability of delay-differential equations, *IEEE Trans. Automat. Contr.*, **26**, 1981, pp. 527–528.

[55] A.A. Suslin, Projectives modules over a polynomial ring are free (in russian), *Dokl. Akad. Nauk. SSSR*, **229**, 1976, pp. 1063-1066 (english translation: *Soviet. Math. Dokl.*, **17**, 1976, pp. 1160-1164).

[56] J.C. Willems, Paradigms and puzzles in the theory of dynamical systems, *IEEE Trans. Automat. Contr.*, **36**, 1991, pp. 259-294.

[57] K. Yosida, *Operational Calculus*, Springer, New York, 1984.

[58] D.C. Youla and G. Gnavi, Notes on n-dimensional system theory, *IEEE Trans. Circuits Syst.*, **26**, 1979, pp. 105–111.

[59] D.C. Youla and P.F. Pickel, The Quillen-Suslin theorem and the structure of n-dimensional elementary polynomial matrices, *IEEE Trans. Circuits Syst.*, **31**, 1984, pp. 513–518.

[60] S. Zampieri, Modellizzazione di Sequenze di Dati Mutlidimensionali, *Tesi*, Università di Padova, 1993.

M.F.: Centre de Mathématiques et de Leurs Applications, École Normale Supérieure de Cachan, 61 avenue du Président Wilson, 94235 Cachan, France. *E-mail*: `fliess@cmla.ens-cachan.fr`

H.M.: Département AXIS, Institut d'Électronique Fondamentale, Bâtiment 220, Université Paris-Sud, 91405 Orsay, France.
E-mail: `mounier@ief.u-psud.fr`

Progress in Systems and Control Theory, Vol. 25

On Canonical Wiener-Hopf Factorizations

P. A. Fuhrmann[1]

1 Introduction

This paper addresses some problems arising in Wiener-Hopf factorizations. The point of interest in this paper is not so much in the results which are slight generalizations of previous results, see Gohberg and Zucker [12, 13] and Zucker [18]. Rather, the main interest is in the different, functional oriented, technique used which allows the establishing of a clearer connection between factorization theory and geometry. That such a connection exists is known to every student of a linear algebra course. Indeed, the computation of a 1-dimensional invariant subspace of a finite dimensional linear transformation via the computation of an eigenvalue and a corresponding eigenvector lead to the factorization of the linear pencil $\lambda I - A$.

The deeper study of the correspondence between factorizations and invariant subspaces began with the work of Livsic and Brodskii on characteristic functions and their factorizations. The next significant result is the work of Sakhnovich [15], see also Bart, Gohberg, Kaashoek and Van Dooren [2], giving a geometric characterization for the existence of a minimal factorization of a rational, normalized biproper function. If $W = \left(\begin{array}{c|c} A & B \\ \hline C & I \end{array}\right)$ is a minimal realization, then a factorization $W = W_1 W_2$ into rational, normalized biproper functions exists if and only if the state space decomposes as $\mathcal{X} = \mathcal{X}_1 \oplus \mathcal{X}_2$, where $\mathcal{X}_1$ is A-invariant and $\mathcal{X}_2$ is $A^\times$-invariant, where $A^\times = A - BC$. If we are interested in the existence of a left canonical Wiener-Hopf factorization, then the subspaces in the previous characterization need to be taken as spectral subspaces and the condition turns out to be the existence of a direct sum representation $\mathcal{X} = \mathcal{X}_+(A) \oplus \mathcal{X}_-(A^\times)$. We note that subspaces of the form $\mathcal{X}_-(A^\times)$ are both controlled as well as conditioned invariant subspaces. Thus it will come as no great surprise that the central results of this paper will be expressed in such terms.

Assuming that a rational matrix function is positive on the imaginary axis, including the point at infinity, its spectral factorization coincides with a Wiener-Hopf factorization. In this case, clearly both factors are invertible at infinity. However, the situation changes when we study canonical factorizations with respect to the unit circle.

[1]Earl Katz Family Chair in Algebraic System Theory. Partially supported by GIF under Grant No. I-526-034.06/97 .

Thus, assuming W is a proper, nonsingular rational function, if $W = W_+W_-$ is a left canonical factorization, with W_-, W_-^{-1} analytic in $|z| > 1$ and W_+, W_+^{-1} analytic in $|z| < 1$, it follows that W_- is invertible at infinity whereas no such restriction is made on W_+. In comparison with the previously mentioned results, one expects a compact geometric condition characterizing the existence of canonical factorizations. This turns out to be the case, however the language of invariant subspaces is not rich enough to express the result. In fact, once we represent W in terms of a state space realization, it is not wise to disregard the rest of system theory. It is the area of system theory, with an emphasis on geometric control theory, that provides the right language and setting in which to express the results. Throughout the paper the method of polynomial models and in particular the shift realization, introduced in Fuhrmann [6, 7, 8], are used consistently. This method proves itself once again in providing a vantage point from which the connections between seemingly different phenomena are most clearly seen.

2 Rational function factorizations

It is well known that the transfer function of a series connection of two systems is the product of the two transfer functions. The inverse problem is considerably more difficult and is yet not completely solved. What we are after is, given a rational function, which can be considered without loss of generality to be the transfer function of a finite dimensional linear system, we want to obtain a characterization of its factorization into the product of two transfer functions. We will assume to begin with that the transfer function is proper with constant term equal to the identity. We will refer to such a function as a **normalized biproper function**. In particular, the inverse of a normalized biproper rational function is also normalized biproper and rational.

In much the same way as factorizations of polynomial matrices are related to invariant subspaces and controlled or conditioned invariant subspaces one expects some such relation in the case of factorizations of transfer functions. The following factorization result due to Sakhnovich [15], see also Bart, Gohberg, Kaashoek and van Dooren [2], is the basis for all that follows.

Theorem 2.1 *Let G be a rational, normalized biproper function and let* $G = \left(\begin{array}{c|c} A & B \\ \hline C & I \end{array}\right)$ *be a minimal realization in the state space $\mathcal{X}$.*

Then a necessary and sufficient condition for G to admit a factorization

$$G = G_2G_1 \tag{1}$$

with G_i normalized, biproper factors is that

$$\mathcal{X} = M_1 \oplus M_2 \tag{2}$$

with M_1 an $A-$invariant subspace, M_2 an $A^\times-$invariant, where $A^\times$ is defined by

$$A^\times = A - BC. \tag{3}$$

Another proof of this result, in the spirit of this paper, can be found in Shamir and Fuhrmann [17]. We sketch the basic idea of that paper for the construction of the factors. Starting from a polynomial left coprime factorization $G = T^{-1}D$, then with respect to the shift realization any A-invariant subspace has the representation $T_1 X_{T_2}$. Similarly, as $G^{-1} = D^{-1}T$, any $A^\times$-invariant subspace has the representation $D_1 X_{D_2}$. Thus the direct sum representation (2) translates into

$$X_T = T_1 X_{T_2} \oplus D_1 X_{D_2}. \tag{4}$$

In order to construct the factors we let E be the l.c.r.m. of T_1 and D_1. Thus $E = T_1\overline{D}_1 = D_1\overline{T}_1$ for a pair $\overline{D}_1, \overline{T}_1$ of right coprime polynomial matrices. In a series of relatively simple steps one can show that, up to some unimodular factors, we have $G_1 = T_2^{-1}\overline{D}_1$ and $G_2 = \overline{T}_1^{-1}D_2$.

The principal results of this paper deals with canonical Wiener-Hopf factorizations with respect to the unit circle. In fact, with respect to any closed contour in the extended complex plane, we can consider the corresponding factorization. We will however concentrate on the two cases of interest to us, namely the unit circle and the imaginary axis. There is another case of interest and that is the factorization which puts all the finite points in one set and the point at infinity in the other. For any, real or complex, rational function this is like taking the factorization with respect to a large enough circle (that contains all the finite poles). This type of factorization can be defined for an arbutrary field and thus covers the case of transfer functions of linear, finite dimensional, discrete time systems over an arbitrary field F.

Definition 2.1 *Let $G \in F^{p\times m}((z^{-1}))$ be rational. A* **right Wiener-Hopf factorization at infinity** *is a factorization of G of the form*

$$G = G_- D G_+ \tag{5}$$

with $G_+ \in F^{m\times m}[z]$ unimodular, $G_- \in F^{p\times p}[z]$ biproper and

$$D(z) = \begin{pmatrix} \Delta(z) & 0 \\ 0 & 0 \end{pmatrix}$$

where $\Delta(z) = \mathrm{diag}\,(z^{\kappa_1}, \ldots, z^{\kappa_r})$. *The integers* κ_i, *assumed decreasingly ordered, are called the* **right factorization indices** *at infinity. A* **left factorization** *and the* **left factorization indices** *are analogously defined with the plus and minus signs in (5) reversed.*

We consider next the case of the real or complex field and the factorization with respect to the imaginary axis or the unit circle.

Definition 2.2 *Let* F *be the real or complex field.*

1. *Let* $G \in F^{p\times m}((z^{-1}))$ *be rational. A* **right Wiener-Hopf factorization with respect to the imaginary axis** *is a factorization of* G *of the form*
$$G = G_- D G_+ \tag{6}$$
with G_+, G_+^{-1} *analytic in the extended left half plane and* G_-, G_-^{-1} *analytic in the extended right half plane.*
$$D(z) = \begin{pmatrix} \Delta(z) & 0 \\ 0 & 0 \end{pmatrix}$$
where $\Delta(z) = \mathrm{diag}\,((\frac{z-1}{z+1})^{\kappa_1}, \ldots, (\frac{z-1}{z+1})^{\kappa_r})$. *The integers* κ_i, *assumed decreasingly ordered, are called the* **right factorization indices**. *A* **left Wiener-Hopf factorization** *and the* **left factorization indices** *are analogously defined with the plus and minus signs in (6) reversed.*

2. *If* G *is square and nonsingular and* $D = I$, *we say that the factorization (6) is right canonical. Left canonical factorizations are similarly defined.*

3. *Let* G *be a* $p \times m$ *rational. and let* α_-, α_+ *be arbitrary points in the interior and exterior of the unit disk respectively. A* **right Wiener-Hopf factorization with respect to the unit circle** *is a factorization of* G *of the form*
$$G = G_- D G_+ \tag{7}$$
with G_+, G_+^{-1} *analytic in the open unit disk and* G_-, G_-^{-1} *analytic in the exterior of the closed unit disk.*
$$D(z) = \begin{pmatrix} \Delta(z) & 0 \\ 0 & 0 \end{pmatrix}$$
where $\Delta(z) = \mathrm{diag}\,((\frac{z-\alpha_-}{z+\alpha_+})^{\kappa_1}, \ldots, (\frac{z-\alpha_-}{z+\alpha_+})^{\kappa_r})$. *The integers* κ_i, *assumed decreasingly ordered, are called the* **right factorization indices**. *A* **left Wiener-Hopf factorization** *and the* **left factorization indices** *are analogously defined with the plus and minus signs in (7) reversed.*

4. *If G is square and nonsingular and $D = I$, we say that the factorization (7) is right canonical. Left canonical factorizations are similarly defined.*

A convenient general reference for Wiener-Hopf theory is Gohberg and Feldman [11]. The following result is due to Bart, Gohberg and Kaashoek [1].

Theorem 2.2 *A normalized, biproper rational function G with minimal realization $\left(\begin{array}{c|c} A & B \\ \hline C & I \end{array}\right)$ in the state space $\mathcal{X}$ has a left canonical factorization $G = G_+G_-$ if and only if $\mathcal{X}_-(A^\times)$ and $\mathcal{X}_+(A)$ are complementary subspaces, i.e.*

$$\mathcal{X} = \mathcal{X}_+(A) \oplus \mathcal{X}_-(A^\times). \tag{8}$$

Proof: Assume $G = G_+G_-$ is a canonical factorization. Such a factorization is automatically minimal. Let $G_+ = T_+^{-1}\overline{D}_+$ and $G_- = \overline{T}_-^{-1}D_-$ be a left coprime factorization.

Let $\overline{D}_+\overline{T}_-^{-1} = T_-^{-1}D_+$ with the last one a left coprime factorization. Clearly T_- is stable and D_+ antistable. Thus

$$G_+G_- = T_+^{-1}T_-^{-1}D_+D_- = (T_-T_+)^{-1}(D_+D_-)$$

By Theorem 2.1 we must have

$$X_{T_-T_+} = T_-X_{T_+} \oplus D_+X_{D_-} \tag{9}$$

But $T_-X_{T_+} = X_+(A)$ and $D_+X_{D_-} = X_-(A^\times)$

Conversely, assume $X = X_+(A) \oplus X_-(A^\times)$. Thus a factorization exists. Let $G = T^{-1}D$. Factor $T = T_-T_+$ and $D = D_+D_-$ such that D_-, T_- are stable and T_+, D_+ are antistable. Such factorizations exist. Also $T_-X_{T_+} = X_+(A)$ and $D_+X_{D_-} = X_-(A^\times)$. By our assumption $X_{T_-T_+} = T_-X_{T_+} \oplus D_+X_{D_-}$ so by Theorem 2.1 this implies a factorization $G = G_+G_-$. ∎

For the rest of the paper we need a few concepts from geometric control which we proceed to introduce.

Given a linear transformation A in X and an A-invariant subspace $\mathcal{V}$, we denote by $A|_{\mathcal{V}}$ the restriction of A to $\mathcal{V}$. By a slight abuse of notation we will denote by $A|_{X/\mathcal{V}}$ the **induced map**, i.e. the map induced by A in the quotient space $X/\mathcal{V}$. This notation extends to conditioned and controlled invariant subspaces. Thus if $\mathcal{V} \subset X$ is a controlled invariant subspace for the pair (A, B) and if F is a feedback such that $(A + BF)\mathcal{V} \subset \mathcal{V}$, then we use the notation $A+BF|_{\mathcal{V}}$ and $A+BF|_{X/\mathcal{V}}$ for the restricted and induced maps respectively.

Our notion of stability depends on whether we consider discrete or continuous time system. In the first case we say that a linear transformation is stable if all its eigenvalues are contained in the open unit disk. For the continuous time case the unit disk is replaced by the open left half plane. Following Hautus [14] and Schumacher [16], we say that a controlled invariant subspace $\mathcal{V}$ for the pair (A, B) is stabilizable, or **inner stabilizable** if there exists a feedback F such that $\mathcal{V}$ is $A+BF$-invariant and $A+BF|_{\mathcal{V}}$ is stable. Analogously, we say that a controlled invariant subspace $\mathcal{V}$ is **outer stabilizable** if there exists a feedback F such that $\mathcal{V}$ is $A + BF$-invariant and $A + BF|_{X/\mathcal{V}}$ is stable. Similarly we define **inner antistabilizable** subspaces.

There are natural dual concepts. Thus a conditioned invariant subspace $\mathcal{V}$ is **inner detectable** if there exists an output injection map H such that $\mathcal{V}$ is $A + HC$-invariant and $A + HC|_{\mathcal{V}}$ is stable. A subspace $\mathcal{V}$ is **outer detectable** if there exists an output injection map H such that $\mathcal{V}$ is $A+HC$-invariant and $A+HC|_{X/\mathcal{V}}$ is stable. Again, the concepts of **inner and outer antidetectability** are naturally defined.

If $\mathcal{V}$ is controlled invariant with respect to the reachable pair (A, B) and K such that $(A + BK)\mathcal{V} \subset \mathcal{V}$, then the pair induced by $(A + BK, B)$ in $X/\mathcal{V}$ is also reachable, hence $\mathcal{V}$ is both outer stabilizable and antistabilizable. From this point of view, it is more interesting to study inner stabilizability and antistabilizability. Similarly, given a conditioned invariant subspace $\mathcal{V}$ with respect to the observable pair (C, A) and an output injection H such that $(A + HC)\mathcal{V} \subset \mathcal{V}$, then $(C, A + HC)$ has a restriction to $\mathcal{V}$ which is observable. Thus if $\mathcal{V}$ is conditioned invariant then it is both inner detectable and antidetectable. By standard duality considerations, we expect the notions of inner stabilizability to be related to outer detectability, and this indeed is the case.

Given a, not necessarily reachable, pair (A, B), there exists a maximal stabilizability subspace. It can be constructed from the sum of the reachability subspace with the spectral subspace corresponding to the stable modes of A. In the same way, given a system $\left(\begin{array}{c|c} A & B \\ \hline C & D \end{array}\right)$, there exists a maximal output nulling, inner stabilizable subspace which we denote by $\mathcal{S}^*_-$. Similarly, $\mathcal{S}^*_+$ denotes the maximal output nulling, inner antistabilizable subspace. In the same fashion, there exist a unique, minimal input containing, outer detectable subspace which we denote by $\mathcal{S}_*^-$ and a unique, minimal input containing, outer antidetectable subspace which we denote by $\mathcal{S}_*^+$.

The following theorem characterizes these subspaces for the case of the shift realization in terms of factorization of polynomial matrices. Since our interest in this paper is in square and nonsingular matrix functions, the results are rather simple. For the general case, as well as the proofs, we refer to Fuhrmann [9]. In the next theorem the curve γ should be interpreted as the unit circle for discrete time systems and the extended imaginary axis for the continuous time case, in both cases we assume it is positive oriented.

Theorem 2.3 *Let G be a proper, $p \times p$, nonsingular rational function with $D^{-1}N$ a left coprime factorization. Assume G has no zeros or poles on γ. Let*

$$N = N_+N_- = \overline{N}_-\overline{N}_+. \tag{10}$$

be factorizations of N into stable and antistable factors. Then, with respect to the shift realization.

1. *We have*

$$\mathcal{S}_-^* = N_+X_{N_-} \tag{11}$$

   and

$$\mathcal{S}_+^* = \overline{N}_-X_{\overline{N}_+}. \tag{12}$$

2. *We have*

$$\mathcal{S}_*^- = X_D \cap \overline{N}_-F^p[z], \tag{13}$$

   and

$$\mathcal{S}_*^+ = X_D \cap N_+F^p[z], \tag{14}$$

3. *We have*

$$\mathcal{X} = \mathcal{S}_-^* \oplus \mathcal{S}_*-\cdot \tag{15}$$

4. *We have the inclusions*

$$\mathcal{S}_+^* \subset \mathcal{S}_*^- \tag{16}$$

   and

$$\mathcal{S}_-^* \subset \mathcal{S}_*^+. \tag{17}$$

5. *G is biproper if and only if we have the equalities*

$$\mathcal{S}_+^* = \mathcal{S}_*^- \tag{18}$$

   and

$$\mathcal{S}_-^* = \mathcal{S}_*^+. \tag{19}$$

We stress once more the fact that these results are for the case that G is nonsingular. The proof of the direct sum representation (15) is based on a codimension formula proved in Fuhrmann and Helmke [10]. Since $\operatorname{codim} X_D \cap \overline{N}_- F^p[z] = \dim X_D - \deg \det \overline{N}$, $\det \overline{N} = \det N$ and $\dim N_+ X_{N_-} = \det N_-$ it suffices to prove that $N_+ X_{N_-} \cap (X_D \cap \overline{N}_- F^p[z]) = 0$, and this is easy as G is assumed nonsingular.
Note that in the case of a biproper rational function we have $X_-(A^\times) = \mathcal{S}_-^*$ and so the direct sum representation (8) can be replaced by

$$\mathcal{X} = \mathcal{X}_+(A) \oplus \mathcal{S}_-^*.$$

3 Left and right canonical factorizations

In the case of the unit circle, the existence of canonical factorizations for a rational matrix W does not imply that it is necessarily biproper. For example for the rational function $W = \dfrac{z}{(z-1/2)(z+3)}$ both left and right canonical factorization exists with the factors $W_- = \dfrac{z}{z-1/2}$ and $W_+ = \dfrac{1}{z+3}$.

Lemma 3.1 *Let $W = D^{-1}N$ be a left coprime factorization of an $m \times m$ nonsingular, proper rational function W having no poles and zeros on the unit circle.*

Let $D = D_- D_+$ and $N = N_+ N_-$ be factorizations into square nonsingular factors with D_+, N_+ antistable, i.e. having their zeros in the exterior of the closed unit disk and D_-, N_- stable, i.e. having their zeros in the interior of the unit disk.

Let E be a least common right multiple of D_- and N_+. Set $E = D_- \overline{N}_+ = N_+ \overline{D}_-$ with $\overline{N}_+, \overline{D}_-$ right coprime.

Then the following statements are equivalent:

1. *We have*
$$X_N \cap E F^m[z] = \{0\}. \tag{20}$$

2. *We have*
$$\operatorname{Ker} \mathcal{T}_{N^{-1}E} = \{0\}. \tag{21}$$

3. $$D_- X_{D_+} \cap N_+ X_{N_-} = \{0\}. \tag{22}$$

4. *All left Wiener-Hopf factorization indices at infinity of $N^{-1}E$ are nonnegative.*

Proof:

We show first the equality

$$D_-X_{D_+} \cap N_+X_{N_-} = X_N \cap EF^m[z]. \tag{23}$$

Note that $D_-F^m[z] \cap N_+F^m[z] = EF^m[z]$, where is a least common right multiple of D_- and N_+.

Since $D^{-1}N$ is proper, it follows that $X_N \subset X_D$. Also, it is clear that $N_+X_{N_-} \subset X_N$. Thus we have $D_-X_{D_+} \cap N_+X_{N_-} \subset X_N \subset X_D$, or

$$\begin{aligned} D_-X_{D_+} \cap N_+X_{N_-} &= (X_D \cap D_-F^m[z]) \cap (X_N \cap N_+F^m[z]) \\ &= X_N \cap (D_-F^m[z] \cap N_+F^m[z]) = X_N \cap EF^m[z]. \end{aligned}$$

Next, we show there is a bijective correspondence between the spaces $X_N \cap EF^m[z]$ and $\operatorname{Ker} \mathcal{T}_{N^{-1}E}$. Assume first that $f \in X_N \cap EF^m[z]$. This means that for some strictly proper h, we have $f = Nh = Eg$, i.e. we have $h = N^{-1}Eg$ and hence $\pi_+N^{-1}Eg = 0$, i.e. $g \in \operatorname{Ker} \mathcal{T}_{N^{-1}E}$.

Conversely, assume $g \in \operatorname{Ker} \mathcal{T}_{N^{-1}E}$. Thus, for some strictly proper h, we have $f = Nh = Eg$, and we have $f \in X_N \cap EF^m[z]$.

It follows that if one of the three spaces is trivial, so are all. Finally, the equivalence of statements 2. and 4. is standard.

∎

Theorem 3.1 *Let $W = D^{-1}N$ be a left coprime factorization of an $m \times m$ nonsingular, proper rational function W having no poles and zeros on the unit circle. we have*

1. *Let $D = D_-D_+$ and $N = N_+N_-$ be factorizations into square nonsingular factors with D_+, N_+ antistable, i.e. having their zeros in the exterior of the closed unit disk and D_-, N_- stable, i.e. having their zeros in the interior of the unit disk.*

   Let E be a least common right multiple of D_- and N_+. Set $E = D_-\overline{N}_+ = N_+\overline{D}_-$ with $\overline{N}_+, \overline{D}_-$ right coprime.

   Then the following statements are equivalent:

   (a) *There exists a left canonical factorization*

   $$W = W_+W_- \tag{24}$$

   into proper factors with W_-, W_-^{-1} analytic in the exterior of the unit disk including the point at infinity, W_+, W_+^{-1} analytic in the interior of the unit disk.

(b) *We have*

$$X_D = D_- X_{D_+} \oplus N_+ X_{N_-}. \tag{25}$$

(c) *Given any minimal realization* $W = \left(\begin{array}{c|c} A & B \\ \hline C & D \end{array}\right)$ *in the state space* $\mathcal{X}$, *let* $\mathcal{X}_+$ *be the antistable spectral subspace of* A *and* $\mathcal{S}_-^*$ *the maximal output nulling, inner stabilizable subspace of* $\mathcal{X}$. *Then we have*

$$\mathcal{X} = \mathcal{X}_+(A) \oplus \mathcal{S}_-^* \tag{26}$$

2. *Let* $D = \hat{D}_+ \hat{D}_-$ *and* $N = \hat{N}_- \hat{N}_+$ *be factorizations into square non-singular factors with* $\hat{D}_+, \hat{N}_+$ *antistable, i.e. having their zeros in the exterior of the closed unit disk and* $\hat{D}_-, \hat{N}_-$ *stable, i.e. having their zeros in the interior of the unit disk.*

   Let E *be a least common right multiple of* D_- *and* N_+. *Set* $E = D_- \overline{N}_+ = N_+ \overline{D}_-$ *with* $\overline{N}_+, \overline{D}_-$ *right coprime.*

   Then the following statements are equivalent:

   (a) *There exists a right canonical factorization*

$$W = W_- W_+ \tag{27}$$

   into proper factors with W_-, W_-^{-1} *analytic in the exterior of the unit disk including the point at infinity,* W_+, W_+^{-1} *analytic in the interior of the unit disk.*

   (b) *We have*

$$X_D = \hat{D}_+ X_{\hat{D}_-} \oplus (X_D \cap \hat{N}_- F^m[z]). \tag{28}$$

   (c) *Given any minimal realization* $W = \left(\begin{array}{c|c} A & B \\ \hline C & D \end{array}\right)$ *in the state space* $\mathcal{X}$, *let* $\mathcal{X}_-$ *be the stable spectral subspace of* A *and* $\mathcal{S}_*^-$ *the minimal input containing, outer detectable subspace of* $\mathcal{X}$. *Then we have*

$$\mathcal{X} = \mathcal{X}_- \oplus \mathcal{S}_*^-. \tag{29}$$

Proof:

1. (a) $\Rightarrow$ (b) Assume $W = W_+ W_-$ is a left canonical Wiener-Hopf factorization. Clearly, W_- is necessarily biproper. Let

$$\left\{ \begin{array}{lcl} W_+ & = & D_+^{-1} \overline{N}_+ \\ W_- & = & \overline{D}_-^{-1} N_- \end{array} \right. \tag{30}$$

with both factorizations left coprime. Thus $W = D_+^{-1}\overline{N}_+\overline{D}_-^{-1}N_-$. Let $\overline{N}_+\overline{D}_-^{-1} = D_-^{-1}N_-$ with the last factorization left coprime. This equality implies

$$E = D_-\overline{N}_+ = N_+\overline{D}_-. \tag{31}$$

In particular, (31) shows that E is a l.c.r.m. of D_-, N_+. The two coprimeness conditions yield the equivalence of $D_-, \overline{D}_-$ and analogously of $N_+, \overline{N}_+$ and in particular the equalities $\deg\det D_- = \deg\det \overline{D}_-$ and $\deg\det N_+ = \deg\det \overline{N}_+$. The biproperness of W_- implies the equality $\deg\det \overline{D}_- = \deg\det N_-$.
Now, clearly we have $D_-X_{D_+} \subset X_D$ and $N_+X_{N_-} \subset X_N \subset X_D$, which taken together imply the inclusion

$$D_-X_{D_+} + N_+X_{N_-} \subset X_D. \tag{32}$$

Since $W_- = \overline{D}_-^{-1}N_-$ is proper its left Wiener-Hopf factorization indices at infinity are nonpositive. This is equivalent to the right Wiener-Hopf factorization indices at infinity of

$$N^{-1}E = N_-^{-1}N_+^{-1}N_+\overline{D}_- = N_-^{-1}\overline{D}_-$$

being nonnegative. In turn, by Lemma 3.1, we get
$D_-X_{D_+} \cap N_+X_{N_-} = \{0\}$. Going back to (32), we have

$$\begin{aligned} & \dim(D_-X_{D_+} + N_+X_{N_-}) \\ = \;& \deg\det D_+ + \deg\det N_- = \deg\det D_+ + \deg\det \overline{D}_- \\ = \;& \deg\det D_+ + \deg\det D_- = \deg\det D = \dim X_D. \end{aligned}$$

Thus the equality $X_D = D_-X_{D_+} + N_+X_{N_-}$ follows.

$(c) \Rightarrow (b)$ Without loss of generality, we can assume that the realization is the shift realization associated with the left coprime factorization $W = D^{-1}N$. In this case we have $\mathcal{X} = X_D$, $\mathcal{X}_+ = D_-X_{D_+}$ and finally

$$\mathcal{S}_-^* = N_+X_{N_-}. \tag{33}$$

Thus (25) follows. Note that

$$\dim \mathcal{S}_-^* = \deg\det N_- = \deg\det D_- = \dim \mathcal{X}_-(A).$$

$(b) \Rightarrow (a)$ Assume (25) holds. We write

$$\begin{aligned} W &= D^{-1}N = (D_-D_+)^{-1}N_+N_- \\ &= D_+^{-1}(D_-^{-1}N_+)N_- = D_+^{-1}(\overline{N}_+\overline{D}_-^{-1})N_- \\ &= (D_+^{-1}\overline{N}_+)(\overline{D}_-^{-1}N_-) = W_+W_-. \end{aligned}$$

From the factorization $D = D_-D_+$ we have

$$\dim X_D = \deg\det D_+ + \deg\det D_-.$$

On the other hand our assumption of the direct sum representation (25) implies

$$\dim X_D = \deg\det D_+ + \deg\det N_-.$$

These two equalities imply $\deg\det D_- = \deg\det N_-$. Next, we compute

$$\begin{aligned} N^{-1}E &= N_-^{-1}N_+^{-1}N_+\overline{D}_- = N_-^{-1}\overline{D}_- \\ &= (U\Delta)\Gamma \end{aligned}$$

where $U\Delta\Gamma$ is the right Wiener-Hopf factorization at infinity of

$$N_-^{-1}\overline{D}_-.$$

Taking determinants it follows that necessarily Δ is trivial. Absorbing the unimodular matrix in N_-, it follows that $W_- = N_-^{-1}\overline{D}_-$ is biproper. Since W is proper so, is W_+.

$(b) \Rightarrow (c)$ With respect to the shift realization associated with the left coprime factorization $W = D^{-1}N$, we have $\mathcal{X} = X_D$, $\mathcal{X}_+ = D_-X_{D_+}$ and finally

$$\mathcal{S}_-^* = N_+X_{N_-}. \tag{34}$$

Thus (26) follows. Note that

$$\dim \mathcal{S}_-^* = \deg\det N_- = \deg\det D_- = \dim \mathcal{X}_-(A). \tag{35}$$

2. $(a) \Rightarrow (b)$ Assume W has a right canonical Wiener-Hopf factorization $W = W_-W_+$. Let

$$\begin{cases} W_- &= D_-^{-1}\overline{N}_- \\ W_+ &= \overline{D}_+^{-1}N_+ \end{cases}$$

Let $D_+^{-1}N_-$ be a left coprime factorization of $\overline{N}_-\overline{D}_+^{-1}$. Then

$$\begin{aligned} W &= W_-W_+ = D_-^{-1}\overline{N}_-\overline{D}_+^{-1}N_+ \\ &= D_-^{-1}D_+^{-1}N_-N_+ = (D_+D_-)^{-1}N_-N_+ = D^{-1}N \end{aligned}$$

With respect to the shift realization corresponding to the coprime factorization $W = (D_+D_-)^{-1}N_-N_+$, we have

$$\begin{aligned} \mathcal{X} &= X_D = X_{D_+D_-} \\ \mathcal{X}_- &= D_+X_{D_-} \\ \mathcal{S}_*^- &= X_{D_+D_-} \cap N_-F^m[z] \end{aligned} \tag{36}$$

The dimensions of these subspaces are given by

$$\begin{aligned}
\dim X_{D_+D_-} &= \deg\det D_+ + \deg\det D_- \\
\dim X_- &= \dim D_+X_{D_-} = \deg\det D_- \\
\dim \mathcal{S}_*^- &= \deg\det D_+ + \deg\det D_- - \deg\det N_- = \deg\det D_+.
\end{aligned} \tag{37}$$

In this computation we used the fact that, by the biproperness of W_-, we have $\deg\det \overline{N}_- = \deg\det D_-$. Also from the equality $N_-\overline{D}_+ = D_+\overline{N}_-$, taken together with the two coprimeness conditions, it follows that $\det N_- = \det \overline{N}_-$. For this see Fuhrmann [6]. Finally, the last dimension formula follows from a codimension result in Fuhrmann and Helmke [10]. The previous formulae show that $\dim \mathcal{X} = \dim \mathcal{X}_- + \dim \mathcal{S}_*^-$. Hence, in order to prove the direct sum representation (28), it suffices to show that

$$D_+X_{D_-} \cap (X_D \cap N_-F^m[z]) = \{0\}.$$

Now

$$D_+X_{D_-} \cap (X_D \cap N_-F^m[z]) = D_+X_{D_-} \cap N_-F^m[z].$$

Let $f \in D_+X_{D_-} \cap N_-F^m[z]$, then we write $f = D_+D_-h_- = N_-g$ and $f_- = D_-h_- = D_+^{-1}N_-g = \overline{N}_-\overline{D}_+^{-1}g$ which in turn implies that we have the representation $g = \overline{D}_+g_1$ for some vector polynomial g_1. Substituting back, we have $f_- = \overline{N}_-g_1$ and $h_- = D_-^{-1}\overline{N}_-g_1$, or equivalently $g_1 = (D_-^{-1}\overline{N}_-)^{-1}h_-$. The left side is polynomial whereas the right side is strictly proper, so both have to be zero. This implies $f_- = f = 0$.

$(b) \Rightarrow (a)$ Assume we have the direct sum representation (28). We write

$$\begin{aligned}
W &= D^{-1}N = (D_+D_-)^{-1}N_-N_+ = D_-^{-1}D_+^{-1}N_-N_+ \\
&= (D_-^{-1}\overline{N}_-)(\overline{D}_+^{-1}N_+) = W_-W_+.
\end{aligned}$$

Our aim is to show that $W_- = D_-^{-1}\overline{N}_-$ is biproper. That would show that the factorization $W = W_-W_+$ is right canonical.

From our assumption it follows that

$$\begin{aligned}
\dim X_D &= \deg\det D_+ + \deg\det D_- \\
&= \deg\det D_- + \dim X_D - \deg\det N_-,
\end{aligned}$$

i.e, $\deg\det D_- = \deg\det N_-$. The coprime factorizations $D_+^{-1}N_- = \overline{N}_-\overline{D}_+^{-1}$ yield $D_+\overline{N}_- = N_-\overline{D}_+$. Now $D_+DX_- \cap N_-F^m[z] = \{0\}$ implies

$$X_{D_-} \cap D_+^{-1}N_-F^m[z] = X_{D_-} \cap \overline{N}_-\overline{D}_+^{-1}F^m[z] = \{0\}.$$

In turn, since $\overline{N}_-F^m[z] \subset \overline{N}_-\overline{D}_+^{-1}F^m[z]$, we have $X_{D_-} \cap \overline{N}_-F^m[z] = \{0\}$. This condition is equivalent to all Wiener-Hopf factorization indices at infinity of W_- being nonpositive. If we take the right Wiener-Hopf factorization at infinity $D_-^{-1}\overline{N}_- = \Gamma\Delta U$, with Γ biproper and U polynomial unimodular, taking determinants and using the equality $\deg\det D_- = \deg\det \overline{N}_-$, we conclude that all the indices are zero. As a consequence, by renormalizing $\overline{N}_-$, we have W_- biproper and we are done.

■

We point out that part 2 of the preceding theorem could have also been derived from the first part using duality theory as developed in Fuhrmann [8].

4 State space analysis

If we are more specific on a minimal realization of W, we can be more specific on the existence of canonical factorizations and the realization of the canonical factors.

Theorem 4.1 *Let $W = D^{-1}N$ be a left coprime factorization of an $m \times m$ nonsingular, proper rational function W having no poles and zeros on the unit circle. Let*

$$W = G_+ + G_- \tag{38}$$

be the partial fraction decomposition of W, where we assume that G_- is analytic in the exterior of the unit disk and, for uniqueness purposes, vanishes at ∞, whereas G_+ is analytic in the open unit disk. Moreover, we assume the minimal realizations

$$G_+ = \left(\begin{array}{c|c} A_+ & B_+ \\ \hline C_+ & D/2 \end{array}\right), \quad G_- = \left(\begin{array}{c|c} A_- & B_- \\ \hline C_- & D/2 \end{array}\right). \tag{39}$$

then

1. *A left canonical factorization exists if and only if there exists a solution Z to the following system of equations*

$$\begin{cases} A_+Z - ZA_- = -\overline{B}_+\overline{C}_- \\ B_+ = \overline{B}_+ - ZB_- \\ C_- = D\overline{C}_- + C_+Z \end{cases} \tag{40}$$

In that case minimal realizations of the left canonical factors are given by

$$W_- = \left(\begin{array}{c|c} A_- & B_- \\ \hline \overline{C}_- & I \end{array}\right) \tag{41}$$

and

$$W_+ = \left(\begin{array}{c|c} A_+ & \overline{B}_+ \\ \hline C_+ & D \end{array}\right). \tag{42}$$

Moreover, we have

$$\begin{aligned} \mathcal{X}_+ &= \operatorname{Im}\begin{pmatrix} 0 \\ I \end{pmatrix} \\ \mathcal{S}_-^* &= \operatorname{Im}\begin{pmatrix} I \\ -Z \end{pmatrix}. \end{aligned} \tag{43}$$

2. *A right canonical factorization exists if and only if there exists a solution Z to the following system of equations*

$$\begin{cases} A_-Y - YA_+ = -\overline{B}_-\overline{C}_+ \\ \overline{B}_-D = B_- + YB_+ \\ \overline{C}_+ = C_+ - C_-Y \end{cases} \tag{44}$$

In that case minimal realizations of the left canonical factors are given by

$$\begin{cases} W_+ = \left(\begin{array}{c|c} A_+ & B_+ \\ \hline \overline{C}_+ & D \end{array}\right) \\ W_- = \left(\begin{array}{c|c} A_- & \overline{B}_- \\ \hline C_- & I \end{array}\right) \end{cases} \tag{45}$$

Moreover, we have

$$\begin{aligned} \mathcal{X}_- &= \operatorname{Im}\begin{pmatrix} I \\ 0 \end{pmatrix} \\ \mathcal{S}_*^- &= \operatorname{Im}\begin{pmatrix} -Y \\ I \end{pmatrix}. \end{aligned} \tag{46}$$

Proof:

1. Assume that in the partial fraction decomposition $W = G_+ + G_-$ the functions have the following coprime factorizations $G_+ = D_+^{-1}E$ and $G_- = FD_-^{-1}$. Invoking the shift realizations, we can assume without

loss of generality that the factors in the left canonical factorization have the minimal realizations of the form

$$W_- = \left(\begin{array}{c|c} A_- & B_- \\ \hline \overline{C}_- & I \end{array}\right) \tag{47}$$

and

$$W_+ = \left(\begin{array}{c|c} A_+ & \overline{B}_+ \\ \hline C_+ & D \end{array}\right). \tag{48}$$

Note that

$$W_-^{-1} = \left(\begin{array}{c|c} A_- - B_-\overline{C}_- & B_- \\ \hline -\overline{C}_- & I \end{array}\right), \tag{49}$$

which implies, by our assumption on W_-, the stability of $A_- - B_-\overline{C}_-$.

Taking the series coupling of the two realizations we obtain the product realization

$$\begin{aligned} W &= W_+W_- = \left(\begin{array}{c|c} A_+ & \overline{B}_+ \\ \hline C_+ & D \end{array}\right) \times \left(\begin{array}{c|c} A_- & B_- \\ \hline \overline{C}_- & I \end{array}\right) \\ &= \left(\begin{array}{cc|c} A_- & 0 & B_- \\ \overline{B}_+\overline{C}_- & A_+ & \overline{B}_+ \\ \hline D\overline{C}_- & C_+ & D \end{array}\right). \end{aligned} \tag{50}$$

Using an old result of Callier and Nahum [3], we note that, by stability considerations, as there are no possible zero pole cancellations, the previous realization is minimal.

Similarly, taking the parallel connection of the two minimal realizations (39), we have

$$\begin{aligned} W &= G_+ + G_- \\ &= \left(\begin{array}{cc|c} A_- & 0 & B_- \\ 0 & A_+ & B_+ \\ \hline C_- & C_+ & D \end{array}\right). \end{aligned} \tag{51}$$

and this too, by Fuhrmann [5], is minimal.

The two realizations (51) and (50) are minimal realizations of W and hence, by the state space isomorphism theorem, are isomorphic. Let

$\begin{pmatrix} Z_{11} & Z_{12} \\ Z_{21} & Z_{22} \end{pmatrix}$ be the intertwining isomorphism, i.e. we have

$$\begin{pmatrix} Z_{11} & Z_{12} \\ Z_{21} & Z_{22} \end{pmatrix}\begin{pmatrix} B_- \\ \overline{B}_+ \end{pmatrix} = \begin{pmatrix} B_- \\ B_+ \end{pmatrix}$$

$$\begin{pmatrix} Z_{11} & Z_{12} \\ Z_{21} & Z_{22} \end{pmatrix}\begin{pmatrix} A_- & 0 \\ \overline{B}_+\overline{C}_- & A_+ \end{pmatrix} = \begin{pmatrix} A_- & 0 \\ 0 & A_+ \end{pmatrix}\begin{pmatrix} Z_{11} & Z_{12} \\ Z_{21} & Z_{22} \end{pmatrix}$$

$$\begin{pmatrix} C_- & C_+ \end{pmatrix}\begin{pmatrix} Z_{11} & Z_{12} \\ Z_{21} & Z_{22} \end{pmatrix} = \begin{pmatrix} D\overline{C}_- & C_+ \end{pmatrix} \tag{52}$$

From the central equation in (52) we get

$$Z_{12}A_+ = A_-Z_{12}.$$

Since A_+, A_- have disjoint spectra, we conclude $Z_{12} = 0$. With this we derive from the first two equations

$$\begin{pmatrix} Z_{11} & 0 \\ Z_{21} & Z_{22} \end{pmatrix}\begin{pmatrix} B_- \\ \overline{B}_+ \end{pmatrix} = \begin{pmatrix} B_- \\ B_+ \end{pmatrix} \tag{53}$$

or $Z_{11}B_- = B_-$. Computing the $1,1$ term we get $Z_{11}A_- = A_-Z_{11}$ and hence by induction $Z_{11}A_-^i = A_-^i Z_{11}$. In particular we get, for all $i \geq 0$, that $Z_{11}A_-^i B_- = A_-^i B_-$. By controllability, we conclude $Z_{11} = I$. Using duality, the last two equations yield $Z_{22} = I$. Moreover, we define $Z = -Z_{21}$.

Rewriting the equations in this notation yields

$$\begin{pmatrix} I & 0 \\ -Z & I \end{pmatrix}\begin{pmatrix} B_- \\ \overline{B}_+ \end{pmatrix} = \begin{pmatrix} B_- \\ B_+ \end{pmatrix}$$

$$\begin{pmatrix} I & 0 \\ -Z & I \end{pmatrix}\begin{pmatrix} A_- & 0 \\ \overline{B}_+\overline{C}_- & A_+ \end{pmatrix} = \begin{pmatrix} A_- & 0 \\ 0 & A_+ \end{pmatrix}\begin{pmatrix} I & 0 \\ -Z & I \end{pmatrix}$$

$$\begin{pmatrix} C_- & C_+ \end{pmatrix}\begin{pmatrix} I & 0 \\ -Z & I \end{pmatrix} = \begin{pmatrix} D\overline{C}_- & C_+ \end{pmatrix} \tag{54}$$

From this we obtain the following

$$\begin{cases} A_+Z - ZA_- = -\overline{B}_+\overline{C}_- \\ B_+ = \overline{B}_+ - ZB_- \\ C_- = D\overline{C}_- + C_+Z \end{cases} \tag{55}$$

We note that the first equation has a unique solution and so, as expected, the matrices $\overline{B}_+, \overline{C}_-$ are uniquely determined.

We show next that

$$\mathcal{S}_-^* = \operatorname{Im}\begin{pmatrix} I \\ -Z \end{pmatrix}. \tag{56}$$

Let us consider the feedback matrix $\begin{pmatrix} \overline{C}_- & 0 \end{pmatrix}$. Then

$$\begin{pmatrix} A_- & 0 \\ 0 & A_+ \end{pmatrix} - \begin{pmatrix} B_- \\ B_+ \end{pmatrix}\begin{pmatrix} \overline{C}_- & 0 \end{pmatrix} = \begin{pmatrix} A_- - B_-\overline{C}_- & 0 \\ -B_+\overline{C}_- & A_+ \end{pmatrix}$$

So

$$\begin{aligned}
&\begin{pmatrix} A_- - B_-\overline{C}_- & 0 \\ -B_+\overline{C}_- & A_+ \end{pmatrix}\begin{pmatrix} I \\ -Z \end{pmatrix} \\
&= \begin{pmatrix} A_- - B_-\overline{C}_- \\ -(B_+ + ZB_-)\overline{C}_- - A_+Z \end{pmatrix} \\
&= \begin{pmatrix} A_- - B_-\overline{C}_- \\ -ZA_- + ZB_-\overline{C}_- \end{pmatrix} \\
&= \begin{pmatrix} I \\ -Z \end{pmatrix}(A_- - B_-\overline{C}_-)
\end{aligned}$$

This shows, using the stability of $A_- - B_-\overline{C}_-$, that $\operatorname{Im}\begin{pmatrix} I \\ -Z \end{pmatrix}$ is an inner stabilizable, controlled invariant subspace. Moreover, from $\begin{pmatrix} C_- & C_+ \end{pmatrix} - D\begin{pmatrix} \overline{C}_- & 0 \end{pmatrix} = \begin{pmatrix} C_+Z & C_+ \end{pmatrix}$, we compute

$$\begin{pmatrix} C_+Z & C_+ \end{pmatrix}\begin{pmatrix} I \\ -Z \end{pmatrix} = 0.$$

This shows that $\operatorname{Im}\begin{pmatrix} I \\ -Z \end{pmatrix}$ is an inner stabilizable, output nulling subspace. Comparing dimensions with $\dim(D_+X_{D_-})$, we get the equality (56).

Since $\begin{pmatrix} I & 0 \\ -Z & I \end{pmatrix}$ is invertible and $\operatorname{Im}\begin{pmatrix} 0 \\ I \end{pmatrix} = \mathcal{X}_+$ we clearly recover

$$\mathcal{X} = \mathcal{X}_+(A) \oplus \mathcal{S}_-^*. \tag{57}$$

2. Assume now that we have the following coprime factorizations

$$\begin{cases} G_- &= D_-^{-1}E \\ G_+ &= FD_+^{-1} \end{cases} \tag{58}$$

Then, in case there exists a right canonical factorization, we have

$$W = G_- + G_+ = D_-^{-1}N_- + N_+D_+^{-1} = D_-^{-1}E_- \cdot E_+D_+^{-1} = W_-W_+. \tag{59}$$

From the previous representation it follows that, without loss of generality, W_- and W_+ have the following, minimal state space realizations

$$\begin{cases} W_+ &= \left(\begin{array}{c|c} A_+ & B_+ \\ \hline \overline{C}_+ & D \end{array}\right) \\ W_- &= \left(\begin{array}{c|c} A_- & \overline{B}_- \\ \hline C_- & I \end{array}\right) \end{cases} \tag{60}$$

Note again that by computing W_-^{-1} and using our assumption, we have the stability of $A_- - \overline{B}_-C_-$. We can compute now

$$\begin{aligned} W &= W_-W_+ = \left(\begin{array}{c|c} A_- & \overline{B}_- \\ \hline C_- & I \end{array}\right) \times \left(\begin{array}{c|c} A_+ & B_+ \\ \hline \overline{C}_+ & D \end{array}\right) \\ &= \left(\begin{array}{cc|c} A_- & \overline{B}_-\overline{C}_+ & \overline{B}_-D \\ 0 & A_+ & B_+ \\ \hline C_- & \overline{C}_+ & D \end{array}\right) \end{aligned} \tag{61}$$

Now we repeat the previous argument, but this time for the realizations (51) and (61). Applying the state space isomorphism theorem, we obtain the following equations.

$$\begin{gathered} \begin{pmatrix} Y_{11} & Y_{12} \\ Y_{21} & Y_{22} \end{pmatrix} \begin{pmatrix} \overline{B}_-D \\ B_+ \end{pmatrix} = \begin{pmatrix} B_- \\ B_+ \end{pmatrix} \\ \begin{pmatrix} Y_{11} & Y_{12} \\ Y_{21} & Y_{22} \end{pmatrix} \begin{pmatrix} A_- & \overline{B}_-\overline{C}_+ \\ 0 & A_+ \end{pmatrix} = \begin{pmatrix} A_- & 0 \\ 0 & A_+ \end{pmatrix} \begin{pmatrix} Y_{11} & Y_{12} \\ Y_{21} & Y_{22} \end{pmatrix} \\ \begin{pmatrix} C_- & C_+ \end{pmatrix} \begin{pmatrix} Y_{11} & Y_{12} \\ Y_{21} & Y_{22} \end{pmatrix} = \begin{pmatrix} \overline{C}_- & C_+ \end{pmatrix} \end{gathered} \tag{62}$$

As before, we conclude that

$$\begin{pmatrix} Y_{11} & Y_{12} \\ Y_{21} & Y_{22} \end{pmatrix} = \begin{pmatrix} I & -Y \\ 0 & I \end{pmatrix}$$

This leads to

$$\begin{cases} A_- Y - Y A_+ = -\overline{B}_- \overline{C}_+ \\ \overline{B}_- D = B_- + Y B_+ \\ \overline{C}_+ = C_+ - C_- Y \end{cases} \tag{63}$$

Consider now the output injection $\begin{pmatrix} H_1 \\ H_2 \end{pmatrix} = \begin{pmatrix} \overline{B}_- \\ 0 \end{pmatrix}$. Then

$$\begin{pmatrix} A_- & 0 \\ 0 & A_+ \end{pmatrix} - \begin{pmatrix} \overline{B}_- \\ 0 \end{pmatrix} \begin{pmatrix} C_- & C_+ \end{pmatrix}$$

$$= \begin{pmatrix} A_- - \overline{B}_- C_- & -\overline{B}_- C_+ \\ 0 & A_+ \end{pmatrix}$$

We compute

$$\begin{pmatrix} A_- - \overline{B}_- C_- & -\overline{B}_- C_+ \\ 0 & A_+ \end{pmatrix} \begin{pmatrix} -Y \\ I \end{pmatrix}$$

$$= \begin{pmatrix} -(A_- - \overline{B}_- C_-)Y - B_- C_+ \\ A_+ \end{pmatrix}$$

$$= \begin{pmatrix} -A_- Y - \overline{B}_-(C_+ - C_- Y) \\ A_+ \end{pmatrix}$$

$$= \begin{pmatrix} -A_- Y - \overline{B}_- \overline{C}_+ \\ A_+ \end{pmatrix} = \begin{pmatrix} -Y \\ I \end{pmatrix} A_+$$

This shows that $\operatorname{Im} \begin{pmatrix} -Y \\ I \end{pmatrix}$ is conditioned invariant and, as $A_- - \overline{B}_- C_-$ is stable, outer detectable. We proceed to show that it is input containing. We compute

$$\begin{pmatrix} B_- \\ B_+ \end{pmatrix} = \begin{pmatrix} \overline{B}_- D - Y B_+ \\ B_+ \end{pmatrix} = \begin{pmatrix} -Y \\ I \end{pmatrix} B_+ + \begin{pmatrix} \overline{B}_- \\ 0 \end{pmatrix} D.$$

This shows that $\operatorname{Im}\left[\begin{pmatrix} B_- \\ B_+ \end{pmatrix} - \begin{pmatrix} B_- \\ 0 \end{pmatrix} D\right] \subset \operatorname{Im}\begin{pmatrix} -Y \\ I \end{pmatrix}$. By a dimensionality argument we have

$$\mathcal{S}_*^- = \operatorname{Im}\begin{pmatrix} -Y \\ I \end{pmatrix}, \tag{64}$$

i.e.

$$\mathcal{X} = \mathcal{X}_- \oplus \mathcal{S}_*^-. \tag{65}$$

∎

Note that, in case D is invertible, the system of equations (40) reduces to a quadratic, Riccati like, equation in Z

$$(A_+ + B_+ D^{-1} C_+)Z - Z(A_- - B_- D^{-1} C_-) + B_+ D^{-1} C_- - Z B_- D^{-1} C_+ Z = 0. \tag{66}$$

A similar remark holds for the system of equations (44).

Having the previous theorem clears the way to the characterization of the existence of both left and right canonical factorizations.

Theorem 4.2 *Let $W = D^{-1}N$ be a left coprime factorization of an $m \times m$ nonsingular, proper rational function W having no poles and zeros on the unit circle.*

Then the following statements are equivalent:

1. *There exist both left and right canonical factorizations of W.*

$$W = W_+ W_- = Y_- Y_+ \tag{67}$$

2. *There exist solutions Z, Y to the systems of equations (40) and (44). In this case all of the following matrices*

$$\begin{pmatrix} I & -Y \\ -Z & I \end{pmatrix}, \quad (I - YZ), \quad (I - ZY) \tag{68}$$

are invertible.

Proof: Follows from Theorem 2.3, the characterizations (43) and (46) and the direct sum representation (15).

∎

References

[1] H. Bart, I. Gohberg, and M.A. Kaashoek, *Minimal Factorization of Matrix and Operator Functions*, Birkhauser, Basel, 1979.

[2] H. Bart, I. Gohberg, M.A. Kaashoek and P. Van Dooren, "Factorizations of transfer functions," *SIAM J. Contr. Optim.*, 18, 675-696, 1979.

[3] F.M. Callier and C.D. Nahum, "Necessary and sufficient conditions for the complete controllability and observability of systems in series using the coprime decomposition of a rational matrix," *IEEE Trans. Circ.and Sys.* 22, 90-95, 1975.

[4] F.M. Callier, "On polynomial matrix spectral factorization by symmetric root extraction," *IEEE Trans. Autom. Contr.* AC-30, 453-464, 1985.

[5] P.A. Fuhrmann, "On controllability and observability of systems connected in parallel," *IEEE Trans. Circuits and Systems* CAS-22, 57, 1975.

[6] P.A. Fuhrmann, "Algebraic system theory: An analyst's point of view," *J. Franklin Inst.*, 301, 521-540, 1976.

[7] P.A. Fuhrmann, "On strict system equivalence and similarity," *Int. J. Contr.*, 25, 5-10, 1977.

[8] P.A. Fuhrmann, "Duality in polynomial models with some applications to geometric control theory," *IEEE Trans. Aut. Control*, AC-26, 284-295, 1981.

[9] P.A. Fuhrmann, "Geometric control revisited," in preparation.

[10] P.A. Fuhrmann and U. Helmke, "On conditioned invariant subspaces and observer theory," *CIM Preprint No. 1*, Coimbra, 1998.

[11] I. Gohberg and I.A. Feldman, *Convolution Equations and Projection Methods for their Solution*, Translations of Mathematical Monographs, vol. 41, Amer. Math. Soc., Providence, 1974.

[12] I. Gohberg and Y. Zucker, "Left and right factorizations of rational matrix functions," *Integral Eq. and Oper. Th.*, 19, 216-239, 1994.

[13] I. Gohberg and Y. Zucker, "On canonical factorizations of rational matrix functions," *Integral Eq. and Oper. Th.*, 25, 73-93, 1996.

[14] M.L.J. Hautus, "(A, B)-invariant and stabilizability subspaces, a frequency domain description," *Automatica*, 16, 703-707, 1990.

[15] L.A. Sakhnovich, "On the factorization of an operator valued transfer function," *Soviet Math. Dokl.*, 17, 203-207, 1976.

[16] J.M. Schumacher, "Dynamic feedback in finite- and infinite-dimensional linear systems," Ph.D. thesis, Free University of Amsterdam, 1981.

[17] T. Shamir and P.A. Fuhrmann, "Minimal factorizations of rational matrix functions in terms of polynomial models," *Lin. Alg. Appl.*, 68, 67-91, 1985.

[18] Y. Zucker, "Constructive factorization and partial indices of rational matrix functions," Ph.D. thesis, Tel-Aviv University, 1998.

Department of Mathematics, Ben-Gurion University of the Negev, Beer Sheva, Israel

Progress in Systems and Control Theory, Vol. 25

State Space Methods for Analysis Problems Involving Rational Matrix Functions

I. Gohberg and M.A. Kaashoek

1 Introduction

During the last twenty years the state space method from systems theory has entered into mathematical analysis and has proved to be an effective tool for solving problems from this area. Today a large number of such applications of the state space method exist (see [1], [6], [9], [10], [13]); in the present paper we review a number of the recent applications. The ones we have chosen deal with the Nevanlinna-Pick interpolation problem, the problem of solving explicitly integral equations of convolution type, direct and inverse spectral problems for canonical systems of differential equations, and with the problem of constructing exact solutions of nonlinear partial differential equations.

2 Nevanlinna-Pick interpolation

Metric constrained interpolation problems from complex analysis have a long and rich history (see, e.g., the books [5], [11]). In this section we shall solve one of the simplest problems of this type, namely a matrix valued version of the classical Nevanlinna-Pick interpolation problem, by using the state space method.

Let $\mathcal{R}H^\infty_{M\times N}$ be the set of all proper stable rational $M \times N$ matrix functions. Thus $F \in \mathcal{R}H^\infty_{M\times N}$ means that F has no poles in the open right half plane (which we will denote by RHP) and

$$\|F\|_\infty = \sup\{\|F(z)\| \mid z \in \mathrm{RHP}\} < \infty.$$

The simplest case of the matrix Nevanlinna-Pick interpolation problem is to find all matrix functions F from $\mathcal{R}H^\infty_{M\times N}$ which satisfy the metric constraint $\|F\|_\infty < 1$ and the following interpolation conditions

$$x_j F(z_j) = y_j \quad \text{for} \quad 1 \leq j \leq n, \tag{2.1}$$

where $z_1, \dots, z_n$ are distinct points in RHP, $x_1, \dots, x_n$ are nonzero row vectors in $\mathbb{C}^M$ and $y_1, \dots, y_n$ are arbitrary row vectors in $\mathbb{C}^N$. We shall refer to this problem as the *tangential Nevanlinna-Pick problem associated with the data set* $w = \{z_j, x_j, y_j \mid 1 \leq j \leq n\}$.

Experience from the classical case suggests that the solutions of the above problem (when they exist) should be parametrized by a linear fractional map

$$F = T_{\Theta}[G] = (\Theta_{11}G + \Theta_{12})(\Theta_{21}G + \Theta_{22})^{-1}, \tag{2.2}$$

where the free parameter G is an arbitrary proper stable rational $M \times N$ matrix function satisfying the constraint $\|G\|_{\infty} < 1$, and

$$\Theta(z) = \begin{pmatrix} \Theta_{11}(z) & \Theta_{12}(z) \\ \Theta_{21}(z) & \Theta_{22}(z) \end{pmatrix}$$

is a stable rational $(M+N) \times (M+N)$ matrix function which is J-inner with respect to the signature matrix

$$J = \begin{pmatrix} I & 0 \\ 0 & -I \end{pmatrix}.$$

The latter means that the stable rational matrix function Θ is J-unitary on the imaginary axis ∂RHP and $\Theta(z)$ is J-contractive for $z \in$ RHP.

Let us assume that such a representation is valid. Then

$$\begin{pmatrix} F(z) \\ I \end{pmatrix} = \Theta(z) \begin{pmatrix} G(z) \\ I \end{pmatrix} (\Theta_{21}(z)G(z) + \Theta_{22}(z))^{-1},$$

and hence

$$\begin{aligned} 0 &= \begin{pmatrix} x_i & -y_i \end{pmatrix} \begin{pmatrix} F(z_i) \\ I \end{pmatrix} \\ &= \begin{pmatrix} x_i & -y_i \end{pmatrix} \Theta(z_i) \begin{pmatrix} G(z_i) \\ I \end{pmatrix} (\Theta_{21}(z_i)G(z_i) + \Theta_{22}(z_i))^{-1}. \end{aligned}$$

This suggests that the stable rational J-inner matrix function Θ should be chosen in such a way that it satisfies the following homogeneous interpolation conditions:

$$\begin{pmatrix} x_i & -y_i \end{pmatrix} \Theta(z_i) = 0, \qquad i = 1, \ldots, n. \tag{2.3}$$

Set B equal to the $n \times (M+N)$ matrix

$$B = \begin{pmatrix} x_1 & -y_1 \\ \vdots & \vdots \\ x_n & -y_n \end{pmatrix}, \tag{2.4}$$

and A equal to the $n \times n$ diagonal matrix

$$A = \operatorname{diag}(z_1, \ldots, z_n). \tag{2.5}$$

Notice that the pair A, B is controllable, and that (2.3) is equivalent to the statement that $(z - A)^{-1}B\Theta(z)$ is stable. Since there are no other constraints on Θ, it is natural to require additionally that the number of zeros of Θ in the RHP (multiplicities taken into account) is precisely equal to n (the order of the matrix A). It turns out that this strategy works. Indeed, assume that Θ is a stable J-inner rational $(M + N) \times (M + N)$ matrix function such that $(z - A)^{-1}B\Theta(z)$ is stable and such that the number of zeros of Θ in the RHP is precisely equal to n. Then the tangential Nevanlinna-Pick interpolation problem considered in the present section has a solution, and any solution F of this problem is given by (2.2), where G is an arbitrary proper stable rational $M \times N$ matrix satisfying $\|G\|_\infty < 1$.

The construction of a stable J-inner rational $(M+N)\times(M+N)$ matrix function Θ such that $(z-A)^{-1}B\Theta(z)$ is stable and the number of zeros of Θ in the RHP is precisely equal to n involves solving the Lyapunov equation

$$SA^* + AS = BJB^*. \tag{2.6}$$

Letting S_{ij} denote the (i, j)-the entry of the unknown matrix S, we see that the latter equation may be reduced to $S_{ij}\overline{z}_j + z_i S_{ij} = x_i x_j^* - y_i y_j^*$ from which we see that $S_{ij} = [x_i x_j^* - y_i y_j^*](z_i + \overline{z}_j)^{-1}$ for $1 \le i, j \le n$.

With the data set $w = \{z_j, x_j, y_j \mid 1 \le j \le n\}$ we associate the matrix

$$\Lambda(w) = \left(\frac{x_i x_j^* - y_i y_j^*}{z_i + \overline{z}_j}\right)_{1\le i,j\le n} \tag{2.7}$$

which solves the Lyapunov equation (2.6). It turns out that $\Lambda(w)$ plays the role of the classical Pick matrix for the problem. The heuristic remarks presented above lead to the following theorem.

Theorem 2.1 [4] *Let $w = \{z_j, x_j, y_j \mid 1 \le j \le n\}$ be the data set for a tangential matrix Nevanlinna-Pick interpolation problem on the* RHP. *Construct matrices B, A and $\Lambda = \Lambda(w)$ according to* (2.4), (2.5) *and* (2.7). *Then there exist functions F in $\mathcal{RH}^\infty_{M\times N}$ with $\|F\|_\infty < 1$ satisfying the interpolation conditions* (2.1) *if and only if Λ is positive definite. Put*

$$\Theta(z) = I - JB^*(zI + A^*)^{-1}\Lambda^{-1}B.$$

Then $F \in \mathcal{RH}^\infty_{M\times N}$ has $\|F\|_\infty < 1$ and satisfies the interpolation conditions (2.1) *if and only if*

$$F = (\Theta_{11}G + \Theta_{12})(\Theta_{21}G + \Theta_{22})^{-1},$$

where

$$\Theta(z) = \begin{pmatrix} \Theta_{11}(z) & \Theta_{12}(z) \\ \Theta_{21}(z) & \Theta_{22}(z) \end{pmatrix}$$

and $G \in \mathcal{RH}^\infty_{M\times N}$ with $\|G\|_\infty < 1$.

For more details about this section we refer to the book [5] which develops the above approach systematically for a large variety of metric constrained interpolation and extension problems.

3 Integral equations of convolution type

In this section we solve integral equations of convolution type by reducing the problem to an inversion problem for systems with boundary conditions. We shall show that this system approach works effectively for equations on a half line as well as for equations on a finite interval. For the case when the symbol of the equations is a rational matrix function the solutions are obtained explicitly in state space form. The results reviewed in this section will play an essential role in the next section.

3.1 Wiener-Hopf equations

We begin with a convolution equation on the half line:

$$\varphi(t) - \int_0^\infty k(t-s)\varphi(s)\,ds = f(t), \quad t \geq 0. \tag{3.1}$$

One usually calls (3.1) a *Wiener-Hopf equation.* Notice that (3.1) stands for a system of equations. Thus the kernel function k is a matrix function of size $m \times m$, say. We shall assume that its entries are integrable on the full line, i.e., $k \in L^1_{m\times m}(\mathbb{R})$. The right hand side f is a given $\mathbb{C}^n$-valued function of which the entries are square integrable, i.e., $f \in L^2_m(0,\infty)$, and the problem is to find $\varphi \in L^2_m(0,\infty)$ such that (3.1) holds.

The usual procedure to solve (3.1) begins by setting $f(t) = 0$ and $\varphi(t) = 0$ for $t < 0$. Then (3.1) takes the form

$$\varphi(t) - \int_{-\infty}^\infty k(t-s)\varphi(s)\,ds = f(t) + \Psi_-(t), \quad t \in \mathbb{R},$$

where $\Psi_-(t)$ is a unknown correction term which has its support on the negative half line. After Fourier transformation this yields:

$$(I_m - \widehat{k}(\lambda))\widehat{\varphi}(\lambda) = \widehat{f}(\lambda) + \widehat{\Psi}_-(\lambda). \tag{3.2}$$

One calls the $m \times m$ matrix function $W(\lambda) := I_m - \widehat{k}(\lambda)$ the *symbol* of the equation (3.1).

The classical method to solve (3.1) is based on factorization of the symbol. Here we shall take another route.

Let us assume that the symbol is rational. The fact that the kernel function is integrable implies that the points on the real line and at infinity cannot be poles of W. It follows from the Kalman realization theory that one can find a square matrix A without eigenvalues on the real line and matrices B and C such that

$$W(\lambda) = I_m + C(\lambda I_n - A)^{-1}B, \ \lambda \in \mathbb{R}. \tag{3.3}$$

Here I_p denotes the identity matrix of order p. Formula (3.3) allows us to rewrite (3.2) as

$$\widehat{\varphi}(\lambda) + C(\lambda I_n - A)^{-1}B\widehat{\varphi}(\lambda) = \widehat{f}(\lambda) + \widehat{\Psi}_-(\lambda). \tag{3.4}$$

Now set $\rho(\lambda) = -(\lambda I_n - A)^{-1}B\widehat{\varphi}(\lambda)$. Using ρ as a new state variable we see that (3.4) can be rewritten in the following equivalent form:

$$\begin{cases} \lambda\widehat{\rho}(\lambda) = A\widehat{\rho}(\lambda) - B\widehat{\varphi}(\lambda), \\ \widehat{f}(\lambda) = -C\widehat{\rho}(\lambda) + \widehat{\varphi}(\lambda) - \widehat{\Psi}_-(\lambda) \end{cases} \quad (\lambda \in \mathbb{R})$$

By applying the inverse Fourier transform we obtain:

$$\begin{cases} \dot{\rho}(t) = -iA\rho(t) + iB\varphi(t) \\ f(t) = -C\rho(t) + \varphi(t) - \Psi_-(t) \end{cases} \quad (t \in \mathbb{R})$$

Recall that the correction term $\Psi_-(t)$ has its support on the negative half line. On the other hand we are interested in solutions on the positive half line only. So in the above system we will restrict the time variable t to the positive time axis. In this way we get rid of the correction term $\Psi_-(t)$. However we do have to insert a boundary condition at zero. In fact the system that has to be considered is the following:

$$\begin{cases} \dot{\rho}(t) = -iA\rho(t) + iB\varphi(t), \quad t \geq 0 \\ f(t) = -C\rho(t) + \varphi(t) \\ (I - P)\rho(0) = 0. \end{cases} \tag{3.5}$$

Here P is spectral (Riesz) projection of A corresponding to eigenvalues of A in open upper half plane, i.e., P is given by

$$P = \frac{1}{2\pi i}\int_\Gamma (\lambda I - A)^{-1}\, d\lambda,$$

where Γ is a positively oriented contour in the open upper half plane around the eigenvalues of A in the open upper half plane. There is also a hidden boundary condition at infinity, namely the state function ρ is square integrable.

The precise connection between (3.1) and the system (3.5) is described in the following theorem.

Theorem 3.1 [7] *Let the symbol of* (3.1) *be given by* (3.3). *The equation* (3.1) *and the linear system* (3.3) *are equivalent in the following sense. If* φ *is a solution of* (3.1) *in* $L^2_m(0,\infty)$, *then system* (3.5) *with input* φ *has output* f; *conversely, if the system* (3.5) *with input* φ *in* $L^2_m(0,\infty)$ *has output* f, *then* φ *is a solution of* (3.1).

According to the above theorem we can solve (3.1) by interchanging in the system (3.5) the role of input and output. The latter yields the following system

$$\begin{cases} \dot{\rho}(t) = -i(A-BC)\rho(t) + iBf(t), \quad t \geq 0 \\ \varphi(t) = C\rho(t) + f(t) \\ (I-P)\rho(0) = 0 \end{cases} \tag{3.6}$$

To obtain the output φ in (3.6) we have to find a square integrable function ρ satisfying the first equation in (3.6) and the boundary condition $(I-P)\rho(0) = 0$. To find such a function ρ is not always possible and requires additional conditions. A sample result is the following inversion theorem.

Theorem 3.2 [7] *Let the symbol of the Wiener Hopf equation* (3.1) *be given by* (3.3), *and put* $A^\times = A - BC$. *Then* (3.1) *is uniquely solvable in* $L^2_m(0,\infty)$ *if and only if*

$$\sigma(A^\times) \cap \mathbb{R} = \emptyset, \quad \mathbb{C}^n = \mathrm{Im}P \dot{+} \mathrm{Ker}P^\times,$$

where $\sigma(A^\times)$ *denotes the set of eigenvalues of* $A^\times$, *and* $P^\times$ *is the Riesz projection of* $A^\times$ *corresponding to the eigenvalues in the open upper half plane* $\mathbb{C}_+$. *Furthermore, in this case the unique solution of* (3.1) *is given by*

$$\varphi(t) = f(t) + \int_0^\infty \gamma(t,s) f(s)\, ds, \quad t \geq 0,$$

where

$$\gamma(t,s) = \begin{cases} iCe^{-itA^\times}\Pi e^{isA^\times}B, & s < t, \\ iCe^{-itA^\times}(\Pi - I)e^{isA^\times}B, & t < s, \end{cases}$$

with Π *being the projection of* $\mathbb{C}^n$ *along* Im P *onto* Ker $P^\times$.

In a similar way (see [7]) one may obtain in this way explicit formulas in terms of A, B and C and associated matrices for Fredholm index and other Fredholm characteristics of the equation (3.1). Also explicit factorization formulas for the symbol may be derived. For these and related results we refer to Chapter XIII in [12].

3.2 Convolution equations on a finite interval

In this subsection we consider the analog of (3.1) on a finite interval:

$$\varphi(t) - \int_0^\tau k(t-s)\varphi(s)\,ds = f(t), \ \ 0 \le t \le \tau. \tag{3.7}$$

As before we assume that the kernel function k is an integrable $m \times m$ matrix function on the full line and that k has a rational Fourier transform. So we may assume that

$$I - \widehat{k}(\lambda) = I_m + C(\lambda I_n - A)^{-1}B, \tag{3.8}$$

where A has no eigenvalue on the real line. In general, the theory of convolution equations on a finite interval is quite different from the theory of the corresponding equations of the half line. However the state space approach sketched in the previous subsection for the half line equation applies equally well to (3.7).

Theorem 3.3 [8] *Let the symbol of* (3.7) *be given by* (3.8), *and let* P *be the spectral projection of* A *corresponding to eigenvalues of* A *in open upper half plane. Then the equation* (3.7) *and the linear system*

$$\begin{cases} \dot{\rho}(t) = -iA\rho(t) + iB\varphi(t), \quad 0 \le t \le \tau, \\ f(t) = -C\rho(t) + \varphi(t) \\ (I-P)\rho(0) = 0, \quad P\rho(\tau) = 0 \end{cases} \tag{3.9}$$

are equivalent in the following sense. If φ *is a solution of* (3.7) *in* $L^2_m(0,\tau)$, *then system* (3.9) *with input* φ *has output* f; *conversely, if the system* (3.9) *with input* φ *in* $L^2_m(0,\tau)$ *has output* f, *then* φ *is a solution of* (3.7).

By interchanging in (3.9) the role of input and output we may derive the following inversion theorem.

Theorem 3.4 [8] *Let the symbol of* (3.7) *be given by* (3.8), *let* P *be the spectral projection of* A *corresponding to eigenvalues of* A *in open upper half plane, and put* $A^\times = A - BC$. *Then the equation* (3.7) *is uniquely solvable in* $L^2_m(0,\tau)$ *if and only if*

$$S_\tau := Pe^{-i\tau A^\times}|_{\mathrm{Im}\ P} : \mathrm{Im}\ P \to \mathrm{Im}\ P$$

is invertible. In this case the unique solution of (3.7) *is given by*

$$\varphi(t) = f(t) + \int_0^\tau \gamma_\tau(t,s)f(s)\,ds,$$

where

$$\gamma_\tau(t,s) = \begin{cases} iCe^{-itA^\times}\Pi_\tau e^{isA^\times}B, & s<t, \\ iCe^{-itA^\times}(\Pi_\tau - I)e^{isA^\times}B, & t<s. \end{cases}$$

Here Π_τ *is the projection of* $\mathbb{C}^n$ *along* Im P *given by*

$$\Pi_\tau x = x - S_\tau^{-1}Pe^{-i\tau A^\times}x, \quad x \in \mathbb{C}^n.$$

Let us note that systems with boundary conditions on the state, like the ones appearing in (3.5), (3.6), and (3.9), are not common in system theory. In fact such systems are not causal; one has to know the past as well the future to determine the state. However, in mathematical physics such systems appear in a natural way and in these cases t is not a time parameter but a distance, e.g., a distance travelled in a medium (see, e.g., [12], Section XIII.9).

We conclude this section with an example. Consider the (scalar) equation:

$$\varphi(t) - \int_0^\tau e^{-|t-s|}\varphi(s)\,ds = f(t), \quad 0 \le t \le \tau. \tag{3.10}$$

The corresponding symbol is given by

$$\begin{aligned} W(\lambda) &= 1 - \int_{-\infty}^{\infty} e^{i\lambda t}e^{-|t|}\,dt = \frac{\lambda^2-1}{\lambda^2+1} \\ &= 1 + C(\lambda I_2 - A)^{-1}B, \end{aligned}$$

where

$$A = \begin{pmatrix} i & 0 \\ 0 & -i \end{pmatrix}, \; B = \begin{pmatrix} 1 \\ -1 \end{pmatrix}, \; C = \begin{pmatrix} i & i \end{pmatrix}$$

Let us compute the operator

$$S_\tau = Pe^{-i\tau A^\times}|_{\text{Im } P} : \text{Im } P \to \text{Im } P.$$

In this case

$$P = \begin{pmatrix} 1 & 0 \\ 0 & 0 \end{pmatrix}, \; A^\times = A - BC = \begin{pmatrix} 0 & -i \\ i & 0 \end{pmatrix}.$$

In particular, Im $P = \mathbb{C}$. To compute the exponential in the formula for S_τ we diagonalize $A^\times$. We have

$$A^\times = \begin{pmatrix} 1 & 1 \\ i & -i \end{pmatrix} \begin{pmatrix} 1 & 0 \\ 0 & -1 \end{pmatrix} \begin{pmatrix} \frac{1}{2} & -\frac{i}{2} \\ \frac{1}{2} & \frac{i}{2} \end{pmatrix}$$

which yields

$$e^{-i\tau A^\times} = \begin{pmatrix} 1 & 1 \\ i & -i \end{pmatrix} \begin{pmatrix} e^{-i\tau} & 0 \\ 0 & e^{i\tau} \end{pmatrix} \begin{pmatrix} \frac{1}{2} & -\frac{i}{2} \\ \frac{1}{2} & \frac{i}{2} \end{pmatrix},$$

and hence

$$\begin{aligned} S_\tau &= \begin{pmatrix} 1 & 0 \end{pmatrix} e^{-i\tau A^\times} \begin{pmatrix} 1 \\ 0 \end{pmatrix} \\ &= \frac{1}{2}(e^{-i\tau} + e^{i\tau}) = \cos\tau. \end{aligned}$$

We conclude that (3.10) is uniquely solvable in $L^2(0,\tau)$ if and only if $\cos\tau$ is different from zero. Assume that $\cos\tau \neq 0$, and let us compute the resolvent kernel using Theorem 3.4. In this case

$$\begin{aligned} \Pi_\tau \begin{pmatrix} x_1 \\ x_2 \end{pmatrix} &= \begin{pmatrix} x_1 \\ x_2 \end{pmatrix} - S_\tau^{-1} P e^{-i\tau A^\times} \begin{pmatrix} x_1 \\ x_2 \end{pmatrix} \\ &= \begin{pmatrix} x_2 \tan\tau \\ x_2 \end{pmatrix}, \end{aligned}$$

and hence

$$\Pi_\tau = \begin{pmatrix} 0 & \tan\tau \\ 0 & 1 \end{pmatrix}.$$

So the resolvent kernel is given by

$$\gamma_\tau(t,s) = \begin{cases} iCe^{-itA^\times}\Pi_\tau e^{isA^\times}B, & s < t, \\ iCe^{-itA^\times}(\Pi_\tau - I)e^{isA^\times}B, & t < s, \end{cases}$$

which yields

$$\begin{aligned} \gamma_\tau(t,s) = \sin|t-s| \quad &+ \quad \cos(t+s) + \\ &+ \quad (\tan\tau)\{\cos(t-s) - \sin(t+s)\}. \end{aligned}$$

4 Canonical systems of differential equations

The most recent application of the state space method concerns direct and inverse spectral problems for canonical systems of differential equations with rational matrix spectral functions.

4.1 The general setting

We first describe the general setting. Consider the system of differential equations:

$$\frac{d}{dx}u(x,\lambda) = ij(\lambda + V(x))u(x,\lambda), \quad x \geq 0. \tag{4.1}$$

Here j and $V(x)$ are $2m \times 2m$ matrices,

$$j = \begin{pmatrix} I_m & 0 \\ 0 & -I_m \end{pmatrix}, \quad V(x) = \begin{pmatrix} 0 & v(x) \\ v(x)^* & 0 \end{pmatrix},$$

the $m \times m$ matrix function $v(\cdot)$ is locally integrable on $(0,\infty)$, and the unknown $u(\cdot,\lambda)$ is a $2m \times 2m$ matrix function which depends on the spectral parameter $\lambda \in \mathbb{C}$. One calls (4.1) a *canonical system*, and one refers to $V(x)$ or to $v(x)$ as the *potential* of (4.1).

The spectral theory of the canonical system (4.1) is considered in terms of a symmetric differential operator associated with (4.1), namely the operator H on $L^2_{2m}(0,\infty)$ defined by

$$(Hf)(x) = -ij\frac{d}{dx}f(x) - V(x)f(x), \quad x \geq 0.$$

The domain $\mathcal{D}(H)$ of H consists of all $f \in L^2_{2m}(0,\infty)$ such that f is absolutely continuous, $f' \in L^2_{2m}(0,\infty)$, and f satisfies the initial condition $(I_m - I_m)f(0) = 0$. A matrix valued measure $d\tau$ on the real line is called a *spectral measure* for (4.1) if there exists a diagonalizing unitary operator U for H such that

$$(UHU^{-1}f)(z) = zf(z) \quad \text{on} \quad L^2_m(\mathbb{R}, d\tau).$$

There are two main spectral problems associated with (4.1):

- the direct problem: given the potential, find the spectral measure;
- the inverse problem: given the spectral measure, find the potential.

The solution of these problems were first obtained by I.M. Gelfand, M.G. Kreĭn, B.M. Levitan, and V.A. Marchenko.

Let us recall here the main elements of M.G. Kreĭn's [17] solution of the inverse problem which concerns the case when $v \in L^1_{m\times m}(0,\infty)$. In [17] it was shown that for such potentials

(i) the spectral measure is absolutely continuous, $d\tau = W(z)\,dz$,

(ii) the spectral density W has the form

$$W(z) = I_m - \widehat{k}(\lambda) \text{ for some } k \in L^1_{m\times m}(\mathbb{R}),$$

(iii) $W(z) \geq \varepsilon I_m$ for $z \in \mathbb{R}$.

Assume now that W is given. Then W defines a unique $k \in L^1_{m\times m}(\mathbb{R})$ such that property (ii) above holds. In other words, W is the symbol of a convolution operator of the type considered in Section 3. Let T_τ be the convolution operator on $L^2_{2m}(0,\infty)$ determined by k, i.e.,

$$(T_\tau f)(x) = f(x) - \int_0^\tau k(x-s)f(s)\,ds. \tag{4.2}$$

By property (iii) the operator T_τ is positive definite on $L^2_{2m}(0,\infty)$; in particular, T_τ is invertible,

$$(T_\tau^{-1} g)(x) = g(x) + \int_0^x \gamma_\tau(x,s)g(s)\,ds. \tag{4.3}$$

M.G. Kreĭn [17] proved that for a matrix function W with the properties (i), (ii), and (iii) mentioned above the solution of the inverse problem is the potential v given by

$$v(x) = -2i\gamma_{2x}(0,2x), \quad x \geq 0.$$

4.2 Positive definite rational spectral densities

Let W be a $m \times m$ matrix function which has the properties (i), (ii), and (iii) described in the previous subsection. In this subsection we assume additionally that W is rational and given by the realization

$$W(\lambda) = I_m + C(\lambda I_n - A)^{-1}B, \quad \lambda \in \mathbb{R}. \tag{4.4}$$

According to property (ii) mentioned above, the integral operator T_τ defined by (4.2) has W as its symbol, and hence the resolvent kernel γ in (4.2) can be computed by using Theorem 3.4. This yields a state space representation for the potential. The following theorem gives the final formula.

Theorem 4.1 [1] *Let W be a $m \times m$ rational matrix functions which is positive definite on the real line and given by the realization* (4.4). *Then $d\tau = W(z)\,dz$ is the spectral measure of the canonical system* (4.1) *with the potential v being given by*

$$v(x) = -2C\big(Pe^{-2ixA^\times}|_{\mathrm{Im}\ P}\big)^{-1}PB, \tag{4.5}$$

where P is the spectral projection of A corresponding to the eigenvalues of A in the open upper half plane, Im P *is the range of P, and $A^\times = A - BC$.*

Again let W be a $m \times m$ rational matrix function which has the properties (i), (ii), and (iii) described in the previous subsection. Notice that so far property (iii), the positivity of W on the real line, has been used only in a mild way (namely to conclude that the integral operator T_τ in (4.2) is invertible) and is not yet reflected in the choice of the realization. To do the latter, let us use the fact that from W is positive definite on the real line it follows that W admits a left spectral factorization,

$$W(z) = g(\bar{z})^* g(z). \tag{4.6}$$

Thus g is a $m \times m$ rational matrix function, which is proper and has the value I_m at infinity, and $g(\cdot)$ and $g(\cdot)^{-1}$ have no poles in open lower half plane $\mathbb{C}_-$ and on the real line. It follows that g admits a minimal realization

$$g(z) = I_m + \gamma_2^*(zI_n - \beta)^{-1}\gamma_1 \quad \text{with} \quad \sigma(\beta) \subset \mathbb{C}_-. \tag{4.7}$$

The minimality of the realization and the fact that $\sigma(\beta) \subset \mathbb{C}_-$ imply that there exists a unique T such that $\beta^*T - T\beta = i\gamma_2\gamma_2^*$. Using the symmetry in the latter equation and the uniqueness of T, we see that $T = T^*$. Furthermore, because of the minimality of thȩ realization, T is invertible. So we can use T as a state space transformation which replaces the realization (4.7) by

$$g(z) = I_m + \widetilde{\gamma}_2^*(zI_n - \widetilde{\beta})^{-1}\widetilde{\gamma}_1,$$

where

$$\widetilde{\beta} = T^{1/2}\beta T^{-1/2}, \ \widetilde{\gamma}_1 = T^{1/2}\gamma_1, \ \widetilde{\gamma}_2 = T^{-1/2}\gamma_2.$$

Notice that $\widetilde{\beta}^* - \widetilde{\beta} = i\widetilde{\gamma}_2\widetilde{\gamma}_2^*$. The above reasoning shows that without loss of generality we may assume that the spectral factor g in (4.6) can be realized as in (4.7) with the additional property that

$$\beta^* - \beta = i\gamma_2\gamma_2^*. \tag{4.8}$$

Using (4.6) and (4.7), we see that W has the realization (4.4) with

$$A = \begin{pmatrix} \beta^* & \gamma_2\gamma_2^* \\ 0 & \beta \end{pmatrix}, \quad B = \begin{pmatrix} \gamma_2 \\ \gamma_1 \end{pmatrix}, \quad C = \begin{pmatrix} \gamma_1^* & \gamma_2^* \end{pmatrix},$$

$$P = \begin{pmatrix} I_n & -iI_n \\ 0 & 0 \end{pmatrix}.$$

But then we can use (4.8) to show that

$$A^\times = A - BC = \begin{pmatrix} \alpha^* & 0 \\ -\gamma_1\gamma_1^* & \alpha \end{pmatrix} \quad \text{with} \quad \alpha = \beta - \gamma_1\gamma_2^*.$$

It follows (see [14]) that in this case the potential v in (4.5) has the following representation:

$$v(x) = 2\gamma_1^* e^{ix\alpha^*} S(x)^{-1} e^{ix\alpha}(i\gamma_1 - \gamma_2),$$

where

$$S(x) = e^{-ix\alpha} S e^{ix\alpha^*} + e^{ix\alpha}(I - S)e^{-ix\alpha^*},$$

with S equal to the unique solution of the equation $\alpha S - S\alpha^* = i\gamma_1{\gamma_1}^*$.

4.3 Pseudo-exponentional potentials

The approach to canonical systems described at the end of the previous section allows for an important generalization. To see this we begin with a lemma.

Lemma 4.2 [14] *Let β, γ_1, and γ_2 be a triple of $m \times m$ matrices such that*

$$\beta^* - \beta = i\gamma_2\gamma_2^*. \tag{4.9}$$

Put

$$A = \begin{pmatrix} \beta^* & \gamma_2\gamma_2^* \\ 0 & \beta \end{pmatrix}, \quad B = \begin{pmatrix} \gamma_2 \\ \gamma_1 \end{pmatrix}, \quad C = \begin{pmatrix} \gamma_1^* & \gamma_2^* \end{pmatrix}, \tag{4.10}$$

$$P = \begin{pmatrix} I_n & -iI_n \\ 0 & 0 \end{pmatrix}, \tag{4.11}$$

and set $A^\times = A - BC$. Then $Pe^{-ixA^\times}|_{\mathrm{Im}\ P} : \mathrm{Im}\ P \to \mathrm{Im}\ P$ is invertible for each $x \geq 0$.

From the above lemma it follows that for $A^\times, B, C$ and P in the above lemma the function

$$v(x) = -2C\big(Pe^{-2ixA^\times}|_{\mathrm{Im}\ P}\big)^{-1}PB, \quad x \geq 0, \tag{4.12}$$

is well-defined. We shall refer to v in (4.12) as the *pseudo-exponential potential determined* by the triple β, γ_1, and γ_2. When we use this phrase it is assumed that (4.9) is satisfied.

The class of pseudo-exponential potentials is larger than the class of potentials corresponding to positive definite rational spectral densities considered in the previous subsection. In fact, in the case of pseudo-exponential potentials

- the matrix β can have real eigenvalues,

- the matrix $A^\times$ can be nilpotent, and hence v can be rational in t.

The following theorem gives the solution of the direct spectral problem for a pseudo-exponential potential.

Theorem 4.3 [14] *Let v be the pseudo-exponential potential determined by the $m \times m$ matrices $\beta, \gamma_1, \gamma_2$, where (4.9) is satisfied. Let $z_1 < z_2 < \cdots < z_p$ be the real eigenvalues of β, and put*

$$\nu_k := \mathrm{Res}_{z=z_k} \gamma_1^*(zI_n - \beta)^{-1}\gamma_1, \quad 1 \le k \le p.$$

Then the spectral measure $d\tau(z)$ of the canonical system (4.1) with potential v is given by

$$\tau(z) = \int_0^z g(t)^* g(t)\, dt + \sum_{z_k < z} 2\pi\nu_k, \quad z \in \mathbb{R}, \tag{4.13}$$

where $g(t) = I_m + \gamma_2^*(tI_n - \beta)^{-1}\gamma_1$.

From the above theorem we see that the absolutely continuous part of the spectral measure corresponding to a pseudo-exponential potential has a rational derivative and the singular part is a step function. Moreover our state space formulas allow us to determine exactly where the jumps in the singular part occur and how big they are.

5 An application to nonlinear equations

It is known that the theory of canonical systems of differential equations can be used to find solutions of nonlinear equations. Here we combine this fact and the state space method to construct new explicit solutions of certain nonlinear partial differential equations.

Let

$$v(x) = -2C\left(Pe^{-2ixA^\times}|_{\mathrm{Im}\ P}\right)^{-1} PB \tag{5.1}$$

be a pseudo-exponential potential determined by the triple of $m \times m$ matrices $\beta, \gamma_1, \gamma_2$. Thus

$$A^\times = A - BC,$$

$$A = \begin{pmatrix} \beta^* & \gamma_2\gamma_2^* \\ 0 & \beta \end{pmatrix}, \quad B = \begin{pmatrix} \gamma_2 \\ \gamma_1 \end{pmatrix}, \quad C = \begin{pmatrix} \gamma_1^* & \gamma_2^* \end{pmatrix},$$

$$\beta^* - \beta = i\gamma_2\gamma_2^*, \qquad P = \begin{pmatrix} I_n & -iI_n \\ 0 & 0 \end{pmatrix}.$$

Introduce an additional time parameter t into v as follows:

$$v(x,t) = -2C\big(Pe^{-2ixA^\times}e^{-2it(A^\times)^k}|_{\operatorname{Im} P}\big)^{-1}PB. \tag{5.2}$$

It turns out that these functions are again pseudo-exponential potentials and they satisfy nonlinear partial differential equations.

Theorem 5.1 [14] *Let v be the pseudo-exponential potential* (5.1). *Then for each t in some neighborhood $[0,\varepsilon)$ the function $v(\cdot,t)$ defined by* (5.2) *is again a pseudo-exponential potential. Furthermore, the function $v(\cdot,\cdot)$ satisfies nonlinear equations, namely*

- *for $k=2$ the nonlinear Schrödinger equation*

$$2\frac{\partial v}{\partial t} + i\frac{\partial^2 v}{\partial x^2} = 2ivv^*v,$$

- *for $k=3$ the matrix modified Korteweg-de Vries equation*

$$4\frac{\partial v}{\partial t} + \frac{\partial^3 v}{\partial x^3} = 3\left(\frac{\partial v}{\partial x}v^*v + vv^*\frac{\partial v}{\partial x}\right).$$

Results analoguous to those obtained in this and the previous section (including solutions of nonlinear partial differential equations) also hold for pseudo-canonical systems with skew selfadjoint potentials V (see [15]) and for Sturm-Liouville systems (see [3], [16]).

References

[1] Alpay D., and I. Gohberg, Inverse spectral problem for differential operators with rational scattering matrix function, *J. Diff. Eqs* **118** (1995), 1–19.

[2] Alpay D., and I. Gohberg, State space methods for inverse spectral problems, in *Systems and Control in the Twenty-First Century* (eds. C.I. Byrnes, B.N. Datta, D.S. Gilliam, C.F. Martin), PSCT **22**, Birkhäuser Verlag, 1997; pp. 1–16.

[3] Alpay D., and I. Gohberg, Inverse problem for Sturm-Liouville operators with rational reflection coefficient, *Integral Equations and Operator Theory* **30** (1998), 317–325.

[4] Ball J.A., I. Gohberg, and L. Rodman, Realization and interpolation of rational matrix functions, in *Topics in interpolation theory of rational matrix-valued functions* (ed. I. Gohberg), OT **33**, Birkhäuser Verlag, Basel, 1988; pp. 1–71.

[5] Ball J.A., I. Gohberg, and L. Rodman, *Interpolation of rational matrix functions*, OT **45**, Birkhäuser Verlag, Basel, 1990.

[6] Ball J.A., I. Gohberg, and L. Rodman, The state space method in the study of interpolation by rational matrix functions, in: *Mathematical System Theory. The influence of Kalman* (ed. A.C. Antoulas), Springer-Verlag, 1991; pp. 503–508.

[7] Bart H., I. Gohberg, and M.A. Kaashoek, Wiener-Hopf integral equations, Toeplitz matrices and linear systems, in: *Toeplitz Centennial* (ed. I. Gohberg), OT **4**, Birkhäuser Verlag, 1982; pp. 85–135.

[8] Bart H., I. Gohberg, and M.A. Kaashoek, Convolution equations and linear systems, *Integral Equations and Operator Theory* **5** (1982), 283–340.

[9] Bart H., I. Gohberg, and M.A. Kaashoek, The state space method in problems of analysis, in: *Proceedings First International Conference on Industrial and Applied Mathematics, Contributions from the Netherlands*, CWI, Amsterdam 1987, pp. 1–16.

[10] Bart H., I. Gohberg, and M.A. Kaashoek, Wiener-Hopf equations and linear systems, in: *Proc. Symposia Appl. Math.* **52**, Amer. Math. Soc. Providence RI, 1996, pp. 115–128.

[11] Foias C., A.E. Frazho, I. Gohberg, and M.A. Kaashoek, *Metric constrained interpolation, commutant lifting and systems*, OT **100**, Birkhäuser Verlag, Basel, 1998.

[12] Gohberg I., S. Goldberg, and M.A. Kaashoek, *Classes of Linear Operators*, Volume I, OT **49**, Birkhäuser Verlag, Basel, 1990.

[13] Gohberg I., and M.A. Kaashoek, The state space method for solving singular integral equations, in: *Mathematical System Theory. The influence of Kalman* (ed. A.C. Antoulas), Springer-Verlag, 1991; pp. 509–523.

[14] Gohberg I., M.A. Kaashoek and A.L. Sakhnovich, Canonical systems with rational spectral densities: explicit formulas and applications, *Mathematische Nachrichten*, to appear.

[15] Gohberg I., M.A. Kaashoek and A.L. Sakhnovich, Pseudo-canonical systems with rational Weyl functions: explicit formulas and applications, *J. Diff. Eqs* **146** (1998), 375–398.

[16] Gohberg I., M.A. Kaashoek and A.L. Sakhnovich, Sturm-Liouville systems with rational Weyl functions: explicit formulas and applications, *Integral Equations and Operator Theory* **30** (1998), 338–377.

[17] On the theory of accelerants and S-matrices of canonical differential systems, *Dokl. Akad. Nauk SSSR* **111** (1956), 1167–1170 [Russian].

School of Mathematical Sciences, Raymond and Beverly Sackler Faculty of Exact Sciences, Tel-Aviv University, 69978 Tel-Aviv, Israel

Department of Mathematics, Faculty of Sciences, Vrije Universiteit, De Boelelaan 1081a, 1081 HV Amsterdam, The Netherlands

Progress in Systems and Control Theory, Vol. 25

Stabilization of Nonlinear Systems Using Output Feedback

A. Isidori[1]

1 Introduction

One of the basic fundamental issues in control theory is the ability to design a feedback law to the purpose of robustly stabilizing a system, in the presence of structured uncertainties, such as parameter variations, and/or unstructured uncertainties, such as unmodeled dynamics.

In the case of finite-dimensional, time-invariant, linear models, a variety of techniques are available, which range from elementary methods based on the properties of the root locus (which lends itself to a very intuitive characterization of the "stability margin" inherent to certain design techniques, such as "small-gain" and/or "high-gain") to more sophisticated methods such as, in the case of parameter uncertainties, those relying upon the analysis of the closed-loop characteristic polynomial (whose uncertain coefficients range over prescribed intervals) or the design of a feedback controller imposing a fixed (parameter independent) Lyapunov function. In the case of unstructured perturbations, a rather common standpoint is to look at the controlled system as to the feedback interconnection of two subsystems, only one of which is accurately modeled, and then to seek a control law able to bring the L_2-gain of the loop below a fixed threshold. In this way a problem of robust stabilization is cast as a problem of lowering the L_2-gain between a disturbance input and a regulated output of an appropriate controlled subsystem.

In the last years, similar methods have been gradually developed also for nonlinear systems. Fundamental in this respect were the methods for robust design based on the theory of differential games and the corresponding L_2-gain analysis developed by Basar and Bernhard [2], Van der Schaft [25], Ball *et al.* [1], the notion and the properties of input-to-state stable system developed by Sontag *et al.* [20] [21], the methods of adaptive output feedback controllers studied by Marino and Tomei [16] [17], the methods for nonlinear adaptive control developed by Krstic, Kanellakopoulos and Kokotovic [15], the methods for robust feedback stabilization proposed by Freeman and Kokotovic [5], [6], and by Sepulchre, Jankovic and Kokotovic [19], the methods for output feedback stabilization developed by Teel and Praly [22] [23], the methods for stabilization of interconnected uncertain systems proposed by Jiang and Mareels [11] [12].

[1]Work supported in part by NSF under grant ECS-9707891, by AFOSR under grant F49620-95-1-0232, and by MURST.

In the case of output feedback, the major achievement in this area of research was the "nonlinear separation principle" proved by Teel and Praly [22], who have shown that global stabilizability via state feedback and a property of uniform observability (previously introduced by Gauthier and and Bornard [7]) imply the possibility of semiglobal stabilization via output feedback. To cope with the restricted information structure, Teel-Praly's stabilization scheme includes an approximate state observer (whose role is actually that of producing approximate estimates of a number of "higher order" derivatives of the output) earlier developed by Khalil [14] to cope with a similar (though more restricted) stabilization problem. Teel and Praly have also shown that, in the presence of parameter uncertainties, semiglobal stabilization via output feedback is still possible if a state-feedback law is known which robustly globally stabilizes the system and its value, at any time, can be expressed as a (fixed) function of the values, at this time, of a fixed number of derivatives of input and output [23].

The separation principle is indeed a cornerstone of linear system theory and this is why we regard its nonlinear version provided by Teel and Praly as one of the finest results achieved in recent years in this domain. In the first part of this paper, we provide a summary of this result, along with an alternative new proof. Then, in the second part of the paper, we explore other methods for constructing (robust) stabilizers via output feedback, which do not necessarily appeal to the principle of "composing a state feedback with (direct or indirect) estimates of the state". In particular, we describe a recursive stabilization scheme, recently introduced in [10], which essentially – at each stage – only uses "small-gain" and/or "high-gain" arguments to determine the values of certain design parameters. For convenience, this method is described first in the case of linear systems, when it is always applicable under no extra hypothesis other than the obvious one of "stabilizability by output feedback", and then in the case of nonlinear systems.

2 Teel-Praly's NL Separation Principle

Consider a system of the form

$$\begin{aligned} \dot{x} &= f(x) + g(x)u \\ y &= h(x)\,, \end{aligned} \tag{1}$$

with state $x \in I\!R^n$, input $u \in I\!R$, output $y \in I\!R$, in which it is assumed that $f(x)$, $g(x)$, $h(x)$ are continuous functions of their arguments, and $f(0) = 0, h(0) = 0$. Consider the sequence of $n+1$ mappings

$$\begin{aligned} \varphi_0 \;:\; I\!R^n &\rightarrow I\!R \\ x &\mapsto \varphi_0(x) \end{aligned}$$

and

$$\begin{array}{cccccc} \varphi_k & : & \mathbb{R}^n \times \mathbb{R}^k & \to & \mathbb{R}^n & \\ & & (x,(v_0,\dots,v_{k-1})) & \mapsto & \varphi_k(x,v_0,\dots,v_{k-1}) & 1 \le k \le n \end{array}$$

defined in the following way

$$\begin{aligned} \varphi_0(x) &= h(x) \\ \varphi_1(x,v_0) &= \frac{\partial \varphi_0}{\partial x}(f(x)+g(x)v_0) \\ &\dots \\ \varphi_k(x,v_0,\dots,v_{k-1}) &= \frac{\partial \varphi_{k-1}}{\partial x}(f(x)+g(x)v_0) + \sum_{i=0}^{k-2} \frac{\partial \varphi_{k-1}}{\partial v_i} v_{i+1} \end{aligned} \tag{2}$$

for $1 < k \le n$.

It is immediate to realize that these mappings, if the input $u(t)$ of (1) is a C^{k-1} function of t, are precisely the mappings which express - for each k and any given time t - the dependence of the k-th derivative $y^{(k)}(t)$ of $y(t)$ on $x(t)$ and $u(t),\dots,u^{(k-1)}(t)$. As a matter of fact,

$$y^{(k)}(t) = \varphi_k(x(t),u(t),\dots,u^{(k-1)}(t)) \,.$$

Suppose now that the first n of these mappings are put together, to define a mapping

$$\begin{array}{ccccc} \Phi & : & \mathbb{R}^n \times \mathbb{R}^{n-1} & \to & \mathbb{R}^n \\ & & (x,v) & \mapsto & w = \Phi(x,v) \end{array} \tag{3}$$

in which

$$v = \begin{pmatrix} v_0 \\ v_1 \\ \cdot \\ v_{n-2} \end{pmatrix}, \qquad \Phi(x,v) = \begin{pmatrix} \varphi_0(x) \\ \varphi_1(x,v_0) \\ \dots \\ \varphi_{n-1}(x,v_0,v_1,\dots,v_{n-2}) \end{pmatrix}.$$

By construction,

$$\Phi \,:\, \left(x(t),u(t),\dots,u^{(n-2)}(t)\right) \mapsto \mathrm{col}(y(t),\dots,y^{(n-1)}(t)) \,.$$

A system of the form (1) is said to be *uniformly observable* (see [7]) if the following conditions are satisfied

(i) the mapping

$$\begin{array}{ccccc} H & : & \mathbb{R}^n & \to & \mathbb{R}^n \\ & & x & \mapsto & \mathrm{col}(h(x), L_f h(x), \dots, L_f^{n-1} h(x)) \end{array}$$

is a global diffeomorphism,

(ii) the rank condition

$$\text{rank}\Big(\frac{\partial \Phi}{\partial x}\Big)(x,v) = n$$

holds for each $(x,v) \in I\!R^n \times I\!R^{n-1}$.

It turns out, that if a system is uniformly observable, the mapping $\Phi(x,v)$ has a global inverse in v (see [7]).

Proposition 2.1 *Suppose (1) is uniformly observable. Then there exists a unique smooth mapping*

$$\begin{array}{rcccc} \Psi & : & I\!R^n \times I\!R^{n-1} & \rightarrow & I\!R^n \\ & & (w,v) & \mapsto & x = \Psi(w,v) \end{array}$$

such that $w = \Phi(\Psi(w,v),v)$ *for all* $(w,v) \in I\!R^n \times I\!R^{n-1}$ *and* $x = \Psi(\Phi(x,v),v)$ *for all* $(x,v) \in I\!R^n \times I\!R^{n-1}$.

Remark. Suppose system (1) is uniformly observable and consider also the function $\varphi_n(x, v_0, \ldots, v_{n-1})$ defined above, which by construction is such that

$$y^{(n)}(t) = \varphi_n(x(t), u(t), \ldots, u^{(n-1)}(t)) \ .$$

Using the function $\Psi(w,v)$, whose existence has been shown in Proposition 2.1, define the system

$$\begin{array}{rcl} \dot{w}_0 & = & w_1 \\ \dot{w}_1 & = & w_2 \\ & \cdots & \\ \dot{w}_{n-2} & = & w_{n-1} \\ \dot{w}_{n-1} & = & \varphi_n(\Psi(w,v), v_0, \ldots, v_{n-1}) \ . \end{array} \tag{4}$$

Then, it is clear from the previous discussion that, if

$$v_i(t) = u^{(i)}(t) \qquad \text{for all } t \geq 0 \text{ and for all } 0 \leq i \leq n-1$$

and

$$w_i(0) = y^{(i)}(0) \qquad \text{for all } 0 \leq i \leq n-1 \ ,$$

then

$$w_i(t) = y^{(i)}(t) \qquad \text{for all } t \geq 0 \text{ and for all } 0 \leq i \leq n-1 \ .$$

In other words, if the initial state (at time $t = 0$) and the inputs to system (4) are appropriately set, the various components of state of this system reproduce the output of (1) and its first $n-1$ derivatives. ◁

Teel and Praly have shown that, if the equilibrium $x = 0$ of

$$\dot{x} = f(x) + g(x)u$$

is *globally asymptotically stabilizable* by means of a *state* feedback and if the system is *uniformly observable*, then the system is *semiglobally stabilizable* by means of a *dynamic output feedback* of the form

$$\begin{aligned} \dot{\xi} &= \eta(\xi, y) \\ u &= \theta(\xi) . \end{aligned} \tag{5}$$

The controller which achieves this goal is constructed as follows. Consider the extended $2n$-dimensional system

$$\begin{aligned} \dot{x} &= f(x) + g(x)v_0 \\ \dot{v}_0 &= v_1 \\ &\cdots \\ \dot{v}_{n-2} &= v_{n-1} \\ \dot{v}_{n-1} &= \bar{u} . \end{aligned} \tag{6}$$

It is well known (see e.g. [3]) that, from the knowledge of a feedback law $u = \alpha(x)$ which globally stabilizes system (1), it is possible to construct a feedback law of the form

$$\bar{u} = \theta(x, v_0, \ldots, v_{n-1}) \tag{7}$$

which globally asymptotically stabilizes system (6).

The interconnection of (6) and (7) can be viewed as the interconnection of the original system (1) and the dynamic feedback

$$\begin{pmatrix} \dot{v}_0 \\ \dot{v}_1 \\ \cdot \\ \dot{v}_{n-2} \\ \dot{v}_{n-1} \end{pmatrix} = \begin{pmatrix} v_1 \\ v_2 \\ \cdots \\ v_{n-1} \\ \theta(x, v_0, \ldots, v_{n-1}) \end{pmatrix} \tag{8}$$

$$u = v_0 .$$

Now, replace x in the controller (8) by $\Psi(\eta, v)$, in which $\Psi(\cdot, \cdot)$ is the mapping defined in Proposition 2.1, v is the state of (8), and η is provided by an "estimator" described by equations of the form

$$\begin{pmatrix} \dot{\eta}_0 \\ \dot{\eta}_1 \\ \cdot \\ \dot{\eta}_{n-2} \\ \dot{\eta}_{n-1} \end{pmatrix} = \begin{pmatrix} \eta_1 \\ \eta_2 \\ \cdots \\ \eta_{n-1} \\ \varphi_n(\Psi(\eta, v), v_0, \ldots, v_{n-1}) \end{pmatrix} + \begin{pmatrix} Lc_{n-1} \\ L^2 c_{n-2} \\ \cdots \\ L^{n-1} c_1 \\ L^n c_0 \end{pmatrix} (y - \eta_0) , \tag{9}$$

in which $L > 0$ is a constant to be determined later and $c_0, c_1, \dots, c_{n-1}$ are coefficients of Hurwitz polynomial

$$p(\lambda) = \lambda^n + c_{n-1}\lambda^{n-1} + \cdots + c_1\lambda + c_0 .$$

By construction, $\eta = \Phi(x, v)$ is a (locally exponentially attractive) invariant manifold for the corresponding closed-loop system. Moreover, by construction, the restriction of the dynamics of the closed-loop system to this invariant manifold is globally asymptotically stable by hypothesis. Therefore, the controller thus found locally asymptotically stabilizes the equilibrium $(x, v, \eta) = (0, 0, 0)$ of the closed-loop system.

In order to achieve "semiglobal stability" (i.e. a region of attraction containing an arbitrarily chosen compact set) one needs to replace, in the dynamic feedback law constructed above, the function $\Psi(\eta, v)$ by another function $\Psi^*(\eta, v)$ defined as follows

$$\Psi^*(\eta, v) = \begin{cases} \Psi(\eta, v) & \text{if } \|\Psi(\eta, v)\| < M \\ \dfrac{\Psi(\eta, v)}{\|\Psi(\eta, v)\|} M & \text{if } \|\Psi(\eta, v)\| \geq M , \end{cases} \tag{10}$$

where $M > 0$ is a design parameter. Note that $\Psi^*(\eta, v)$ is a function which coincides with $\Psi(\eta, v)$ for all (η, v) such that the norm of $\Psi(\eta, v)$ is less than a fixed number M, and bounded (in norm) by M elsewhere.

This yields a control law, for (1) described by equations of the form

$$\begin{pmatrix} \dot{v}_0 \\ \dot{v}_1 \\ \cdot \\ \dot{v}_{n-2} \\ \dot{v}_{n-1} \end{pmatrix} = \begin{pmatrix} v_1 \\ v_2 \\ \cdots \\ v_{n-1} \\ \theta(\Psi^*(\eta, v), v_0, \dots, v_{n-1}) \end{pmatrix} , \tag{11}$$

$$\begin{pmatrix} \dot{\eta}_0 \\ \dot{\eta}_1 \\ \cdot \\ \dot{\eta}_{n-2} \\ \dot{\eta}_{n-1} \end{pmatrix} = \begin{pmatrix} \eta_1 \\ \eta_2 \\ \cdots \\ \eta_{n-1} \\ \varphi_n(\Psi^*(\eta, v), v_0, \dots, v_{n-1}) \end{pmatrix} + \begin{pmatrix} Lc_{n-1} \\ L^2 c_{n-2} \\ \cdots \\ L^{n-1} c_1 \\ L^n c_0 \end{pmatrix} (y - \eta_0) , \tag{12}$$

and

$$u = v_0 . \tag{13}$$

Note that the system thus defined is a feedback law of the form (5), with

$$\xi = \mathrm{col}(v_0, v_1, \dots, v_{n-1}, \eta_0, \eta_1, \dots, \eta_{n-1}) .$$

This feedback law is such that the following result holds (see [22]).

Theorem 2.1 *For each $R > 0$ there exist numbers $R' > 0$, $M^* > 0$ and, for each $M > M^*$, a number $L^*_M > 0$ such that, if $M \geq M^*$ in (10) and*

$L \geq L_M^*$ in (12), the equilibrium $(z,\xi) = (0,0)$ of the closed loop system (1)-(11)-(12)-(13) is locally asymptotically stable and, moreover,

$$\|x(0)\| \leq R, \|\xi(0)\| \leq R' \quad \Rightarrow \quad \left\{ \begin{array}{l} \lim_{t\to\infty} x(t) = 0 \\ \lim_{t\to\infty} \xi(t) = 0 \,. \end{array} \right.$$

Proof. Consider the closed-loop system and change η into e defined as

$$e = D_L(\Phi(x,v) - \eta)$$

where

$$D_L = \text{diag}\{L^{n-1}, \ldots, L, 1\} \,.$$

This yields equations of the form

$$\begin{aligned} \dot{z} &= F(z) + p_1(z,e) \\ \dot{e} &= LAe + p_2(z,e) \end{aligned} \tag{14}$$

in which $z = \text{col}(x,v)$ and

$$p_1(z,e) = \begin{pmatrix} 0 \\ \cdots \\ \phi_1(z,e) \end{pmatrix}, \quad p_2(z,e) = \begin{pmatrix} 0 \\ \cdots \\ \phi_2(z,e) \end{pmatrix},$$

and

$$\begin{aligned} \phi_1(z,e) &= \theta(\Psi^*(\Phi(x,v) - D_L^{-1}e, v_0, \ldots, v_{n-1}) - \theta(x, v_0, \ldots, v_{n-1}) \\ \phi_2(z,e) &= \varphi_n(\Psi^*(\Phi(x,v) - D_L^{-1}e, v_0, \ldots, v_{n-1}) - \varphi_n(x, v_0, \ldots, v_{n-1}). \end{aligned}$$

By construction, system

$$\dot{z} = F(z)$$

has a globally asymptotically stable equilibrium at $z = 0$. Thus, there exists a smooth real-valued function $V(z)$ satisfying

$$\begin{aligned} &\alpha_1(\|z\|) \leq V(z) \leq \alpha_2(\|z\|) \\ &\frac{\partial V}{\partial z} F(z) \leq -\alpha_3(\|z\|) \end{aligned}$$

for all z, where the $\alpha_i(\cdot)$'s are class $\mathcal{K}_\infty$ functions.

By hypothesis,

$$\{z(0), \eta(0)\} \in \mathcal{S}_z \times \mathcal{S}_\eta$$

where $\mathcal{S}_z$ and $\mathcal{S}_\eta$ are fixed compact sets. Choose c such that

$$\Omega_c = \{z : V(z) \leq c\} \supset \mathcal{S}_z$$

and then choose the parameter M in the definition of the function $\Psi^*(\eta, v)$ as

$$M = \max_{z \in \Omega_{c+1}} \|x\|.$$

With this choice, it is easy to check that the two vectors $p_1(z, e)$ and $p_2(z, e)$ defined before vanish at $e = 0$, so long as $z \in \Omega_{c+1}$. In fact, recall that $x = \Psi(\Phi(x, v), v)$. Thus, if $z \in \Omega_{c+1}$, we have $\|\Psi(\Phi(x, v), v)\| \leq M$. As a consequence, if $z \in \Omega_{c+1}$

$$[\Psi^*(\eta, v)]_{e=0} = \Psi^*(\Phi(x, v), v) = \Psi(\Phi(x, v), v) = x,$$

and this shows that if $z \in \Omega_{c+1}$,

$$p_1(z, 0) = 0, \quad p_2(z, 0) = 0.$$

Note that, for all $(z, e) \in \Omega_{c+1} \times \mathbb{R}^n$,

$$\|\Psi^*(\Phi(x, v) - D_L^{-1} e, v)\| \leq M,$$

Thus, there exist positive numbers β_1, β_2, independent of L, such that

$$\|p_i(z, e)\| \leq \beta_i, \text{ for all } (z, e) \in \Omega_{c+1} \times \mathbb{R}^n. \tag{15}$$

Moreover, using the fact that $p_1(z, e)$ and $p_2(z, e)$ vanish at $e = 0$ (if $z \in \Omega_{c+1}$), it is easy to see that, by construction, there exists a positive nondecreasing function $\gamma(\cdot)$, with $\gamma(0) = 0$, which is independent of L (if $L > 1$), such that

$$\|p_1(z, e)\| \leq \gamma(\|e\|), \text{ for all } (z, e) \in \Omega_{c+1} \times \mathbb{R}^n. \tag{16}$$

Let k_1 be such that

$$\|\frac{\partial V(z)}{\partial z}\| \leq k_1 \text{ for all } z \in \Omega_{c+1}.$$

Then

$$\frac{\partial V(z)}{\partial z}(F(z) + p_1(z, e)) \leq -\alpha_3(V(z)) + k_1\beta_1 \tag{17}$$

for all $(z, e) \in \Omega_{c+1} \times \mathbb{R}^n$.

Let P be the positive definite solution of $PA + A^T P = -I$ and observe that, for all $(z, e) \in \Omega_{c+1} \times \mathbb{R}^n$, the function $Q(e) = e^T P e$ satisfies

$$\begin{aligned} \frac{\partial Q}{\partial e}\dot{e} &\leq -L\|e\|^2 + 2|e^T P|\beta_2 \leq -\left(L - \frac{k_2}{\mu}\right)\|e\|^2 + \mu\beta_2^2 \\ &\leq -\left(L - \frac{k_2}{\mu}\right) k_3 Q(e) + \mu\beta_2^2 \end{aligned}$$

where $k_2 > 0$ and $k_3 > 0$ are numbers depending only on P and μ is any positive number. Set

$$a = \left(L - \frac{k_2}{\mu}\right) k_3,$$

to obtain

$$\frac{\partial Q}{\partial e}(LAe + p_2(z,e)) \leq -aQ(e) + \mu\beta_2^2 \tag{18}$$

for all $(z, e) \in \Omega_{c+1} \times I\!R^n$.

Inequalities (17) and (18) show that, if $z(t) \in \Omega_{c+1}$, and L is large enough (so that $a > 0$),

$$\frac{dV(z(t))}{dt} \leq k_1\beta_1, \quad \frac{dQ(e(t))}{dt} \leq \mu\beta_2^2.$$

¿From these it can be seen that there is a fixed time $T > 0$ (independent of L) such that, for every initial state $z(0)$, $\eta(0) \in \mathcal{S}_z \times \mathcal{S}_\eta$, the solution $z(t)$, $\eta(t)$ is defined for all $t \in [0, T]$ and, in particular, $z(t) \in \Omega_{c+1}$ for all $t \in [0, T]$. In fact, choosing $T = (2k_1\beta_1)^{-1}$, it is immediate to conclude that $z(t)$, $e(t)$ is defined for all $t \in [0, T]$ and $V(z(t)) \leq c + \frac{1}{2}$, because otherwise the inequalities

$$V(z(t)) - V(z(0)) \leq k_1\beta_1 t$$
$$Q(e(t)) - Q(e(0)) \leq \mu\beta_2^2 t$$

would be contradicted.

We prove now the following Lemma.

Lemma 2.1 *For any ϵ there exists a number $L^* > 0$ (independent of $z(0)$, $\eta(0)$) such that, if $L > L^*$,*

$$\|e(T)\| \leq \epsilon,$$

and, moreover, for all $T' > T$

$$z(t) \in \Omega_{c+1} \text{ for all } t \in [T, T'] \Rightarrow \|e(t)\| \leq \epsilon \text{ for all } t \in [T, T'].$$

Proof of the Lemma. Consider again (18). By standard comparison arguments, it is deduced that

$$\|e(t)\|^2 \leq k_4 \left[e^{-at}\|e(0)\|^2 + \frac{1 - e^{-at}}{a}\mu\beta_2^2\right],$$

where $k_4 > 0$ is a number depending only on P. Fix ϵ and choose μ to satisfy $2k_4\mu\beta^2 \leq \epsilon^2$, so that, if $a > 1$,

$$\|e(t)\|^2 \leq k_4 e^{-at}\|e(0)\|^2 + \frac{\epsilon^2}{2}.$$

Observe that

$$\lim_{L\to\infty} e^{-aT}\|e(0)\|^2 = 0,$$

because $\|e(0)\| = \|D_L(\Phi(x(0), v(0)) - \eta(0)\|$ is bounded by a polynomial of order $n-1$ in L. In particular, since $x(0)$, $v(0)$ and $\eta(0)$ range over a compact set, there is L^*,

$$k_4 e^{-aT}\|e(0)\|^2 \leq \frac{\epsilon^2}{2}$$

for every $L > L^*$ and every $x(0)$, $v(0)$, $\eta(0)$ and the result follows. ◁

Using this, it can be proven that the trajectories of the system are bounded and, actually, converge to an arbitrarily small neighborhood of the origin. To this end, let $\delta > 0$ be any (small) number, satisfying $\delta < \|z(0)\|$ and such that

$$B_\delta = \{z : \|z\| < \delta\} \subset \text{int}(\Omega_{c+1}).$$

Then, choose ϵ so that

$$\Omega_{a_3^{-1}(k_1\gamma(\epsilon))+d} \subset B_\delta,$$

for some $d > 0$ and consider the set

$$\Gamma = \{z \in \Omega_{c+1} : z \notin \Omega_{a_3^{-1}(k_1\gamma(\epsilon))}\}.$$

By construction, so long as

$$z(t) \in \Gamma, \quad \|e(t)\| \leq \epsilon$$

the function $V(z(t))$ satisfies

$$\frac{dV(z(t))}{dt} \leq -\alpha_3(V(z(t)) + k_1\gamma(\epsilon) < 0 \tag{19}$$

i.e. is decreasing.

We have already shown that, for all $t \in [0, T]$, the solution $z(t)$, $\eta(t)$ is defined and $z(t) \in \text{int}(\Omega_{c+1})$. If $L > L^*$, we know from the Lemma that $\|e(T)\| \leq \epsilon$. Then, using again this Lemma, we see that $z(t)$ cannot leave the set Ω_{c+1} and $\|e(t)\|$ is bounded by ϵ for all $t \geq T$. In fact, $z(t)$ cannot reach the boundary of the set Ω_{c+1}, where $V(z) > V(z(T))$ without contradicting (19).

It can also be shown that, in finite time, $z(t)$ enters the set $\Omega_{a_3^{-1}(k_1\gamma(\epsilon))+d}$ and remains there for all subsequent times. For, suppose this is not true. Then, $V(z(t))$ is always decreasing and converges to a nonnegative limit $V_0 \geq \alpha_3^{-1}(k_1\gamma(\epsilon)) + d$. Let γ^+ denote the ω-limit set of the trajectory in question. It is well-known that $V(z) = V_0$ at each point of γ^+. Pick any initial condition in γ^+ and observe that the function $V(z)$ is constant along the corresponding trajectory. Thus we have

$$0 \leq -\alpha_3(V_0) + k_1\gamma(\epsilon),$$

i.e.

$$V_0 \leq \alpha_3^{-1}(k_1\gamma(\epsilon))$$

which is a contradiction. Once $z(t)$ has entered $\Omega_{\alpha_3^{-1}(k_1\gamma(\epsilon))+d}$, it can never leave this set, because $\dot{V}(z(t))$ is negative at each point of its boundary.

Having proven that, in finite time, both $z(t)$ and $e(t)$ enter an arbitrarily small neighborhood of the equilibrium $(z, e) = (0, 0)$, the proof can be completed by showing that the latter is locally asymptotically stable.

To this end, observe that if ϵ and δ are sufficiently small,

$$\Psi^*(\Phi(x, v) - D_L^{-1}e, v) = \Psi(\Phi(x, v) - D_L^{-1}e, v).$$

Thus, $p_2(z, e)$ is a smooth function of z, e which vanishes at $e = 0$ and can be expressed as

$$p_2(z, e) = g_2(z, e)e.$$

Then, the function $Q(e)$ satisfies

$$\frac{\partial Q}{\partial e}\dot{e} \leq -L\|e\|^2 + 2\|e\|^2 k_5$$

where k_5 is such that $\|Pg_2(z, e)\| < k_5$ for all $\|z\| \leq \delta$ and all $\|e\| \leq \epsilon$.

This shows that, if L is large enough,

$$\lim_{t\to\infty} e(t) = 0.$$

¿From this, standard arguments show that also

$$\lim_{t\to\infty} z(t) = 0$$

and this completes the proof. ◁

3 Stabilization via output feedback without separation principle

In this section we illustrate a simple recursive design method for stabilizing a linear system by output feedback, which does not use any technique for pole assignment nor the separation principle, but rather uses concepts reminiscent of the classical design methods based on root-locus properties. Of course, this does not lead to any specific breakthrough in the case of linear systems, where the solution of a diophantine equation would suffice, but its nonlinear version, described in the next section, opens a totally new perspective in the problem of robustly stabilizing, via output feedback, a possibly unstable and non-minimum phase nonlinear system.

Consider a single-input single-output linear system with no input-output feed-through. Let n denote the dimension of its state space and let r denote its relative degree. It is well known that this system can always be modeled by equations of the form

$$\begin{aligned} \dot{z} &= Fz + Gv \\ \dot{v} &= Hz + Jv + bu \\ y &= Kz\,, \end{aligned} \tag{20}$$

in which $z \in I\!R^{n-1}$ and in which v coincides with $y^{(r-1)}$, the $(r-1)$-th derivative of the output with respect to time.

With this system, we associate an *auxiliary subsystem*, with input u_{a}, output y_{a}, and state z, defined as follows

$$\begin{aligned} \dot{z} &= Fz + Gu_{\mathrm{a}} \\ y_{\mathrm{a}} &= Hz + Ju_{\mathrm{a}}\,. \end{aligned} \tag{21}$$

The main purpose of this section is to show that, if the auxiliary subsystem thus defined is stabilizable by output feedback, so is the full system (20). This result is the direct consequence of two Lemmas. The first Lemma shows how to stabilize (20) using a dynamic output feedback which uses v as input.

Lemma 3.1 *Suppose system*

$$\begin{aligned} \dot{\eta} &= L\eta + My_{\mathrm{a}} \\ u_{\mathrm{a}} &= N\eta \end{aligned}$$

stabilizes system (21). Then there is a number k^* such that, for all $k > k^*$ the feedback law

$$\begin{aligned} \dot{\eta} &= L\eta + Mk(v - N\eta) \\ u &= \frac{1}{b}\Big[N[L\eta + Mk(v - N\eta)] - k(v - N\eta)\Big] \end{aligned} \tag{22}$$

stabilizes system (20).

Proof. Consider the interconnection of (20) with (22) and change the state variable v into a state variable w defined as

$$w = k(v - N\eta) - Hz - Jv\,.$$

This change of variable is well defined if k is large enough, and

$$v = a(\varepsilon)w + a(\varepsilon)Hz + b(\varepsilon)N\eta$$

where

$$\varepsilon = \frac{1}{k}, \quad a(\varepsilon) = \frac{\varepsilon}{1 - \varepsilon J}, \quad b(\varepsilon) = \frac{1}{1 - \varepsilon J}\,.$$

This yields

$$\begin{aligned}\dot{z} &= (F + a(\varepsilon)GH)z + b(\varepsilon)GN\eta + a(\varepsilon)Gw \\ \dot{\eta} &= b(\varepsilon)MHz + (L + b(\varepsilon)MJN)\eta + b(\varepsilon)Mw \\ \dot{w} &= (-\frac{1}{\varepsilon} - a(\varepsilon)HG - J[b(\varepsilon)NM - 1])w \\ &\quad - (H[F + a(\varepsilon)GH] - b(\varepsilon)JNMH)z \\ &\quad - (b(\varepsilon)HGN + JN[L + b(\varepsilon)MJN])\eta\end{aligned} \tag{23}$$

The latter can be viewed as the feedback interconnection of

$$\begin{aligned}\begin{pmatrix}\dot{z}\\ \dot{\eta}\end{pmatrix} &= \begin{pmatrix}F + a(\varepsilon)GH & b(\varepsilon)GN \\ b(\varepsilon)MH & L + b(\varepsilon)MJN\end{pmatrix}\begin{pmatrix}z\\ \eta\end{pmatrix} + \begin{pmatrix}a(\varepsilon)G \\ b(\varepsilon)M\end{pmatrix} w \\ \zeta &= -(H[F + a(\varepsilon)GH] - b(\varepsilon)JNMH)z \\ &\quad - (b(\varepsilon)HGN + JN[L + b(\varepsilon)MJN])\eta\end{aligned} \tag{24}$$

and

$$\dot{w} = (-\frac{1}{\varepsilon} - a(\varepsilon)GH - J[b(\varepsilon)NM - 1])w + \zeta\,. \tag{25}$$

Noting that system (24) is stable by hypothesis at $\varepsilon = 0$, let γ denote its H_∞-gain. By continuity, for any $\delta > 0$, there is an ε^* such that, for all $0 < \varepsilon < \varepsilon^*$, system (24) is stable and its H_∞-gain does not exceed $\gamma + \delta$. On the other hand, for small ε, system (25) is stable and has an H_∞-gain $\gamma(\varepsilon)$ which tends to 0 as ε tends to 0. As a consequence, by the small gain theorem, if ε is small enough the feedback interconnection of (24) and (25) is stable. $\triangleleft$

The second Lemma shows how, in the previous stabilizer, v can be replaced by an estimate generated by a filter which receives y as input. Define

$$P = \begin{pmatrix}-gc_{r-1} & 1 & 0 & \cdots & 0\\ -g^2c_{r-2} & 0 & 1 & \cdots & 0\\ \cdot & \cdot & \cdot & \cdots & \cdot\\ -g^{r-1}c_1 & 0 & 0 & \cdots & 1\\ -g^rc_0 & 0 & 0 & \cdots & 0\end{pmatrix}, \qquad Q = \begin{pmatrix}-gc_{r-1}\\ -g^2c_{r-2}\\ \cdot\\ -g^{r-1}c_1\\ -g^rc_0\end{pmatrix}$$

$$R = \begin{pmatrix}0 & 0 & 0 & \cdots & 1\end{pmatrix},$$

where $c_0, c_1, \ldots, c_{r-1}$ are the coefficients of fixed Hurwitz polynomial, and $g > 0$ is a parameter to be determined.

Lemma 3.2 *Suppose system*

$$\begin{aligned}\dot{\eta} &= A\eta + Bv \\ u &= C\eta + Dv\end{aligned}$$

stabilizes system (20). Then, there is a number g^* such that, for all $g > g^*$ the feedback law

$$\begin{aligned} \dot{\xi} &= P\xi + Qy \\ \dot{\eta} &= A\eta + BR\xi \\ u &= C\eta + DR\xi \end{aligned} \tag{26}$$

stabilizes system (20).

Proof. Define $e_i = g^{r-i}(y^{(i-1)} - \xi_i)$, for $i = 1, 2, \ldots, r$ and observe that the interconnection of (20) and (26) becomes

$$\begin{aligned} \dot{z} &= Fz + Gv \\ \dot{v} &= Hz + Jv + b(C\eta + Dv) - bDe_r \\ \dot{\eta} &= A\eta + Bv - Be_r \\ \dot{e} &= g\tilde{A}e + \tilde{B}(Hz + Jv + b(C\eta + Dv) - bDe_r)\,. \end{aligned} \tag{27}$$

where $\tilde{A}$ is a Hurwitz matrix (independent of g) and $\tilde{B} = \text{col}(0, 0, \ldots, 0, 1)$. This can be viewed as the feedback interconnection of two subsystems. One of these, with input e_r, internal state (z, v, η) and output $Hz + Jv + b(C\eta + Dv)$ is stable by construction and has an H_∞-gain bounded by a fixed number independent of g, if $g > 1$. The other one, with input $Hz + a_r v + b(C\eta + Dv)$, internal state e and output e_r is stable for large g and, as an elementary calculation shows, has an H_∞-gain that decreases to 0 as $g \to \infty$. Thus, again by the small gain theorem, the interconnection is stable for large g. ◁

Of course, the question naturally arises of how general is the approach outlined above to the problem of stabilization by output feedback, or – what is the same – how much one can count on the hypothesis, basic to the whole construction, that the auxiliary subsystem (21) is stabilizable by output feedback. Actually, this question can be answered in positive terms, by showing that this hypothesis *must necessarily hold* (modulo a memoryless output feedback) if the original system is stabilizable by output feedback.

Proposition 3.1 *Suppose the system*

$$\begin{aligned} \dot{x} &= Ax + Bu \\ y &= Cx \end{aligned} \tag{28}$$

is stabilizable by output feedback. Then, for every $k \in \mathbb{R}$ except possibly one single value k_0, the auxiliary subsystem associated to

$$\begin{aligned} \dot{x} &= (A + kBC)x + Bu \\ y &= Cx \end{aligned} \tag{29}$$

is stabilizable by output feedback.

Remark. System (29) is obtained from system (28) via a memoryless output feedback of the form

$$u \mapsto u + ky$$

and this indeed leaves the property of stabilizability by output feedback unchanged. The previous Proposition shows that, in case the auxiliary system associated with (28) *is not* stabilizable by output feedback, for every nonzero k the auxiliary system associated with (29) *is* stabilizable by output feedback. $\triangleleft$

Proof. Suppose, without loss of generality, that

$$\begin{pmatrix} A & B \\ C & 0 \end{pmatrix} = \begin{pmatrix} F_0 & G_0 & 0 & \cdots & 0 & 0 \\ 0 & 0 & 1 & \cdots & 0 & 0 \\ 0 & 0 & & \cdots & 0 & 0 \\ \cdot & \cdot & \cdot & \cdots & \cdot & \cdot \\ 0 & 0 & 0 & \cdots & 1 & 0 \\ H_0 & a_1 & a_2 & \cdots & a_r & b \\ 0 & 1 & 0 & \cdots & 0 & 0 \end{pmatrix}.$$

Using standard tests, show first that stabilizability and detectability of the triplet (A, B, C) implies stabilizability and detectability of the triplet (F_0, G_0, H_0). Bearing in mind this property, show that the triplet $(A + kBC, B, C)$ is stabilizable and detectable if and only if

$$\det \begin{pmatrix} F_0 & G_0 \\ H_0 & a_1 + kb \end{pmatrix} \neq 0 .$$

Observe that

$$\det \begin{pmatrix} F_0 & G_0 \\ H_0 & a_1 + kb \end{pmatrix} = (a_1 + kb)\det(F_0) + \det \begin{pmatrix} F_0 & G_0 \\ H_0 & 0 \end{pmatrix} .$$

The right-hand side of this identity, a polynomial of first degree in k, is not identically vanishing because otherwise the stabilizability and detectability of the triplet (F_0, G_0, H_0) would be contradicted. From this the result follows. $\triangleleft$

In other words, there is no loss of generality in approaching the problem of stabilization via output feedback in the manner suggested above, i.e. by "reducing" the problem to that of stabilizing an auxiliary lower-dimensional subsystem. Indeed, the process can be iterated all the way down to the case in which the auxiliary subsystem has dimension one (where a simple gain would suffice to stabilize). This yields a systematic method for stabilization via output feedback, which does not use any technique for pole assignment nor the idea of solving separately a problem of state-feedback stabilization and a problem of asymptotic state reconstruction. Rather, it is based on the recursive update of a sequence of "dynamic" output feedback stabilizers for a sequence of subsystem of increasing dimension.

4 Robust output feedback stabilization of nonlinear systems

A number of methods for stabilizing a nonlinear system using output feedback rely upon the hypothesis that the system has a stable zero dynamics (we recall that the *zero dynamics* of a system are the internal dynamics which arise when input and initial conditions are chosen in such a way as to constrain the output to remain identically zero), i.e. the system belong to a class which is the analogue of the class of linear minimum phase systems. The main reason why this occurs is that most of the methods in question use feedback in order to overcome certain unwanted terms in the dynamics and this in turn tends to render the system poorly observable, with the consequence of enforcing an internal behavior which is very close to that of the zero dynamics. As a result, stabilization may occur only if the latter is asymptotically stable, i.e. if the system is minimum phase. For example, in the stabilization of a system of the form

$$\begin{aligned} \dot{z} &= f_0(z, y, p) \\ \dot{y} &= f_1(z, y, p) + u \end{aligned}$$

robust stability (with respect to the uncertain parameter p) is usually achieved by choosing the control u in such a way as to offset the (possibly unpleasant) presence on the term $f_1(x, y, p)$ in the derivative of a candidate Lyapunov function of the form

$$U(z, y) = V(z) + (y - y^*(z))^2 .$$

If only y is available for feedback, this can always be achieved (in semiglobal terms) whenever $y^*(z) = 0$ and $V(z)$ is a Lyapunov function for the zero dynamics

$$\dot{z} = f_0(z, 0, p) ,$$

by means of a control of the form $u = -ky$, where k is a sufficiently large number. If, on the contrary, the zero dynamics are unstable, some (output-based) estimate of a nontrivial $y^*(z)$ is needed, and this might not be compatible with a control action which, by dominating the term $f_1(x, y, p)$, would render z poorly observable from y.

On the other hand, the method described in the previous section does not rely at all upon a similar procedure, but rather aims at taking explicit advantage of the term $f_1(z, y, p)$, to the purpose of determining a stabilizer for what was called the "auxiliary subsystem", in this case system

$$\begin{aligned} \dot{z} &= f_0(z, u_a, p) \\ y_a &= f_1(z, u_a, p) . \end{aligned}$$

In this section we describe a nonlinear version of that method.

Consider a nonlinear system modeled by equations of the form

$$
\begin{aligned}
\dot{z} &= f_0(z, x_1, \ldots, x_{r-1}, x_r, p) \\
\dot{x}_1 &= x_2 \\
&\cdots \\
\dot{x}_{r-1} &= x_r \\
\dot{x}_r &= h_0(z, x_1, \ldots, x_{r-1}, x_r, p) + b(x_1, \ldots, x_r)u \\
y &= x_1 \,,
\end{aligned}
\tag{30}
$$

in which $z \in I\!R^{n-r}$ and p is a (possibly vector-valued) unknown parameter, ranging over a compact set $\mathcal{P}$.

Remark. Note that any system modeled by equations of the form

$$
\begin{aligned}
\dot{z} &= f_0(z, x_1, \ldots, x_{r-1}, x_r, p) \\
\dot{x}_1 &= x_2 + f_1(z, x_1, p) \\
&\cdots \\
\dot{x}_{r-1} &= x_r + f_{r-1}(z, x_1, \ldots, x_{r-1}, p) \\
\dot{x}_r &= h_0(z, x_1, \ldots, x_{r-1}, x_r, p) + b(x_1)u \\
y &= x_1 \,,
\end{aligned}
\tag{31}
$$

is globally diffeomorphic to a system of the form (30). The global diffeomorphism changing (31) into (30) is parameter-dependent, and thus useless for the purpose of designing state-feedback laws, but this is not a problem in the present setup, where output-feedback laws are sought. $\triangleleft$

With this system, we associate an *auxiliary plant*

$$
\begin{aligned}
\dot{x}_{\mathrm{a}} &= f_{\mathrm{a}}(x_{\mathrm{a}}, u_{\mathrm{a}}, p) \\
y_{\mathrm{a}} &= h_{\mathrm{a}}(x_{\mathrm{a}}, u_{\mathrm{a}}, p)
\end{aligned}
\tag{32}
$$

in which

$$
x_{\mathrm{a}} = \begin{pmatrix} z \\ x_1 \\ \cdots \\ x_{r-2} \\ x_{r-1} \end{pmatrix}, \qquad f_{\mathrm{a}}(x_{\mathrm{a}}, u_{\mathrm{a}}, p) = \begin{pmatrix} f_0(z, x_1, \ldots, x_{r-1}, u_{\mathrm{a}}, p) \\ x_2 \\ \cdots \\ x_{r-1} \\ u_{\mathrm{a}} \end{pmatrix},
$$

and

$$
h_{\mathrm{a}}(x_{\mathrm{a}}, u_{\mathrm{a}}, p) = h_0(z, x_1, \ldots, x_{r-1}, u_{\mathrm{a}}, p) \,.
$$

Motivated by the arguments presented in the previous section, we assume the following.

Assumption A. *The auxiliary system (32) is globally robustly stabilizable by means of a dynamic output feedback controller of the form*

$$\begin{aligned} \dot{\eta} &= L(\eta) + M y_{\mathrm{a}} \\ u_{\mathrm{a}} &= N(\eta) \,. \end{aligned} \tag{33}$$

Consider now, for the original plant (30), the dynamic feedback law

$$\begin{aligned} \dot{\eta} &= L(\eta) + Mk(x_r - N(\eta)) := \varphi(\eta, x_r) \\ u &= \frac{1}{b(x_1, \ldots, x_r)} \Big[\frac{\partial N}{\partial \eta} [L(\eta) + Mk(x_r - N(\eta))] - k[x_r - N(\eta)] \Big] \\ &:= \frac{1}{b(x_1, \ldots, x_r)} \gamma(\eta, x_r) \end{aligned} \tag{34}$$

where k is a positive number.

We claim that a controller with this structure is able to *robustly semiglobally practically* stabilize the plant (30) (in the sequel we denote, regardless of the actual value of the integer k, by $\mathcal{B}_R$ the closed cube $\mathcal{B}_R = \{x \in \mathbb{R}^k : |x_i| \leq R, 1 \leq i \leq k\}$).

Theorem 4.1 *For any $R > 0$ and any $\varepsilon > 0$, there is a value k^* such that, if $k > k^*$, in the feedback interconnection of (30) and (34), namely in the system*

$$\begin{aligned} \dot{x}_{\mathrm{a}} &= f_{\mathrm{a}}(x_{\mathrm{a}}, x_r, p) \\ \dot{x}_r &= h_{\mathrm{a}}(x_{\mathrm{a}}, x_r, p) + \gamma(\eta, x_r) \\ \dot{\eta} &= \varphi(\eta, x_r) \,, \end{aligned} \tag{35}$$

any initial condition satisfying $\{x_{\mathrm{a}}(0), x_r(0), \eta(0)\} \in \mathcal{B}_R$ produces a trajectory that converges, in finite time, to the set $\mathcal{B}_\varepsilon$.

Proof. Set

$$\theta = x_r - N(\eta)$$

and observe that the closed-loop system (35) can be viewed as the system resulting from the interconnection of a system with output θ and input v, modeled by equations of the form

$$\begin{aligned} \dot{x}_{\mathrm{a}} &= f_{\mathrm{a}}(x_{\mathrm{a}}, \theta + N(\eta), p) \\ \dot{\eta} &= L(\eta) - Mv \\ \dot{\theta} &= h_{\mathrm{a}}(x_{\mathrm{a}}, \theta + N(\eta), p) + v \end{aligned} \tag{36}$$

and a memoryless system (with input θ and output v)

$$v = -k\theta \,. \tag{37}$$

System (36) has uniform relative degree one between input v and output θ, and its "high-frequency gain" is equal to one. Its zero dynamics, found by setting $\theta = 0$, i.e. $x_r - N(\eta)$, are those of

$$\begin{array}{rcl} \dot{x}_{\rm a} & = & f_{\rm a}(x_{\rm a}, N(\eta), p) \\ \dot{\eta} & = & L(\eta) + M h_{\rm a}(x_{\rm a}, N(\eta), p) \, . \end{array} \tag{38}$$

These dynamics have by hypothesis a globally asymptotically stable equilibrium at $(x_{\rm a}, \eta) = (0, 0)$. As a consequence, standard arguments (see e.g. [23]) prove that system (36) can be semiglobally practically stabilized by means of a feedback law of the form (37), i.e. prove the Theorem.

Basically, these arguments consists in the following. Let $V(x_{\rm a}, \eta)$ be a Lyapunov function for (38) and consider the candidate Lyapunov function

$$W(x_{\rm a}, \eta, \theta) = V(x_{\rm a}, \eta + M\theta) + \theta^2$$

for the closed-loop system (36)-(37). Let Ω_c denote the set

$$\Omega_c = \{(x_{\rm a}, \eta, \theta) : W(x_{\rm a}, \eta, \theta) \leq c\}$$

and choose $c > 0, d > 0$ such that $\Omega_d \subset \mathcal{B}_\varepsilon \subset \mathcal{B}_R \subset \Omega_c$. Then, there exists k^* such that, if $k > k^*$, the derivative of $W(x_{\rm a}, \eta, \theta)$ along the vector field (35) is negative at each point of the compact set

$$\{(x_{\rm a}, \eta, \theta) : d \leq W(x_{\rm a}, \eta, \theta) \leq c\} \, .$$

This proves that all trajectories with initial conditions in $\mathcal{B}_R$ remain in Ω_c for all $t \geq 0$. Thus, these trajectories are bounded, and using standard ω-limit set arguments, it can also be seen that these trajectories ultimately enter the set $\mathcal{B}_\varepsilon$. ◁

The controller (34) uses the state variables $x_1, \ldots, x_r$ as inputs. In case these are not available for feedback, they must be replaced by appropriate estimates $\xi_1, \ldots, \xi_r$ provided by a system of the form

$$\dot{\xi} = P\xi + Qy$$

with Q and P having the form discussed earlier. However, if this is the case, a saturation is needed, in order to avoid the occurrence of finite escape times for large values of g. In what follows, we describe the construction of the stabilizing controller in the particular case in which $b(x_1, \ldots, x_r)$ is a function of x_1 only (and still denoted by $b(x_1)$). In the controller (34), the term

$$k(x_r - N(\eta))$$

must be replaced by the term

$$\sigma_\rho(k\xi_r - kN(\eta)) \, ,$$

in which $\sigma_\rho(\cdot)$ is a saturation function

$$\sigma_\rho(a) = \begin{cases} a & \text{if } |a| < \rho \\ \dfrac{a}{|a|}\rho & \text{if } |a| \geq \rho \,. \end{cases}$$

This yields a controller modeled by equations of the form

$$\begin{aligned} \dot{\eta} &= \tilde{\varphi}(\eta, \xi_r) \\ u &= \frac{1}{b(y)}\tilde{\gamma}(\eta, \xi_r) \\ \dot{\xi} &= P\xi + Qy \,. \end{aligned} \tag{39}$$

A controller of this type is able to robustly semiglobally practically stabilize the plant (30).

Theorem 4.2 *For any $R > 0$ and any $\varepsilon > 0$, there are values of ρ, k, such that, in the feedback interconnection of (30) and (39), namely in the system*

$$\begin{aligned} \dot{x}_a &= f_a(x_a, x_r, p) \\ \dot{x}_r &= h_a(x_a, x_r, p) + \tilde{\gamma}(\eta, \xi_r) \\ \dot{\eta} &= \tilde{\varphi}(\eta, \xi_r) \\ \dot{\xi} &= P\xi + Qx_1 \,, \end{aligned} \tag{40}$$

any initial condition satisfying $\{x_a(0), x_r(0), \eta(0), \xi(0)\} \in \mathcal{B}_R$ produces a trajectory that converges, in finite time, to the set $\mathcal{B}_\varepsilon$.

The proof of this result follows a pattern similar to that of the proof of Theorem 4.1 and will not be repeated here.

We conclude the section with some remarks about the structure of the controller used to stabilize the "auxiliary subsystem" (32). The previous theorem considers the case in which the stabilizer for (32) is a strictly proper system, namely a system with no feed-through between y_a and u_a, and has the special structure (33). To deal with the more general case in which the stabilizer for the auxiliary system is a proper system, non-affine in y_a, one can use the following result.

Consider two subsystems

$$\begin{aligned} \dot{x} &= f(x, u) \\ y &= h(x, u) \,, \end{aligned} \tag{41}$$

$$\begin{aligned} \dot{\eta} &= \theta(\eta, v) \\ u &= \kappa(\eta, v) \,, \end{aligned} \tag{42}$$

in which $f(0,0) = 0$, $h(0,0) = 0$, $\theta(0,0) = 0$ and $\kappa(0,0) = 0$. Suppose that, for each x, η, the equation

$$h(x, \kappa(\eta, v)) = v$$

as a unique solution $v = \alpha(x, \eta)$, which is a smooth function. In this case, the feedback interconnection of (41) and (42) via $v = y$ is well-defined. Suppose it is globally asymptotically stable at the equilibrium $(x, \eta) = (0, 0)$. Suppose also that

$$\xi[\xi + \alpha(x, \eta) - h(x, \kappa(\eta, \xi + \alpha(x, \eta)))]$$

is non negative and zero only at $\xi = 0$.

Consider now the strictly proper system

$$\begin{aligned} \dot{\eta} &= \theta(\eta, v) \\ \dot{v} &= -Kv + Ky \\ u &= \kappa(\eta, v) \end{aligned} \tag{43}$$

(note that its structure is exactly the structure of the controller (33)). This system is able to semiglobally practically stabilize system (41).

Note that the feedback interconnection of (41) and (43) differs from the feedback interconnection of (41) and (42) by the "insertion" of a (small, if K is large) time constant between y and v. To prove the result, change v into

$$\xi = v - \alpha(x, \eta)$$

to obtain

$$\begin{aligned} \dot{x} &= f(x, \kappa(\eta, \xi + \alpha(x, \eta)) \\ \dot{\eta} &= \theta(\eta, \xi + \alpha(x, \eta)) \\ \dot{\xi} &= -K[\xi + \alpha(x, \eta) - h(x, \kappa(\eta, \xi + \alpha(x, \eta)))] - \dot{\alpha} \,. \end{aligned} \tag{44}$$

For $\xi = 0$ the top two equations reduce to the feedback interconnection of (41) and (42), globally asymptotically stable by hypothesis. Then, standard arguments, similar to those used earlier, show that there exists a positive definite proper function $V(x, \eta, \xi)$ whose derivative along the vector field (44), for any choice of arbitrarily small $d' > 0$ and arbitrarily large $c' > 0$, can be rendered negative at each point of the set

$$\{(x, \eta, \xi) : d' \leq V(x, \eta, \xi) \leq c'\}$$

by choosing K large enough.

References

[1] J. Ball, J.W. Helton, and M.L. Walker, H_∞ control for nonlinear systems via output feedback, *IEEE Trans. Autom. Control*, **38**: pp. 546-559, 1993

[2] T. Basar and P. Bernhard, H_∞*-optimal control and related Minimax design problems*, Birkhauser, 1990.

[3] C.I. Byrnes, A. Isidori, Asymptotic stabilization of minimum-phase nonlinear systems, *IEEE Trans. Aut. Contr.*, **AC-36**: pp. 1122-1137, 1991.

[4] F. Esfandiari and H.K. Khalil, Output feedback stabilization of fully linearizable systems, *Int. J. Contr.*, **56**: pp. 1007–1037, 1992.

[5] R.A. Freeman and P.V. Kokotović, Design of 'softer' robust nonlinear control laws, *Automatica*, **29**: pp. 1425–1437, 1993.

[6] R.A. Freeman and P.V. Kokotović, Tracking controllers for systems linear in the unmeasured states, *Automatica*, **32**: pp. 735–746, 1996.

[7] J.P. Gauthier and G. Bornard, Observability for any $u(t)$ of a class of nonlinear systems, *IEEE Trans. Automat. Contr.*, **AC-26**: pp. 922–926, 1981.

[8] J. Imura, T. Sugie, and T. Yoshikawa, Global robust stabilization of nonlinear cascaded systems, *IEEE Trans. Automat. Contr.*, **AC-39**: pp. 1084–1089, 1994.

[9] A. Isidori, *Nonlinear Control Systems*, Springer-Verlag, New York, third edition, 1995.

[10] A. Isidori, Semiglobal practical stabilization of uncertain non-minimum-phase nonlinear systems via output feedback, *4th IFAC Symp. Nonlinear Contr. Syst. Design*, 1998, to appear.

[11] Z.P. Jiang and I. Mareels, A small gain control method for nonlinear cascaded systems with dynamic uncertainties, *IEEE Trans. Automat. Contr.*, **AC-42**: pp. 292–308, 1997.

[12] Z.P. Jiang, I.M. Mareels and Y. Wang, A Lyapunov formulation of the nonlinear small-gain theorem for interconnected ISS systems, *Automatica*, **32**: pp. 1211–1215, 1996.

[13] Z.P. Jiang, A. Teel and L. Praly, Small-gain theorem for ISS systems and applications, *Mathematics of Control, Signals and Systems*, **7**: pp. 95–120, 1994.

[14] H.K. Khalil and F. Esfandiari, Semiglobal stabilization of a class of nonlinear systems using output feedback, *IEEE Trans. Automat. Contr.*, **AC-38**: pp. 1412–1415, 1993.

[15] M. Krstic, I. Kanellakopoulos, P. Kokotovic, *Nonlinear Adaptive Control Design*, J.Wiley (New York), 1995.

[16] R. Marino, P. Tomei, Global adaptive output feedback control of nonlinear systems, part I: linear parametrization, *IEEE Trans. Autom. Control* **AC-38**: pp. 17–32, 1993.

[17] R. Marino, P. Tomei, Global adaptive output feedback control of nonlinear systems, part II: nonlinear parametrization, *IEEE Trans. Autom. Control* **AC-38**: pp. 33–48, 1993.

[18] F. Mazenc and L. Praly, Adding integrations, saturated controls and stabilization for feedforward systems, *IEEE Trans. Automat. Contr.*, **AC-41**: pp. 1559–1577, 1996.

[19] R. Sepulchre. M. Jankovic, P.V. Kokotovic, *Constructive Nonlinear Control*, Springer Verlag, 1996.

[20] E.D. Sontag, On the input-to-state stability property, *European J. Contr.*, **1**: pp. 24–36, 1995.

[21] E. Sontag, Y. Wang, On characterizations of the input-to-state stability property, *Syst. Contr. Lett.* **24**: pp. 351–359, 1995.

[22] A. Teel and L. Praly, Global stabilizability and observability imply semi-global stabilizability by output feedback, *Systems Contr. Lett.*, **22**: pp. 313–325, 1994.

[23] A. Teel and L. Praly, Tools for semiglobal stabilization by partial state and output feedback, *SIAM J. Control & Optimization*, **33**: pp. 1443–1488, 1995.

[24] A. Tornambe, Output feedback stabilization of a class of non-minimum phase nonlinear systems, *Systems Contr. Lett.*, **19**: pp. 193–204, 1992.

[25] A.J. Van der Schaft, L_2-gain analysis of nonlinear systems and nonlinear H_∞ control, *IEEE Trans. Autom. Control*, **AC-37**: pp.770–784, 1992.

Department of System Science and Mathematics, Washington University, St. Louis, MO 63130

Dipartimento di Informatica e Sistemistica, Università di Roma "La Sapienza", 00184 Rome, Italy

Progress in Systems and Control Theory, Vol. 25

Towards a System Theory for Model Set: Chain-Scattering Approach

Hidenori Kimura

1 Introduction

Recent progress of robust control has shifted the target of control theory from a single model to *a set of models*. In robust control, model is no longer a precise description of the plant to be controlled. It is rather a crude description of the mechanism that explains behaviors of the plant subject to various kinds of uncertainties [14]. The uncertainty precludes the unique representation of the plant. Thus, the plant is actually characterized by a set of models that incorporates all possible perturbations to the nominal model. Robust control is expected to control this set of models, rather than the nominal model itself [13] . There, the set of models in naturally introduced to represent the uncertainty in the nominal model. The uncertainty is ascribable to various reasons. First, it is ascribed to the lack of sufficient knowledge and/or information for constructing a precise model. Second, it is ascribed to the need for simplifying the model. Third, it is ascribed to the complexity of the real system that generates intrinsic indeterminability. Discussions about the source of uncertainties in models and modeling would require a number of pages and are omitted here. We only note that the uncertainty is not an accidental feature of models. It is rather an intrinsic characteristic feature of model that tries to describe the real system by an abstract framework with a finite number of words [7]. Robust control is no doubt the first systematic attempt to take this important feature of models seriously into account.

In the early stage of its development, the role of robust control theory was to construct a robust controller that covers the uncertainty which was given *a priori*. Then, some people began to realize that the design of robust control systems was not completed unless the plant uncertainty are fully taken into account in the design. Modeling is now a part of design, rather than a separate task. The new field which is sometimes called *identification for robust control* emerged as a natural consequence of the development of robust control. The main focus of identification for robust control is to estimate the uncertainty of the model quantitatively to facilitate the design of robust control, especially to reduce excessive conservatism in the design. The identification for robust control is then the *identification of a set of models*, rather than that of a single model.

Now, we have two fields that deal with model set: control and identification. If the notion of model set is a natural way of describing the real system, there is no reason to confine our scope of theory of model set within control and identification. A lot of theoretical issues on model set exist. We could say almost all achievements of system theory which deals only with a single model can be extended to model set. This motivates *system theory of model set* as an extension of system theory of single model.

The purpose of this expository paper is to give a perspective of a system theoretic approach to model set based on the chain-scattering representation of the plant. The chain-scattering representation was used in circuit theory and signal processing. It is believed to be the plant appropriate framework for dealing with model sets.

Since we have quite a few nice expository papers and books on control and identification of model set (e.g., [6] [15]), we exclude these topics in this paper. Currently model set system theory is at the starting level of maturity, so that this paper contains more problems than solutions.

Notations:

$\boldsymbol{R}$: The set of real numbers

$\boldsymbol{R}^{n\times m}$: The set of $n \times m$ matrices whose elements are in $\boldsymbol{R}$.

$\boldsymbol{C}$: The set of complex numbers

$\boldsymbol{C}^{n\times m}$: The set of $n \times m$ matrices whose elements are in $\boldsymbol{C}$

$\bar{z}$: The complex conjugate of z.

$A^{\sim}(s) = A^T(-s), \quad A^* = \bar{A}^T(s)$

$\boldsymbol{D}$: $\{z \in \boldsymbol{C} : \ |z| < 1\}$

$\boldsymbol{H}^{\infty} = \{f(s) : \text{analytic on } Res \geq 0, \|f\|_{\infty} = \sup_{\omega} |f(j\omega)| < \infty\}.$

$\boldsymbol{B}^{n\times m} = \{\Delta : \ \Delta \in \boldsymbol{C}^{n\times m}, \ \|\Delta\| < 1\}$

$\boldsymbol{BH}^{\infty} = \{f(s) : \ f \in \boldsymbol{H}^{\infty}, \ \|f\|_{\infty} < 1\}.$

$\overline{\boldsymbol{BH}}^{\infty}$ is the closure of $\boldsymbol{BH}^{\infty}$*i.e.,* $\overline{\boldsymbol{BH}}^{\infty} = \{f(s) : \ f \in \boldsymbol{H}^{\infty}, \ \|f\|_{\infty} \leq 1\}.$

2 Chain-Scattering Representation

2.1 Chain-Scattering Representation of Dynamical Systems

Consider a linear dynamical system of Fig.1 with two inputs $b_1 \in \boldsymbol{R}^r$, $b_2 \in \boldsymbol{R}^p$ and two outputs $a_1 \in \boldsymbol{R}^m$, $a_2 \in \boldsymbol{R}^r$ represented in terms of a transfer function P, i.e.,

$$\begin{bmatrix} a_1 \\ a_2 \end{bmatrix} = P \begin{bmatrix} b_1 \\ b_2 \end{bmatrix} = \begin{bmatrix} P_{11} & P_{12} \\ P_{21} & P_{22} \end{bmatrix} \begin{bmatrix} b_1 \\ b_2 \end{bmatrix}. \tag{2.1.1}$$

Assuming that P_{21} is invertible, the above relation is solved with respect to b_1, i.e.,

$$\begin{bmatrix} a_1 \\ b_1 \end{bmatrix} = \begin{bmatrix} P_{12} - P_{11}P_{21}^{-1}P_{22} & P_{11}P_{21}^{-1} \\ -P_{21}^{-1}P_{22} & P_{21}^{-1} \end{bmatrix} \begin{bmatrix} b_2 \\ a_2 \end{bmatrix} \tag{2.1.2}$$

The above representation is called *the chain-scattering representation* of P, and the above relation is represented as

$$\begin{aligned} \begin{bmatrix} a_1 \\ b_1 \end{bmatrix} &= CHAIN(P) \begin{bmatrix} b_2 \\ a_2 \end{bmatrix} \\ CHAIN(P) &:= \begin{bmatrix} P_{12} - P_{11}P_{21}^{-1}P_{22} & P_{11}P_{21}^{-1} \\ -P_{21}^{-1}P_{22} & P_{21}^{-1} \end{bmatrix} \end{aligned} \tag{2.1.3}$$

A schematic diagram of the chain-scattering representation is shown in Fig.2, which indicates *scattering* between the $(b_2,\ a_1)$-wave travelling from right to left and the $(b_1,\ a_2)$-wave travelling oppositely.

The advantage of using the chain-scattering representation lies in its cascade property. Consider a feedback connection of two systems P and Q shown in Fig.3. The resulting system is represented in the input/output form

$$\begin{bmatrix} a_1 \\ c_2 \end{bmatrix} = f(P,\ Q) \begin{bmatrix} b_1 \\ d_2 \end{bmatrix}.$$

The transfer function $f(P,\ Q)$ of the connected system is computed from the relation (2.1.1) and

$$\begin{bmatrix} c_1 \\ c_2 \end{bmatrix} = Q \begin{bmatrix} d_1 \\ d_2 \end{bmatrix}$$

using the identities $a_2 = d_1$, $b_2 = c_1$. The functional form of $f(P,\ Q)$ is quite complicated and is usually refered to as a *star product* of P and Q [18]. The feedback connection of Fig.3 is nothing but a cascade connection in the chain-scattering formalism. The chain-scattering matrix of the resulting system is just a product of the two chain-scattering matrix, as shown in Fig.4, e.g.,

$$CHAIN\text{star product of } P \text{ and } Q = CHAIN(P) \cdot CHAIN(Q). \tag{2.1.4}$$

This simplifies the arguments concerning the feedback connection greatly. The notion of chain-scattering matrix is of course not new. It was introduced originally in circuit theory and then used extensively in signal processing [4]. In control community, it was introduced in the early stage of development of H^{∞} control. An extensive discussion of the chain-scattering representation is found in [12].

2.2 Algebraic Properties of Chain-Scattering Representation

It is easily seen that the representation (2.1.3) allows a factorization

$$CHAIN(P) = \begin{bmatrix} P_{12} & P_{11} \\ 0 & I \end{bmatrix} \begin{bmatrix} I & 0 \\ P_{22} & P_{21} \end{bmatrix}^{-1}. \tag{2.2.1}$$

Each factor of the right-hand side is a chain-scattering representation, i.e.,

$$\begin{aligned} \begin{bmatrix} P_{12} & P_{11} \\ 0 & I \end{bmatrix} &= CHAIN\left(\begin{bmatrix} P_{11} & P_{12} \\ I & 0 \end{bmatrix}\right) \\ \begin{bmatrix} I & 0 \\ P_{22} & P_{21} \end{bmatrix}^{-1} &= \begin{bmatrix} I & 0 \\ -P_{21}^{-1}P_{22} & P_{21}^{-1} \end{bmatrix} = CHAIN\left(\begin{bmatrix} 0 & I \\ P_{21} & P_{22} \end{bmatrix}\right). \end{aligned}$$

Thus, the relation (2.2.1) is represented as

$$CHAIN(P) = CHAIN\left(\begin{bmatrix} P_{11} & P_{12} \\ I & 0 \end{bmatrix}\right) \cdot CHAIN\left(\begin{bmatrix} 0 & I \\ P_{21} & P_{22} \end{bmatrix}\right). \tag{2.2.2}$$

(2.1.4), the cascade property of the chain-scattering representation, the above relation implies that P is the star-product of the two plants $\begin{bmatrix} P_{11} & P_{12} \\ I & 0 \end{bmatrix}$ and $\begin{bmatrix} 0 & I \\ P_{21} & P_{22} \end{bmatrix}$. This is indeed the case as shown in Fig.5.

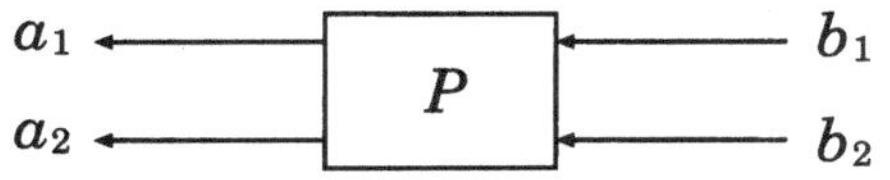

Figure 1: A Linear Dynamical System

Figure 2: Chain-Scattering Representation

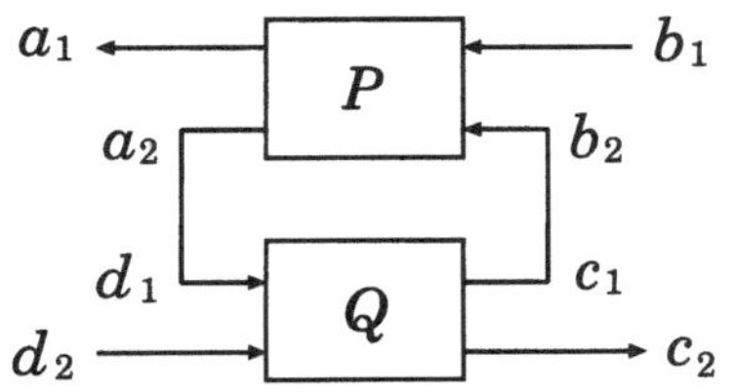

Figure 3: A Star-Product

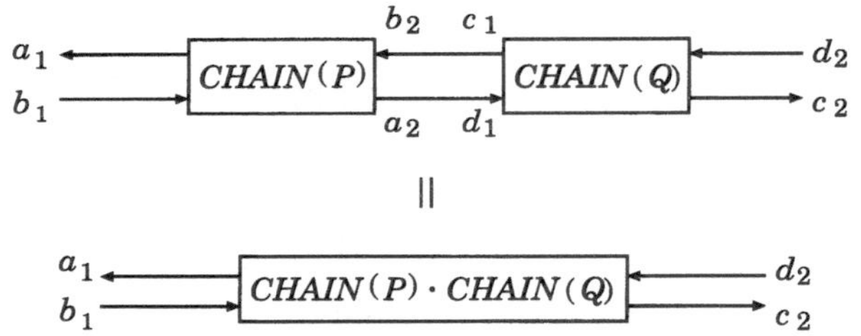

Figure 4: Cascade Property of Chain-Scattering Representation

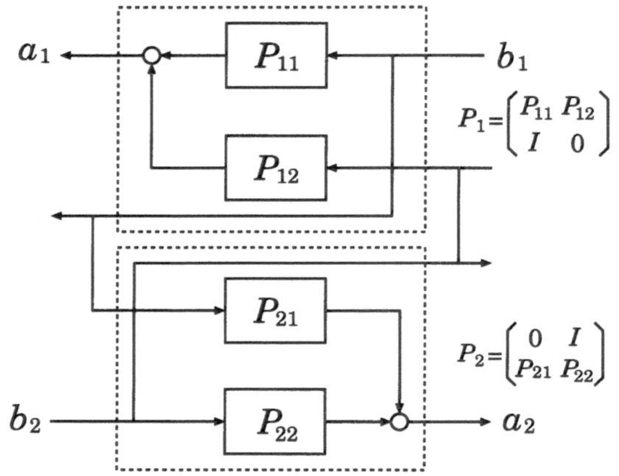

Figure 5: The Representation of Plant (2.1.1) as a Star-Product of the Two Subsystems P_1 and P_2 (Illustration of Equation (2.2.2))

2.3 Chain-Scattering Representation of LFT

The feedback configuration shown in Fig.6 is frequently used in the literature of robust control. The closed-loop transfer function is given by a linear fractional transformation

$$LFT(P:\ \Delta) := P_{11} + P_{12}\Delta(I - P_{22}\Delta)^{-1}P_{21} \tag{2.3.1}$$

In the chain-scattering formalism, Fig.6 is represented as Fig.7, and the closed-loop transfer function is given by a *Homographic transformation.*

$$\begin{aligned} HM(G:\ \Delta) &= (G_{11}\Delta + G_{12})(G_{21}\Delta + G_{22})^{-1}, \\ G &= CHAIN(P) \end{aligned} \tag{2.3.2}$$

Note that the representation (2.3.2) is valid only when $CHAIN(P)$ exists, i.e., P_{21} is invertible. The representation (2.3.2) is sometimes called a *Möbius transformation.* The following lemma lists up some properties of $HM(G;\ \Delta)$. Proofs are selfevident, and omitted here.

Lemma 2.3.1 *HM(G; Δ) satisfies the following properties:*

(1) $LFT(P: \Delta) = HM(CHAIN(P); \Delta)$

(2) $HM(G_1: HM(G_2: \Delta)) = HM(G_1G_2: \Delta)$

(3) $HM(I: \Delta) = \Delta$

(4) $HM(G: \Delta) = \Delta_1$ *implies* $\Delta = HM(G^{-1}: \Delta_1)$ *when* G^{-1} *exists.*

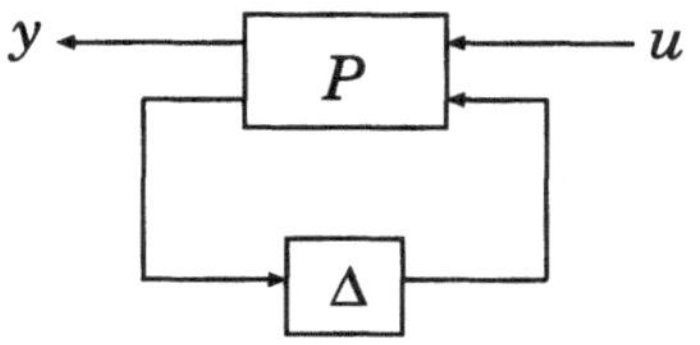

Figure 6: Feedback Configuration

Figure 7: Chain-Scattering Representation of Fig.6

The transformation (2.3.2) is a matrix extension of the linear fractional transformation $\boldsymbol{C} \longrightarrow \boldsymbol{C}$ given by

$$f(z) = \frac{g_{11}z + g_{12}}{g_{21}z + g_{22}} \tag{2.3.3}$$

which plays a central role in complex analysis(e.g.,[1]). The transformation (2.3.3) enjoys many interesting properties. The most important property of $f(z)$ is that it transforms lines in $\boldsymbol{C}$ to lines and circles in $\boldsymbol{C}$ to circles. The matrix version (2.3.2) of (2.3.3) was extensively investigated by Siegel [19] from the viewpoint of function theory of several complex variables.

One of the familiar facts in complex function theory states that every conformal mapping that maps the unit disk onto itself is a linear fractional transformation (2.3.3) with

$$\begin{bmatrix} \bar{g}_{11} & \bar{g}_{21} \\ \bar{g}_{12} & \bar{g}_{22} \end{bmatrix} \begin{bmatrix} 1 & 0 \\ 0 & -1 \end{bmatrix} \begin{bmatrix} g_{11} & g_{12} \\ g_{21} & g_{22} \end{bmatrix} = \begin{bmatrix} 1 & 0 \\ 0 & -1 \end{bmatrix}, \tag{2.3.4}$$

under some normalization (e.g., $g_{11}g_{22} - g_{12}g_{21} = 1$). Its extension to matrix cases was obtained by Siegel [19]. An important generalization was derived by Phillips [17] which is quite relevant to our problem. It should be noted that the matrix version of the unit disk is $\boldsymbol{B}^{m \times r}$.

Theorem 2.3.1 *If $f(\Delta)$ is a bianalytic function that maps $\boldsymbol{B}^{m\times r}$ onto $\boldsymbol{B}^{m\times r}$, then $f(\Delta)$ is a linear fractional transformation $HM(G:\ \Delta)$ given by (2.3.2) where G satisfies, under some normalization,*

$$\begin{bmatrix} G_{11}^* & G_{21}^* \\ G_{12}^* & G_{22}^* \end{bmatrix} \begin{bmatrix} I & 0 \\ 0 & -I \end{bmatrix} \begin{bmatrix} G_{11} & G_{12} \\ G_{21} & G_{22} \end{bmatrix} = \begin{bmatrix} I & 0 \\ 0 & -I \end{bmatrix} \tag{2.3.5}$$

Before concluding this subsection, we compute the inverse transform of *CHAIN*. Let

$$H = CHAIN(P) = \begin{bmatrix} H_{11} & H_{12} \\ H_{21} & H_{22} \end{bmatrix}$$

Then, due to (2.1.3), we have

$$P = \begin{bmatrix} H_{12}H_{22}^{-1} & H_{11} - H_{12}H_{22}^{-1}H_{21} \\ H_{22}^{-1} & -H_{22}^{-1}H_{21} \end{bmatrix}, \tag{2.3.6}$$

assuming that H_{22} is invertible. This relation is represented in a factorized form

$$P = \begin{bmatrix} H_{11} & H_{12} \\ 0 & I \end{bmatrix} \begin{bmatrix} H_{21} & H_{22} \\ I & 0 \end{bmatrix}^{-1} \tag{2.3.7}$$

2.4 Lossless Systems

A chain-scattering matrix $\Theta(s)$ with $(p+r)$ rows and $(m+r)$ columns is said to be *J-unitary*, if

$$\Theta^{\sim}(s)J\Theta(s) = J, \qquad \forall s \tag{2.4.1}$$

where

$$J := \begin{bmatrix} I_m & 0 \\ 0 & -I_r \end{bmatrix}. \tag{2.4.2}$$

If a J-unitary $\Theta(s)$ satisfies

$$\Theta^*(s)J\Theta(s) \leq J \qquad \forall Re[s] \geq 0, \tag{2.4.3}$$

then, it is said to be *J-lossless.*

Assume that $\Theta(s) = CHAIN(P(s))$ for some transfer function $P(s)$. Then, using the relation (2.2.1), we see that (2.4.1) implies

$$P^{\sim}(s)P(s) = \begin{bmatrix} I_r & 0 \\ 0 & I_m \end{bmatrix} \tag{2.4.4}$$

The above relation implies that $P(s)$ is unitary. In a similar way, we can show easily that (2.4.3) implies

$$P^*(s)P(s) \leq \begin{bmatrix} I_r & 0 \\ 0 & I_m \end{bmatrix}, \qquad \forall Re[s] \geq 0. \tag{2.4.5}$$

The relation (2.4.4) and (2.4.5) implies that $P(s)$ is *bounded* or *lossless.* Thus, the J-losslessness is a chain-scattering version of losslessness. We note this fact as follows:

Proposition 2.4.1 *$\Theta = CHAIN(P)$ is J-unitary, iff P is unitary. $\Theta = CHAIN(P)$ is J-lossless, iff P is lossless.*

Let Θ be a J-lossless matrix. Assume that

$$\Theta \begin{bmatrix} U \\ V \end{bmatrix} = \begin{bmatrix} X \\ Y \end{bmatrix}.$$

From the definition of *HM* in (2.3.2), we have

$$XY^{-1} = HM(\Theta : \; UV^{-1}),$$

provided that both V^{-1} and Y^{-1} exist. Due to (2.4.2) and (2.4.4), we have

$$\begin{aligned} U^{\sim}U - V^{\sim}V &= X^{\sim}X - Y^{\sim}Y, \quad \forall s, \\ U^{*}U - V^{*}V &\geq X^{*}X - Y^{*}Y, \quad \forall Re[s] \geq 0. \end{aligned}$$

Thus, we have the following important properties of J-lossless matrix.

Proposition 2.4.2 *Assume that Θ is J-lossless. Then, $HM(\Theta : \Delta)$ maps $\boldsymbol{BH}^{\infty}$ to $\boldsymbol{BH}^{\infty}$ one to one and onto.*

Another important property of J-lossless matrix in relation to J-unitary matrix is given, the proof of which is given in [12].

Proposition 2.4.3 *Let Θ be a J-unitary matrix. If $HM(\Theta : \Delta)$ maps $\boldsymbol{BH}^{\infty}$ into $\boldsymbol{BH}^{\infty}$, then it must be J-lossless.*

The following result is also well-known [12].

Proposition 2.4.4 *Assume that both Θ and Θ^{-1} are J-lossless. Then, Θ is a constant J-unitary matrix.*

3 Representation of Model Set

3.1 Uniqueness of Representation

In robust control theory, the most general class of model set is represented in Fig.6, where P denotes the known part of the plant and Δ represents the uncertainty included in the model. For instance, consider a model with additive uncertainty given by

$$G = G_0 + W\Delta \qquad \Delta \in \boldsymbol{B\bar{H}}^{\infty}, \tag{3.1.1}$$

where G_0 denotes the nominal model, Δ the uncertainty and W a frequency dependent magnitude of uncertainty. The equation (3.1.1) is represented by linear fractional transformation (2.3.1) with the identification

$$P = \begin{bmatrix} G_0 & W \\ I & 0 \end{bmatrix}. \tag{3.1.2}$$

Similarly, a model with multiplicative uncertainty given by

$$G = G_0(I + W\Delta), \qquad \Delta \in \boldsymbol{BH}^{\infty} \tag{3.1.3}$$

is represented by (2.3.1) with

$$P = \begin{bmatrix} G_0 & G_0 W \\ I & 0 \end{bmatrix}. \tag{3.1.4}$$

These uncertainty models are also represented by chain-scattering formalism. For (3.1.2), application of (2.1.3) yields

$$CHAIN(P) = \begin{bmatrix} W & G_0 \\ 0 & I \end{bmatrix}. \tag{3.1.5}$$

Also, for (3.1.4), we have

$$CHAIN(P) = \begin{bmatrix} G_0 W & G_0 \\ 0 & I \end{bmatrix}. \tag{3.1.6}$$

Now, we consider the model set

$$\mathcal{M} = \{M : \ M = LFT(P : \ \Delta); \ \Delta \in \overline{\boldsymbol{BH}}^{\infty}\}. \tag{3.1.7}$$

We may call P a *generator* of the model set $\mathcal{M}$. We assume that P has a chain-scattering representation

$$H = CHAIN(P). \tag{3.1.8}$$

Then, due to (1) of Lemma 2.3.1, the model set (3.1.7) is represented as

$$\mathcal{M} = \{M : \ M = HM(H : \ \Delta); \ \Delta \in \overline{\boldsymbol{BH}}^{\infty}\}. \tag{3.1.9}$$

The uncertainty Δ may not be just an arbitrary element of $\overline{\boldsymbol{BH}}^{\infty}$, but rather subject to various constraints that are known a priori. Such a *structured uncertainty* is much harder to deal with than the *unstructured* case represented by (3.1.9) [5]. In this paper, we only deal with the simplest unstructured case.

Now, we consider the uniqueness of the representation (3.1.9). Our problem is whether H uniquely determine the set $\mathcal{M}$ in the representation (3.1.9). As is easily seen from the definition (2.3.2) of HM, there is an inherent non-uniqueness in the representation (3.1.9) of model sets, in the sense that, for any scalar λ,

$$HM(H : \ \Delta) = HM(\lambda H : \ \Delta). \tag{3.1.10}$$

This implies that multiplication by an arbitrary scalar to the generater H does not change the model set. In order to account for this indeterminability of the representation (3.1.9), we identify the two generators H_1 and H_2, if $H_1 = \lambda H_2$ for some scalar λ. We may use the notation.

$$H_1 = H_2 \qquad (\text{mod}\lambda). \tag{3.1.11}$$

The same identerminability occurs in LFT given by (2.3.1). Indeed,

$$LFT(G_1 : \Delta) = LFT(G_2 : \Delta)$$

if

$$G_1 = \begin{bmatrix} \lambda I & 0 \\ 0 & I \end{bmatrix} G_2 \begin{bmatrix} \lambda^{-1} I & 0 \\ 0 & I \end{bmatrix}. \tag{3.1.12}$$

for any scalar λ. This indeterminacy is much more complicated than that of HM given in (3.1.10). This is another reason why we use the chain-scattering formalism.

Apart from this obvious non-uniqueness, how the set $\mathcal{M}$ is determined by the generator H in the representation (3.1.9)? The following result answers this question.

Theorem 3.1.1 *Assume that* $\mathcal{M}_1 = \mathcal{M}_2$ *for*

$$\mathcal{M}_i = \{H : H = HM(H_i : \Delta); \Delta \in \overline{\boldsymbol{BH}}^{\infty}\}, \quad i = 1, 2,$$

where both H_1 *and* H_2 *are invertible. Then,* $H_1 = H_2 \Pi$ *(mod* λ*) where* Π *is a constant* J*-unitary matrix, i.e.,* $\Pi^* J \Pi = J$.

Proof. Obviously, $\mathcal{M}_1 = \mathcal{M}_2$ implies $\mathcal{M}_1 \subset \mathcal{M}_2$ and $\mathcal{M}_2 \subset \mathcal{M}_1$. The inclusion $\mathcal{M}_1 \subset \mathcal{M}_2$ implies that for any $\Delta_1 \in \overline{\boldsymbol{BH}}^{\infty}$, there exists a $\Delta_2 \in \overline{\boldsymbol{BH}}^{\infty}$ such that

$$HM(H_1 : \Delta_1) = HM(H_2 : \Delta_2). \tag{3.1.13}$$

Now, we fix $s = j\omega$. Due to (4) of Lemma 2.3.1, we have

$$\Delta_2(j\omega) = HM(H_2^{-1}(j\omega)H_1(j\omega) : \Delta_1(j\omega)) \in \overline{\boldsymbol{B}}^{p\times r}$$

This implies that the function $f(\Delta_1) := HM(H_2^{-1}(j\omega)H_1(j\omega) : \Delta_1(j\omega))$ maps $\boldsymbol{B}^{m\times r}$ into $\boldsymbol{B}^{m\times r}$.

On the other hand, the opposite inclusion $\mathcal{M}_2 \subset \mathcal{M}_1$ implies that for any $\Delta_2 \in \boldsymbol{B}^{m\times r}$, there exists $\Delta_1 \in \boldsymbol{B}^{m\times r}$ such that (3.1.13) holds. This implies that $f(\Delta_1) = HM(H_2^{-1}H_1 : \Delta_1)$ maps $\boldsymbol{B}^{m\times r}$ onto $\boldsymbol{B}^{m\times r}$. Since this holds for each $s = j\omega$, we have, due to Theorem 2.3.1, $H_2^{-1}H_1$ is J-unitary, i.e.,

$$H_2^{-1}H_1 = \Theta,$$

where Θ is J-unitary.

Since $HM(H_2^{-1}H_1 : \Delta_1) \in \overline{\boldsymbol{BH}}^{\infty}$, Proposition 2.4.3 implies that Θ is J-lossless.

The same argument holds if we replace H_1 with H_2, which yields that $\Theta^{-1} = H_1^{-1}H_2$ is also J-lossless. According to Proposition 2.4.4, Θ must be a constant J-unitary matrix. This complete the proof. □

This theorem states that in the representation (3.1.9), the generator H is essentially unique.

We apply the above result for special cases. The model set with additive uncertainty (see (3.1.1)) is given by

$$\mathcal{M}_a = \{G : \ G = G_0 + W\Delta, \ \Delta \in \overline{\boldsymbol{BH}}_\infty\}. \tag{3.1.14}$$

It was shown that $\mathcal{M}_a$ is represented in the form (3.1.9) with the generator H given by (3.1.5), i.e.,

$$H = \begin{bmatrix} W & G_0 \\ 0 & I \end{bmatrix}. \tag{3.1.15}$$

If there exist another pair G_0' and W' such that

$$\mathcal{M}_a = \{G : \ G = G_0' + W'\Delta, \ \Delta \in \overline{\boldsymbol{BH}^\infty}\},$$

Theorem 3.1.1 implies that

$$\begin{bmatrix} W & G_0 \\ 0 & I \end{bmatrix} = \gamma \begin{bmatrix} W' & G_0' \\ 0 & I \end{bmatrix} \begin{bmatrix} \Pi_{11} & \Pi_{12} \\ \Pi_{21} & \Pi_{22} \end{bmatrix},$$

where γ is a scalar and Π is a constant J-unitary matrix. The above relation implies

$$\begin{aligned} \Pi_{21} &= 0, \qquad \gamma\Pi_{22} = I \\ W &= \gamma W'\Pi_{11} \qquad G_0 = \gamma(W'\Pi_{12} + G_0'\Pi_{22}) \end{aligned}$$

Since Π is J-unitary, we have $\Pi_{22}^{\sim}\Pi_{22} = I$. Hence, $\Pi_{22} = I$ and $\gamma = 1$. Therefore, $\Pi_{12} = 0$ and $\Pi_{11}^{\sim}\Pi_{11} = I$. Thus, we obtain the following result.

Corollary 3.1.1 *If*

$$\{G : G = G_0 + W\Delta, \ \Delta \in \overline{\boldsymbol{BH}^\infty}\} = \{G : G = G_0' + W'\Delta, \ \Delta \in \overline{\boldsymbol{BH}^\infty}\},$$

then $G_0 = G_0'$ *and* $W = W'\Pi_{11}$ *for some unitary matrix* Π_{11}.

In exactly the same way, we can show a similar result to the multiplicative uncertainty model set:

Corollary 3.1.2 *If*

$$\{G : G = G_0(I+W\Delta), \ \Delta \in \overline{\boldsymbol{BH}^\infty}\} = \{G : G = G_0'(I+W'\Delta), \ \Delta \in \overline{\boldsymbol{BH}^\infty}\},$$

then $G_0 = G_0'$ *and* $W = W'\Pi_{11}$ *for some unitary matrix* Π_{11}.

We can prove the uniqueness of the hybrid representation.

Corollary 3.1.3 *If*

$$\{G : G = G_0 + W\Delta,\ \Delta \in \overline{\boldsymbol{BH}^{\infty}}\} = \{G : G = G_0'(I + W'\Delta),\ \Delta \in \overline{\boldsymbol{BH}^{\infty}}\},$$

then $G_0 = G_0'$ *and* $W = G_0 W' \Pi_{11}$ *for some unitary matrix.*

The proof is done based on the identity

$$\begin{bmatrix} W & G_0 \\ 0 & I \end{bmatrix} = \gamma \begin{bmatrix} G_0'W' & G_0' \\ 0 & I \end{bmatrix} \begin{bmatrix} \Pi_{11} & \Pi_{12} \\ \Pi_{21} & \Pi_{22} \end{bmatrix},$$

for some J-unitary Π.

3.2 Topological Aspects of Model Set

The representation of model set is not completed unless we give a proper topology to the set. This problem is quite difficult and so far we do not have many results to state here. We only give some preliminary results on this problem and suggest future work.

The fundamental issue is the *metrization*, which is closely related to the purpose of using the model set. We may use for the set of transfer functions the usual H^p norm regarding the set as a subset of H^p space. H^∞-norm is particularly convenient because it is an induced norm which is consistent with the input/output view of the system. It is also useful for robust control. The other popular metrics used in the field of robust control are the Hankel norm and gap metric [8]. It is worth noting that interesting research on the set of systems is going on in the framework of information geometry initiated by Amari and his coworkers [2].

Topological aspects of model sets necessarily involves stability consideration. If a model set contains both stable models and unstable ones, its topology becomes difficult to formulate. Therefore, it is usual to assume that the model set contains only stable models. This assumption entails some restrictions for the generator H in the representation (3.1.9) of the model set. In the representation (3.1.7) of the model set in terms of LFT it is known that the model set $\mathcal{M}$ contains only stable models iff P is stable and $\|P_{22}\|_\infty \leq 1$. In that case, we call the model set *stable*. This restriction is converted to chain-scattering formalism through the relation (2.3.7).

Proposition 3.2.1 *The model set* $\mathcal{M}$ *given by (3.1.7) is stable, iff*

(i)

$$\begin{bmatrix} H_{11} & H_{12} \\ 0 & I \end{bmatrix} \begin{bmatrix} H_{21} & H_{22} \\ 0 & I \end{bmatrix}^{-1}$$

is stable,

(ii)

$$H_{22}^{-1}H_{21} \in \boldsymbol{BH}^{\infty}$$

An important application of metrization is the quantitative measure of the largeness of the model set. The most usual notion for that purpose is the *diameter* defined by

$$diam(\mathcal{M}) := \sup_{G\in\mathcal{M}} \sup_{H\in\mathcal{M}} \|G - H\|. \tag{3.2.1}$$

Also, the *radius* defined by

$$rad(\mathcal{M}) := \inf_{G\in\mathcal{M}} \sup_{H\in\mathcal{M}} \|G - H\| \tag{3.2.2}$$

is also used. The model that attains the infimum of the above definition

$$c(\mathcal{M}) := \arg \min_{G\in\mathcal{M}} \sup_{H\in\mathcal{M}} \|G - H\| \tag{3.2.3}$$

is called the *center* of $\mathcal{M}$ [20].

The above quantities are usually not easy to compute except very special cases. For instance, the diameter of the additive uncertainty model (3.1.14) is given by

$$d(\mathcal{M}) = 2\|W\|_{\infty}, \tag{3.2.4}$$

if the metric is $\boldsymbol{H}^{\infty}$ norm.

It should be noted that in the representations of the model set in terms of perturbation Δ, the center does not coincide with the nominal model for which $\Delta = 0$. For instance, the model set (3.1.9) for which the generator H is J-lossless is $\boldsymbol{BH}^{\infty}$ whose center is obviously $c(\mathcal{M}) = 0$. However, the nominal model of the set (3.1.9) is given by

$$c(\mathcal{M}) = G_{12}G_{22}^{-1},$$

which is not zero usually.

If G is a constant matrix in $\boldsymbol{C}^{2m\times 2m}$, then

$$f(\Delta) = HM(G: \ \Delta)$$

defines a map from $\Delta \in \boldsymbol{C}^{m\times m}$ to $\boldsymbol{C}^{m\times m}$. In [19], Siegel introduced an interesting metric in $\boldsymbol{C}^{m\times m}$ based on the *cross-ratio*

$$R(\Delta,\ \Delta') := (\Delta - \Delta')(\Delta - \bar{\Delta}')^{-1}(\bar{\Delta} - \bar{\Delta}')(\Delta - \Delta')^{-1}.$$

This metric leads to a matrix generalization of the Poincaré non-euclidian metric which was used by et.al. in the early development of robust control theory [11]. In our case, the problem is much more involved, because G and

Δ are not constant matrices but functions of s. Nevertheless, the metric introduced in [19] gives some insight to the metrization of the model set given in (3.1.7).

The following result gives an upper bound of the diameter of the model set of (3.1.9) under the condition of Proposition 3.2.1.

Proposition 3.2.2 *Assume that the generator H of the model set (3.1.9) is invertible and satisfies the conditions of Proposition 3.2.1. Then,*

$$diam(\mathcal{M}) \leq 2\|H_{22}^{-1}\|_\infty \|\hat{H}_{11}^{-1}\|_\infty (1-\|H_{22}^{-1}H_{21}\|_\infty)^{-1}(1-\|\hat{H}_{11}^{-1}\hat{H}_{21}\|_\infty)^{-1} \tag{3.2.5}$$

where $\hat{H} := H^{-1}$

Proof: Write

$$M(\Delta) := HM(H:\Delta) = (H_{11}\Delta + H_{12})(H_{21}\Delta + H_{22})^{-1}$$

Due to the relation

$$[I \quad -\Delta]\hat{H}H\begin{bmatrix}\Delta \\ I\end{bmatrix} = 0,$$

we have an alternative representation of $M(\Delta)$ as

$$M(\Delta) = -(\hat{H}_{11} - \Delta\hat{H}_{21})^{-1}(H_{21} - \Delta\hat{H}_{22}) \tag{3.2.6}$$

The condition (ii) of Proposition 3.2.1 is given in terms of $\hat{H}$ as

$$\hat{H}_{21}\hat{H}_{11}^{-1} \in \boldsymbol{BH}^\infty. \tag{3.2.7}$$

From the definition (3.2.1) of the diameter, we have

$$diam(\mathcal{M}) = \sup_{\Delta_1 \in \overline{\boldsymbol{BH}}^\infty} \sup_{\Delta_2 \in \overline{\boldsymbol{BH}}^\infty} \|M(\Delta_1) - M(\Delta_2)\|_\infty$$

From (3.2.6) and $\hat{H}H = I$, it follows that

$$\begin{aligned} M(\Delta_1) - M(\Delta_2) &= (H_{11}\Delta_1 + H_{12})(H_{21}\Delta_1 + H_{22})^{-1} \\ &+ (\hat{H}_{11} - \Delta_2\hat{H}_{21})^{-1}(\hat{H}_{12} - \Delta_2\hat{H}_{22}) \\ &= (\hat{H}_{11} - \Delta_2\hat{H}_{21})^{-1}(\Delta_1 - \Delta_2)(H_{21}\Delta_1 + H_{22})^{-1}. \end{aligned}$$

Thus, we have

$$\|M(\Delta_1) - M(\Delta_2)\|_\infty \leq \|\Delta_1 - \Delta_2\|_\infty \|(\hat{H}_{11} - \Delta_2\hat{H}_{21})^{-1}\|_\infty \|(H_{21}\Delta_1 + H_{22})^{-1}\|_\infty$$

Since $\|\Delta_1 - \Delta_2\|_\infty \leq \|\Delta_1\|_\infty + \|\Delta_2\|_\infty \leq 2$ for each $\Delta_1,\ \Delta_2 \in \overline{\boldsymbol{BH}}^\infty$, we have

$$diam(\mathcal{M}) \leq 2 \sup_{\Delta \in \overline{\boldsymbol{BH}}^\infty} \|(\hat{H}_{11} - \Delta\hat{H}_{21})^{-1}\|_\infty \cdot \sup_{\Delta \in \overline{\boldsymbol{BH}}^\infty} \|(H_{21}\Delta - H_{22})^{-1}\|_\infty \tag{3.2.8}$$

Now, we have

$$\sup_{\Delta \in \overline{BH}^{\infty}} \|(H_{21}\Delta + H_{22})^{-1}\|_\infty \leq \|H_{22}^{-1}\|_\infty \sup_{\Delta \in \overline{BH}^{\infty}} \|(I + H_{22}^{-1}H_{21}\Delta)^{-1}\|_\infty$$

Note that $H_{22}^{-1}H_{21} \in \boldsymbol{BH}^\infty$ due to Proposition 3.2.1. It is easy to see that for each ω,

$$\sup_{\Delta \in \bar{B}} \|(I + H_{22}^{-1}(\omega)H_{21}(j\omega)\Delta)^{-1}\|_\infty = (1 - \|H_{22}^{-1}(j\omega)H_{21}(j\omega)\|)^{-1}$$

Therefore, we have

$$\sup_{\Delta \in \overline{BH}^{\infty}} \|(I + H_{22}^{-1}H_{21}\Delta)^{-1}\| \leq (1 - \|H_{22}^{-1}H_{21}\|_\infty)^{-1}.$$

In the same way, we can show that

$$\sup_{\Delta \in \overline{BH}^{\infty}} \|(\hat{H}_{11} - \Delta\hat{H}_{21})^{-1}\|_\infty \leq \|\hat{H}_{11}^{-1}\|_\infty \cdot (1 - \|\hat{H}_{21}\hat{H}_{11}^{-1}\|_\infty)^{-1}$$

Substituting these inequalities in (3.2.7), we have (3.2.5). □

As an application of the above proposition, consider the model set with additive uncertainty given by (3.1.14). The corresponding chain-scattering generator is given by (3.1.15) and its inverse is given by

$$H^{-1} = \begin{bmatrix} W^{-1} & -W^{-1}G_0 \\ 0 & I \end{bmatrix} = \begin{bmatrix} \hat{H}_{11} & \hat{H}_{12} \\ \hat{H}_{21} & \hat{H}_{22} \end{bmatrix}$$

assuming that W^{-1} exists. Since $\hat{H}_{21} = 0$, $H_{21} = 0$ and $H_{22} = I$, the inequality (3.2.5) becomes in this case

$$diam(\mathcal{M}_a) \leq 2\|W\|_\infty \tag{3.2.9}$$

Actually, due to (3.2.4), the equality holds in (3.2.9). Therefore, the upper bound of the diameter is the diameter itself in this case, which suggests that the upper bound (3.2.5) is quite tight.

In system theory of model set, it is important to consider the distance between two different sets, as well as the one between two models in the same set. If the two sets $\mathcal{M}_1$ and $\mathcal{M}_2$ are embedded in a metric space, then a usual definition of the distance between $\mathcal{M}_1$ and $\mathcal{M}_2$ is the Hausdorf distance

$$\begin{aligned} d(\mathcal{M}_1, \mathcal{M}_2) &:= \max(\vec{d}(\mathcal{M}_1, \mathcal{M}_2), \vec{d}(\mathcal{M}_2, \mathcal{M}_1)) \\ \vec{d}(\mathcal{M}_1, \mathcal{M}_2) &:= \sup_{M_1 \in \mathcal{M}_1} \inf_{M_2 \in \mathcal{M}_2} \|M_1 - M_2\|. \end{aligned}$$

If $\mathcal{M}_1 \subset \mathcal{M}_2$, $\vec{d}(\mathcal{M}_1, \mathcal{M}_2) = 0$. Hence, $d(\mathcal{M}_1, \mathcal{M}_2) = \vec{d}(\mathcal{M}_2, \mathcal{M}_1)$. The distance between the model sets becomes important in approximation of model set.

4 Other Problems

4.1 Connections

In system theory, connection of systems is an important issue. This gives rise connections of model sets in the present context. The connection of model sets is usually reduced to separating uncertainties from known parts of the original model sets. This sort of manipulation is quite popular and useful in control system design (e.g. [16]). We show how to represent the connections of model sets in the chain-scattering formalism (3.1.9). After the connection, the uncertainty of the resulting model set becomes always structured.

4.1.1 Cascade Connection

The cascade connection of two model sets

$$\mathcal{M}_1 = \{M : M = CHAIN(G : \Delta_1);\ \Delta_1 \in \overline{\boldsymbol{BH}}^{\infty}\} \quad (4.1.1)$$
$$\mathcal{M}_2 = \{M : M = CHAIN(H : \Delta_2);\ \Delta_2 \in \overline{\boldsymbol{BH}}^{\infty}\} \quad (4.1.2)$$

is illustrated in Fig.8. Define

$$\Delta = \begin{bmatrix} \Delta_1 & 0 \\ 0 & \Delta_2 \end{bmatrix}. \quad (4.1.3)$$

We have

$$\begin{bmatrix} a_2 \\ c_2 \end{bmatrix} = \Delta \begin{bmatrix} b_2 \\ d_2 \end{bmatrix}.$$

Eliminating a_2, c_2, b_2 and d_2, from the relations

$$\begin{bmatrix} a_1 \\ b_1 \end{bmatrix} = H \begin{bmatrix} a_2 \\ b_2 \end{bmatrix}, \qquad \begin{bmatrix} c_1 \\ d_1 \end{bmatrix} = G \begin{bmatrix} c_2 \\ d_2 \end{bmatrix},$$

we have

$$\begin{pmatrix} a_1 \\ d_1 \end{pmatrix} = HM(R : \Delta) \begin{pmatrix} b_1 \\ c_1 \end{pmatrix}$$

where

$$R := \begin{bmatrix} H_{11} & 0 & H_{12} & 0 \\ 0 & G_{21} & 0 & G_{22} \\ H_{21} & 0 & H_{22} & 0 \\ 0 & G_{11} & 0 & G_{12} \end{bmatrix}. \quad (4.1.4)$$

Since $b_1 = c_1$, we have

$$a_1 = HM(HM(R : \Delta) : I) d_1.$$

Thus, the model set generated by the cascade connection of the two model set (4.1.1) and (4.1.2) are given by

$$\mathcal{M} = \{M : M = HM(HM(R : \Delta) : I);\ \Delta = \begin{bmatrix} \Delta_1 & 0 \\ 0 & \Delta_2 \end{bmatrix}, \Delta_i \in \overline{\boldsymbol{BH}}^{\infty}\}$$

4.1.2 Paralell Connection

Paralell connection of two model sets (4.1.1) and (4.1.2) is illustrated in Fig.9. In the same way as in the preceding section, the model set of Fig.9 is given by

$$\mathcal{M} = \{M : M = HM(HM(Q : \Delta) : I);\ \Delta = \begin{bmatrix} \Delta_1 & 0 \\ 0 & \Delta_2 \end{bmatrix}, \Delta_i \in \overline{\boldsymbol{BH}}^{\infty}\},$$

where

$$Q := \begin{bmatrix} G_{11} & H_{11} & G_{12} & H_{12} \\ G_{21} & 0 & G_{22} & 0 \\ G_{21} & 0 & G_{22} & 0 \\ 0 & H_{21} & 0 & H_{22} \end{bmatrix}$$

4.1.3 Feedback Connection

The feedback connection comes from the usual plant/controller connection shown in Fig.10. Taking $G = CHAIN(P)$ and $H = CHAIN(C)$ with Δ_1 and Δ_2 being uncertainties contained in the plant and controller, respectively, the feedback connection is illustrated in Fig.11. It is not difficult to see that

$$\begin{bmatrix} e_1 \\ w_1 \\ w_2 \\ e_2 \end{bmatrix} = F \begin{bmatrix} q_1 \\ q_2 \\ r_1 \\ r_2 \end{bmatrix}$$

$$F := \begin{bmatrix} G_{21} & 0 & G_{22} & 0 \\ G_{21} & -H_{11} & G_{22} & -H_{12} \\ -G_{11} & H_{21} & -G_{12} & H_{22} \\ 0 & H_{21} & 0 & H_{22} \end{bmatrix}$$

Since

$$\begin{bmatrix} q_1 \\ q_2 \end{bmatrix} = \Delta \begin{bmatrix} r_1 \\ r_2 \end{bmatrix}, \quad \Delta = \begin{bmatrix} \Delta_1 & 0 \\ 0 & \Delta_2 \end{bmatrix},$$

we have

$$\begin{bmatrix} e_1 \\ w_1 \end{bmatrix} = HM(F : \Delta) \begin{bmatrix} w_2 \\ e_2 \end{bmatrix}.$$

This is the chain-scattering representation of the familiar relation

$$\begin{bmatrix} e_1 \\ e_2 \end{bmatrix} = \begin{bmatrix} I & -C \\ -P & I \end{bmatrix}^{-1} \begin{bmatrix} w_1 \\ w_2 \end{bmatrix}.$$

Therefore, the model set generated by the feedback connection of $\mathcal{M}_1$ and $\mathcal{M}_2$ given by (4.1.1) and (4.1.2), respectively, is represented as

$$\mathcal{M} = \{M : M = HM(F : \Delta),\ \Delta = \begin{bmatrix} \Delta_1 & 0 \\ 0 & \Delta_2 \end{bmatrix}, \Delta_i \in \overline{\boldsymbol{BH}}^{\infty}\}. \quad (4.1.5)$$

A fundamental problem of feedback control is the reduction of uncertainty by feedback [21]. From this viewpoint, it is desirable that $diam(\mathcal{M})$ of (4.1.5) is smaller than the sum $diam(\mathcal{M}_1) + diam(\mathcal{M}_2)$, where $\mathcal{M}_1$ and $\mathcal{M}_2$ are given in (4.1.1) and (4.1.2), respectively.

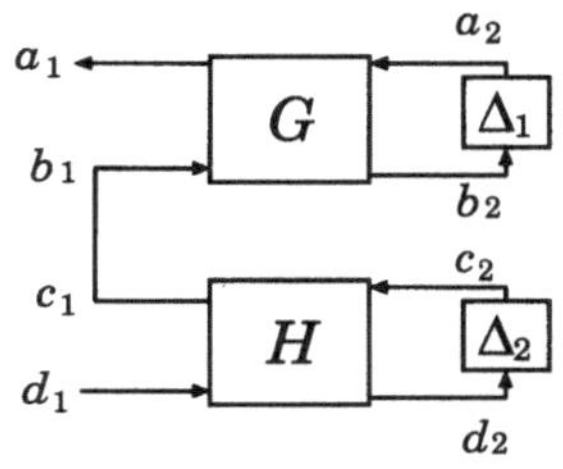

Figure 8: Cascade Connection

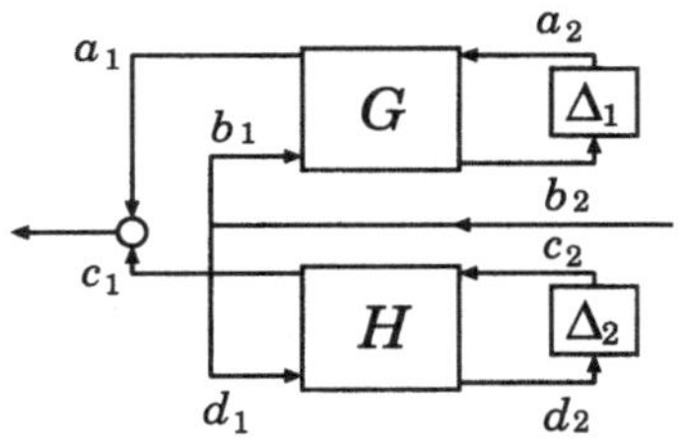

Figure 9: Parallel Connection

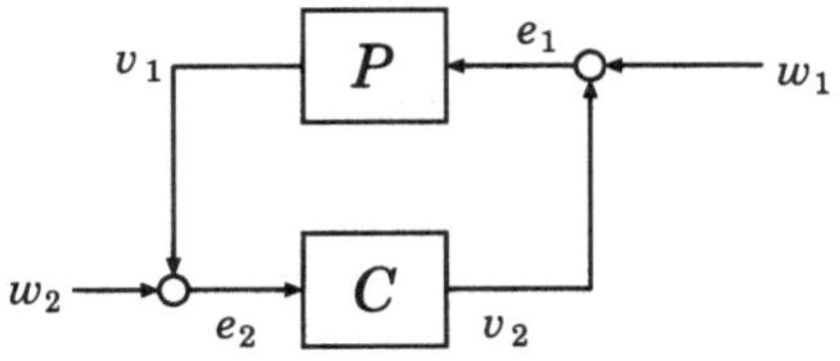

Figure 10: Plant/Controller Connection

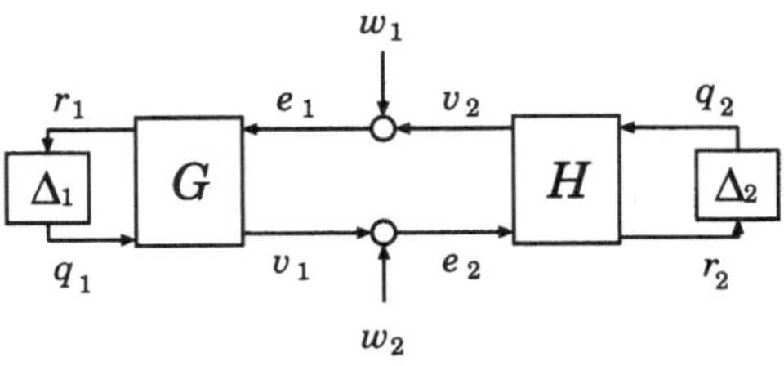

Figure 11: Feedback Connection Corresponding to Fig.10, $G = CHAIN(P)$, $H = CHAIN(C)$

4.2 Model Set Reduction

Model reduction has been an important area of research in system theory which has produced many remarkable results from both theoretical and practical points of view (see,e.g., [9]). Corresponding to model reduction, there is of course the model set reduction which is also quite important in system theory.

Consider the model set (3.1.14) of additive uncertainty. The traditional model reduction is to get a simplified approximation of the nominal model G_0. On the other hand, the model set reduction aims to reduce not only the nominal model G_0, but also the frequency dependent uncertainty band function W. This is the sharp point of departure of the model set reduction from model reduction. In the context of chain-scattering formalism, the problem of model set reduction is formulated as follows: *Given* H, *find a simpler* H' *such that* $d(\mathcal{M},\ \mathcal{M}')$ is small where $\mathcal{M} = \{M :\ M = CHAIN(H :\ \Delta);\ \Delta \in \overline{\boldsymbol{BH}_\infty}\}$, $\mathcal{M}' = \{M :\ M = CHAIN(H' :\ \Delta);\ \Delta \in \overline{\boldsymbol{BH}_\infty}\}$.

The model set $\mathcal{M}'$ is a reduced approximation of $\mathcal{M}$. It is natural to assume that $\mathcal{M}'$ contains the original $\mathcal{M}$, i.e.,

$$\mathcal{M} \subset \mathcal{M}'. \tag{4.2.1}$$

In that case,

$$\begin{aligned} d(\mathcal{M},\ \mathcal{M}') &= \vec{d}(\mathcal{M},\ \mathcal{M}') \\ &= \sup_{M_1 \in \mathcal{M}} \inf_{M_2 \in \mathcal{M}'} \|M_1 - M_2\|. \end{aligned} \tag{4.2.2}$$

The relation (4.2.1) implies that the any model that is considered to be true is contained in the reduced set. This is almost a necessary condition for $\mathcal{M}'$ to be useful as an approximation of $\mathcal{M}$. The relation (4.2.1) can be interpreted that, due to simplification of the generator, the set is expanded, i.e., the uncertainty is increased. This is also comparable with the natural phenomenon of traditional model reduction that the more the reduction is, the more is caused the approximation error. One might think that since the model set reduction is boiled down to the reduction of the generator, it is just a special case of traditional model reduction. This is not the case. The constraint (4.2.1) gives a very different feature of model set reduction from the traditional one. The four block submatrices of H play different roles in the representation of the model set (3.1.9).

Some interesting results were derived in a pioneering work by Beck et.al. [3]. Their results does not satisfy the inclusion condition (4.2.1). As for the model set reduction problem with the inclusion condition (4.2.1), we do not have many results except a very preliminary one [10].

It should be noted that the model set reduction has much larger flexibility in the selection of the approximation than the usual model reduction, because of the uncertainty. It may be possible to reduce the model set significantly with a slight increase of uncertainty. For instance, a model set with order more than one hundred can be represented by a model set with order three by slight increase of uncertainty. This sort of observation may be able to explain why controllers of very low order (e.g., PID controllers) can control complex systems successfully.

5 Conclusion

An overview of system theoretic approach to model set is presented in the framework of chain-scattering representation of models. Model set approach to deal with real systems was initiated in robust control. It was naturally extended to identification problems. The purpose of this expository paper is to show that the system theoretic approach to model set is important and fruitful not only in control and identification, but also in other problems such as uniqueness of representation, metrization, connection and reduction. Learning is another issue of importance in which system theoretic approach to model set plays fundamental role. This topics is omitted due to space limitation, and will be fully discussed elsewhere.

The current status of system theory of model set is at immatured level except control and identification. Even in the areas of control and identification, there is wide room for future research if we take the model set point of view more clearly and radically. The chain-scattering approach which is used in this paper seems to be an efficient way of dealing with model set, though there may be many other approaches to model set.

We believe that system theory of model set will entail a paradigm shift in the area of systems and control.

ACKNOWLEDGEMENT

The author is grateful to Professor Helton for variable discussions about the system theoretic implication of Möbius transformation. He is also grateful to Dr. Tsumura for helpful comments.

References

[1] L. Ahlfors, *Complex Analysis*, McGraw-Hill, (1979).

[2] S. Amari, *Differential Geometric Methods in Statistics,* Springer, (1985).

[3] C. Beck, J. Doyle and K. Glover, "Model reduction of multidimensional and uncertain systems," *IEEE Trans. Auto. Control*, **AC-41**, 1466-1477, (1996).

[4] P. Devilde and H. Dym, "Lossless chain-scattering matrices and optimum linear prediction; the vector case," *Int. J. of Circuit Theory and Appl.,* **9**, 135-175, (1981).

[5] J. C. Doyle, "Analysis of feedback systems with structured uncertainties," *IEEE Proc.* **133**, Prt D, 45-56, (1982).

[6] J. C. Doyle, B. A. Francis and A. Tannenbaum, *Feedback Control Theory,* Macmillan Publishing Company, (1992).

[7] J. C. Doyle, "Theory in modeling and simulation," Reprint, (1989).

[8] T. T. Georgiou and M. C. Smith, "Optimal robustness in the gap metric," *Proc. CDC,* Tampa, 2331-2336, (1992).

[9] K. Glover, "All optimal Hankel-norm approximations of linear multivariable systems and their L_∞-error bounds," *Int. J. Control,* **39**, 1115-1193, (1984).

[10] I. Jikuya and H. Kimura, "Reduction of model set with inclusion," *Proc. CDC,* Tampa, (1998), to appear.

[11] P. P. Khargonekar and A. Tannenbaum, "Non-Euclidian metrics and the robust stabilization of systems with parameter uncertainties," *IEEE Trans. Auto. Control,* **AC-30**, 1005-1013, (1985).

[12] H. Kimura, *Chain-Scattering Approach to H^∞ Control,* Birkheuser, (1996).

[13] H. Kimura, "How the model gets reality ?," *Proc. 2nd Asian Control Conference*, Seoul, **2**, 3-10, (1997).

[14] H. Kimura, "Non-uniqueness, uncertainty and complexity in modeling," *J. of Appl. Computation, Control, Signals and Circuits*, **1**, 455-485, (1998).

[15] M. Milanese, J. Norton, H. Piet-Lahanier and Eric Walter (eds.), *Bounding Approaches to System Identification*, Plenum Press, (1996).

[16] A. Packard. "Gain scheduling via linear fractional transformations," *Systems and Control Letters*, **22**, 79-92, (1994).

[17] R. S. Phillips, "On symplectic mappings of contraction operators," *Studia Mathematica*, **T. XXX1**, 15-27, (1968).

[18] R. M. Redheffer, "On a certain linear fractional transformation," *J. of Mathematics and Physics*, **39**, 269-286, (1960).

[19] C. L. Siegel, *Symplectic Geometry,* Academic Press, (1964).

[20] J. F. Traub, G. W. wasilkowski and H. Woznlakowski, *Information-based Complexity*, Academic Press, (1988).

[21] G. Zames, "Feedback and Optimal sensitivity: Model reference transformations, multiplicative seminorms, and approximate inverses," *IEEE Trans. Auto. Control*, **26**, 301-320, (1981).

The University of Tokyo, Japan

Progress in Systems and Control Theory, Vol. 25

The Role of the Hamiltonian in the Solution of Algebraic Riccati Equations

Peter Lancaster

1 Introduction: Riccati's time

Beginning with the time of Riccati himself, we trace the origin of the Hamiltonian matrix and developments on the theme (in the context of the two basic algebraic Riccati equations) from about two hundred years ago. Our main emphasis is on developments in the last fifty years and the extension or adaption of the Hamiltonian from a block matrix to pencils which retain either the Hamiltonian or symplectic properties, as appropriate.

In this expository paper the reader is introduced to the development of the part played by the Hamiltonian in the solution of algebraic Riccati equations. However, as this paper was prepared for a special meeting to honour Count Jacopo Riccati himself, an attempt is made to make some connection, however tenuous, with the mathematical interests of Riccati, and to survey briefly the development of relevant notions over the subsequent three-hundred years or so. As a consequence, we begin with some basic ideas about ordinary differential equations (rather than algebraic equations).

In 1723 Riccati investigated real nonlinear differential equations of the form

$$x'(t) = c(t) - d(t)x^2(t), \tag{1}$$

where the accent refers to differentiation with respect to t, $d(t) = t^{-n}$, $c(t) = -nt^{n+m-1}$, and n and m are constants. A burning question for Riccati was: For which integer values of n and m is this equation solvable in *finite terms*? This means that solutions can be expressed as finite expressions (i.e. no limiting processes) of elementary functions of t. For example, this is possible when $n = 0$.

Note that Riccati's problem is already phrased in modern terminology and. to expand a little on this point, it is worth spending some time on the context of Riccati's work. We can conveniently begin with the publication of Newton's "Principia Mathematica" in 1687. Riccati himself lived from 1676 to 1754 and was self-educated in mathematics. It is a measure of his stature that he was an early reader and commentator on the Principia. Thus, his research work was done not so long after the discovery of a comprehensive differential and integral calculus. Of course, there are several other very important names in that enterprise, notably Leibniz and the Bernoulli family, and Riccati was in correspondence with several of these extraordinary mathematicians. See Chapter 1 of Watson [17], and Appendix A of Ince [8] for more details.

Equations of the above type play a role in the theory of special functions. For example, solutions of Riccati equations of the form $x'(t) = t^a - t^{-a-1}x^2(t)$ are expressed in terms of Bessel functions in Section 4.15 of [17] (and this helps to explain why Riccati's work is discussed in a treatise on Bessel functions).

The eighteenth century saw developments in analysis and differential equations by such intellectual giants as Euler, Lagrange, and Laplace, and the age of rigorous mathematics began with Gauss (1777-1855). Note also that it was only within a few years of 1800 that the complex number system and some of its ramifications were well understood. The "fundamental theorem of algebra" was proved by Gauss in 1799, and the geometric picture of the "Argand diagram" was published in 1806. Thus, Riccati's work was done in ignorance of these ideas, which might now be described as high-school mathematics. In particular, these developments paved the way for the study of differential equations over the complex numbers. Then the fundamental theorem of Cauchy (1789-1857) concerning analytic functions was published in 1814 and opened the door for the wonders of analysis of functions of a complex variable. For more information on this period of mathematics see E.T.Bell's "Men of Mathematics" [2] and the biographies [18], [7], for example.

2 Linearization

The significance of *linear* equations was already understood in Riccati's time. In contrast, equation (1) is a prototypical nonlinear equation, with more difficult questions of existence and uniqueness of solutions. In 1728 Euler published a technique for transforming certain nonlinear to linear equations by using clever transformations of the variables (see [8]). For us, a different linearization process to a linear *system* is illuminating. Such a technique was certainly clear to Sturm and Liouville a hundred years after Riccati.

It is convenient to first make a generalization of equation (1). According to Watson, it was D'Alembert who first named the equation

$$x'(t) = c(t) - 2a(t)x(t) - d(t)x^2(t), \tag{2}$$

with a general t-dependent quadratic term on the right, *Riccati's generalized equation*. This was in 1763. Consider also the system

$$\begin{aligned} -v'(t) &= -c(t)u(t) + a(t)v(t), \\ u'(t) &= a(t)u(t) + d(t)v(t), \end{aligned} \tag{3}$$

and assume that $a(t)$, $c(t)$, $d(t)$ are continuous on an interval I. Then the following statement is easily verified and involves little more than the quotient rule for differentiation (see Chapter 1 of [15]):

Theorem 1 *Equation* (2) *has a solution* $x(t)$ *on* I *if and only if there exist functions* $u(t)$ *and* $v(t)$ *on* I *satisfying* (3), *for which* $u(t) \neq 0$ *on* I *and for which* $x(t) = v(t)/u(t)$.

With our twentieth century conditioning we are led to write the system (3) in matrix form:

$$\begin{bmatrix} 0 & -1 \\ 1 & 0 \end{bmatrix} \begin{bmatrix} u' \\ v' \end{bmatrix} = \begin{bmatrix} -c & a \\ a & d \end{bmatrix} \begin{bmatrix} u \\ v \end{bmatrix},$$

(for convenience, we suppress the t-dependence) and then

$$\begin{bmatrix} u' \\ v' \end{bmatrix} = \begin{bmatrix} a & d \\ c & -a \end{bmatrix} \begin{bmatrix} u \\ v \end{bmatrix}. \tag{4}$$

We now see the first appearance of a Hamiltonian in this discussion. A 2×2 matrix A is said to be Hamiltonian if HA is real and symmetric where

$$H = \begin{bmatrix} 0 & 1 \\ -1 & 0 \end{bmatrix}, \tag{5}$$

and the coefficient matrix of equation (4) has this property. Equation (4) may now be described as a Hamiltonian system.

From the historical point of view we have taken quite a lot for granted. The powerful matrix notation emerged in the nineteenth century through the work of Sylvester (1814-1897), Cayley (1821-1895), and several others and, of course, Hamilton (1805-1865) recognized the importance of the symmetry displayed here. As with complex numbers and the calculus, one recognises the importance and power of good notation.

3 Matrix equations

We now formulate matrix generalizations of the ideas in Section 2. There is some room for manoeuvre here, but we take matrix equations which arise in several important applications (familiar to an MTNS audience). Let $X(t)$ take values in the complex $n \times n$ matrices, i.e. in $\mathbb{C}^{n\times n}$, and consider the following generalization of (2):

$$X'(t) = C(t) - X(t)A(t) - A^*(t)X(t) - X(t)D(t)X(t) \tag{6}$$

where $C(t)$, $D(t)$ take Hermitian values, all three coefficient matrices are $n \times n$ and, in view of the symmetry of the equation, it is natural to investigate Hermitian solutions $X(t)$.

We associate with this matrix Riccati equation a *linear* first order system generalizing that of equations (3). Here, $U(t)$ and $V(t)$ take values in $\mathbb{C}^{n\times n}$ (we suppress t once more):

$$\begin{aligned} U' &= AU + DV, \\ V' &= CU - A^*V. \end{aligned} \tag{7}$$

Although this generalization of the scalar equation may seem like a great leap forward, it is surprising that some important properties generalize with just some elementary matrix algebra. For example, if equation (6) has a Hermitian solution at some t_0, it is not immediately clear that it will evolve as a Hermitian function of t in a neighbourhood of t_0. However, we have the following nice result whose proof is quite easy and attractive (cf. Chapter 8 of [1]):

Theorem 2 *Let U, V satisfy equations* (7) *and let VU^{-1} exist and be Hermitian for at least one t. Then VU^{-1} is Hermitian whenever it exists. Furthermore, $X := VU^{-1}$ satisfies equation* (6).

Proof We have $(U^*V - V^*U)' = (U^*)'V + U^*V' - (V^*)'U - V^*U'$, and substituting on the right-hand-side from (7) it is found that this derivative is zero. Consequently,

$$U^*(t)V(t) - V^*(t)U(t) = M,$$

a constant matrix.

Now, if VU^{-1} exists at some t and is Hermitian, then $(U^{-1})^*V^* = VU^{-1}$, or $U^*V = V^*U$ so that, in fact, we have $M = 0$. It follows that $V(t)U^{-1}(t)$ is Hermitian where it exists.

The verification that X satisfies the Riccati equation is straightforward, and requires only that we know about the derivative of an inverse.

$$\begin{aligned} (VU^{-1})' &= V'U^{-1} - VU^{-1}U'U^{-1} \\ &= (CU - A^*VU^{-1} - VU^{-1}(AU + DV)U^{-1} \\ &= C - A^*X - XA - XDX. \qquad \square \end{aligned}$$

We now extend the definition of (5) and write

$$H = \begin{bmatrix} 0 & I \\ -I & 0 \end{bmatrix}, \tag{8}$$

where I is the $n \times n$ identity matrix. Then the coefficient matrix

$$N = \begin{bmatrix} A & D \\ C & -A^* \end{bmatrix} \tag{9}$$

is Hamiltonian in the sense that HN is Hermitian. This implies the similarity $(-N^*) = HNH^{-1}$. Consequently, at any fixed t, the eigenvalues of N are distributed in the complex plane symmetrically with respect to the imaginary axis. If, in addition, N is real, then the eigenvalues are distributed symmetrically with respect to *both* axes in $\mathbb{C}$.

Equation (6) plays an important role in the calculus of variations, which blossomed in the seventeenth century and has attracted the attention of analysts ever since that time including Morse [13] and Birkhoff and Hestenes [3] in 1935. Any account of this connection is beyond the scope of the present exposition (but see Chapter 5 of [15] for a quick introduction). More recently, the seminal work of Kalman on linear-quadratic regulator problems beginning in the late 1950's has led a vigorous development of control and filtering theory in which equations of this type play a central part. It is this area which motivates the subsequent discussion - on the main topic of this article.

4 The algebraic Riccati equation

An algebraic equation is obtained from a time-invariant equation of the form (6) (i.e. A, C, and D do not depend on t) when we seek steady-state solutions. Thus, we obtain the equation

$$XDX + XA + A^*X - C = 0, \tag{10}$$

with $D^* = D$, $C^* = C$, and we seek Hermitian solutions X. To get some idea of the possible form of the solution sets of such equations consider the following examples.

EXAMPLES

1. Scalar equations of the form $dx^2 + 2ax - c = 0$, where $a, d, c \in \mathbb{R}$ and $d > 0$, $a \geq 0$. The Hermitian property of solutions means just real solutions here, and they may or may not exist. The existence of a nonnegative solution requires $c \geq 0$.

2. The equation (10) with
$$D = \begin{bmatrix} 0 & 1 \\ 1 & 0 \end{bmatrix}, \quad A = 0, \quad C = \begin{bmatrix} 1 & 0 \\ 0 & 0 \end{bmatrix}$$
has no solution (either real or complex).

3. The equation (10) with
$$D = \begin{bmatrix} 1 & 0 \\ 0 & 0 \end{bmatrix}, \quad A = \begin{bmatrix} 0 & 0 \\ 1 & 0 \end{bmatrix}, \quad C = \begin{bmatrix} -2 & 0 \\ 0 & 1 \end{bmatrix}$$
has the one Hermitian solution $X = \begin{bmatrix} 0 & -1 \\ -1 & 0 \end{bmatrix}$ and two non-Hermitian solutions.

4. The equation (10) with

$$D = \begin{bmatrix} 0 & 0 \\ 0 & 1/3 \end{bmatrix}, \quad A = \begin{bmatrix} 0 & -1 \\ 2/3 & 0 \end{bmatrix}, \quad C = \begin{bmatrix} 5/3 & 0 \\ 0 & 1 \end{bmatrix}$$

has four Hermitian solutions including the maximal solution $X = \begin{bmatrix} 3 & 1 \\ 1 & 3 \end{bmatrix}$, and two non-Hermitian solutions.

5. The equation (10) with

$$D = \begin{bmatrix} 1 & 0 \\ 0 & 0 \end{bmatrix}, \quad A = \begin{bmatrix} 0 & 0 \\ 0 & 1 \end{bmatrix}, \quad C = \begin{bmatrix} 1 & 0 \\ 0 & 0 \end{bmatrix}$$

has an isolated solution $X = \begin{bmatrix} 1 & 0 \\ 0 & 0 \end{bmatrix}$, and a continuum of Hermitian solutions depending on a parameter $a \in \mathbb{C}$:

$$X_a = \begin{bmatrix} -1 & a \\ \bar{a} & -\frac{1}{2}|a|^2 \end{bmatrix},$$

and also a continuum of non-Hermitian solutions.

Our first general observation makes use of the Hamiltonian N of (9). If there is a solution X of (6) and we define $Z = A + DX$, then

$$\begin{bmatrix} A & D \\ C & -A^* \end{bmatrix} \begin{bmatrix} I \\ X \end{bmatrix} = \begin{bmatrix} I \\ X \end{bmatrix} Z. \tag{11}$$

This is very easily verified. Another way to say this is that the *graph* of X,

$$\mathcal{G}(X) := \operatorname{Im} \begin{bmatrix} I \\ X \end{bmatrix} \subseteq \mathbb{C}^{2n}$$

is invariant under N. Using this notion a converse statement is soon verified and we have the important statement:

Proposition 1 *A matrix* $X \in \mathbb{C}^{n\times n}$ *is a solution of* (10) *if and only if the graph of* X *is* T*-invariant.*

Note that, more generally, a graph subspace has the form

$$\operatorname{Im} \begin{bmatrix} X_1 \\ X_2 \end{bmatrix},$$

where X_1 is nonsingular, because

$$\operatorname{Im} \begin{bmatrix} X_1 \\ X_2 \end{bmatrix} = \operatorname{Im} \begin{bmatrix} X_1 \\ X_2 \end{bmatrix} X_1^{-1} = \operatorname{Im} \begin{bmatrix} I \\ X \end{bmatrix}, \tag{12}$$

where $X = X_2 X_1^{-1}$.

The importance of the last proposition lies in the fact that it admits the solution of algebraic Riccati equations by standard processes of numerical linear algebra for the eigenvalue-eigenvector problem. As before, the Hamiltonian is serving to replace a nonlinear problem by a linear one.

The next question is: When is a solution X Hermitian? This can be described geometrically in terms of the fundamental matrix H of (8). It is easily seen that X is Hermitian if and only if the graph of X is H-neutral. This is because

$$\begin{bmatrix} I & X^* \end{bmatrix} H \begin{bmatrix} I \\ X \end{bmatrix} = X - X^*.$$

Thus, we are armed with two important ideas and are to find n-dimensional N-invariant subspaces which are also H-neutral. IF such a subspace is also a *graph* subspace then it determines a Hermitian solution of equation (10). However, further conditions are required on the coefficient matrices before the existence of such a subspace can be guaranteed. It turns out that sufficient conditions arise from control theory, and admit a general theory for problems of that kind. Here is an existence theorem of Lancaster and Rodman [9] published in 1980. Generalizations have been published since then. The notion of a *controllable* pair of matrices is needed and we assume the reader is familiar with this (or see [10], for example).

Theorem 3 *Assume that $D \geq 0$ and the pair (A, D) is controllable. Then* (10) *admits Hermitian solutions if and only if there exists an n-dimensional, N-invariant, H-neutral subspace $\mathcal{S}$. Moreover, the formula*

$$\mathcal{S} = \operatorname{Im} \begin{bmatrix} I \\ X \end{bmatrix}$$

establishes a one-to-one correspondence between the set of Hermitian solutions X and the set of n-dimensional, N-invariant, H-neutral subspaces.

As such an invariant subspace $\mathcal{S}$ has a basis of n eigenvectors and generalized eigenvectors from a $2n$ dimensional space, it follows that, generically, the solution set will have cardinality $(2n!)/(n!)^2$. (Thus, Example 4 above is generic.)

Returning to equation (11) observe that, for a solution X of the CARE,

$$\begin{bmatrix} A & D \\ C & -A^* \end{bmatrix} \begin{bmatrix} I \\ X \end{bmatrix} = \begin{bmatrix} I \\ X \end{bmatrix} (A + DX). \tag{13}$$

We see that $(A + DX)$ is the restriction of N to the graph subspace, and must have as its spectrum a corresponding portion of the spectrum of N. Thus, (counting algebraic multiplicities) n of the $2n$ eigenvalues of N are inherited by $A + DX$. The neutral property of the graph subspace demands that, among these n eigenvalues there must be NO pairs which are symmetric with respect to the imaginary axis (recall the symmetry of the spectrum of the Hamiltonian). Furthermore, the description of such sets is easy in the physically significant case when N has no purely imaginary eigenvalues. In particular, the existence of a Hermitian solution is guaranteed in this case.

Furthermore, when N has no purely imaginary eigenvalues and the theorem holds, the invariant subspace of N associated with the right half of the complex plane will generate a solution X_+ for which $A + DX_+$ has its spectrum in the open right half plane. Furthermore, X_+ is the maximal solution in the sense that $X_+ > X$ for all Hermitian solutions of (10). With the same hypotheses, there is also a minimal solution X_- constructed in an analogous way and for which $A - DX_-$ is stable, i.e. has all its eigenvalues in the open left half-plane. This solution is said to be *stabilizing*.

5 The concerns of numerical analysis

First of all, general purpose algorithms for solving the CARE are designed for equations with *real* matrix coefficients and are designed to compute the *stabilizing* solution, when it exists. The algorithms of choice are usually based on so-called "subspace" methods relying on the theoretical ideas developed in the preceding section. Conceivably, this could be done by direct calculation of eigenvectors and, possibly, generalized eigenvectors. However, this strategy presents difficult problems of numerical stability and is generally avoided. A class of more stable algorithms originates with Laub [11]. They are known as Schur methods because they involve reduction to triangular (or quasi-triangular) form by orthogonal similarity - following a technique of Schur from 1909. The use of *orthogonal* similarity is the basis of more reliable numerical performance.

Thus, a real orthogonal matrix $U = [U_1\ U_2]$ is to be found so that

$$[U_1\ U_2]^T \begin{bmatrix} A & D \\ C & -A^T \end{bmatrix} [U_1\ U_2] = \begin{bmatrix} S_1 & S_{12} \\ 0 & S_2 \end{bmatrix}, \tag{14}$$

where U_1 and U_2 are $2n \times n$, and S_1, S_2 are real $n \times n$ quasi-triangular matrices. Their diagonal blocks of size one determine real eigenvalues, and those of size two determine conjugate pairs. The presence of the zero block on the right shows immediately that $\mathrm{Im} U_1$ is N-invariant. One must therefore ensure that the eigenvalues of S_1 are those required to generate the stabilizing subspace (i.e. associated with the eigenvalues of the Hamiltonian in the left half-plane). When this is the case, we write

$$U_1 = \begin{bmatrix} X_1 \\ X_2 \end{bmatrix} \quad \text{and} \quad \begin{bmatrix} A & D \\ C & -A^T \end{bmatrix} \begin{bmatrix} X_1 \\ X_2 \end{bmatrix} = \begin{bmatrix} X_1 \\ X_2 \end{bmatrix} S_1,$$

so that $X = X_2 X_1^{-1}$ (see equation (12)). Notice also that numerical difficulties are to be anticipated when the Hamiltonian has purely imaginary eigenvalues, because they they are necessarily multiple eigenvalues.

Another (more substantial) difficulty is the design of an algorithm which uses orthogonal similarities but also retains the Hamiltonian symmetry at each stage. This can be resolved by further restricting the matrix U to be symplectic, i.e. satisfying $U^T HU = H$. There is a remarkable theorem of Paige and van Loan [14] which has played an important role in this connection. Recall first that a real matrix A has Hamiltonian structure if and only if $(HA) = (HA)^T$. Then, if A is Hamiltonian and U is symplectic, a little manipulation shows that UAU^{-1} is also Hamiltonian.

The next two lemmas are easy preparations for Theorem 4, which takes more proof.

Lemma 1 *A real orthogonal matrix U of size $2n$ is symplectic if and only if*

$$U = \begin{bmatrix} Q_1 & Q_2 \\ -Q_2 & Q_1 \end{bmatrix}$$

where Q_1 and Q_2 are in $\mathbb{R}^{n\times n}$.

Lemma 2 *If A is Hamiltonian and U is orthogonal and symplectic, then UAU^T is Hamiltonian.*

Theorem 4 *If N has no eigenvalues on the imaginary axis, $D^T = D$ and $C^T = C$, then there exists a real matrix U which is both orthogonal and symplectic for which*

$$U^T \begin{bmatrix} A & D \\ C & -A^T \end{bmatrix} U = \begin{bmatrix} R & G \\ 0 & -R^T \end{bmatrix},$$

where all matrices are real, R is upper quasi-triangular, and $G^T = G$.

Work in this direction combined with the matrix pencil formulation continues (see Mehrmann [12] and several subsequent papers).

6 Dilation to matrix pencils

Several problems of control and filtering theory give rise to Riccati equations (10) in which the coefficients D, A, C appear in composite form. Consider the equation

$$XDX + X\hat{A} + \hat{A}^*X - \hat{Q} = 0, \tag{15}$$

with

$$\hat{A} = -A + BR^{-1}C, \quad \hat{Q} = Q - C^*R^{-1}C, \quad D = BR^{-1}B^*, \tag{16}$$

A, Q are $n\times n$, B, C^* are $n\times m$, and R is $m\times m$, Hermitian and nonsingular. The objective now is to re-formulate the problem in such a way as to avoid the inversion of R required in the direct evaluation of the coefficients in (15), and this is done at the expense of the Hamiltonian. This will improve the condition of the problem when R is nonsingular but ill-conditioned. The eigenvalue problem for N is now replaced by an eigenvalue problem for a dilated problem involving a *matrix pencil* rather than the single matrix N. This idea originates with Van Dooren in 1981 [16].

The Hamiltonian is now

$$N = \begin{bmatrix} \hat{A} & D \\ \hat{Q} & -\hat{A}^* \end{bmatrix} = \begin{bmatrix} -A & 0 \\ Q & A^* \end{bmatrix} - \begin{bmatrix} -B \\ C^* \end{bmatrix} R^{-1}[C \ \ B^*],$$

and the right hand side is recognized as a Schur complement. This clue leads one to the dilation (and strict equivalence):

$$\begin{bmatrix} \lambda I - N & 0 \\ 0 & -R \end{bmatrix} = E_1(\lambda G - F)E_2 \tag{17}$$

where

$$G = \begin{bmatrix} H & 0 \\ 0 & 0 \end{bmatrix}, F = \begin{bmatrix} Q & A^* & C^* \\ A & 0 & B \\ C & B^* & R \end{bmatrix},$$

and the nonsingular transforming matrices are

$$E_1 = \begin{bmatrix} 0 & -I & BR^{-1} \\ I & 0 & -C^*R^{-1} \\ 0 & 0 & I \end{bmatrix}, \quad E_2 = \begin{bmatrix} I & 0 & 0 \\ 0 & I & 0 \\ -R^{-1}C & -R^{-1}B^* & I \end{bmatrix}.$$

Strict equivalence is important for us because it preserves the set of all eigenvalues, finite and infinite, and their partial multiplicities (see [4], for example).

Observe that equation (17) allows us to work now with the *Hermitian* pencil $\mu(iG)-F$ where $\mu = -i\lambda$. The symmetry is apparent, but the leading term iG is singular. However, since R is nonsingular, the pencil is *regular* and quite tractable. (By definition, $\lambda G - F$ is *regular* if $\det(\lambda G - F) \not\equiv 0$). Solutions of (15) are now associated with *deflating subspaces* of $\lambda G - F$. (A subspace $\mathcal{S}$ of $\mathbb{C}^{2n}$ is said to be deflating for the pencil $\lambda G - F$ if there is a subspace $\mathcal{T}$ of the same dimension as $\mathcal{S}$ such that $G\mathcal{S} \subseteq \mathcal{T}$ and $F\mathcal{S} \subseteq \mathcal{T}$.)

We will not pursue this line of argument further, but note the more general formulation of the Hamiltonian, and the need to adapt analysis and numerical algorithms to this new formulation.

7 A second algebraic Riccati equation

There is a second nonlinear matrix equation which arises when one considers *difference* equations in the role of the differential equations of Sections 2 and 3. It also appears in the discrete forms of the linear-quadratic regulator problem, and so on. We will treat a simplified form of the equation and sketch the role of the matrix (matrix pencil) analogous to the Hamiltonian in investigations of (10). The simplifed equation in question is:

$$X = A^*XA + Q - A^*XB(R + B^*XB)^{-1}B^*XA, \tag{18}$$

where A and Q are $n \times n$, B is $n \times m$, R is $m \times m$, and Q and R are Hermitian and nonnegative (or positive) definite. The solutions of interest are Hermitian matrices X for which $R + B^*XB$ is nonsingular. To the writer's knowledge, equations of this type have a relatively short history.

A single matrix T playing the role of the Hamiltonian in a closely comparable way can be formulated at the expense of two unfortunate hypotheses: *that both R and A are nonsingular.* In this case we may define

$$T = \begin{bmatrix} A + (BR^{-1}B^*)A^{*-1}Q & -(BR^{-1}B^*)A^{*-1} \\ -A^{*-1}Q & A^{*-1} \end{bmatrix}, \tag{19}$$

and it is easily verified that $T^*HT = H$. In other words, T is unitary with respect to H (of equation (8)). This ensures that the spectrum of T is symmetric with respect to the unit circle in $\mathbb{C}$. If, in addition, T is real, then T is said to be *symplectic* and its spectrum is symmetric with respect to both the unit circle and the real axis.

First, it is reassuring that there is a direct analogue of Proposition 1 plus the H-neutral property of invariant subspaces of T associated with solutions of (18), but it takes a little more proof (see Section 12.2 of [10], for example).

Theorem 5 *Equation* (18) *has an Hermitian solution X if and only if the graph subspace of X is n-dimensional, H-neutral, and T-invariant.*

An analogue of the important Theorem 3 (not in the most general form) guaranteeing the existence of a subspace with the properties decribed in Theorem 5 is now:

Theorem 6 *Assume that $R > 0$, $Q \geq 0$, (A, B) is controllable, and A is invertible. Then equation* (18) *has an Hermitian solution if and only if there is an n-dimensional, H-neutral and T-invariant subspace $\mathcal{S}$. Moreover, the formula*

$$\mathcal{S} = \operatorname{Im} \begin{bmatrix} I \\ X \end{bmatrix}$$

establishes a one-to-one correspondence between Hermitian solutions X and n-dimensional, H-neutral, T-invariant subspaces $\mathcal{S}$.

The proof of a more general form of this result uses the fact that, if we define $N = (T + I)(T - I)^{-1}$, then N is Hamiltonian (see Section 12.2 of

[10]). Also, if

$$T\begin{bmatrix} I \\ X \end{bmatrix} = \begin{bmatrix} I \\ X \end{bmatrix} T_1,$$

then we need to know about T_1, because it tells us about the restriction of T to $\mathcal{S}$. At the expense of some computation it can be shown that $T_1 = A + BZ$, where

$$Z = Z(X) = -(R + B^*XB)^{-1}B^*XA, \tag{20}$$

(cf. equation (13)).

Analogues of the results discussed at the end of Section 4 can now be developed, including the existence of maximal and minimal solutions for equation (18), as well as the existence of a stabilizing solution when T has no unimodular eigenvalues. Here, a solution X is said to be stabilizing when the eigenvalues of $A + BZ$ are all in the open unit disc.

For discussion of the numerical solution of equation (18) we assume (as in Section 5) that all matrices are real. Then matrix T is symplectic, and an analogue of Therem 4 (going from real Hamiltonian to symplectic matrices) is required. This can be obtained from Theorem 4 with an argument using a Cayley transformation (see [14] and Section 7 of [12]).

Theorem 7 *Let T be symplectic and have no unimodular eigenvalues. Then there is a real matrix U which is both orthogonal and symplectic such that*

$$U^TTU = \begin{bmatrix} S & K \\ 0 & (S^{-1})^T \end{bmatrix},$$

where S, $K \in \mathbb{R}^{n\times n}$ and S is upper quasi-triangular.

Again, this result is just the beginning of an extensive literature developed in recent years.

8 Matrix pencils again

The matrix T of (19) is difficult to work with, and the presence of A^{-1} and R^{-1} in its formulation is a serious limitation and may lead to numerical difficulties, even when A and R are both nonsingular. In this section we make two successive dilations from T to matrix pencils. The first admits singular matrices A. In the second we unfold further to remove reference to R^{-1} (an analogue of the process used in Section 4). Consider the $2n \times 2n$ matrix pencil $\lambda F - G$ where

$$F = \begin{bmatrix} I & BR^{-1}B^* \\ 0 & A^* \end{bmatrix}, \quad G = \begin{bmatrix} A & 0 \\ -Q & I \end{bmatrix}, \tag{21}$$

and note that, when A is nonsingular, $F^{-1}G = T$. It is found that

$$FHF^* = GHG^* = \begin{bmatrix} 0 & A \\ -A^* & 0 \end{bmatrix}. \tag{22}$$

To describe the first of these equations, we say that the pencil $\lambda F^* - G^*$ is unitary in the skew-symmetric H-scalar product. This implies that the spectrum is symmetric about the unit circle. The generalized form of Theorem 5 is now:

Theorem 8 *Let the pencil $\lambda F - G$ defined by* (21) *be regular. Then an $n \times n$ Hermitian matrix X is a solution of* (18) *if and only if the graph of X is H-neutral and deflating for the pencil $\lambda F - G$.*

Now consider a further dilation to a pencil $\lambda F_e - G_e$, where

$$F_e = \begin{bmatrix} I & 0 & 0 \\ 0 & A^* & 0 \\ 0 & -B^* & 0 \end{bmatrix}, \quad G_e = \begin{bmatrix} A & 0 & B \\ -Q & I & 0 \\ 0 & 0 & R \end{bmatrix}. \tag{23}$$

When R^{-1} (and hence F) exist, $\lambda F_e - G_e$ is strictly equivalent to a simple dilation of $\lambda F - G$. Thus, if we write $\hat{B} = [0 \; -B^*]$, there exist nonsingular P_1 and P_2 such that

$$P_1 \left(\lambda \begin{bmatrix} 0 & \hat{B} \\ 0 & F \end{bmatrix} - \begin{bmatrix} R & 0 \\ 0 & G \end{bmatrix} \right) P_2 = \lambda F_e - G_e,$$

(see Section 15.2 of [10]). Concerning symmetry, it is easily seen that $\lambda F_e - G_e$ is no longer unitary in any real skew scalar product. However, the *finite* spectrum retains symmetry with respect to the unit circle (and symplectic symmetry when all matrices are real).

We now need to generalize the graph-subspace idea. A subspace $\mathcal{S}$ of $\mathbb{C}^{2n+m}$ is called an *extended graph subspace* if it can be written in the form

$$\mathcal{S} = \operatorname{Im} \begin{bmatrix} I_n \\ X \\ Z \end{bmatrix}$$

for some matrices X and Z with sizes $n \times n$ and $m \times n$, respectively. Such a subspace is necessarily n-dimensional and (as in equation (12)) could also be written in the form

$$\mathcal{S} = \operatorname{Im} \begin{bmatrix} X_1 \\ X_2 \\ X_3 \end{bmatrix}$$

where X_1 is nonsingular. It turns out that, indeed X is a candidate for the solutions set of our second Riccati equation (18) and, furthermore, Z is just the matrix required in establishing the stabilizing property (see equation (20)).

Theorem 9 *Let the pencil $\lambda F_e - G_e$ be regular. Then* (18) *has a solution X if and only if there is a deflating subspace for this pencil of the form*

$$\mathcal{S}_e = \mathrm{Im}\begin{bmatrix} X_1 \\ X_2 \\ X_3 \end{bmatrix}$$

which is an extended graph subspace and $R + B^T X_2 X_1^{-1} B$ is nonsingular. Then $X = X_2 X_1^{-1}$ is a solution of (18), *and $X_3 X_1^{-1} = Z$ of equation* (20).

We conclude by noting another useful strict equivalence, and this one admits the possibility that R is singular (see [5]). Namely, $\lambda F_e - G_e$ and $\lambda M_e - N_e$ are strictly equivalent where, with Z as defined in (20),

$$M_e = \begin{bmatrix} I & 0 & 0 \\ 0 & (A+BZ)^* & 0 \\ 0 & -B^* & 0 \end{bmatrix}, \quad N_e = \begin{bmatrix} A+BZ & 0 & B \\ 0 & I & 0 \\ 0 & 0 & R+B^*XB \end{bmatrix}.$$

Some essential properties of the spectrum of $\lambda F_e - G_e$ are apparent from this result.

9 Concluding remarks

Beginning with some comments on the investigations of Jacopo Riccati, a connection has been made via generalized differential Riccati equations with the pervasive algebraic Riccati equation of (10) and, from there, with (18). In each case there is (under simplifying assumptions) a single block matrix which is (in the real case) either Hamiltonian or symplectic, respectively, whose invariant subspaces can be used to characterize the solutions (if any) of the corresponding nonlinear matrix equation. It has been shown that there are good reasons to replace the eigenvalue problems for these block matrices with eigenvalue problems for matrix pencils which retain the vital symmetries of the simpler problems.

Our comments on the numerical solution of algebraic Riccati equations suggest that the difficult cases numerically are those for which the underlying Hamiltonian or symplectic matrix (or pencil generalization of one of these) has eigenvalues on the imaginary axis, or on the unit circle. This is is indeed the case, and it may be that quite a different line of attack is needed to resolve these problems satisfactorily. One approach is developed by Guo for equations including (18) in [5], and this builds on earlier work dealing with (10) in [6].

References

[1] Atkinson, F.V., *Discrete and Continuous Boundary Problems*, Academic Press, 1964.

[2] Bell, E.T., *Men of Mathematics*, Penguin Books, Harmondsworth, 1953. (First edition, 1937.)

[3] Birkhoff, G.D. and Hestenes, M.R., *Natural isoperimetric conditions in the calculus of variations* Duke Math. Journal, **1**, 1935, 198-286.

[4] Gantmacher, F.R., *The Theory of Matrices*, vols. 1 and 2. Chelsea, New York, 1959.

[5] Guo, C.H., *Newton's method for the discrete algebraic Riccati equation when the closed-loop matrix has eigenvalues on the unit circle*, SIAM Journal on Matrix Analysis and Applications, (to appear).

[6] Guo, C.H. and Lancaster, P., *Analysis and modification of Newton's method for algebraic Riccati equations*, Mathematics of Computation, **67**, 1998, 1089-1105.

[7] Hall, T., *Carl Friedrich Gauss*, MIT Press, Cambridge, Mass., 1970.

[8] Ince, E.L., *Ordinary Differential Equations* Dover, New York, 1956. (First published by Longmans, Green and Co. in 1926.)

[9] Lancaster, P. and Rodman, L., *Existence and uniqueness theorems for the algebraic Riccati equation*, International Journal of Control, **32**, 1980, 285-309.

[10] Lancaster, P. and Rodman, L., *Algebraic Riccati Equations*, Oxford University Press, 1995.

[11] Laub, A.J., *A Schur method for solving algebraic Riccati equations*, IEEE Trans. on Automatic Control, **AC-24**, 1979, 913-921.

[12] Mehrmann, V., *The Autonomous Linear Quadratic Control Problem*, Springer Verlag, Berlin, 1991. (Lecture Notes in Control and Information Sciences, Vol.163.)

[13] Morse, M., *Calculus of Variations in the Large*, American Math. Soc. Colloquium Pubs., vol. 18, New York, 1935.

[14] Paige, C.C. and Van Loan, C.F., *A Schur decomposition for Hamiltonian matrices*, Linear Algebra and its Applications, **41**, 1981, 11-32.

[15] Reid, W.T., *Riccati Differential Equations*, Academic press, New York, 1972.

[16] Van Dooren, P., *A generalized eigenvalue problem approach for solving Riccati equations* SIAM Jour. Scientific and Statistical Computing, **2**, 1981, 121-135.

[17] Watson, G.N., *The Theory of Bessel Functions*, Cambridge University Press, 1958. (First edition, 1922.)

[18] Westfall, R.S., *Never at Rest. A Biograhy of Isaac Newton*, Cambridge University Press, 1980.

[19] Wonham, W.M., *Linear Multivariable Control*, Springer Verlag, Berlin, 1979. (First edition, 1970)

Department of Mathematics and Statistics, University of Calgary, Calgary, AB T2N 1N4, Canada

Progress in Systems and Control Theory, Vol. 25

The Control and Mechanics of Human Movement Systems

Clyde F. Martin and Lawrence Schovanec [1]

1 Introduction

Issues that are central to the modeling and analysis of a human movement system include (1) musculotendon dynamics, (2) the kinetics and kinematics of the biomechanical system, and (3) the relationship between neurological control and the formulation of the system as an open or closed loop process. This paper will address these problems in the context of two particular movement systems. The first to be addressed is the human ocular system. Eye movement systems are ideal for studying human control of movement since they are of relatively low dimension and easier to control than other neuromuscular systems. By scrutinizing the trajectories of eye movements it is possible to infer the effects of motoneuronal activity, deduce the central nervous system's control strategy, and systematically observe the effects of perturbations in the controls. An application of the locomotory-control system will also be presented in this paper. In particular, a model of human gait is developed for the purpose of relating neural controls to the state of stress in a skeletal member. This is achieved by modeling the human body as an ensemble of articulating rigid-body segments controlled by a minimal muscle set. Neurological signals act as the input into the musculotendon dynamics and from the resulting muscular forces, the joint moments and resulting motion of the segmental model are derived. At fixed moments in the gait cycle, the joint torques and joint reaction forces are incorporated into an equilibrium analysis of the segmental elements, modeled as elastic bodies undergoing biaxial bending. Both movement systems that are discussed here emphasize a forward or direct dynamic approach that results in a natural flow of neural-to-muscular-to-movement events while utilizing physiologically realistic models of the musculotendon actuators that faithfully reproduce trajectories and muscle tension.

1.1 Inverse versus Forward Dynamics

If a movement system has n degrees of freedom, then the equations of motion governing the system can be written in the form

$$[M(\theta)]\ddot{\theta} = C(\theta, \dot{\theta}) + G(\theta) + F_m(\theta) \tag{1.1.1}$$

[1] This research was supported by NSF grants ECS-9720357, DMS-9628558 and Texas Advanced Research Program Grant No. 003644-123.

where $\theta, \dot{\theta}, \ddot{\theta}$ are $n \times 1$ vectors of displacement, velocity and acceleration, $[M(\theta)]$ is the $n \times n$ inertia matrix, $C(\theta, \dot{\theta})$ is an $n \times 1$ vector of coriolis and centrifugal terms, $G(\theta)$ is an $n \times 1$ vector of gravitational terms, and $F_m(\theta)$ is an $n \times 1$ vector of applied moments. There are are essentially two approaches to analyzing such a movement system.

Inverse dynamics is an approach that proceeds from known kinematic data and external forces and moments to arrive at expressions for the resultant forces and moments. For this method, motion acts as the input and torques are the output. This approach requires experimental and kinematic data and observed motions. If these variables are assumed to be known functions of time, (1.1.1) becomes an algebraic equation for $F_m(\theta)$. It is a difficult problem however, to determine the forces of the individual muscles that result in an applied moment since there are typically more unknown muscle forces than can be determined from mechanical relations alone. This is referred to as the *redundancy problem.* To address this problem the muscle set that contributes to a specific motion may be reduced by grouping muscles of similar function or by using EMG activity as a guide in determining which muscles were used during a specific movement [4, 20, 24]. If the problem is still indeterminate, a static optimization scheme is employed. For these types of optimization methods the selection of appropriate optimization criterion is somewhat arbitrary and the schemes do not take into account musculotendon dynamics [8, 26, 13]. Consequently, the static optimization approach often results in discontinuous muscle force histories [29]. Another limitation associated with the inverse method is that it is not predictive in nature in that one is limited to studying motion which is produced by monitored subjects.

In contrast, the forward or direct-dynamic approach provides the motion of the system over a given time period as a consequence of the applied forces and given initial conditions. Solution of the forward dynamics problem makes it possible to simulate and predict motion as a result of the forces that produce it. In a forward analysis the torques or the muscle forces that generate the moments are the inputs and the body motion is the output. This relationship is emphasized by writing equation (1.1.1) in the form

$$\ddot{\theta} = [M(\theta)]^{-1}(C(\theta, \dot{\theta}) + G(\theta) + F_m(\theta)).$$

Since neural input activates the muscles, i.e., the actuators of the system, the true input into the system is indeed neural input. Because controls for each muscle are needed, the redundancy problem reoccurs. In the forward approach, muscle dynamics are incorporated into the optimization techniques used to determine the controls. For instance, when human gait is to be simulated, the dynamic optimization methods employ cost functions that usually involve both a tracking error term and a term influencing the distribution of muscle force [9]. Once controls are achieved, the system of differential equations for the body segments and the muscle groups can be integrated forward in time to obtain the motion trajectories. In this sense, a direct-dynamic analysis is self-validating in that the controls specified do indeed result in the observed motion.

A complete development of a direct dynamic model needs to include a representation of the musculotendon complex, anatomical geometry, and kinematic models and inertial characteristics of the underlying movement system. In the next section we provide an overview of musculotendon dynamics. In subsequent sections we utilize these dynamics in developing movement systems that describe ocular motion and human gait. Within the context of these specific applications we will present the relevant geometrical, kinematical, and inertial information.

2 Musculotendon Dynamics

2.1 Functional Properties

Muscles are the actuators of the neuro-musculo-skeletal control system that produce movement. In the analysis of a control system as complex as that governing movement it is essential to have a clear understanding of the physical nature of the actuators and a tractable mathematical representation of their dynamics. The muscle models utilized in this investigation are referred to as Hill-type models. A phemenological representation of the musculotendon complex as idealized mechanical objects is presented in Figure 1. This model has been shown to incorporate enough complexity while remaining computationally practical.

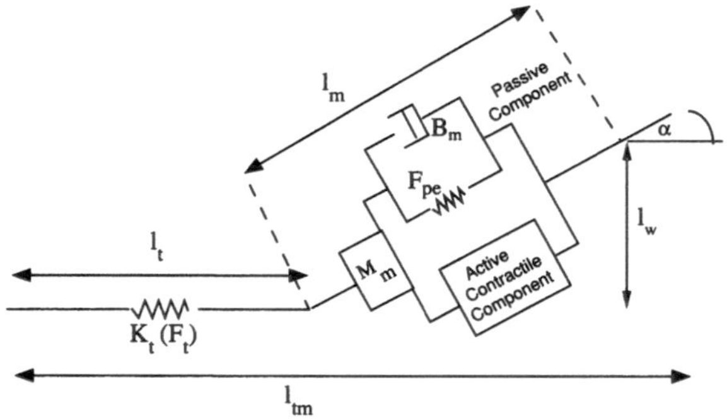

Figure 1. Hill Type Model of the musculotendon complex.

The muscle of length l_m is in series and off-axis by a pennation angle α with the tendon of length l_t. The total pathlength of the musculotendon complex is denoted by l_{tm}. The muscle is assumed to consist of two components: an active force generator and a parallel passive component. The passive component includes a parallel elastic element (F_{pe}) that describes the passive muscle elasticity and a damping component which corresponds to the passive muscle viscosity (B_m). The model for the active contractile component is based on the generally accepted notion that the active muscle force is the product of three factors: (1) a length-tension relation $f_l(l_m)$, (2) a velocity-tension relation $f_v(\dot{l}_m)$, and (3) the activation level $a(t)$. In this paper, the curves utilized in the modeling of these relations are developed by two methods. In the case that sufficient data is available, a natural cubic spline will be fit to the data. As an alternative approach, analytical expressions that capture the qualitative properties of the curves will be used. The parameters that appear in these expressions will be determined by imposing smoothness conditions in combination with a fit of experimental data.

For multiple muscle systems, such as that used in the simulation of human gait, it is advantageous to develop curves describing the attributes of a generic muscle. This curve can then be scaled with appropriate parameters to reflect the dynamics of a particular muscle. We will see that the scale parameters needed for each musculotendon group include: (1) maximal isometric active muscle force F_o, (2) optimal muscle length, l_o, (3) pennation angle α_o when $l_m = l_o$, and (4) tendon slack length l_{ts}. In developing nondimensional representations for these curves the approach of [31] is implemented and all forces and lengths are scaled as $\tilde{F} = F/F_o$ and $\tilde{l}_m = l_m/l_o$. Another quantity used to specify a muscle specific force-velocity relation is the maximum speed of shortening defined as $v_o \equiv l_o/\tau_c$. This quantity scales time and varies for fast and slow muscles. In the case of the lower extremities a standard value of $\tau_c = .1s$ is used for all muscles types.

Muscle force is easily measured at various lengths under isometric conditions to produce force-length relationships. The curve produced when muscle is not stimulated is the passive force-length curve, $F_{pe}(l_m)$. When muscle is activated the curve that results represents both passive and active contributions. The difference in these two curves is the active force-length relation, $f_l(l_m)$. The length at which the maximum active muscle force, F_o, is developed is called the optimal muscle length, l_o. Figure 2 displays the qu

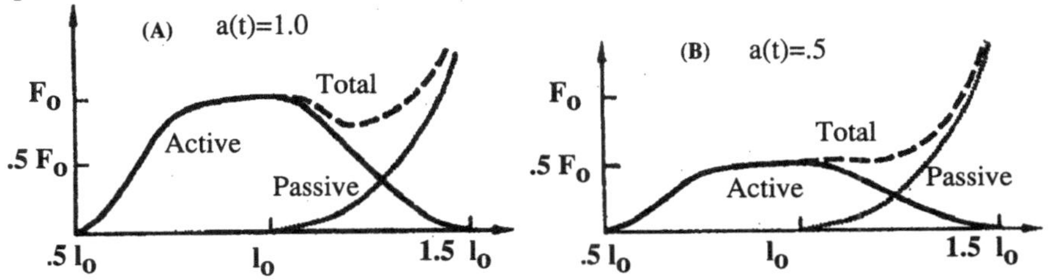

Figure 2. Isometric Force-length Relation for Muscle: (A) Full activation. (B) Active force scales with activation but passive is force is unaffected (adapted from [31]).

A theoretical explanation for the active f_l relation is based on the microscopic nature of muscle and is explained by the Sliding Filament Theory [23]. This theory offers an explanation for the generally accepted notion that when a muscle is completely tetanized, the active force displays a parabolic dependence on length in a nominal region, $.5 \leq \tilde{l}_m \leq 1.5$ with a maximum value of F_o when $l_m = l_o$. At less than full activation, the force-length dependence is obtained by scaling the fully activated f_l curve [31]. In this paper, the active force length relation for muscles of the lower extremities are constructed as a natural cubic spline that fits data reported by [10]. This curve is then scaled to provide a description for specific muscle. For the ocular system, an analytical model of the force length effect is utilized. Several approaches have been suggested, for example [14], but a simple normalized form that is utilized here is

$$f_l(\bar{l_m}) = F_o\left(1 - ((\bar{l_m} - 1)/0.5)^2\right).$$

The nonlinear passive dependence of muscle force on length is described by the function $F_{pe}(l_m)$. Just as in the case of the active force length, the passive force length curves are constructed as cubic splines or modeled by analytical expressions. A commonly used form for F_{pe} and that which is employed here is given in [15] and is expressed as

$$F_{pe}(l_m) = \begin{cases} \left(\frac{k_{ml}}{k_{me}}\right)\left[\exp(k_{me}(l_m - l_{ms})) - 1\right] & l_{ms} \leq l_m < l_{mc} \\ k_{pm}(l_m - l_{mc}) + F_{mc} & l_m > l_{mc} \\ 0 & \text{otherwise.} \end{cases} \qquad (2.1.1)$$

The passive muscle slack length is l_{ms} and corresponds to a length at which no force is generated. The transition length from the linear to nonlinear regime is l_{mc} corresponding to a force of F_{mc}. The specific methods by which these parameters as well as the stiffness and shape parameters k_{me}, k_{pm}, k_{ml} are determined from data relevant to a specific application.

Active muscle force is also dependent on muscle velocity. When a muscle actively shortens, it produces less force than it would under isometric conditions. A.V. Hill [17] was the first to quantify this result with an empirical hyperbolic relationship when a muscle is shortening as

$$f_v(v_m) = F_o\frac{v_o - v_m}{v_o + cv_m}$$

where $v_m = \dot{l_m}$ and shortening corresponds to $v_m > 0$. In contrast to a concentric contraction, when a muscle is actively lengthening it is able to produce forces above the maximal isometric force. Experimental data reveals that this relationship is not an extension of Hill's equation and exhibits a threshold which limits the amount of tension muscle can withstand, approximately $1.8F_o$. The f_v curve is also thought to scale with activation and is displayed in Figure 3. For the muscles involved in the gait model the force velocity curve is constructed as a natural cubic spline fit to data collected while the muscle lengthened and shortened [10]. For the ocular system, an analytical model of f_v is utilized. Due to the way the curve is utilized in the formulation of the dynamics, it is convenient to express this relation in terms of the inverse $f_v^{-1}(F)$. The output of the inverse is normalized with respect to v_o, the maximum speed of shortening. In particular if $\tilde{F}$ denotes a normalized active muscle force,

$$\tilde{v}^m = v^m/v^o = f_v^{-1}(\tilde{F}).$$

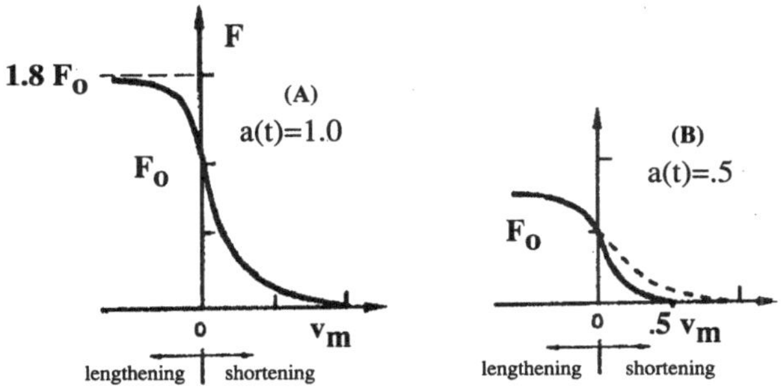

Figure 3. Force-velocity relation for muscles: (A) Full activation when $l_m = l_{mo}$. (B) The force-velocity scales with activation.

There is evidence to suggest that total active force generation is best described by a force-length-velocity relationship which is usually quantified as the product of the force-length and force-velocity curves [31] where the resulting surface is scaled by muscle activation. Consequently, it is convenient to visualize force generation as a collections of surfaces described in terms of nondimensionalzed force velocity and force length curve,

$$F_{act} = a(t)F_0 f_l(\tilde{l_m}) f_v(\tilde{v_m}).$$

Muscle activation, $a(t)$, is related to the neural neural input, $n(t)$, by a process known as contraction dynamics. Both quantities, $n(t)$ and $a(t)$, can be related to experimental data. In particular $n(t)$ is related to rectified EMG while $a(t)$ is related to filtered, rectified EMG [31]. The process through which neural input is transformed into activation is known to be mediated through a calcium diffusion process and is represented by the first order differential equation

$$\frac{da(t)}{dt} + \left[\frac{1}{\tau_{act}} \; (\beta + (1-\beta)u(t))\right] \; a(t) = \frac{1}{\tau_{act}} \; u(t)$$

where $0 < \beta < 1$ and τ_{act} is an activation time constant that varies with fast and slow muscle.

The series elastic element in Figure 1 corresponds to the muscle tendon. More precisely, the tendon is assumed to behave non-linearly under minimal extension and then to become linear with stiffness constant k_s beyond a given length l_{tc} associated with a particular level of resisting force, F_{tc}. A common approach to tendon dynamics (see, for example [15]) is to assume a model of the form

$$\dot{F}_t = K_t(F_t)\dot{l}_t \tag{2.1.2}$$

where

$$K_t(F_t) = \begin{cases} k_{te}F_t + k_{tl} & 0 \leq F_t < F_{tc} \\ k_s & F_t \geq F_{tc} \end{cases} .$$

By integrating the above equation the tendon force can be alternatively expressed as a function of tendon length l_t as

$$F_t(l_t) = \begin{cases} \left(\frac{k_{tl}}{k_{te}}\right) \left[\exp\left(k_{te}(l_t - l_{ts})\right) - 1\right] & l_{ts} \leq l_t < l_{tc} \\ k_s(l_t - l_{tc}) + F_{tc} & l_t > l_{tc} \\ 0 & \text{otherwise} \end{cases} \tag{2.1.3}$$

where l_{ts} denotes tendon slack length. By imposing smoothness conditions on this curve and some notion of fit to experimental data, the shape and stiffness parameters can be specified. However, for the gait model it is convenient to use a generic force-length relationship for tendon derived by a method discussed in [31]. In particular, define tendon strain by $\varepsilon^t = (l_t - l_{ts})/l_{ts}$ and normalized tendon force as $\tilde{F}_t = F_t/F_o$. We assume a generic force-strain curve ($\tilde{F}_t$ vs ε^t) based on the assumptions that a nominal stress-strain curve can be formulated that represents all tendon and that the strain in a tendon when force in the tendon equals the maximal isometric muscle force is independent of the musculotendon unit. By scaling the generic force-strain relationship by F_o and l_{ts}, a force-length function is found for a specific tendon. If we adopt the notion that the tendon behaves as an exponential spring and fit an analytical model as in equation (2.1.3) to data reported by [10], a generic force-strain relationship may be obtained in the form

$$F_t(\varepsilon^t) = \begin{cases} .10377\left(e^{91\varepsilon^t} - 1\right) & 0 \leq \varepsilon^t < .01516 \\ 37.526\varepsilon^t - .26029 & .01516 \leq \varepsilon^t < .1 \end{cases} \tag{2.1.4}$$

If the strain in tendon reach values beyond .1, the tendon is known to rupture [31]. Since such an extreme value of strain should not occur during normal locomotion, this part of the curve need not be included in our analysis. From (2.1.4) it follows that tendon stiffness defined by $K_t(F_t) = dF_t/dl_t$ is given by

$$\begin{aligned} K_t(F_t) &= \frac{dF_t}{d\tilde{F}_t} \cdot \frac{d\tilde{F}_t}{d\varepsilon^t} \cdot \frac{d\varepsilon^t}{dl_t} \\ &= \frac{F_o}{l_{ts}} \cdot \frac{d\tilde{F}_t}{d\varepsilon^t} \\ &= \begin{cases} \left(\frac{F_o}{l_{ts}}\right) 91(\tilde{F}_t + .10377) & 0 \leq \tilde{F}_t \leq .3086 \\ \left(\frac{F_o}{l_{ts}}\right) 37.526 & \tilde{F}_t \geq .3086 \end{cases} \\ &= \left(\frac{F_o}{l_{ts}}\right) K_t(\tilde{F}_t). \end{aligned} \tag{2.1.5}$$

2.2 Contraction Dynamics

From Figure 1 it readily follows that the total force of a muscle is the sum of the passive and the active forces, $F_m = F_{pe} + F_{act} + B_m \dot{l}_m$. Muscle is known to maintain a constant volume and so l_w is constant. With this observation and since

$$l_{mt} = l_t + l_m \cos \alpha,$$

it follows

$$\dot{\alpha} = -\frac{\dot{l}_m}{l_m} \tan \alpha.$$

and

$$\dot{l}_t = \dot{l}_{mt} - \frac{\dot{l}_m}{\cos \alpha}.$$

The tendon dynamics may now be expressed as

$$\dot{F}_t = K_t(F_t)(\dot{l}_{tm} - \dot{l}_m / \cos \alpha). \tag{2.2.6}$$

The equation of motion for the muscle mass is

$$M_m \frac{d^2(l_m \cos \alpha)}{dt^2} = F_t - [F_{act} + F_{pe} + B_m \dot{l}_m] \cos \alpha$$

and with some simple manipulations, the muscle dynamics take the form

$$M_m \ddot{l}_m = F_t \cos \alpha - \cos^2 \alpha [F_{act} + F_{pe} + B_m \dot{l}_m] + \frac{M_m \dot{l}_m^2 \tan^2 \alpha}{l_m}. \tag{2.2.7}$$

Two state variables are required to describe the contraction dynamics of the musculotendon actuator as given by (2.2.6) and (2.2.7). For multiple muscle systems such as that needed to describe gait, it is desirable to reduce the system dimension. This can be achieved by eliminating the muscle mass. In this case, (2.2.7) becomes a statement of force balance

$$F_t = F^m \cos\alpha = \left(F_o a(t) f_l(\tilde{l}_m) f_v(\tilde{v}^m) + F_{pe}(\tilde{l}_m) + B_m \dot{l}_m\right) \cos\alpha.$$

If it is assumed that the passive muscle viscosity effect is small, then $\dot{l}_m$ can be computed from

$$\dot{l}_m = v^m = v_o f_v^{-1}\left(\frac{(F_t/\cos\alpha) - F_{pe}(\tilde{l}_m)}{F_0 a(t) f_l(\tilde{l}_m)}\right).$$

It now follows from (2.2.6) and that the differential equation describing the contraction dynamics of the musculotendon is

$$\frac{dF_t}{dt} = \frac{F_0}{l_{ts}} K^t(\tilde{F}_t)\left[\dot{l}_{mt} - \frac{v_o}{\cos\alpha} f_v^{-1}\left(\frac{(F_t/\cos\alpha) - F_0 f_p(\tilde{l}_m)}{F_0 a(t) f_l(\tilde{l}_m)}\right)\right] \tag{2.2.8}$$

where

$$v_o = (1/\tau_c) l_0$$

$$\tilde{l}_m = l_m/l_0 = \sqrt{(l_{mt} - l_t)^2 + (l_w)^2}/l_0,$$

$$l_t = \begin{cases} l_{ts}\left(1 + \ln\left(\tilde{F}_t/.10377 + 1\right)/91\right) & 0 \le \tilde{F}_t \le .3086 \\ l_{ts}\left(1 + (\tilde{F}_t + .26029)/37.526\right) & .3086 \le \tilde{F}_t \end{cases},$$

$$\cos\alpha = (l_{mt} - l_t)/l_m,$$

$$\tilde{F}_t = F_t/F_0.$$

3 Ocular Dynamics

In human binocular vision the movement of each eye is controlled by a set of six muscles. When the eyes are fixed on an object two things occur. First, the eye rotates so that the image of the object formed by the eyes' lens system is projected onto the fovea of the retina. This is the area of the retina of greatest ocular acuity. Secondly, the eye lenses adjust to bring the object into focus. This section is concerned with the first part of this process: how the rotation develops and how it is controlled. The brain and central nervous system process information obtained by the retina, and then transmit signals to the extraocular muscles. These muscles work in three agonist-antagonist pairs to exert forces on the eye causing it to rotate.

The three muscle pairs consist of the medial and lateral recti, the superior and inferior recti, and the superior and inferior obliques. The lateral and medial recti produce primarily horizontal rotation. The superior and inferior recti work mainly to control vertical rotation. The superior oblique controls the intorsional rotation (toward the nose) of the eye while the inferior oblique controls mainly extorsional rotation (toward the temple). The eye has three degrees of freedom but experimental evidence shows that for any horizontal rotation (θ) and vertical rotation (ϕ), the amount of torsional rotation (ψ) is determined by a phenomena known as Listing's Law,

$$\psi = \arccos\left(\frac{\sin\theta\sin\phi}{1+\cos\theta\cos\phi}\right).$$

The motion of the eye is a result of moments produced by the six extraocular muscles and a passive moment due to the orbit which restrains the rotation of the globe. If ω_x, ω_y, ω_z denote the components of angular velocity with respect to a fixed inertial reference frame with coordinates x, y, z, then the equation of motion for the globe is of the form

$$\begin{pmatrix} \dot{\omega_x} \\ \dot{\omega_y} \\ \dot{\omega_z} \end{pmatrix} = \frac{1}{J_G}\Big[(\vec{r_{lr}} \times \vec{F_{lr}}) + (\vec{r_{mr}} \times \vec{F_{mr}}) + (\vec{r_{sr}} \times \vec{F_{sr}}) \tag{3.0.1}$$

$$+ (\vec{r_{ir}} \times \vec{F_{ir}}) + (\vec{r_{so}} \times \vec{F_{so}}) + (\vec{r_{io}} \times \vec{F_{io}}) + (\vec{rp} \times \vec{F_p})\Big].$$

Here $\vec{r_{lr}}$ and$\vec{F_{lr}}$ denote the moment arm and force associated with the lateral rectus, with the obvious interpretation of the other terms. Three dimensional simulations carried out in [22] support the claim that horizontal eye movement may be accurately modeled by including only the medial and lateral rectus muscles. We will assume that the points of attachment are such that the moment these muscles generate is in the direction of the z axis (see Figure 4).

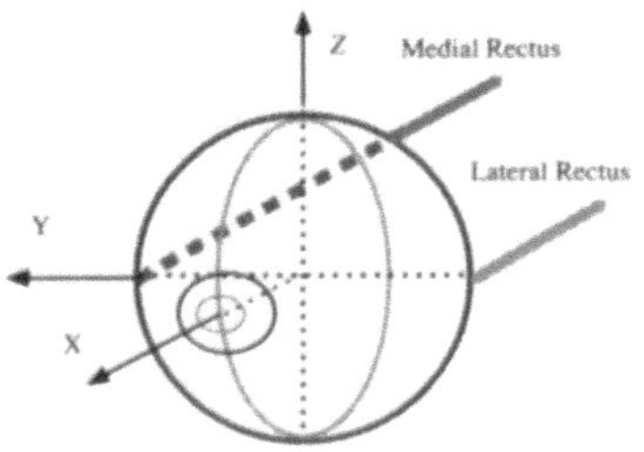

Figure 4. Left globe with medial and lateral rectus muscles.

More specifically, the model presented here will be restricted to horizontal saccadic eye movements. The purpose of saccadic eye movement is to position the high-resolution fovea, the central part of the retina, on the important features of a scene. This is achieved by altering the direction and magnitude of the saccade so as to correct for position error. Because saccades are among the fastest muscle movements that are performed, vision is suppressed during this motion and it is generally accepted that the the oculomotor system operates in an open-loop mode when performing a saccadic movement. The model proposed herein builds upon earlier models of the oculomotor system [2, 6]. However, the model presented here incorporates more general formulations of the musculotendon dynamics as presented in the previous section and allows for the inclusion of muscle mass. There are several reasons that a meaningful model of the horizontal saccadic movement system can be developed. Experimental data has been gathered that provides values for many of the parameters that arise in the description of the model. Most importantly in this regard is information regarding the time course of innervation that is specific to saccadic movement. Experimental data provides fairly conclusive information regarding the nature of the inputs, or controls, that are appropriate to saccades. More specifically, the input into the oculomotor plant is derived from experimental records of raw EMG corresponding to the firing rate of motor units. It is accepted as 'sufficiently realistic as to be useful' [25] that the rate of discharge corresponding to a saccadic movement can be described by filtered EMG that corresponds to a rectangular pulse followed by a step.

Figure 5 provides a mechanical representation of the eye plant model in which the elements of the plant are displayed as if the muscles and the globe are undergoing linear motion. One must distinguish between the lateral rectus, (the agonist) and the medial rectus (the antagonist). All model components that pertain to the agonist will be denoted as $F_{t1}, l_{tm1}, B_{m1}, M_{m1}$, etc, while the corresponding quantities for the agonist are indicated by $F_{t2}, l_{tm2}, B_{m2}, M_{m2}$, etc.

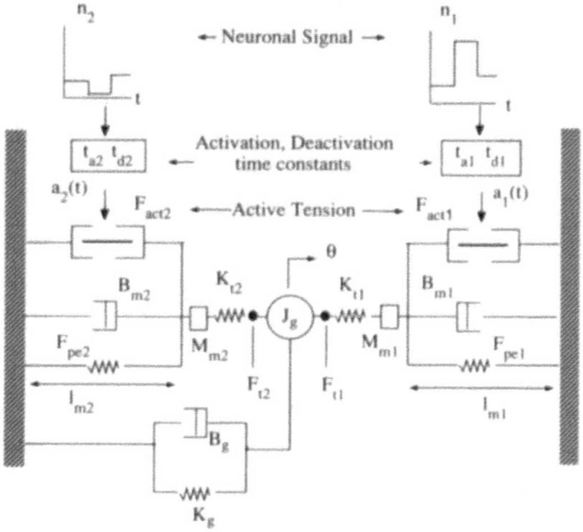

Figure 5. The model of the eye plant.

Because of conventions adopted in the ocular literature, it is convenient to report rotations of the eye in degrees (°) and to write the equation of motion for the globe and the musculotendon dynamics in terms of pure tendon force in units of grams tension. To this end let J_G, B_G, and K_G denote the parameters for globe inertia, globe viscosity, and globe elasticity, respectively, in terms of cgs units and radians. Let r denote the radius of the globe and define

$$J_g = \frac{J_G}{980r(180/\pi)}, \quad \text{in gt} \cdot \text{s}^2/°$$

and similarly for B_g, K_g. If F_{t1} and F_{t2} are reported in gt, then equation (3.0.1), when expressed in degrees and units of gt, becomes

$$J_g\ddot{\Theta} + B_g\dot{\Theta} + K_g\Theta = F_{t1} - F_{t2}.$$

If pennation effects are ignored, the equation of motion for the muscle mass of the agonist and antagonist, previously expressed in (2.2.70, now takes the form for $i = 1, 2$

$$\frac{M}{980}\ddot{l}_{mi} + B_{pm}\left(\frac{180}{\pi r}\right)\dot{l}_{m1} = F_{ti} - F_{acti} - F_{pe1}(l_{mi}).$$

In a similar fashion, the tendon dynamics and the description of the passive muscle elasticity must be amended to account for the change in units. In particular, the modified forms of (2.1.1),(2.1.2) are

$$F_{pe}(l_m) = \begin{cases} \left(\frac{k_{ml}}{k_{me}}\right)\left[\exp(k_{me}\left(\frac{180}{\pi r}\right)(l_m - l_{ms})) - 1\right] & l_{ms} \leq l_m < l_{mc} \\ k_{pm}\left(\frac{180}{\pi r}\right)(l_m - l_{mc}) + F_{mc} & l_m > l_{mc} \\ 0 & \text{otherwise} \end{cases}$$

and

$$\dot{F}_t = K_t(F_t)\left(\frac{180}{\pi r}\right)\dot{l}_t$$

with the obvious change in (2.1.3). For the agonist, the tendon dynamics become,

$$\dot{F}_{t1} = K_t(F_{t1})\left[-\dot{\Theta} - \left(\frac{180}{\pi r}\right)\dot{l}_{m1}\right]$$

while for the agonist,

$$\dot{F}_{t2} = K_t(F_{t2})\left[\ \dot{\Theta} - \left(\frac{180}{\pi r}\right)\dot{l}_{m2}\right].$$

If a state vector is selected as

$$\mathbf{x}^T(t) = [\ \Theta \quad \dot{\Theta} \quad l_{m1} \quad \dot{l}_{m1} \quad l_{m2} \quad \dot{l}_{m2} \quad F_{t1} \quad F_{t2} \quad a_1 \quad a_2\ ],$$

the state equations for the system are given by

$$
\dot{\mathbf{x}}(t) = \begin{bmatrix}
x_2 \\
\frac{1}{J_g}(x_7 - x_8 - B_g x_2 - K_g x_1) \\
x_4 \\
\frac{980}{M}(x_7 - F_{act}(x_3, x_4, x_9) - F_{pe}(x_3) - B_{pm}(\frac{180}{\pi r})x_4) \\
x_6 \\
\frac{980}{M}(x_8 - F_{act}(x_5, x_6, x_{10}) - F_{pe}(x_5) - B_{pm}(\frac{180}{\pi r})x_6) \\
K_t(x_7)\left[-x_2 - \left(\frac{180}{\pi r}\right)x_4\right] \\
K_t(x_8)\left[\ x_2 - \left(\frac{180}{\pi r}\right)x_6\right] \\
\frac{1}{\tau_1(t)}[n_1(t) - x_9] \\
\frac{1}{\tau_2(t)}[n_2(t) - x_{10}]
\end{bmatrix}
$$

The simulations illustrated in Figure 6 originate from the primary position and the initial conditions for muscle length, tendon force, and activation are taken from [25]. The vector of initial data is then $\mathbf{x}^\mathsf{T}(0) = [0\ 0\ 4\ 0\ 4\ 0\ 20\ 20\ .17\ .17]$.

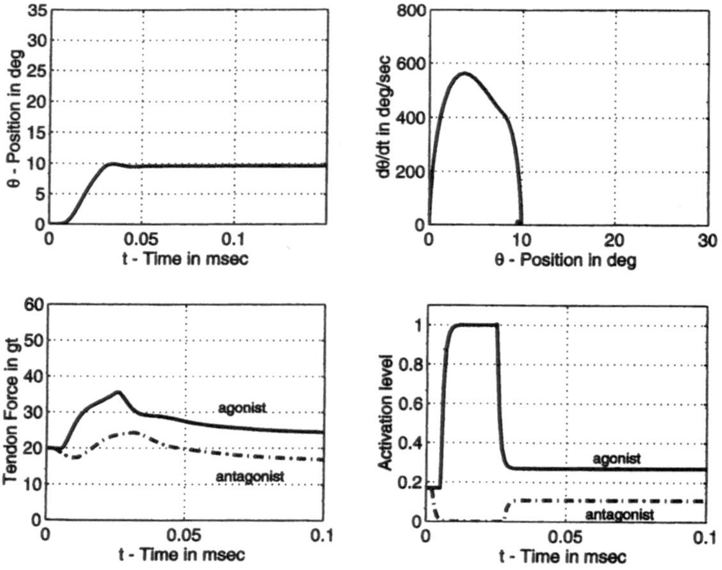

Figure 6. Model simulation of 10 degree saccade.

The simulated trajectory, velocity, and tendon force are in close agreement with the empiric data of [7, 25]. The results of Figure 7 illustrate the effects of pulse-width and pulse-height mismatch in glissadic overshoot and undershoot. The results show that the pulse-height errors are associated with unusually large peak velocities. In contrast, pulse width mismatches produce saccades with reasonable predictions of velocities. These results are in qualitative agreement with experimental results of [3] which suggest that when glissadic overshoot is associated with low peak velocities, the error is caused by the erroneous neural input in computing the pulse width, not the height.

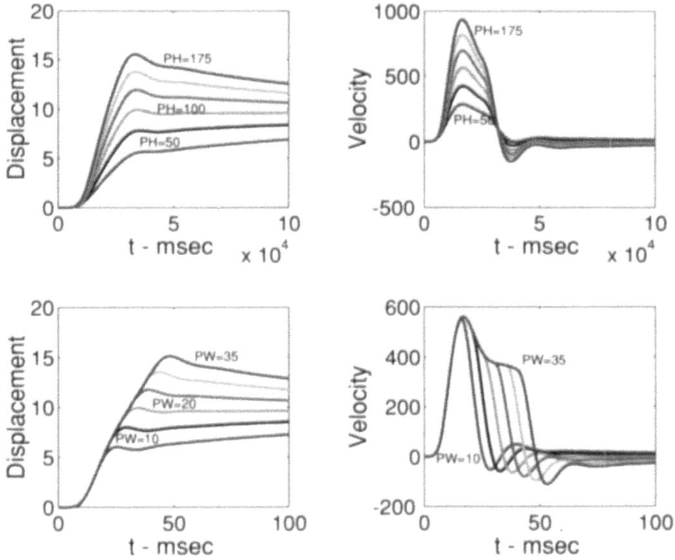

Figure 7. Glissadic overshoot due to pulse width and height errors.

4 A Forward Dynamic Model of Gait

In order to generate the loading conditions during gait, it is necessary to develop a model driven by musculotendon actuators that simulates normal gait. Although many gait models have been built, few include the complexity which is needed for realistic dynamic simulations. The approach that is adopted here builds on a model developed by [30]. The model constrains seven rigid-body segments which represent the feet, shanks, thighs and trunk to 8 degrees of freedom. All joints are assumed to be revolute having one degree of freedom with the exception of the stance hip which has two degrees of freedom. This allows the hip to ab/adduct, a condition which reduces the degree of coupling between the trunk and the swing leg. Figure 8 shows the generalized coordinates used to describe the configuration of the body. Joint angles q_1, q_2, and q_3 are measured with respect to the horizontal or transverse plane and the rotation of these joints occur about an axes parallel to $\vec{n}_2$. Movement of the stance leg is confined to the sagittal plane, but due to the extra degree of freedom granted to the stance hip, the swing leg and trunk can also move in the frontal or coronal plane through pelvic list. Joint angle q_4 tilts the trunk as well as the swing leg about an axis parallel to $\vec{n}_1 = \vec{d}_1$. Joint angles q_5 through q_8 are measured in the tilted plane and these rotations occur about an axis parallel to $\vec{d}_2$.

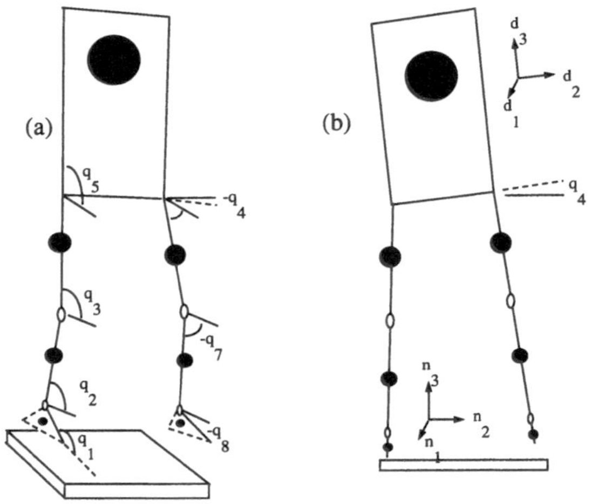

Figure 8. 3-D, 8 DOF Model Showing Segment Angle Definitions: (a) The stance hip has two DOF, while all other joints are revolute. (b) Front view showing pelvis list. Stance angles are specified with respect to the inertial frame, $\vec{n}$, while the swing angles are respect to the titled trunk reference frame, $\vec{d}$ (Adapted from [30]).

This musculoskeletal model represents a normal male with mass totaling 76 kilograms. Segmental dimensions and inertial parameters used in the model are presented in [11]. A key element of the model is that only one-half (14%, approximately left-toe-off, to 62%, approximately left-foot-flat) of the gait cycle is simulated in this analysis. The complete gait cycle can be reconstructed under the assumption of bilateral symmetry. This assumption simplifies the modeling in that the stance leg is always the stance leg and the swing leg is always the swing leg. As a result, muscles in the swing and stance leg can vary according to the function of that leg in the simulation. With this assumption, it is also valid to have the stance toe constrained to the ground which eliminates one more degree of freedom. This simplification requires fewer muscles to be modeled and reduces the system dimension.

The range of each joint angle is limited to normal ranges through the use of ligamentous constraints. If the joint angle stays within a nominal range, then the effects of the passive structure are minimal, but when the nominal range is exceeded, the passive torques grow exponentially. The general form of the passive moments is given by

$$M_{pass} = k_1\, e^{-k_2(\theta-\theta_2)} - k_3\, e^{-k_4(\theta_1-\theta)} - c\,\dot{\theta}.$$

Here θ is the joint angle measured in radians, $\dot{\theta}$ is the joint velocity measured in radian per second and M_{pass} is measured in Newton-meters. Note that $\theta_2 < \theta < \theta_1$ represents a nominal range for that joint. Initial estimates of the parameters k_j, θ_j, and c were taken from [9] and then modified in the course of validating the model. The inclusion of the passive term $(-c\,\dot{\theta})$ is vital since the passive viscosity was excluded from the the musculotendon model when mass was eliminated. A list of the passive parameters and definitions of the joint angles in terms of the generalized coordinates is given in [11].

Although the stance toe is constrained throughout the simulation, it is necessary to incorporate additional constraints in order to prevent the stance heel and swing foot from penetrating the "ground" and to eliminate excess sliding of the swing foot during double support. These additional constraints are considered to be soft [16]. The vertical ground reaction forces which act on the heels of both the stance and swing leg are modeled as highly-damped, stiff linear springs

$$F_{normal} = \begin{cases} 0 & zheel \geq 0 \\ -(1.5 \times 10^5) zheel - (1 \times 10^3) zheel & zheel < 0 \end{cases}$$

where $zheel$ is the height of the heel above the ground. On the swing heel an additional frictional force applied in a direction parallel to the ground in the sagittal plane is used to prevent excess sliding. This force is proportional to the normal force applied to the swing heel,

$$F_{friction} = \begin{cases} 0 & zheel_{swing} \geq 0 \\ -.5|F_{normal}(zheel_{swing})| & zheel_{swing} < 0 \end{cases}.$$

Once flat-foot is achieved by the swing leg, a counterclockwise torque is applied to prevent the foot from penetrating the ground. The torque is model as a damped, torsional spring which resists plantarflexion of the foot. This torque is given by

$$T = \begin{cases} 0 & zheel_{swing} > 0 \text{ or } \delta \geq 0 \\ -5696\delta - 38.19\dot{\delta} & zheel_{swing} \leq 0 \text{ and } \delta < 0 \end{cases}$$

where δ is the angle the bottom of the foot makes with the ground. These soft constraints allow the same model to be used during the single and double support phases of gait. If the ground had been modeled as a "hard" constraint, then one would lose a degree of freedom on the swing leg when the toe touches the ground. This would mean two models would be needed to simulate this phase of gait and a switch between the two systems would occur at heel strike.

The model that is developed here incorporates the same muscle sets as that utilized in [30]. As a result, ten musculotendon units are incorporated into our model, five on each leg. On the stance leg, the relevant muscle groups are the soleus, gastrocnemius, vasti, gluteus medius and minimus, and the iliopsoas. The swing leg utilizes the dorsiflexors, hamstring, vasti, gluteus medius and minimus and the iliopsoas. Thus seven different musculotendon groups need to be specified in this model. The constituent muscles composing each musculotendon group, and the parameters which are used to distinguish the dynamics of each particular musculotendon unit, that is, tendon slack length, optimal muscle length, maximal isometric force and pinnation angle, are listed in [11].

In order to incorporate the musculotendon actuators into the dynamics, it is necessary to place the muscles geometrically on the body segment so that the length of the musculotendon, l_{mt}, and the velocity of the musculotendon, v_{mt}, can be derived as a functions of the state variables, i.e. the generalized coordinates, q_i. The attachment of the musculotendon to bone is specified by the defining an origin or proximal attachment, and an insertion or distal attachment. Effective origins and effective insertions are specified when the straight path from the actual origin to actual insertion passes through bone or out of anatomical range during certain body configurations. Origins and insertions are specified with respect to coordinate systems which are directed along the bones and are fixed with respect to the foot, shank, thigh and pelvis. The origin and insertion points for the 7 muscle groups used in this analysis were based on from [5, 18] and are summarized in [11]

The total length and velocity of the musculotendon complex, which is needed as input into the musculotendon dynamics, can be derived for most muscles through vector addition. In this case, the length of the musculotendon is given by

$$l_{mt} = |\overrightarrow{O_a O_e}| + |\overrightarrow{O_e I_e}| + |\overrightarrow{I_e I_a}|$$

where O and I refer to origin and insertion, and the subscripts e and a refer to actual and effective coordinates (Figure 9). The velocity of the musculotendon is then just the time derivative of the length. Musculotendon forces were incorporated into the body dynamics as joints torques. When a musculotendon spans a joint a torque is realized across that joint. The joint torque due to musculotendons which do not span the knee are computed using standard vector cross product methods. In this case, the moment $\overrightarrow{M}$ acting on the proximal segment is defined as

$$\overrightarrow{M} = \pm F^t \left(\vec{p} \times \frac{\overrightarrow{O_e I_e}}{|\overrightarrow{O_e I_e}|} \right). \tag{4.0.1}$$

Here $\vec{p}$ is a vector from the joint to a point on the line of action of the musculotendon. The line of action is defined by the unit vector $\overrightarrow{O_e I_e}/|\overrightarrow{O_e I_e}|$, and the tension in the musculotendon complex, F^t, is produced according to the muscle dynamics (2.2.8). The sign in equation (4.0.1) depends on whether the musculotendon acts to extend or flex the joint it spans. The cross product,

$$\overrightarrow{m_e} = \vec{p} \times \frac{\overrightarrow{O_e I_e}}{|\overrightarrow{O_e I_e}|},$$

represents the effective moment arm associated with the musculotendon at a certain body configuration.

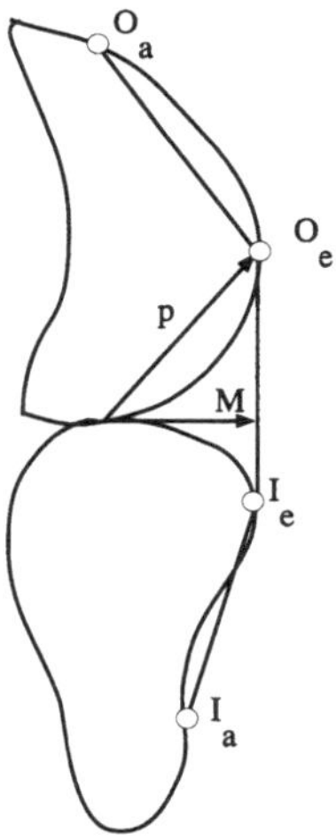

Figure 9. Muscle pathway and effective moment arm.

Vector addition works well for all muscles which do not span the knee. The complication which arises for those musculotendons which do span the knee (vasti, hamstring, and gastrocnemius) occurs because of the complexity of the knee joint. In short, as the knee flexion angle varies, this produces both a change in the location of the knee joint center and in the location of the patella. Since this complexity is not accounted for in the simple segmental model formulated here, an alternative to the vector method is implemented for the knee. We define the effective moment arms of the vasti, hamstring and gastrocnemius according to curves formulated in [29]. The joint moments produced by these musculotendons which span the knee are given by

$$\vec{M} = \pm F^t \cdot m_e(\theta_f)$$

where F^t is calculated from (2.2.8) and m_e is the effective moment arm as a function of the knee flexion angle, θ_f. The length of these musculotendons is calculated by integration of the moment arm as in [18]. In this method the relationship between the length of the musculotendon, the effective moment arm, and the joint angle is given by

$$v^{mt} = \frac{d\,L^{mt}}{dt} = m_e(\theta_f)\frac{d\,\theta_f}{dt}.$$

Consequently, three more differential equations are added to the system.

When a direct-dynamic analysis of gait is to be simulated, controls for the musculotendon actuators must be derived. Developing these controls constitutes one of the more difficult aspects of a forward analysis. Some form of dynamic optimization is usually utilized in developing the controls. The controls used in the gait simulations of this paper are derived through a two-phase process: (1) a coarse formulation based on a dynamic optimization scheme, and, (2) a fine-tuning via trial-and error. The coarse controls are derived as in [30] by employing a cost function that consists of a error tracking term which penalizes deviations from a desired trajectory and a term related to muscle fatigue [8],

$$J_i(k) = \sum_{l=1}^{16} w_{x,l}(x_l - x_{l,des})^2 + \sum_{l=1}^{m} w_{u,l}\left(\frac{F_{tl}}{A_l}\right)^3. \qquad (4.0.2)$$

In equation (4.0.2), x_l is one of the elements of the state variable, $\vec{X} = (q_1, q_2, \ldots, \dot{q_1}, \dot{q_2}, \dot{q_3}, \ldots)$ (refer to Figure 8), and F_{tl} and A_l are the force and physiological cross-sectional area of muscle l. The nominal gait trajectory is described by $x_{l,des}$ and is specified according to data recorded in [28]. Once crude activation controls are formulated, simulations were run and these controls were fine-tuned by ad-hoc methods to meet the specific needs of our model. The final formulation of the control laws derived for the model varied slightly from those of [30]. The resulting muscle activation is illustrated in Figure 10.

For this study, the algebraic manipulation program AUTOLEV [19] was utilized to derive the equations of motion for the gait simulation. This program implements Kane's method as a means of deriving Euler's equations of motion. A detailed discussion of the procedure by which the equations are derived is given in [11]. It suffices here to point out that AUTOLEV generates the equations of motion as well as FORTRAN or C code for integrating these equation forward in time using a fourth-order Runge Kutta scheme.

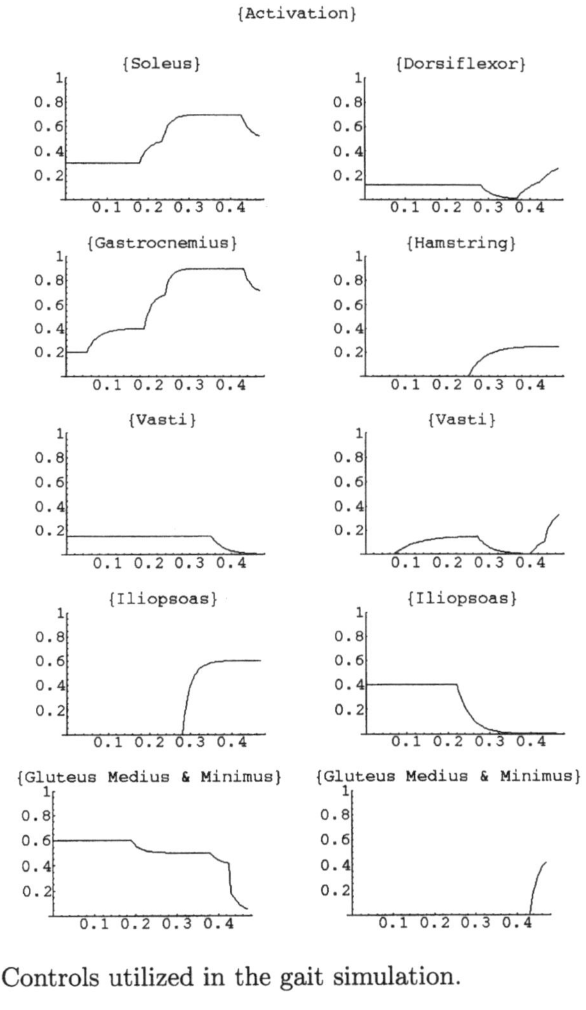

Figure 10. Controls utilized in the gait simulation.

Necessary initial data to run a simulations included the initial segmental orientations $(q_1, ..., q_8)$, the initial angular speeds $(\dot{q}_1, ..., \dot{q}_8)$, the initial force in each musculotendon, and the initial lengths of the musculotendons. Precise values of this data is not readily available and was approximated by several methods. For instance, the initial force in each musculotendon unit was found by running simulations to find the steady state muscle force achieved by the initial control when the motion of the model was constrained. Initial lengths of the musculotendon were estimated so that the muscle maintained an appropriate length throughout the gait cycle. A summary of initial conditions that were utilized in running gait simulations can be found in [11].

A standard means of comparing and validating gait simulations is achieved by displaying the joint torques which drive the system and the resulting joint trajectories. Figure 11 depicts the standard definitions of the joint angles. Notice that these joint angles are distinct from the generalized coordinates used in the analysis. Figure 12 displays the resulting joint trajectories in the simulation. These trajectories provide good qualitative agreement with the trajectories reported by other researchers (see for example [28]). Joint torques and power trajectories, which reflect the rate at which the muscles and tendons are expending and absorbing energy, were in general accordance with the data reported by other researchers [13, 9].

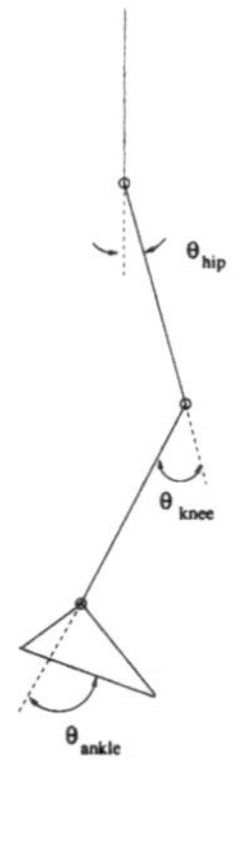

Figure 11. Joint angle definitions.

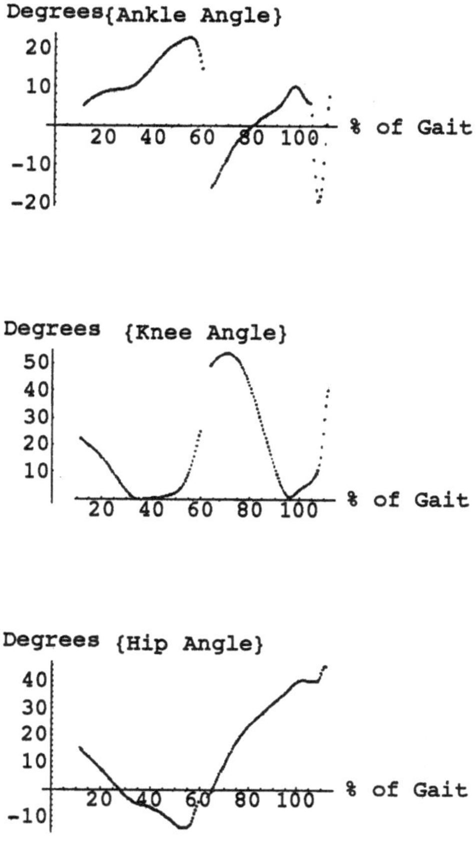

Figure 12. Joint Trajectories.

Figure 13 reports the normalized force realized in the tendons of all ten musculotendon actuators. These normalized forces, F_t/F_0, were obtained through the dynamics in 2.2.8 and are utilized in specifying the loading conditions for the stress analysis. Additional boundary conditions needed for the stress analysis were obtained by appropriately constraining generalize coordinates so that joint reaction forces could be determined.

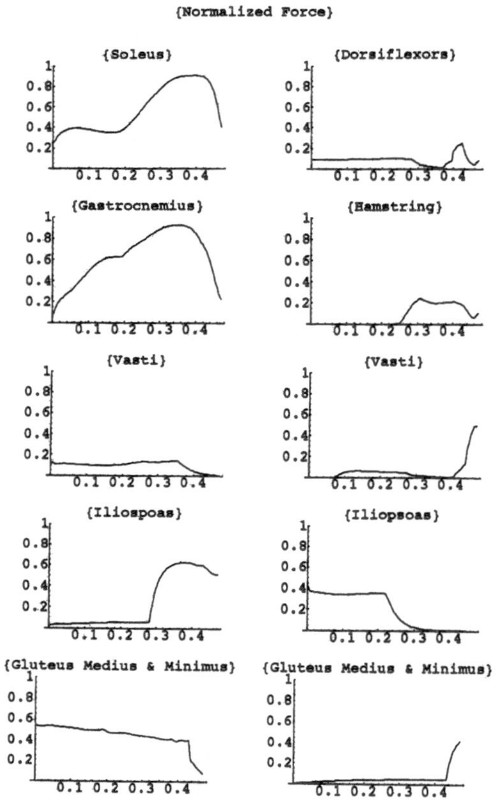

Figure 13. Normalized Force Histories for each musculotendon. Left column refers to stance muscle while the right column refers to the swing side muscles.

4.1 The Stress Analysis

The stress distribution in the long bones of the lower extremities is calculated from equilibrium considerations by 'freezing' the gait model at a fixed instant in time and regarding the segmental elements as linear isotropic elastic bodies. The method that is implemented here follows the approach given in [27]. The applied loads for the stress analysis are the joint reaction forces and the muscle forces as calculated in the gait analysis. The joint reaction forces and joint moments due to musculature effects are resolved into forces and moments at each cross section of the tibia. The components of the stress tensor are calculated in terms of these internal forces and moments.

The relevant geometry is indicated in Figure 14. Assign to the *ith* cross section a local set of orthogonal axes parallel to the unit vectors ζ_i, η_i, ξ_i with the origin at the center of gravity. The following measures of anthropometric data for the each segment for each segmental link must be collected or approximated. Let A_i denote the area of the ith cross section, $(I_\xi)_i$, $(I_\eta)_i$ the statical moments of inertia around the ξ_i and η_i axes, and $(I_{\xi\eta})_i$ the statical product of inertia of the cross sectional area. In the ith section, let w_{ijk} denote the width of the cross section at the point (j,k) parallel to the ξ_i axis, $(W_\xi)_{ijk}$ the moment about the ξ axes of the area bounded by the segment of length w_{ijk} and the perimeter of the cross section, and $(W_\eta)_{ijk}$ the moment about the η axis. (see Figure 14). Similarly, let h_{ijk} denote the height of the cross section at the point (j,k) parallel to the η_i axis and $(H_\xi)_{ijk}$ the moment about the ξ axis of the area bounded by the segment of length h_{ijk} and the perimeter of the cross section with the obvious interpretation of $(H_\eta)_{ijk}$.

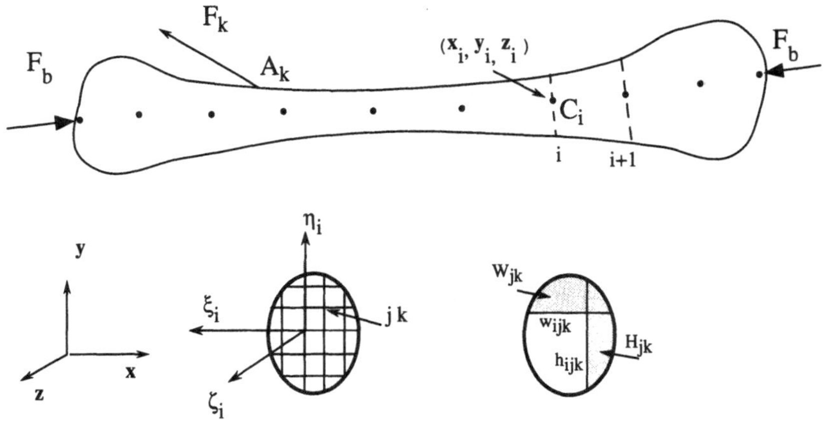

Figure 14. Local coordinates and cross sections of bone.

From equilibrium considerations, the joint reaction forces must be collinear and give rise to an internal force vector on each cross section whose components in terms of the local coordinates are equal and opposite to those of F_b. The component of this internal force acting on the ith cross section is obtained by projection onto the local coordinates. Thus for the ζ direction,

$$(P_\zeta)_i = -F_b \cdot \zeta_i.$$

and similarly for $(P_\eta)_i$ and $(P_\xi)_i$. If $\overrightarrow{C_iC_1}$ denotes the vector from (x_i, y_i, z_i) to (x_1, y_1, z_1), then the component of the reactive moment due to F_b the ζ direction is given by

$$(M_\zeta)_i = -(M)_i \cdot \zeta_i$$

where

$$(M)_i = \overrightarrow{C_iC_1} \times F_b$$

with $(M_\eta)_i$ and $(M_\xi)_i$ defined in a similar way.

If a muscle force acts on the bone, reactive forces and moments at the joints must be introduced to maintain the state of equilibrium. These joint forces and moments will contribute to the internal forces and moments on the cross sections. In particular, suppose a muscle force F_k^m is attached to the bone a point A_k. Introduce joint reaction forces $R_{k1} = (-1/2)F_k^m$ at the points (x_1, y_1, z_1) and (x_n, y_n, z_n). The moment at (x_n, y_n, z_n) due to R_{k1} is

$$M_{k1} = \overrightarrow{C_nC_1} \times R_{k1}$$

and the moment due to the muscle force is

$$M_{k2} = \overrightarrow{C_nA_k} \times F_k^m.$$

To maintain equilibrium in the segment we introduce a reactive moment given by

$$R_{k2} = -(M_{k1} + M_{k2}).$$

The internal forces on a cross section due to R_{k1} are computed in the same manner as for the joint reaction force by simply replacing F_b with R_{k1}. The corresponding internal moment on a section C_i between C_1 and A_k is

$$(M_k)_i = (\overrightarrow{C_iC_1} \times R_{k1}) - R_{k2}.$$

If the section is between A_k and C_n, replace $\overrightarrow{C_nA_k}$ by $\overrightarrow{C_1A_k}$. When several muscles act on the segment, the computations are repeated for each muscle and the effect of all muscles is then obtained by summing contributions of each force. The total internal forces due to the combined effect of the joint reaction force and all muscle attachments is

$$(P_\zeta)_i = -F_b \cdot \zeta_i - \sum_k R_{k1} \cdot \zeta_i$$

and similarly for $(P_\eta)_i$ and $(P_\xi)_i$. The total internal moments are

$$(M_\zeta)_i = -M_i \cdot \zeta_i - \sum_k (M_k)_i \cdot \zeta_i$$

with the obvious expressions for $(M_\eta)_i$ and $(M_\xi)_i$.

If the bone is assumed to be in a state of biaxial bending, the stress tensor has the general form

$$S_{ijk} = \begin{bmatrix} (\sigma_\zeta)_{ijk} & (\tau_{\zeta\eta})_{ijk} & (\tau_{\zeta\xi})_{ijk} \\ (\tau_{\zeta\eta})_{ijk} & 0 & 0 \\ (\tau_{\zeta\xi})_{ijk} & 0 & 0 \end{bmatrix}.$$

The normal stresses due to the compressive and bending forces are computed from well-known formulas concerning biaxial bending and are given by

$$(\sigma_\zeta)_{ijk} = (\sigma_c)_{ijk} + (\sigma_b)_{ijk},$$

where

$$(\sigma_c)_{ijk} = \frac{(P_\zeta)_i}{A_i},$$

$$(\sigma_b)_{ijk} = \frac{[(M_\eta)_i(I_\xi)_i + (M_\xi)_i(I_{\eta\xi})_i]\xi_{ijk} - [(M_\eta)_i(I_{\eta\xi})_i + (M_\xi)_i(I_\eta)_i]\eta_{ijk}}{B_i},$$

$$B_i = (I_\eta)_i(I_\xi)_i - (I_{\eta\xi})_i,$$

and ξ_{ijk}, η_{ijk} are the distances from (j,k) to the ξ and η axes respectively. The shearing stresses are given by

$$(\tau_{\zeta\eta})_{ijk} = (\tilde{\tau}_{\zeta\eta})_{ijk} + (\hat{\tau}_{\zeta\eta})_{ijk}, \qquad (\tau_{\zeta\xi})_{ijk} = (\tilde{\tau}_{\zeta\xi})_{ijk} + (\hat{\tau}_{\zeta\xi})_{ijk}$$

where

$$(\hat{\tau}_{\zeta\eta})_{ijk} = \frac{(\tau_\zeta)_{ijk}\xi_{ijk}}{\sqrt{(\eta_{ijk})^2 + (\xi_{ijk})^2}}, \quad (\hat{\tau}_{\zeta\xi})_{ijk} = -\frac{(\tau_\zeta)_{ijk}\eta_{ijk}}{\sqrt{(\eta_{ijk})^2 + (\xi_{ijk})^2}},$$

$$(\tau_\zeta)_{ijk} = \frac{(M_\zeta)_i\sqrt{(\eta_{ijk})^2 + (\xi_{ijk})^2}}{(I_\eta)_i + (I_\xi)_i},$$

$$(\tilde{\tau}_{\zeta\xi})_{ijk} = \frac{[(P_\eta)_i(I_\eta)_i - (P_\xi)_i(I_{\eta\xi})_i](H_\xi)_{ijk} + [(P_\xi)_i(I_\xi)_i - (P_\eta)_i(I_{\eta\xi})_i](H_\eta)_{ijk}}{h_{ijk}B_i},$$

and

$$(\tilde{\tau}_{\zeta\eta})_{ijk} = \frac{[(P_\eta)_i(I_\eta)_i - (P_\xi)_i(I_{\eta\xi})_i](W_\xi)_{ijk} + [(P_\xi)_i(I_\xi)_i - (P_\eta)_i(I_{\eta\xi})_i](W_\eta)_{ijk}}{w_{ijk}B_i}.$$

In Figures 15-16 the normal stress, (σ_ζ), at heel strike and flat foot is computed on the medial, lateral, anterior, and posterior cortex of the tibia, and plotted versus length of the tibia, from the proximal to distal end. The stress in the absence of muscular effects is indicated by (NM). The results show that muscular loads have a dramatic effect on the stress distribution. Computations based on joint reaction forces alone would underestimate stresses by an order of magnitude. With the inclusion of muscular attachment, the predicted stress is in qualitative agreement with the results of [21] in that tensile stresses are generally greater at flat foot as compared to heel strike. The results are not surprising in view of the fact that muscular forces often exceed body weight and thus dominate the loading on bones both at the joints and at the point of muscle attachment.

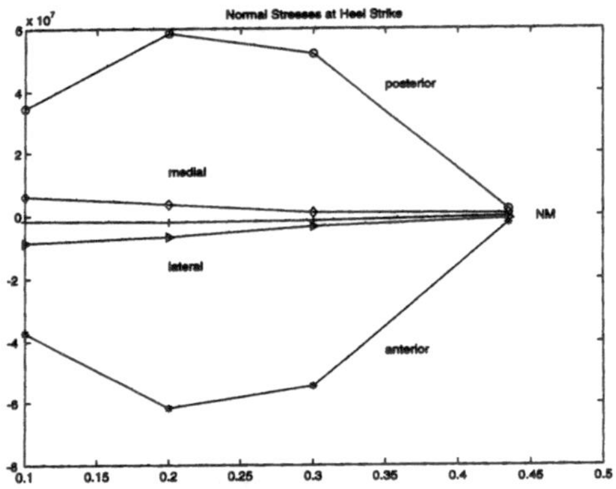

Figure 15. The normal component of stress in the axial direction of the tibia at heel strike.

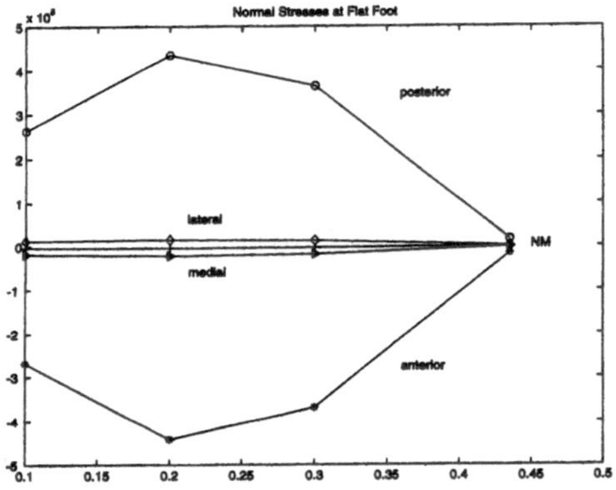

Figure 16. The normal component of stress in the axial direction of the tibia at flat foot.

References

[1] Audu, M.l., and Davy, D.T. The influence of muscles model complexity in musculoskeltal motion modeling. *J. Biomech. Engrg.*, **107**: 147-157 (1985).

[2] Bahill, A., Latimer, J., and Troost, B. Linear homeomorphic model for human movement. *IEEE Trans. Biomed. Eng.*, Vol. BME-27, No. 11, (1980).

[3] Baloh, R. W., etal. Internuclear Opthalmoplegia. *Archives of Neurology*, Vol. 35, pp. 484-493, (1978).

[4] Biewener, A.A. *Biomechanics Structures and Systems: A Practical Approach*, Oxford University Press, New York (1992).

[5] Brand, R.A., Crowninshield, R.D., Wittstock, C.E., Pedersen, D.R., Clark, C.R., and Van Krieken, F.M. A model of lower extremity muscular anatomy.*J. Biomech. Engrg.*, **100**: 88-92 (1982).

[6] Clark, M. R., and Stark, L. Control of human eye movements: I. Modelling of extraocular muscles; II. A model for the extraocular plant mechanism; III. Dynamic characteristics of the eye tracking mechanism. *Mathematical Biosciences*, Vol. 20, pp. 91-265, (1974).

[7] Collins, C. Orbital Mechanics. In P. Bach-y-Rita, C. Collins, Eds., *The Control of Eye Movements*, Academic Press, New York, p. 283 (1973).

[8] Crowninshield, R.D., Brand, R.A. A physiologically based criterion of muscle force prediction on locomotion.*J. Biomechanics.*, **14**: 793-801, (1981).

[9] Davy, D.T., and Audu, M.l. A dynamic optimization technique for predicting forces in the swing phase of gait.*J. Biomechanics.*, **20**: 187-202 (1987).

[10] Delp, S.L., Loan, J.P., Hoy, M.G., Zajac, F.E., Topp E.L., Rosen, J.M. An interactive graphics-based model of the lower extremity to study orthopaedic surgical procedures. *IEEE Transactions on Biomedical Engineering.* **37**: 757-767 (1990).

[11] DeWoody, Y., The role of muscular skeletal dynamics and neuromuscular control in stress development in bone. *Ph.D. Dissertation*, Department of Mathematics and Statistics, Texas Tech University (1998).

[12] Gordon, A.M., Huxley, A.F., and Julian, F.J. Tension development in highly stretched vertebrate muscle fibres. *J. Physiol.*, **184**: 143-169 (1966).

[13] Hardt, D.E. Determining muscle forces in the leg during normal human walking – An application and evaluation of optimization methods.*J. Biomech. Eng.*, **100**: 72-78 (1978).

[14] Hatze, H. The complete optimization of human motion. *Math. Biosci.*, **28**: 99 (1976).

[15] He, J. A feedback control analysis of the neuro-musculo-skeletal control system of a cat hindlimb. *Ph.D. Dissertation*, Department of Electrical Engineering, University of Maryland (1988).

[16] Hemami, H., Jaswa, V.C., and McGhee, R.B. Some alternative formulations of manipulator dynamics for computer simulation studies. *Prc. 13th Allerton Conf. on Circuit Theory,* University of Illinois, (1975).

[17] Hill, A.V. The heat of shortening and dynamic constants of muscle. *Proc. Roy. Soc. B.*, **126**: 136-195 (1938).

[18] Hoy, M.G., Zajac, F.E., and Gordon, M.E. A musculoskeletal model of the human lower extremity: The effects of muscle, tendon, and moment arm on the moment-angle relationship of musculotendon actuators at the hip, knee, and ankle. *J. Biomechanics*, **23**: 157-169 (1990).

[19] Kane, T.R. and Levinson, D.A. *Dynamics Online: Theory and Implementation with* AUTOLEV. Online Dynamics, Inc. Sunnyvale, CA. (1996).

[20] King, A.I. A review of biomechanical models *J. Biomech. Eng.*, **106**: 97-104 (1984).

[21] Lanyon, L.E., Hampson, W.G.J., Goodship, A.E., and Shah, J.S. Bone deformation recorded in vivo from strain gauges attached to the human tibia shaft. *Acta Ortho. Scand.*, **46:** 256-268 (1974).

[22] Lockwood, P. A three dimensional model of dynamic human eye movement. *Ph.D. Dissertation*, Department of Mathematics and Statistics, Texas Tech University (1998).

[23] McMahon, T.A. *Muscles, Reflexes, and Locomotion*, Princeton University Press, Princeton, New Jersey (1984).

[24] Patriarco, A.G., Mann, R.W., Simon, S.R., and Mansour, J.M., An evaluation of the approaches of optimization models in the prediction of muscle forces during human gait. *J. Biomechanics*, **14**: 513-525 (1981).

[25] Robinson, D.A., The mechanics of human saccadic eye movement, *J. of Physiol.*, **174**: 245-264, (1964).

[26] Seireg, A., and Arvikar, R.J. The prediction of muscular load sharing anf joint forces in the lower extremities during walking. *J. Biomechanics*, **8**: 89-102 (1975).

[27] Toridis, T. Stress analysis of the femur. *J. Biomechanics*, **2**: 163-174 (1964).

[28] Winter, D.A. *The Biomechanics and Motor Control of Human Gait.* University of Waterloo Press, Waterloo, Ontario, Canada (1987).

[29] Yamaguchi, G.T., and Zajac, F.E. Restoring unassisted natural gait to paraplegics via functional neuromuscular stimulation: A computer simulation study. *IEEE Trans. Biomed. Eng.*, **37**: 886-902 (1990).

[30] Yamaguchi, G.T. Feasibility and Conceptual design of functional neuromuscular stimulation systems for the restoration of natural gait to paraplegics based on dynamic musculoskeletal models. *Ph.D. Thesis,* Department of Mechanical Engineering, Stanford University, Stanford, CA Aug. (1989).

[31] Zajac, F.E. Muscle and tendon: Properties, models, scaling, and application to biomechanics and motor control. *In CRC Critical Reviews in Biomechanical Engineering.* **17**: 359-410 (1989).

Department of Mathematics, Texas Tech University, Lubbock, Tx 79409

Department of Mathematics, Texas Tech University, Lubbock, Tx 79409

Progress in Systems and Control Theory, Vol. 25

Numerical Methods for Linear Quadratic and H_∞ Control Problems

Peter Benner, Ralph Byers[1], Volker Mehrmann[2] and Hongguo Xu[2]

1 Introduction and preliminaries

The numerical solution of linear quadratic control problems and H_∞ control problems is of great importance in the design of controllers, in particular when robust controllers are desired, [19, 27, 38, 33, 40].

The continuous time linear quadratic control problem has the following form. Minimize

$$\mathcal{S}(x,u) = \int_{t_0}^{\infty} \begin{bmatrix} x(t) \\ u(t) \end{bmatrix}^T \begin{bmatrix} Q & S \\ S^T & R \end{bmatrix} \begin{bmatrix} x(t) \\ u(t) \end{bmatrix} dt \tag{1}$$

subject to the differential-algebraic system

$$E\dot{x} = Ax + Bu, \qquad x(t_0) = x^0. \tag{2}$$

Application of the maximum principle [39, 33] leads to the two-point boundary value problem of Euler-Lagrange equations

$$\mathcal{E}_c \begin{bmatrix} \dot{x} \\ \dot{\mu} \\ \dot{u} \end{bmatrix} = \mathcal{A}_c \begin{bmatrix} x \\ \mu \\ u \end{bmatrix}, \qquad x(t_0) = x^0, \qquad \lim_{t\to\infty} E^T\mu(t) = 0, \tag{3}$$

with the matrix pencil

$$\alpha\mathcal{E}_c - \beta\mathcal{A}_c := \alpha \begin{bmatrix} E & 0 & 0 \\ 0 & -E^T & 0 \\ 0 & 0 & 0 \end{bmatrix} - \beta \begin{bmatrix} A & 0 & B \\ Q & A^T & S \\ S^T & B^T & R \end{bmatrix}. \tag{4}$$

The optimal solution $x(t)$ is required to be stable. If both E and R are nonsingular, then with $\eta := -E^T\mu$, (3) reduces to the two-point boundary value problem

$$\begin{bmatrix} \dot{x} \\ \dot{\eta} \end{bmatrix} = \mathcal{H} \begin{bmatrix} x \\ \eta \end{bmatrix}, \qquad x(t_0) = x^0, \qquad \lim_{t\to\infty} \eta(t) = 0 \tag{5}$$

[1]This author was partially supported by National Science Foundation awards CCR-9404425, DMS-9628626, CCR-9732671, and by the NSF EPSCoR/K*STAR program through the Center for Advanced Scientific Computing.

[2]Supported by *Deutsche Forschungsgemeinschaft*, Research Grant Me 790/7-2.

with the *Hamiltonian matrix*

$$\mathcal{H} = \begin{bmatrix} F & G \\ H & -F^T \end{bmatrix} := \begin{bmatrix} E^{-1}(A - BR^{-1}S) & E^{-1}BR^{-1}B^TE^{-T} \\ Q - SR^{-1}S^T & -(E^{-1}(A - BR^{-1}S))^T \end{bmatrix}. \tag{6}$$

The solution of the boundary value problem can be obtained in many different ways. For example, let Y be the stabilizing solution of the associated algebraic Riccati equation

$$0 = H + YF + F^TY - YGY. \tag{7}$$

Multiplying (6) from the left by

$$\begin{bmatrix} I & 0 \\ Y & I \end{bmatrix}$$

and changing the variables to

$$\begin{bmatrix} x \\ \xi \end{bmatrix} = \begin{bmatrix} I & 0 \\ Y & I \end{bmatrix} \begin{bmatrix} x \\ \eta \end{bmatrix}$$

one obtains the decoupled Hamiltonian system

$$\begin{bmatrix} \dot{x} \\ \dot{\xi} \end{bmatrix} = \begin{bmatrix} F - GY & G \\ 0 & -F^T + YG \end{bmatrix} \begin{bmatrix} x \\ \xi \end{bmatrix}, \; x(t_0) = x_0, \; \lim_{t\to\infty} \xi(t) = 0, \tag{8}$$

from which the solution $[\, x^T, \, \xi^T \,]^T$ may be obtained by one reverse time and one forward time integration.

If E is singular and R is nonsingular then the system (3) represents a differential-algebraic system. Using $u(t) = -R^{-1}(S^Tx(t) + B^T\mu(t))$, system (3) reduces to

$$\mathcal{S} \begin{bmatrix} \dot{x} \\ \dot{\mu} \end{bmatrix} = \mathcal{H} \begin{bmatrix} x \\ \mu \end{bmatrix}, \qquad x(t_0) = x^0, \qquad \lim_{t\to\infty} E^T\mu(t) = 0, \tag{9}$$

with the reduced pencil

$$\alpha\mathcal{S} - \beta\mathcal{H} := \alpha \begin{bmatrix} E & 0 \\ 0 & -E^T \end{bmatrix} - \beta \begin{bmatrix} A - BR^{-1}S & -BR^{-1}B^T \\ Q - SR^{-1}S^T & (A - BR^{-1}S)^T \end{bmatrix} \tag{10}$$

In this case it is possible to write down a generalized Riccati equation but the relationship with solutions of the optimal control problem is lost or hidden. See [5, 25, 26, 33] for details.

If R is singular and E is nonsingular then the situation becomes more complicated. Although the boundary value problem remains well defined, the Riccati equation does not. The analysis of this case has been recently studied in [36, 24, 23] and numerical methods have been introduced in [4, 35, 41].

The case in which both E and R are singular has not been analyzed in full generality yet.

Note that the reduction to the form (5) may be still very ill-conditioned even if E or R are invertible. Hence it may happen that the transformed coefficient matrices in (6) are so corrupted by rounding errors that the solution obtained from them is of limited value. The same may happen if the solution is computed via the Riccati equation (7).

Hamiltonian matrices and Riccati equations of a similar structure occur in the H_∞ control problem. See, e.g., the recent monographs [21, 43]. The extended Hamiltonian pencils typically take the form

$$\alpha\mathcal{E}_h - \beta\mathcal{A}_h := \alpha \begin{bmatrix} E & 0 & 0 \\ 0 & -E^T & 0 \\ 0 & 0 & 0 \end{bmatrix} - \beta \begin{bmatrix} A & \gamma^{-2}B_1B_1^T & B_2 \\ C^TC & A^T & 0 \\ 0 & B_2^T & I \end{bmatrix}. \quad (11)$$

In particular if E is nonsingular, then the reduced order system has the form (5). The main difference though is that in the linear quadratic control problem the matrix G is positive semidefinite, while in the H_∞ case it may be indefinite.

The Euler-Lagrange and Riccati equations, their solvability and their numerical solution has been the subject of numerous publications in recent years. See, e.g., [33, 11, 27, 40].

It was observed in [41] that it suffices to study the deflating subspaces of the pencil $(\mathcal{E}_c, \mathcal{A}_c)$ in (4) to solve the control problems. Suppose $(\mathcal{E}_c, \mathcal{A}_c)$ has an n-dimensional deflating subspace associated with eigenvalues in the left half plane. Let this subspace be spanned by the columns of a matrix $\mathcal{U}$, partitioned analogous to the pencil as

$$\mathcal{U} = \begin{bmatrix} U_1 \\ U_2 \\ U_3 \end{bmatrix}.$$

Then, if U_1 is invertible, the optimal control is a linear feedback of the form $u(t) = U_3U_1^{-1}x(t)$ and the solution of the associated Riccati equation is $Y = U_2U_1^{-1}E^{-1}$. See [33] for details.

Unfortunately, if E is singular, then such an n-dimensional deflating subspace in general does not exist. Under certain restrictions [33] we can complete the subspace to an n-dimensional subspace by adding appropriate eigenvectors and principal vectors associated with the eigenvalue ∞.

A feature of the pencils associated with the two-point boundary value problems is that they have algebraic structures which reflect the model and lead to a certain symmetry in the spectrum. Roundoff errors can destroy this symmetry leading to physically meaningless results unless the numerical method also preserves the algebraic structure of the pencil. Preserving algebraic structure also leads to more efficient as well as more accurate numerical methods.

Definition 1 *Let* $J := \begin{bmatrix} 0 & I_n \\ -I_n & 0 \end{bmatrix}$, *where* I_n *is the* $n \times n$ *identity matrix.*

a) A matrix $\mathcal{H} \in \mathbf{R}^{2n\times 2n}$ *is* Hamiltonian *if* $(\mathcal{H}J)^T = \mathcal{H}J$. *The Lie Algebra of Hamiltonian matrices in* $\mathbf{R}^{2n\times 2n}$ *is denoted by* $\mathcal{H}_{2n}$.

b) A matrix $\mathcal{H} \in \mathbf{R}^{2n\times 2n}$ *is* skew-Hamiltonian *if* $(\mathcal{H}J)^T = -\mathcal{H}J$. *The Jordan algebra of skew-Hamiltonian matrices in* $\mathbf{R}^{2n\times 2n}$ *is denoted by* $\mathcal{SH}_{2n}$.

c) A matrix $\mathcal{S} \in \mathbf{R}^{n\times n}$ *is* symplectic *if* $\mathcal{S}J\mathcal{S}^T = J$. *The Lie group of symplectic matrices in* $\mathbf{R}^{n\times n}$ *is denoted by* $\mathcal{S}_{2n}$.

d) A matrix $\mathcal{U}_d \in \mathbf{R}^{2n\times 2n}$ *is* orthogonal symplectic *if* $\mathcal{U}_d J \mathcal{U}_d^T = J$ *and* $\mathcal{U}_d \mathcal{U}_d^T = I_{2n}$. *The compact Lie group of orthogonal symplectic matrices in* $\mathbf{R}^{n\times n}$ *is denoted by* $\mathcal{US}_{2n}$.

The reduced Euler-Lagrange equations (5) involve a Hamiltonian matrix and (9) involves a pencil with one skew-Hamiltonian and one Hamiltonian matrix.

The pencil (4) does not have this structure but many of the properties of Hamiltonian matrices carry over. We will discuss this in the next section.

Let us close the introductory remarks with some historical background on the numerical solution of the eigenvalue problems for matrices and pencils involving the structures in Definition 1.

The eigenproblem for Hamiltonian matrices has been a topic of research, since the landmark papers of Laub [28] and Paige/Van Loan [37]. While the Schur method proposed in [28] ignores the Hamiltonian structure and uses the standard QR algorithm to obtain the desired deflating subspace, the results in [37] suggest how to use the Hamiltonian structure. In [41] it was then discussed how to effectively use a staircase algorithm to treat the extended matrix pencil (4). But despite these important results and many other contributions, see [14, 29, 33] and the references therein, a completely satisfactory method is still an open problem. Such a method would be a numerically backward stable method, that has a complexity of $\mathcal{O}(n^3)$ and at the same time preserves the Hamiltonian structure.

There are two main reasons why this problem resisted solution. First of all one would need a triangular-like form under orthogonal symplectic similarity transformations from which the desired deflating subspaces can be read off. Such a Hamiltonian Schur form was first suggested in [37] but it is clear that not every Hamiltonian matrix has such a condensed form. The exact characterization when such a form exists was first proposed in [30] and finally proved in [34]. We will give a brief overview of these results in Section 3. The second difficulty arises from the fact that even if a Hamiltonian Schur form exists, it is not clear how to construct a method with the desired features to compute it numerically. It has been shown in [1] that a modification of standard QR-like methods to solve this problem is (except for special cases [15, 16]), in general hopeless, due to the missing reduction to a Hessenberg-like form. For this reason other methods like the multishift-method of [2] or the structured method of [9] were developed that do not follow the direct line of a standard QR-like method. Although these methods still do not fulfill all the requirements to a full extend, they come quite close to the optimal methods. We will review the method of [9] and indicate how it can be extended to skew-Hamiltonian/Hamiltonian pencils.

The outline of the paper is as follows. In Section 2 we describe a way to embed the extended Hamiltonian pencils into pencils with a skew-Hamiltonian/Hamiltonian structure and how this embedding can be interpreted from a system theoretic point of view. In Section 3 we briefly review the results on the existence of Hamiltonian Schur forms and in Section 4 we present numerical methods for the computation of the eigenvalues of Hamiltonian matrices as well as skew-Hamiltonian/Hamiltonian pencils. In Section 5 we then show how we can determine the deflating subspaces that we are interested in via structure preserving methods. and we discuss how the presented results can be extended to the complex case in Section 6.

2 Embedding of extended pencils

In this section we will show how we to endow the extended Hamiltonian pencil (4) with the structure of the pencil (10) by embedding the Euler-Lagrange equations (3) into a larger system. Introducing $\tilde{B} \in \mathbf{R}^{m \times n}$, $\tilde{R} \in \mathbf{R}^{m \times m}$, and an additional control vector $v \in \mathbf{R}^m$, we consider the extended dynamical system

$$E\dot{x} = Ax + Bu + \tilde{B}v, \qquad x(t_0) = x^0. \tag{12}$$

That is, the new control vector is given by $\begin{bmatrix} u \\ v \end{bmatrix}$. We also introduce a new cost functional

$$\mathcal{S}_e(x,u) = \int_{t_0}^{\infty} \begin{bmatrix} x(t) \\ u(t) \\ v(t) \end{bmatrix}^T \begin{bmatrix} Q & S & 0 \\ S^T & R & 0 \\ 0 & 0 & \tilde{R} \end{bmatrix} \begin{bmatrix} x(t) \\ u(t) \\ v(t) \end{bmatrix} dt. \tag{13}$$

For this embedded system the Euler-Lagrange equations written in the appropriate order take the form

$$\mathcal{E}_e \begin{bmatrix} \dot{x} \\ \dot{u} \\ \dot{\mu} \\ \dot{v} \end{bmatrix} = \mathcal{A}_e \begin{bmatrix} x \\ u \\ \mu \\ v \end{bmatrix}, \qquad x(t_0) = x^0, \qquad \lim_{t\to\infty} E^T \mu(t) = 0, \tag{14}$$

with the extended skew-Hamiltonian/ Hamiltonian pencil

$$\alpha\mathcal{E}_e - \beta\mathcal{A}_e := \alpha \left[\begin{array}{cc|cc} E & 0 & 0 & 0 \\ 0 & 0 & 0 & 0 \\ \hline 0 & 0 & E^T & 0 \\ 0 & 0 & 0 & 0 \end{array}\right] - \beta \left[\begin{array}{cc|cc} A & B & 0 & \tilde{B} \\ 0 & 0 & \tilde{B}^T & \tilde{R} \\ \hline -Q & -S & -A^T & 0 \\ -S^T & -R & -B^T & 0 \end{array}\right]. \quad (15)$$

In this embedding there is a lot of freedom. In principle we can choose $\tilde{R}$ and $\tilde{B}$ arbitrarily, but it seems appropriate to follow certain general rules. First of all we should choose them in such a way that the resulting pencil is a regular pencil, so that the associated two-point boundary value problem of differential-algebraic equations (14) has a unique solution for all consistently chosen initial values x^0, see [26]. If the original extended Hamiltonian pencil (4) has this property then this is easily achieved. Note that this regularity property can always be guaranteed via an appropriate preprocessing of the system, see [17, 26, 33]. Secondly, we should ensure that the embedded pencil has a structured Schur-like form as it was introduced for skew-Hamiltonian/Hamiltonian pencils in [32, 31]. Furthermore the problem of computing the desired invariant subspace should not become more ill-conditioned than that for the pencil (4). For a detailed discussion of the choice of $\tilde{B}, \tilde{R}$ see [7].

If we obey these principles then the embedding just means that we have added some eigenvalues at infinity to the system and increased the associated deflating subspace.

There is a certain philosophy behind this embedding trick. First of all we can view this extension as the converse operation to the reduction of the extended problem (3) to the problem (9). Furthermore if we consider a behavior approach, i.e., if we do not distinguish state and input variables, then the same global structure occurs for singular E of the form $\begin{bmatrix} \tilde{E} & 0 \\ 0 & 0 \end{bmatrix}$, if in the reduced system (10) we partition x and μ according to the partitioning in E. In view of these observations, the embedding seems a natural approach from a system theoretic point view. It allows better use of structure, and it avoids the inversion of matrices, which may lead to large numerical errors.

3 Hamiltonian Schur forms

In this section we briefly review the results on the existence of structured Schur forms for Hamiltonian matrices and skew-Hamiltonian/Hamiltonian pencils.

To simplify notation we use *eigenvalue* for eigenvalues of matrices and also for pairs $(\alpha, \beta) \neq (0,0)$ for which the determinant of a matrix pencil $\alpha E - \beta A$ vanishes. These pairs are not unique. If $\beta \neq 0$ then we identify

(α, β) with $(\frac{\alpha}{\beta}, 1)$ or $\lambda = \frac{\alpha}{\beta}$. Pairs $(\alpha, 0)$ with $\alpha \neq 0$ are called *infinite eigenvalues*.

A *quasi-triangular* matrix A is triangular with 1×1 or 2×2 blocks on the diagonal. We call a real matrix *Hamiltonian quasi-triangular* if it is Hamiltonian and has the form

$$\begin{bmatrix} F & G \\ 0 & -F^T \end{bmatrix},$$

where F is quasi-triangular in real Schur form [20]. Similarly we call a real matrix *skew-Hamiltonian quasi-triangular* if it is skew-Hamiltonian and has the form

$$\begin{bmatrix} F & G \\ 0 & F^T \end{bmatrix},$$

where F is quasi-triangular. If a Hamiltonian (skew-Hamiltonian) matrix $\mathcal{H}$ can be transformed into Hamiltonian quasi-triangular from by a similarity transformation with a matrix $\mathcal{U} \in \mathcal{US}_{2n}$, then we say that $\mathcal{U}^T\mathcal{H}\mathcal{U}$ has *Hamiltonian Schur form* (*skew-Hamiltonian Schur form*).

For Hamiltonian matrices that have no purely imaginary eigenvalues the existence of a Hamiltonian Schur form was proved in [37]. The general result was suggested in [30] and a proof based on a structured Hamiltonian Jordan form was recently given in [34]. Since the general result is quite technical, we only give here parts of the result proved in [34].

Theorem 2 *[34]*

Let $\mathcal{H}$ be a real Hamiltonian matrix, let $i\alpha_1, \ldots, i\alpha_\nu$ be its pairwise distinct nonzero purely imaginary eigenvalues and let U_k, $k = 1, \ldots, \nu$, be the associated invariant subspaces. Then the following are equivalent.

i) There exists a real symplectic matrix $\mathcal{S}$ such that $\mathcal{S}^{-1}\mathcal{H}\mathcal{S}$ is real Hamiltonian quasi-triangular.

ii) There exists a real orthogonal symplectic matrix $\mathcal{U}$ such that $\mathcal{U}^T\mathcal{H}\mathcal{U}$ is real Hamiltonian quasi-triangular.

iii) $U_k^H J U_k$ is congruent to J for all $k = 1, \ldots, \nu$, where J is always of the appropriate dimension.

This theorem gives necessary and sufficient conditions for the existence of a real Hamiltonian Schur form under orthogonal symplectic similarity transformations. On the other hand, there are Hamiltonian matrices for which these conditions do not hold, but nevertheless there exists a *nonsymplectic* similarity transformation to Hamiltonian quasi-triangular form. A class of such matrices are the matrices J of a size that is divisible by 4. Orthogonal symplectic similarity transformations do not change these matrices, hence they have no Hamiltonian quasi-triangular form under symplectic similarity transformations. But such matrices are similar to a Hamiltonian quasi triangular form under nonsymplectic transformations. As an example consider $J \in \mathbf{R}^{4\times 4}$. Set $V = [e_1, e_3, e_2, e_4]$, then $V^H JV = \mathrm{diag}(\begin{bmatrix} 0 & 1 \\ -1 & 0 \end{bmatrix}, \begin{bmatrix} 0 & 1 \\ -1 & 0 \end{bmatrix})$ is Hamiltonian triangular.

Necessary and sufficient conditions for the existence of a Hamiltonian Schur form under nonsymplectic similarity transformations are given in the following theorem.

Theorem 3 *[34] A real Hamiltonian matrix $\mathcal{H}$ is similar to a real Hamiltonian triangular form if and only if the algebraic multiplicities of all purely imaginary eigenvalues with positive imaginary parts are even.*

A similar theorem applies to skew-Hamiltonian/Hamiltonian pencils.

Theorem 4 *[32, 31] Let $\alpha\mathcal{S} - \beta\mathcal{H}$ be a regular skew-Hamiltonian/ Hamiltonian pencil, let $i\alpha_1, \ldots, i\alpha_\nu$ be its pairwise distinct nonzero purely imaginary eigenvalues with algebraic multiplicities $p_1, \ldots, p_\nu$ and let U_k, $k = 1, \ldots, \nu$, be the associated deflating subspaces. Furthermore let p_∞ be the algebraic multiplicity of the eigenvalue infinity and let U_∞ be the associated deflating subspace. Then the following are equivalent.*

i) There exists a nonsingular matrix $\mathcal{P}$, such that

$$J\mathcal{P}^T J(\alpha\mathcal{S} - \beta\mathcal{H})\mathcal{P} = \alpha \begin{bmatrix} S_{11} & S_{12} \\ 0 & S_{11}^T \end{bmatrix} - \beta \begin{bmatrix} H_{11} & H_{12} \\ 0 & -H_{11}^T \end{bmatrix} \quad (16)$$

where S_{11} is upper triangular and H_{11} is quasi upper triangular.

ii) There exists a real orthogonal matrix $\mathcal{U}$ such that $J\mathcal{U}^T J(\alpha\mathcal{S} - \beta\mathcal{H})\mathcal{U}$ has the form (16).

iii) $U_k^H JSU_k$ is congruent to J of appropriate dimension for all $k = 1, \ldots, \nu$. Furthermore if $p_\infty \neq 0$ then $U_\infty^T JHU_\infty$ is congruent to iJ of appropriate dimension.

Note that a necessary condition for iii) to hold is that all purely imaginary eigenvalue of $\mathcal{H}$ have even algebraic multiplicities. Similar results also exist for complex matrices and pencils, see [34] and [32, 31]. The results also demonstrate that whenever a structured triangular form exists, then it also exists under orthogonal transformations. This fact gives hope that these forms and therefore also the eigenvalues and deflating subspaces can be computed with structure preserving numerically stable methods. We discuss such methods for the computation of eigenvalues of Hamiltonian matrices and skew-Hamiltonian/Hamiltonian pencils in the remaining sections.

4 Eigenvalue computation

We have seen that (possibly by an appropriate embedding) the solution of our robust control problems leads to the problem of computing eigenvalues and deflating subspaces for Hamiltonian matrices or skew-Hamiltonian/ Hamiltonian pencils. It is well-known that if $\mathcal{H}$ is a Hamiltonian matrix, $\mathcal{H}^2$ is a skew-Hamiltonian matrix. It is easier to compute the eigenvalues of a real skew-Hamiltonian matrix than those of a Hamiltonian matrix [42]. This suggests computing the eigenvalues of $\mathcal{H}$ by taking square roots of the eigenvalues of $\mathcal{H}^2$. This method was proposed in [42]. Unfortunately, in a worst case scenario one might obtain only half of the possible accuracy in the computed eigenvalues [15, 42]. An example demonstrating this was given in [42]. A way out of this dilemma was recently presented in [10].

Theorem 5 *[10] Let $\mathcal{H} \in \mathcal{H}_{2n}$. Then there exist $Q_1, Q_2 \in \mathcal{US}_{2n}$, such that*

$$Q_1^T \mathcal{H} Q_2 = \begin{bmatrix} H_{11} & H_{12} \\ 0 & H_{22} \end{bmatrix}, \tag{17}$$

with H_{11} upper triangular and H_{22}^T quasi upper triangular. Furthermore the eigenvalues of $\mathcal{H}$ are the square roots of the eigenvalues of $-H_{11}H_{22}^T$.

Note that the resulting matrix in (17) is neither Hamiltonian nor similar to $\mathcal{H}$, but a simple calculation shows that both $Q_1^T \mathcal{H}^2 Q_1$ and $Q_2^T \mathcal{H}^2 Q_2$ are real skew-Hamiltonian quasi-triangular.

For skew-Hamiltonian/Hamiltonian pencils $\alpha\mathcal{S} - \beta\mathcal{H}$ of the form (10) or (15), we can construct similar methods. Roughly, the idea is to factor $\mathcal{S} = \mathcal{S}_1\mathcal{S}_2$ with $\mathcal{S}_1 = J\mathcal{S}_2^T J^T$ (e.g., for (10), $\mathcal{S}_1 = \begin{bmatrix} I & 0 \\ 0 & E^T \end{bmatrix}$) and to apply the previous procedure formally to the Hamiltonian matrix

$$\mathcal{S}_1^{-1}\mathcal{H}\mathcal{S}_2^{-1}$$

without ever forming the product or the inverses.

Theorem 6 *[7] If the skew-Hamiltonian/Hamiltonian pencil*

$$\alpha\mathcal{S}_1\mathcal{S}_2 - \beta\mathcal{H} := \alpha \begin{bmatrix} I & 0 \\ 0 & E^T \end{bmatrix} \begin{bmatrix} E & 0 \\ 0 & I \end{bmatrix} - \beta \begin{bmatrix} F & G \\ H & -F^T \end{bmatrix} \tag{18}$$

is regular, then there exist orthogonal matrices Q_3, Q_4 and orthogonal symplectic matrices Q_1, Q_2, such that

$$\begin{aligned} Q_3^T \mathcal{S}_1 Q_1 &= \begin{bmatrix} S_{11} & S_{12} \\ 0 & S_{22} \end{bmatrix}, \\ Q_2^T \mathcal{S}_2 Q_4 &= \begin{bmatrix} T_{11} & T_{12} \\ 0 & T_{22} \end{bmatrix}, \\ Q_3^T \mathcal{H} Q_4 &= \begin{bmatrix} H_{11} & H_{12} \\ 0 & H_{22} \end{bmatrix}, \end{aligned} \tag{19}$$

where $S_{11}, T_{11}, H_{11}, S_{22}^T, T_{22}^T$ are upper triangular and H_{22}^T is quasi upper triangular. Furthermore, the finite eigenvalues of $\alpha\mathcal{S} - \beta\mathcal{H}$ are

1. *the square roots of the finite eigenvalues of*

$$\alpha S_{11} S_{22}^T + \beta H_{11} T_{11}^{-1} T_{22}^{-T} H_{22}^T;$$

2. *or, equivalently, the eigenvalues of*

$$\alpha \begin{bmatrix} T_{22}^T T_{11} & 0 \\ 0 & S_{11} S_{22}^T \end{bmatrix} - \beta \begin{bmatrix} 0 & -H_{22}^T \\ H_{11} & 0 \end{bmatrix}.$$

The proof of this result as well as the proof of Theorem 5 amount to algorithms for the computation of (17) and (19). The reduction procedures are based on a finite elimination procedure that brings all of the diagonal blocks except H_{22} to triangular form. The block H_{22} reduces to lower Hessenberg form. This initial reduction is then followed by the periodic QR-algorithm or QZ-algorithm [12, 22] applied to $-H_{11}H_{22}^T$ or

$$-S_{11}^{-1} H_{11} T_{11}^{-1} T_{22}^{-T} H_{22}^T S_{22}^{-T}, \tag{20}$$

respectively.

The periodic QR-algorithm applied to $-H_{11}H_{22}^T$ yields real orthogonal transformation matrices $U, V \in \mathbf{R}^{n \times n}$ such that $U^T H_{11} V$ is upper triangular and $(U^T H_{22} V)^T$ is quasi upper triangular. Analogously the periodic QZ-algorithm applied to (20) yields real orthogonal transformation matrices $P, Q, U, V, Y, Z \in \mathbf{R}^{n\times n}$, such that $P^T S_{11} Q$, $P^T H_{11} U$, $V^T T_{11} U$, $W^T T_{22}^T V$, $Q^T S_{22}^T Y$ are upper triangular, and $W^T H_{22}^T Y$ is quasi upper triangular. The 2×2 blocks are associated only with nonsingular blocks in S_{11}, S_{22}, T_{11}, T_{22}.

After these forms have been computed, we can compute the eigenvalues of $\mathcal{H}$ or $\alpha\mathcal{S} - \beta\mathcal{H}$, respectively by solving 1×1 or 2×2 eigenvalue problems and taking square roots. For algorithmic details and a detailed error analysis see [7, 10].

To demonstrate the efficiency of this approach we present numerical examples for the Hamiltonian matrix case. The numerical tests were performed using IEEE double precision arithmetic with machine precision $\varepsilon \approx 2.2204 \times 10^{-16}$ on a HP Model 712/60 workstation with operating system HP-UX 9.0. We used the HP-UX Fortran 77 compiler invoked by `f77`. The programs were compiled using standard optimization.

We compared the following methods:

- **URVPSD**, the method based on Theorem 5 as suggested in [10],
- **SQRED**, Van Loan's square reduced method [42] as implemented in [6],

- **LAPACK**, the nonsymmetric eigenproblem solver DGEEVX from LAPACK [3].

All subroutines use the BLAS and LAPACK [3] compiled from Fortran source with `f77 -O`. The implementations of URVPSD and SQRED are not block-oriented.

Example 1 [42, Example 2] Let $F = \text{diag}(1, 10^{-2}, 10^{-4}, 10^{-6}, 10^{-8})$, and let H be the Hamiltonian matrix obtained by

$$H = U^T \begin{bmatrix} F & 0 \\ 0 & -F^T \end{bmatrix} U,$$

with $U \in \mathcal{US}_{2n}$ randomly generated by five symplectic rotations and five reflectors. Thus,

$$\sigma(H) = \{\pm 1, \pm 10^{-2}, \pm 10^{-4}, \pm 10^{-6}, \pm 10^{-8}\}.$$

Table 1 shows the absolute errors in the eigenvalue approximations computed by the three methods.

λ	URVPSD	SQRED	LAPACK
1	0	0	7.8×10^{-16}
10^{-2}	5.5×10^{-16}	5.5×10^{-16}	5.0×10^{-17}
10^{-4}	1.6×10^{-18}	1.6×10^{-14}	2.6×10^{-18}
10^{-6}	1.0×10^{-18}	1.5×10^{-11}	8.4×10^{-18}
10^{-8}	3.1×10^{-17}	2.2×10^{-9}	4.7×10^{-17}

Table 1: Example 1, absolute errors $|\lambda - \tilde{\lambda}|$.

Table 1 demonstrates that SQRED calculates large magnitude eigenvalues to full precision (apart from the effects of eigenvalue ill-conditioning) but that small magnitude eigenvalues to only half precision. LAPACK and URVPSD [10] calculate all eigenvalues to full precision.

Example 2 We also tested the three methods for randomly generated Hamiltonian matrices with entries distributed uniformly in the interval $[-1, 1]$. Since the eigenvalue distribution for these examples usually behaves "nicely", the eigenvalues computed by either of the methods are computed to almost the same accuracy. We give the CPU times for $2n \times 2n$ examples for several sizes of n. For each size of n, we computed 100 examples. The values given in Table 2 are the mean values of the CPU times measured on a HP Model 712/60 work station.

n	URVPSD	SQRED	LAPACK
25	0.092	0.061	0.142
50	0.56	0.34	0.77
75	1.72	1.03	2.36
100	3.95	2.41	5.30
125	7.36	4.66	10.07
150	12.33	7.99	17.36
175	19.52	12.53	27.79
200	28.61	18.51	41.44

Table 2: Example 2, CPU times in seconds.

Table 2 shows that URVPSD and SQRED are much faster than the standard QR algorithm. The speed-ups are roughly proportional to the flop counts. There is a little overhead which causes both methods to be slightly slower than to be expected from the flop counts, though. This is due to the fact that these methods are more complex as far as index handling, memory access, and subroutine calls are concerned.

We have seen that it is possible to use the algebraic structure of Hamiltonian matrices and skew-Hamiltonian/Hamiltonian pencils effectively to speed up the computation of eigenvalues while still achieving full possible accuracy. Unfortunately this new approach is not perfect. We would like to have the Hamiltonian Schur form, since it would give us the eigenvalues and also the deflating subspaces. Currently the only other candidate for an optimal algorithm, the multishift algorithm of [2], sometimes has convergence problems.

For the computation of the deflating subspaces we now use another embedding procedure. These ideas are presented in the next section.

5 Invariant subspace computation for Hamiltonian matrices

In this section we discuss structure preserving methods to compute the invariant subspaces of Hamiltonian matrices. This approach can also be applied to general matrices, so we present it in general and then show how it specializes for Hamiltonian matrices. The description of the treatment of skew-Hamiltonian/Hamiltonian pencils is similar, but technically involved.

For this reason we present here only the method for the Hamiltonian matrix case and refer the reader to the forthcoming paper [7] for the pencil case.

Let $\lambda_-(A), \lambda_+(A), \lambda_0(A)$ denote the spectra in the open left half plane, in the open right half plane and on the imaginary axis, of a matrix A, respectively. The associated invariant subspaces are denoted by $\mathrm{Inv}_-(A), \mathrm{Inv}_+(A), \mathrm{Inv}_0(A)$, respectively.

Let $A \in \mathbf{R}^{n\times n}$. If

$$B = \begin{bmatrix} 0 & A \\ A & 0 \end{bmatrix}, \tag{21}$$

then

$$\begin{aligned} \lambda(B) &= \lambda(A) \cup \lambda(-A), \\ \lambda_0(B) &= \lambda_0(A) \cup \lambda_0(A), \\ \lambda_+(B) &= \lambda_+(A) \cup \lambda_+(-A) = \lambda_+(A) \cup (-\lambda_-(A)), \\ \lambda_-(B) &= \lambda_-(A) \cup \lambda_-(-A) = (-\lambda_+(A)) \cup \lambda_-(A) = -\lambda_+(B). \end{aligned} \tag{22}$$

Furthermore we obtain the following relations for the invariant subspaces of A and B.

Theorem 7 *[9] Let $A \in \mathbf{R}^{n\times n}$ and $B \in \mathbf{R}^{2n\times 2n}$ be related as in (21) and let $\begin{bmatrix} Q_1 \\ Q_2 \end{bmatrix} \in \mathbf{R}^{2n\times n}$, $Q_1, Q_2 \in \mathbf{R}^{n\times n}$, have orthonormal columns, such that*

$$B \begin{bmatrix} Q_1 \\ Q_2 \end{bmatrix} = \begin{bmatrix} Q_1 \\ Q_2 \end{bmatrix} R, \tag{23}$$

where

$$\lambda_+(B) \subseteq \lambda(R) \subseteq \lambda_+(B) \cup \lambda_0(B). \tag{24}$$

Then

$$\mathrm{range}\{Q_1 + Q_2\} = \mathrm{Inv}_+(A) + \mathcal{N}_1, \quad \text{where} \quad \mathcal{N}_1 \subseteq \mathrm{Inv}_0(A), \tag{25}$$

$$\mathrm{range}\{Q_1 - Q_2\} = \mathrm{Inv}_-(A) + \mathcal{N}_2, \quad \text{where} \quad \mathcal{N}_2 \subseteq \mathrm{Inv}_0(A). \tag{26}$$

Moreover, if we partition R as

$$R = \begin{bmatrix} R_{11} & R_{12} \\ 0 & R_{22} \end{bmatrix}, \text{ where } \lambda(R_{11}) = \lambda_+(B), \tag{27}$$

and, accordingly, $Q_1 = \begin{bmatrix} Q_{11} & Q_{12} \end{bmatrix}$, $Q_2 = \begin{bmatrix} Q_{21} & Q_{22} \end{bmatrix}$, then

$$B \begin{bmatrix} Q_{11} \\ Q_{21} \end{bmatrix} = \begin{bmatrix} Q_{11} \\ Q_{21} \end{bmatrix} R_{11}, \tag{28}$$

and there exists an orthogonal matrix Z such that

$$\begin{aligned} \frac{\sqrt{2}}{2}(Q_{11} + Q_{21}) &= \begin{bmatrix} 0 & P_+ \end{bmatrix} Z, \\ \frac{\sqrt{2}}{2}(Q_{11} - Q_{21}) &= \begin{bmatrix} P_- & 0 \end{bmatrix} Z, \end{aligned} \tag{29}$$

where P_+, P_- are orthogonal bases of $\mathrm{Inv}_+(A)$, $\mathrm{Inv}_-(A)$, *respectively.*

In the case of a Hamiltonian matrix $\mathcal{H} = \begin{bmatrix} F & G \\ H & -F^T \end{bmatrix}$ again consider the block matrix

$$\mathcal{B} = \begin{bmatrix} 0 & \mathcal{H} \\ \mathcal{H} & 0 \end{bmatrix}. \tag{30}$$

If

$$\mathcal{P} = \begin{bmatrix} I_n & 0 & 0 & 0 \\ 0 & 0 & I_n & 0 \\ 0 & I_n & 0 & 0 \\ 0 & 0 & 0 & I_n \end{bmatrix}, \tag{31}$$

then

$$\tilde{\mathcal{B}} := \mathcal{P}^T \mathcal{B} \mathcal{P} = \begin{bmatrix} 0 & F & 0 & G \\ F & 0 & G & 0 \\ 0 & H & 0 & -F^T \\ H & 0 & -F^T & 0 \end{bmatrix} \tag{32}$$

is again Hamiltonian.

Theorem 8 *[9] Let $\mathcal{H} \in H_{2n}$ and let $\mathcal{B}$ be as in (30). Then there exists $\mathcal{U} \in U_{4n}$, such that*

$$\mathcal{U}^T \mathcal{B} \mathcal{U} = \begin{bmatrix} R & D \\ 0 & -R^T \end{bmatrix} =: \mathcal{R} \tag{33}$$

is in Hamiltonian quasi-triangular form and $\lambda_-(R) = \emptyset$. Moreover, $\mathcal{U} = \mathcal{P}\mathcal{W}$ with $\mathcal{W} \in US_{4n}$, and

$$\mathcal{R} = \mathcal{W}^T \tilde{\mathcal{B}} \mathcal{W}, \tag{34}$$

i.e., $\mathcal{R}$ is the Hamiltonian quasi-triangular form of the Hamiltonian matrix $\tilde{\mathcal{B}}$. Furthermore, if $\mathcal{H}$ has no purely imaginary eigenvalues, then R has only eigenvalues with positive real part.

A constructive proof leading to the following algorithm for this result is given in [9], but the theorem also follows from Theorems 3 and 2.

Algorithm 1

Input: *A Hamiltonian matrix $\mathcal{H} \in H_{2n}$ having an n-dimensional Lagrangian invariant subspace.*

Output: *$Y \in \mathbf{R}^{2n\times n}$, with $Y^T Y = I_n$, such that the columns of Y span a Lagrange invariant subspace.*

Step 1 *Apply Algorithm 2 of [10] to $\mathcal{H}$ and compute orthogonal symplectic matrices $Q_1, Q_2 \in \mathcal{US}_{2n}$ such that*

$$Q_1^T \mathcal{H} Q_2 = \begin{bmatrix} H_{11} & H_{12} \\ 0 & H_{22} \end{bmatrix}$$

is the decomposition (17).

Step 2 *Determine an orthogonal matrix Q_3, such that*

$$Q_3^T \begin{bmatrix} 0 & -H_{22}^T \\ H_{11} & 0 \end{bmatrix} Q_3 = \begin{bmatrix} T_{11} & T_{12} & T_{13} \\ 0 & T_{22} & T_{23} \\ 0 & 0 & T_{33} \end{bmatrix}$$

is in real Schur form ordered such that the eigenvalues of T_{11} have positive real part, the eigenvalues of T_{22} have zero real part, and the eigenvalues of T_{33} have negative real part.

Step 3 *Use the orthogonal symplectic reordering scheme of [16] to determine an orthogonal symplectic matrix $V \in \mathcal{US}_{4n}$ such that with*

$$U = \begin{bmatrix} U_{11} & U_{12} \\ U_{21} & U_{22} \end{bmatrix} := \begin{bmatrix} Q_1 Q_3 & 0 \\ 0 & Q_2 Q_3 \end{bmatrix} V.$$

At this point we have the Hamiltonian quasi-triangular form

$$U^T \mathcal{B} U = \begin{bmatrix} F_{11} & F_{12} & G_{11} & G_{12} \\ 0 & F_{22} & G_{21} & G_{22} \\ 0 & 0 & -F_{11}^T & 0 \\ 0 & 0 & -F_{12}^T & -F_{22}^T \end{bmatrix},$$

where F_{11}, F_{22} are quasi upper triangular with eigenvalues only in the closed right half plane.

Step 4 *Set $\hat{Y} := \frac{\sqrt{2}}{2}(U_{11} - U_{21})$. Compute Y, an orthogonal basis of* range$\{\hat{Y}\}$, *using any numerically stable orthogonalization scheme, for example a rank-revealing QR-decomposition; see, e.g., [18].*

End

The estimated computational cost for this algorithm is given in Table 3.

Step	1	2	3	4	total
flops	$103\ n^3$	$9\ n^3$	$9\ n^3$	$21\ n^3$	$142\ n^3$

Table 3: Flop counts for Algorithm 1

These numbers compare with $203n^3$ flops for the computation of the same invariant subspace via the standard QR-algorithm as suggested in [28].

Clearly we can obtain the desired solution of the Riccati equation (if it exists) from the invariant subspace but it is also possible to get it directly from $\hat{Y}$. See [9] for details.

A detailed description of this algorithm, an error and perturbation analysis as well as a comparison of different Riccati solvers are given in [9].

6 Complex problems

Complex versions of Theorems 3 and 6 are given in [34], and a unitary-symplectic method for the Hamiltonian eigenproblem is presented in [13]. The algorithms described in the previous sections do not directly carry over to the case of complex Hamiltonian matrices. Another simple embedding, however, yields methods for the complex case [8].

Since for every complex Hamiltonian matrix $\mathcal{H}$, $i\mathcal{H}$ is complex skew-Hamiltonian, it suffices to study skew-Hamiltonian matrices. Let $N = N_1 + iN_2$ be a complex skew-Hamiltonian matrix with a real skew-Hamiltonian matrix N_1 and a real Hamiltonian matrix N_2 with

$$N_1 = \begin{bmatrix} F_1 & D_1 \\ G_1 & F_1^T \end{bmatrix}, \quad N_2 = \begin{bmatrix} F_2 & D_2 \\ G_2 & -F_2^T \end{bmatrix}.$$

Then with the unitary matrix

$$Q_{2n} := \frac{\sqrt{2}}{2} \begin{bmatrix} I_{2n} & iI_{2n} \\ I_{2n} & -iI_{2n} \end{bmatrix}, \tag{35}$$

and the permutation matrix $\mathcal{P}$ of (31) we obtain the real skew-Hamiltonian matrix

$$\begin{aligned} \mathcal{N} &:= \mathcal{P}^H Q_{2n}^H \operatorname{diag}(N, \overline{N}) Q_{2n} \mathcal{P} = \begin{bmatrix} F_1 & -F_2 & D_1 & -D_2 \\ F_2 & F_1 & D_2 & D_1 \\ G_1 & -G_2 & F_1^T & F_2^T \\ G_2 & G_1 & -F_2^T & F_1^T \end{bmatrix} \\ &=: \begin{bmatrix} \mathcal{F} & \mathcal{D} \\ \mathcal{G} & \mathcal{F}^T \end{bmatrix}. \end{aligned} \tag{36}$$

for which we can easily, see [42], obtain the real skew-Hamiltonian quasi-triangular form

$$W^T \mathcal{N} W = \begin{bmatrix} R & T \\ 0 & R^T \end{bmatrix} =: \mathcal{R}, \tag{37}$$

where $R \in \mathbf{R}^{2n \times 2n}$ is quasi upper triangular, $T = -T^T$, and $W \in \mathcal{US}_{4n}$ is real orthogonal symplectic. As in Section 5, we can then determine the desired subspaces. See [8] for details.

7 Conclusion

We have given a survey on recent results concerning the existence of Schur like forms for Hamiltonian matrices and skew-Hamiltonian/Hamiltonian pencils. Furthermore we have discussed structure preserving numerical methods for the computation of invariant and deflating subspaces for these matrices and pencils. The methods can also be used for problems with purely imaginary eigenvalues, although there are still some open problems to be settled. The presented ideas allow a universal treatment of linear quadratic and H_∞ problems and analogous results and methods are also available for discrete-time problems.

References

[1] G. S. Ammar and V. Mehrmann. On Hamiltonian and symplectic Hessenberg forms. *Linear Algebra Appl.*, 149:55–72, 1991.

[2] G.S. Ammar, P. Benner, and V. Mehrmann. A multishift algorithm for the numerical solution of algebraic Riccati equations. *Electr. Trans. Num. Anal.*, 1:33–48, 1993.

[3] E. Anderson, Z. Bai, C. Bischof, J. Demmel, J. Dongarra, J. Du Croz, A. Greenbaum, S. Hammarling, A. McKenney, S. Ostrouchov, and D. Sorensen. *LAPACK Users' Guide.* SIAM, Philadelphia, PA, second edition, 1995.

[4] W.F. Arnold, III and A.J. Laub. Generalized eigenproblem algorithms and software for algebraic Riccati equations. *Proc. IEEE*, 72:1746–1754, 1984.

[5] D.J. Bender and A.J. Laub. The linear-quadratic optimal regulator for descriptor systems. *IEEE Trans. Automat. Control*, AC-32:672–688, 1987.

[6] P. Benner, R. Byers, and E. Barth. HAMEV and SQRED: Fortran 77 subroutines for computing the eigenvalues of Hamiltonian matrices using Van Loan's square reduced method. Technical Report SFB393/96-06, Fak. f. Mathematik, TU Chemnitz–Zwickau, 09107 Chemnitz, FRG, 1996. Available from `http://www.tu-chemnitz.de/sfb393/sfb96pr.html`.

[7] P. Benner, R. Byers, V. Mehrmann, and H. Xu. Numerical computation of deflating subspaces of embedded Hamiltonian and symplectic pencils. In preparation.

[8] P. Benner, V. Mehrmann, and H. Xu. A note on the numerical solution of the complex Hamiltonian and skew-Hamiltonian eigenvalue problems. In preparation.

[9] P. Benner, V. Mehrmann, and H. Xu. A new method for computing the stable invariant subspace of a real Hamiltonian matrix. *J. Comput. Appl. Math.*, 86:17–43, 1997.

[10] P. Benner, V. Mehrmann, and H. Xu. A numerically stable, structure preserving method for computing the eigenvalues of real Hamiltonian or symplectic pencils. *Numer. Math.*, 78(3):329–358, 1998.

[11] S. Bittanti, A. Laub, , and J. C. Willems, editors. *The Riccati Equation.* Springer-Verlag, Berlin, 1991.

[12] A. Bojanczyk, G.H. Golub, and P. Van Dooren. The periodic Schur decomposition; algorithms and applications. In *Proc. SPIE Conference, vol. 1770*, pages 31–42, 1992.

[13] A. Bunse-Gerstner and H. Faßbender. A Jacobi-like method for solving algebraic Riccati equations on parallel computers. *IEEE Trans. Automat. Control*, 42(8):1071–1084, 1997.

[14] A. Bunse-Gerstner, R. Byers, and V. Mehrmann. Numerical methods for algebraic Riccati equations. In S. Bittanti, editor, *Proc. Workshop on the Riccati Equation in Control, Systems, and Signals*, pages 107–116, Como, Italy, 1989.

[15] R. Byers. *Hamiltonian and Symplectic Algorithms for the Algebraic Riccati Equation.* PhD thesis, Cornell University, Dept. Comp. Sci., Ithaca, NY, 1983.

[16] R. Byers. A Hamiltonian QR-algorithm. *SIAM J. Sci. Statist. Comput.*, 7:212–229, 1986.

[17] R. Byers, T. Geerts, and V. Mehrmann. Descriptor systems without controllability at infinity. *SIAM J. Cont.*, 35(2):462–479, 1997.

[18] T. Chan. Rank revealing QR factorizations. *Linear Algebra Appl.*, 88/89:67–82, 1987.

[19] L. Dai. *Singular Control Systems.* Number 118 in Lecture Notes in Control and Information Sciences. Springer-Verlag, Berlin, 1989.

[20] G.H. Golub and C.F. Van Loan. *Matrix Computations.* Johns Hopkins University Press, Baltimore, third edition, 1989.

[21] M. Green and D.J.N Limebeer. *Linear Robust Control.* Prentice-Hall, Englewood Cliffs, NJ, 1995.

[22] J.J. Hench and A.J. Laub. Numerical solution of the discrete-time periodic Riccati equation. *IEEE Trans. Automat. Control*, 39:1197–1210, 1994.

[23] V. Ionescu, C. Oară, and M. Weiss. *Generalized Riccati theory. A Popov function approach.* John Wiley & Sons, Chichester, 1998.

[24] V. Ionescu and M. Weiss. Continuous and discrete-time Riccati theory: A Popov function approach. *Linear Algebra Appl.*, 193:173–209, 1993.

[25] P. Kunkel and V. Mehrmann. Numerical solution of Riccati differential algebraic equations. *Linear Algebra Appl.*, 137/138:39–66, 1990.

[26] P. Kunkel and V. Mehrmann. The linear quadratic control problem for linear descriptor systems with variable coefficients. *Math. Control, Signals, Sys.*, 10:247–264, 1997.

[27] P. Lancaster and L. Rodman. *The Algebraic Riccati Equation.* Oxford University Press, Oxford, 1995.

[28] A.J. Laub. A Schur method for solving algebraic Riccati equations. *IEEE Trans. Automat. Control*, AC-24:913–921, 1979. (See also *Proc. 1978 CDC (Jan. 1979)*, pp. 60-65).

[29] A.J. Laub. Invariant subspace methods for the numerical solution of Riccati equations. In S. Bittanti, A.J. Laub, and J.C. Willems, editors, *The Riccati Equation*, pages 163–196. Springer-Verlag, Berlin, 1991.

[30] W.-W. Lin and T.-C. Ho. On Schur type decompositions for Hamiltonian and symplectic pencils. Technical report, Institute of Applied Mathematics, National Tsing Hua University, Taiwan, 1990.

[31] C. Mehl. *Compatible Lie and Jordan algebras and applications to structured matrices and pencils.* Dissertation, Fakultät für Mathematik, TU Chemnitz, 09107 Chemnitz (FRG), 1998.

[32] C. Mehl. Condensed forms for skew-Hamiltonian/Hamiltonian pencils. Technical Report SFB393/98-8, Fak. f. Mathematik, TU Chemnitz, 09107 Chemnitz, FRG, 1998. Available from `http://www.tu-chemnitz.de/sfb393/sfb98pr.html`.

[33] V. Mehrmann. *The Autonomous Linear Quadratic Control Problem, Theory and Numerical Solution.* Number 163 in Lecture Notes in Control and Information Sciences. Springer-Verlag, Heidelberg, July 1991.

[34] V. Mehrmann and H. Xu. Canonical forms for Hamiltonian and symplectic matrices and pencils. Technical Report SFB393/98-7, Fak. f. Mathematik, TU Chemnitz, 09107 Chemnitz, FRG, 1998. Available from `http://www.tu-chemnitz.de/sfb393/sfb98pr.html`.

[35] C. Oară. Proper deflating subspaces: Properties, algorithms and applications. *Numer. Algorithms*, 7:355–373, 1994.

[36] C. Oară. *Generalized Riccati Theory: A Ppopov Function Approach.* PhD thesis, Polytechnic University Bucharest, Bucharest, Romania, 1995.

[37] C.C. Paige and C.F. Van Loan. A Schur decomposition for Hamiltonian matrices. *Linear Algebra Appl.*, 14:11–32, 1981.

[38] P.H. Petkov, N.D. Christov, and M.M. Konstantinov. *Computational Methods for Linear Control Systems.* Prentice-Hall, Hertfordshire, UK, 1991.

[39] L.S. Pontryagin, V. Boltyanskii, R. Gamkrelidze, and E. Mishenko. *The Mathematical Theory of Optimal Processes.* Interscience, New York, 1962.

[40] V. Sima. *Algorithms for Linear-Quadratic Optimization*, volume 200 of *Pure and Applied Mathematics.* Marcel Dekker, Inc., New York, NY, 1996.

[41] P. Van Dooren. A generalized eigenvalue approach for solving Riccati equations. *SIAM J. Sci. Statist. Comput.*, 2:121–135, 1981.

[42] C.F. Van Loan. A symplectic method for approximating all the eigenvalues of a Hamiltonian matrix. *Linear Algebra Appl.*, 61:233–251, 1984.

[43] K. Zhou, J.C. Doyle, and K. Glover. *Robust and Optimal Control.* Prentice-Hall, Upper Saddle River, NJ, 1996.

Zentrum für Technomathematik, Fachbereich 3/Mathematik und Informatik, Universität Bremen, D-28334 Bremen, FRG.

Department of Mathematics, University of Kansas, Lawrence, KS 66045-2142, USA.

Fakultät für Mathematik, TU Chemnitz, D-09107 Chemnitz, FRG.

Fakultät für Mathematik, TU Chemnitz, D-09107 Chemnitz, FRG.

Progress in Systems and Control Theory, Vol. 25

Nonlinear Feedback Stabilization Revisited

Eduardo D. Sontag*

1 Introduction

Methods of feedback design are undergoing an exceptionally rich period of progress and maturation, fueled to a great extent by the discovery of new conceptual notions as well as by the systematic application of certain ideas such as that of control-Lyapunov functions (clf's). This paper, which can be seen as an updated version of [22], discusses the problem of state stabilization, the understanding of which is a fundamental prerequisite to the solution of control problems such as tracking, disturbance rejection, output feedback, or adaptive and robust design.

It is known that, in general, in order to control nonlinear systems one must use switching (discontinuous) mechanisms of various types. Of course, time-optimal solutions for even linear systems often involve such discontinuities, see for instance [23], Chapter 10. But, whereas for linear systems most control problems often admit also (perhaps suboptimal) continuous solutions, when dealing with arbitrary systems discontinuities are unavoidable, even when no optimality objectives are imposed. As in [22], we begin therefore by discussing the necessity of such discontinuities, and explain the characterization of regular stabilizability in terms of differentiable clf's.

Among other results which only became available after [22] was written, we will mention the use of differentiable clf's as a tool in the characterization of robustness with respect to small observation noise. Another major way in which this paper extends the material from [22] is in its treatment of *non*smooth clf's and, associated to them, techniques of discontinuous stabilization. This leads us into the subject of precisely defining what we mean by "solution" of a discontinuous differential equation, and makes contact with the literature on differential games as well as nonsmooth analysis.

There is often a tradeoff between generality and clarity of exposition. In this paper, we have opted for clarity, not necessarily presenting results in their most general formulations. The citations should be consulted for generalizations as well as for the omitted technical details. (The web site: **http://www.math.rutgers.edu/~sontag** contains several of the papers referenced.) However, we have included technical proofs of a few minor extensions of results, which were not available in the literature in the form needed for this exposition.

One subject which was covered in the lecture, but which we cannot include here because of space limitations, is that dealing with the study of

*The research was supported in part by Grant F49620-98-1-0242.

the effect of "large" disturbances on the behavior of feedback systems. This study leads one to the very active area of input to state stability (ISS) and related notions (output to state stability as a model of detectability, input to output stability for the study of regulation problems, and so forth), and constitutes, in the author's view, the most exciting area of current work in nonlinear control. The web site referenced above may be consulted for several recent papers as well as expository articles on that subject.

2 Preliminaries

In this paper, we consider exclusively continuous-time systems evolving in finite-dimensional spaces $\mathbb{R}^n$, and we suppose that controls take values in $\mathcal{U} = \mathbb{R}^m$. A control or input is any measurable locally essentially bounded map $u(\cdot) : [0, \infty) \to \mathcal{U} = \mathbb{R}^m$. In general, we use the notation $|x|$ for Euclidean norms, and use $\|u\|$, or $\|u\|_\infty$ for emphasis, to indicate the essential supremum of a function $u(\cdot)$. For basic terminology and facts about control systems, we rely upon [23]. Given a map $f : \mathbb{R}^n \times \mathbb{R}^m \to \mathbb{R}^n$ which is locally Lipschitz and satisfies $f(0,0) = 0$, we consider the system

$$\dot{x}(t) \;=\; f(x(t), u(t)) \tag{1}$$

and, when f does not depend on u (for instance, when we substitute, later, a feedback law $u = k(x)$, leading to $\dot{x} = f(x, k(x))$), we have a system with no inputs

$$\dot{x}(t) \;=\; f(x(t))\,. \tag{2}$$

All definitions for such systems are implicitly applied as well to systems with inputs (1) by setting $u \equiv 0$; for instance, we define the global asymptotic stability (GAS) property for (2), but we say that (1) is GAS if $\dot{x} = f(x, 0)$ is. The maximal solution $x(\cdot)$ of (1), corresponding to a given initial state $x(0) = x^0$, and to a given control u, is defined on some maximal interval $[0, t_{\max}(x^0, u))$ and is denoted by $x(t, x^0, u)$. For systems with no inputs (2) we write just $x(t, x^0)$.

2.1 Stability and Asymptotic Controllability

The use of "comparison functions" has become widespread in stability analysis, as this formalism allows elegant formulations of most concepts. We recall the relevant definitions here. The class of $\mathcal{K}_\infty$ functions consists of all $\alpha : \mathbb{R}_{\geq 0} \to \mathbb{R}_{\geq 0}$ which are continuous, strictly increasing, unbounded, and satisfy $\alpha(0) = 0$, cf. Figure 1. The class of $\mathcal{KL}$ functions consists of those $\beta : \mathbb{R}_{\geq 0} \times \mathbb{R}_{\geq 0} \to \mathbb{R}_{\geq 0}$ with the properties that (1) $\beta(\cdot, t) \in \mathcal{K}_\infty$ for all t, and (2) $\beta(r, t)$ decreases to zero as $t \to \infty$. (It is worth remarking, cf. [24], that for each $\beta \in \mathcal{KL}$, there exist two functions $\alpha_1, \alpha_2 \in \mathcal{K}_\infty$ so that $\beta(r, t) \leq \alpha_2\left(\alpha_1(r)e^{-t}\right)$ for all s, t. Thus, as every function of the latter form

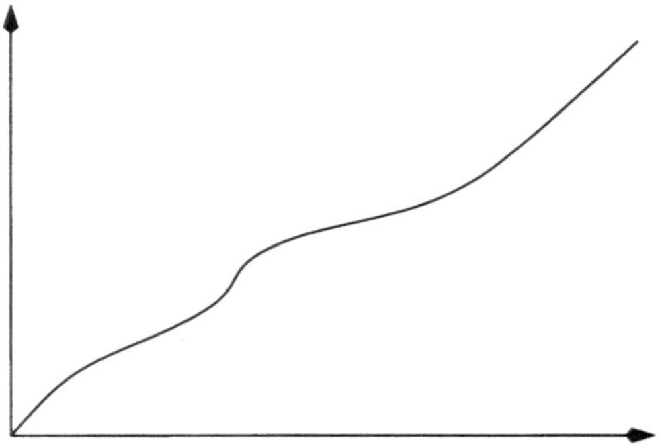

Figure 1: A function of class $\mathcal{K}_\infty$

is in $\mathcal{KL}$, and since $\mathcal{KL}$ functions are only used in order to express upper bounds, in a sense there is no need to introduce the class $\mathcal{KL}$.) We will also use $\mathcal{N}$ to denote the set of all nondecreasing functions $\sigma : \mathbb{R}_{\geq 0} \to \mathbb{R}_{\geq 0}$.

Expressed in this language, the property of *global asymptotic stability (GAS)* for a system with no inputs (2) becomes:

$$(\exists \beta \in \mathcal{KL}) \quad |x(t,x^0)| \;\leq\; \beta\left(|x^0|,t\right) \quad \forall x^0 \,,\; \forall t \geq 0\,.$$

It is an easy exercise to show that this definition is equivalent to the usual "$\varepsilon - \delta$" definition; for one implication, simply observe that

$$|x(t,x^0)| \;\leq\; \beta\left(|x^0|,0\right)$$

provides the stability (or "small overshoot") property, while

$$|x(t,x^0)| \;\leq\; \beta\left(|x^0|,t\right) \underset{t\to\infty}{\longrightarrow} 0$$

gives attractivity.

More generally, we define what it means for the system with inputs (1) to be (open loop, globally) *asymptotically controllable (AC)*. The definition amounts to requiring that for each initial state x^0 there exists some control $u = u_{x^0}(\cdot)$ defined on $[0,\infty)$, such that the corresponding solution $x(t,x^0,u)$ is defined for all $t \geq 0$, and converges to zero as $t \to \infty$, with "small" overshoot. Moreover, we wish to rule out the possibility that $u(t)$ becomes unbounded for x near zero. The precise formulation is as follows.

$$(\exists \beta \in \mathcal{KL})\,(\exists \sigma \in \mathcal{N}) \quad \forall x^0 \in \mathbb{R}^n \;\; \exists u(\cdot)\,,\; \|u\|_\infty \leq \sigma(|x^0|)\,,$$

$$|x(t,x^0,u)| \;\leq\; \beta\left(|x^0|,t\right) \quad \forall t \geq 0\,.$$

Finally, we say that $k : \mathbb{R}^n \to \mathcal{U}$ is a *feedback stabilizer* for the system with inputs (1) if k is locally bounded (that is, k is bounded on each bounded subset of $\mathbb{R}$), $k(0) = 0$, and the closed-loop system

$$\dot{x} \;=\; f(x, k(x)) \tag{3}$$

is GAS, i.e. there is some $\beta \in \mathcal{KL}$ so that $|x(t)| \leq \beta(|x(0)|, t)$ for all solutions and all $t \geq 0$. Obviously, if there exists a feedback stabilizer for (1), then (1) is also AC (just use $u(t) := k(x(t, x^o))$ as u_{x^o}). A most natural question is to ask if the converse holds as well, namely: *is every asymptotically controllable system also feedback stabilizable?* We will see that the answer is "yes", provided that we allow discontinuous feedbacks k, which in turn leads to the technical problem of defining precisely what one means by a "solution" of an initial value problem for (1) when $f(x, k(x))$ is not continuous (since, in that case, the standard theorems on existence do not apply). This is a question that is very much central to the rest of the paper.

2.2 Regularity of Feedback

As already mentioned, one of the central issues with which we will be concerned is that of dealing with possibly discontinuous feedback laws k. Before addressing that subject, however, we will study what can be done with continuous feedback.

It turns out that requirements away from 0, say asking whether k is continuous or smooth, are not very critical; it is often the case that one can "smooth out" a continuous feedback (or, even, make it real-analytic, via Grauert's Theorem) away from the origin. So, in order to avoid unnecessary complications in exposition due to nonuniqueness, let us call a feedback k *regular* if it is locally Lipschitz on $\mathbb{R}^n \setminus \{0\}$. For such k, solutions of initial value problems $\dot{x} = f(x, k(x))$, $x(0) = x^o$, are well defined (at least for small time intervals $[0, \varepsilon)$) and, provided k is a stabilizing feedback, are unique (cf. [23], Exercise 5.9.9).

On the other hand, behavior at the origin cannot be "smoothed out" and, at zero, the precise degree of smoothness plays a central role in the theory. For instance, consider the system

$$\dot{x} = x + u^3 .$$

The continuous (and, in fact, smooth away from zero) feedback $u = k(x) := -\sqrt[3]{2x}$ globally stabilizes the system (the closed-loop system becomes $\dot{x} = -x$). However, there is no possible stabilizing feedback which is differentiable at the origin, since $u = k(x) = O(x)$ implies that

$$\dot{x} = x + O(x^3)$$

about $x = 0$, which means that the solution starting at any positive and small point moves to the right, instead of towards the origin. (A general result, assuming that A has no purely imaginary eigenvalues, cf. [23], Section 5.8, is that if –and only if– $\dot{x} = Ax + Bu + o(x, u)$ can be locally asymptotically stabilized using a feedback which is differentiable at the origin, the linearization $\dot{x} = Ax + Bu$ must be itself stabilizable. In the example that we gave, this linearization is just $\dot{x} = x$, which is not stabilizable.)

3 Nonexistence of Regular Feedback

We now turn to the question of existence of regular feedback stabilizers. We first study a comparatively trivial case, namely systems with one state variable and one input. After that, we turn to multidimensional systems.

3.1 The Special Case $n = m = 1$

There are algebraic obstructions to the stabilization of $\dot{x} = f(x,u)$ if the input u appears nonlinearly in f. Ignoring the requirement that there be a $\sigma \in \mathcal{N}$ so that controls can be picked with $\|u\| \leq \sigma(|x^{\mathrm{o}}|)$, asymptotic controllability is, for $n = m = 1$, equivalent to:

$$(\forall\, x \neq 0)\,(\exists\, u)\;\; xf(x,u) < 0 \tag{4}$$

(this is proved in [25]; it is fairly obvious, but some care must be taken to deal with the fact that one is allowing arbitrary measurable controls; the argument proceeds by first approximating such controls by piecewise constant ones). Let us introduce the following set:

$$\mathcal{O} \;:=\; \{(x,u) \mid xf(x,u) < 0\}\,,$$

and let $\pi : (x,u) \mapsto x$ be the projection into the first coordinate in $\mathbb{R}$. Then, (4) is equivalent to:

$$\pi\mathcal{O} \;=\; \mathbb{R} \setminus \{0\}\,.$$

(One can easily include the requirement "$\|u\| \leq \sigma(|x^{\mathrm{o}}|)$" by asking that for each interval $[-K,K] \subset \mathbb{R}$ there must be some compact set $C_K \subset \mathbb{R}^2$ so that $[-K,K] \subseteq \pi(C_K)$. For simplicity, we ignore this technicality.) In these terms, a stabilizing feedback is nothing else than a locally bounded map $k : \mathbb{R} \to \mathbb{R}$ such that $k(0) = 0$ and so that k is a section of π on $\mathbb{R} \setminus \{0\}$:

$$(x, k(x)) \in \mathcal{O} \;\; \forall\, x \neq 0\,.$$

For a regular feedback, we ask that k be locally Lipschitz on $\mathbb{R} \setminus \{0\}$.

Clearly, there is no reason for Lipschitz, or for that matter, just continuous, sections of π to exist. As an illustration, take the system

$$\dot{x} \;=\; x\left[(u-1)^2 - (x-1)\right]\left[(u+1)^2 + (x-2)\right]\,.$$

Let

$$\mathcal{O}_1 = \{(u+1)^2 < (2-x)\} \quad \text{and} \quad \mathcal{O}_2 = \{(u-1)^2 < (x-1)\}$$

(see Figure 2). Here, $\mathcal{O}$ has three connected components, namely $\mathcal{O}_2$ and $\mathcal{O}_1$ intersected with $x > 0$ and $x < 0$. It is clear that, even though $\pi\mathcal{O} = \mathbb{R}$, there is no continuous curve (graph of $u = k(x)$) which is always in $\mathcal{O}$

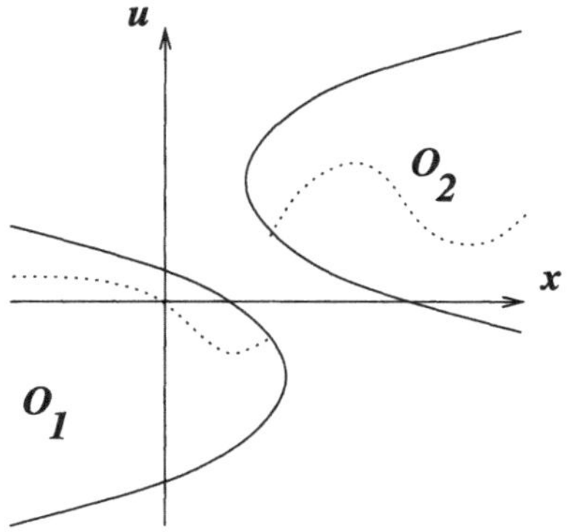

Figure 2: Two regions

and projects onto $\mathbb{R} \setminus \{0\}$. On the other hand, there exist many possible feedback stabilizers provided that we allow one discontinuity (e.g., dotted curve in figure). Although our interest here is primarily in time-invariant feedback $u = k(x)$, it is worth pointing out that it is often possible to overcome obstructions to regular feedback stabilization by means of the use of *time-varying* feedback $u = k(t, x)$. A general result in that direction was proved in [25], which established that every one-dimensional system can be stabilized using $k(t, x)$ continuous (it is not difficult to see that the proof can be adapted to obtain $k(t, x)$ periodic in t, for each x). More recently, a very different construction of time-varying feedbacks was accomplished by Coron, cf. [6], who established a general result valid in any dimension, but restricted to systems with no drift, and by Coron and Rosier, cf. [7], for systems for which smooth clf's (see later) exist.

3.2 Obstructions

When feedback laws are required to be continuous at the origin, new obstructions arise. The case of systems with $n = m = 1$ is also a good way to introduce this subject. The first observation is that stabilization about the origin (even if just local) means that we must have, near zero:

$$f(x, k(x)) \begin{cases} > 0 & \text{if } x < 0 \\ < 0 & \text{if } x > 0 \\ = 0 & \text{if } x = 0 \,. \end{cases}$$

In fact, all that we need is that $f(x_1, k(x_1)) < 0$ for some $x_1 > 0$ and $f(x_2, k(x_2)) > 0$ for some $x_2 < 0$. This guarantees, via the intermediate-value theorem that, if k is continuous, the projection

$$(-\varepsilon, \varepsilon) \to \mathbb{R}, \quad x \mapsto f(x, k(x))$$

is onto a neighborhood of zero, for each $\varepsilon > 0$, see Figure 3. It follows, in

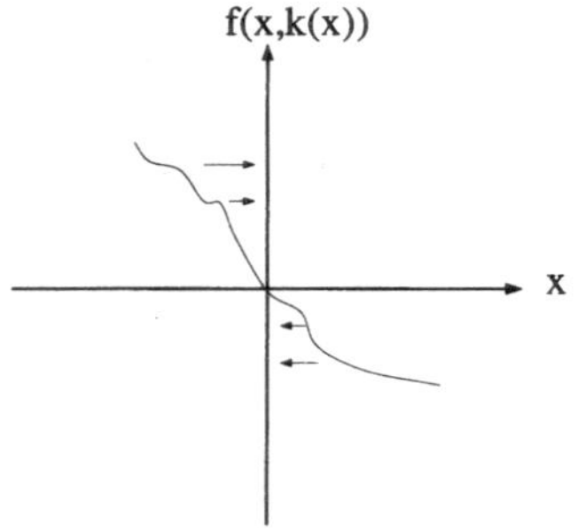

Figure 3: Onto projection, for case $n = m = 1$

particular, that

$$(-\varepsilon, \varepsilon) \times (-\varepsilon, \varepsilon) \to \mathbb{R}, \quad (x, u) \mapsto f(x, u)$$

also contains a neighborhood of zero, for any $\varepsilon > 0$ (that is, the map $(x, u) \mapsto f(x, u)$ is open at zero). This last property is intrinsic, being stated in terms of the original data $f(x, u)$ and not depending upon the feedback k. Brockett's condition, to be described next, is a far-reaching generalization of this argument; in its proof, degree theory replaces the use of the intermediate value theorem.

Logical Decisions are Often Necessary

If there are *global obstacles* in the state-space (that is, if the state-space is a proper subset of $\mathbb{R}^n$), discontinuities in feedback laws cannot in general be avoided. Even if it is in principle possible to reach the origin, it may not be possible to find a regular feedback stabilizer. Actually, this is fairly obvious, and is illustrated in intuitive terms by Figure 4. We think of the position of the (immobile) cat as the origin. In deciding in which way to move, as a function of its current position, the dog must at some point in the state-space make a discontinuous decision: move to the left or to the right of the obstacle (represented by the shaded rectangle)? Formally, this setup can be modeled as a problem in which the state-space is the complement in $\mathbb{R}^2$ of the obstacle, and the fact that discontinuities are necessary is a particular case of a general fact (a theorem of Milnor's), namely that the domain of attraction of an asymptotically stable vector field must be diffeomorphic to Euclidean space (which the complement of the rectangle is not); see [23] for more on the subject.

The interesting point is that even if, as in this exposition, we assume that states evolve in Euclidean spaces, similar obstructions may arise. These are due not to the topology of the state space, but to "virtual obstacles"

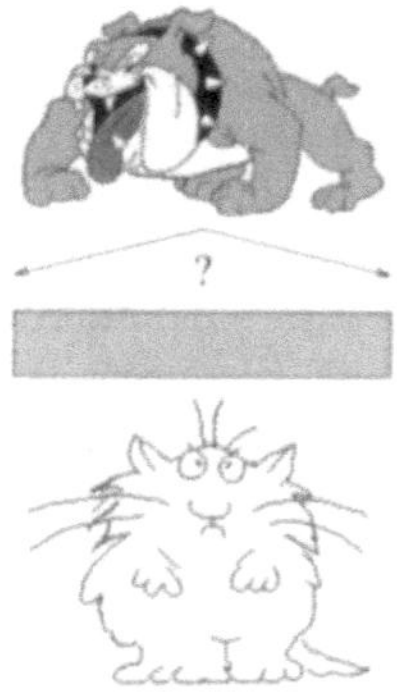

Figure 4: At some point, a discontinuous decision is necessary

implicit in the form of the system equations. These obstacles occur when it is impossible to move *instantaneously* in certain directions, even if it is possible to move *eventually* in every direction ("nonholonomy").

Nonholonomy and Brockett's Theorem

As an illustration, let us consider a model for the "shopping cart" shown in Figure 5 ("knife-edge" or "unicycle" are other names for this example). The state is given by the orientation θ, together with the coordinates x_1, x_2 of the midpoint between the back wheels. The front wheel is a castor, free

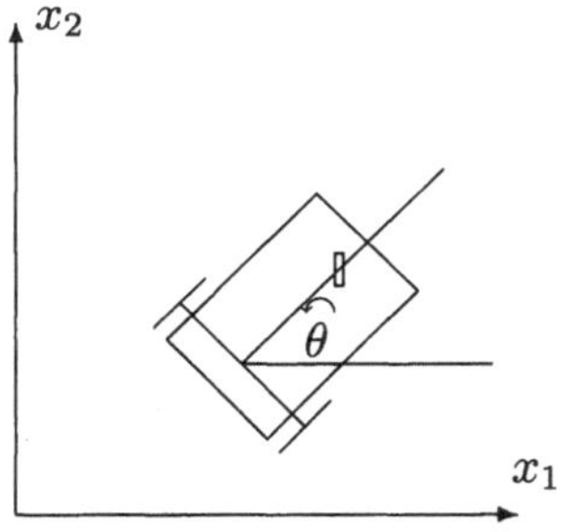

Figure 5: Shopping cart

to rotate. There is a non-slipping constraint on movement: the velocity $(\dot{x}_1, \dot{x}_2)'$ must be parallel to the vector $(\cos\theta, \sin\theta)'$. This leads to the following equations:

$$\begin{aligned}\dot{x}_1 &= u_1 \cos\theta\\ \dot{x}_2 &= u_1 \sin\theta\\ \dot{\theta} &= u_2\end{aligned}$$

where we may view u_1 as a "drive" command and u_2 as a steering control; in practice, one would implement these controls by means of differential forces on the two back corners of the cart. The feedback transformation $z_1 := \theta$, $z_2 := x_1 \cos\theta + x_2 \sin\theta$, $z_3 := x_1 \sin\theta - x_2 \cos\theta$, $v_1 := u_2$, and $v_2 := u_1 - u_2 z_3$ brings the system into the system with equations $\dot{z}_1 = v_1$, $\dot{z}_2 = v_2$, $\dot{z}_3 = z_1 v_2$ known as "Brockett's example" or "nonholonomic integrator" (yet another change can bring the third equation into the form $\dot{z}_3 = z_1 v_2 - z_2 v_1$). We view the system as having state space $\mathbb{R}^3$. Although a physically more accurate state space would be the manifold $\mathbb{R}^2 \times \mathbb{S}^1$, the necessary condition to be given is of a local nature, so the global structure is unimportant.

This system is (obviously) completely controllable (in any case, controllability can be checked using the Lie algebra rank condition, as in e.g. [23], Exercise 4.3.16), and in particular is AC. But we may expect that discontinuities are unavoidable due to the non-slip constraint, which does not allow moving from, for example the position $x_1 = 0$, $\theta = 0$, $x_2 = 1$ in a straight line towards the origin. Indeed, we have:

Theorem A (Brockett [2]) *If there is a stabilizing feedback which is regular and continuous at zero, then the map $(x, u) \mapsto f(x, u)$ is open at zero.*

The test fails here, since no points of the form $(0, \varepsilon, *)$ belong to the image of the map

$$\mathbb{R}^5 \to \mathbb{R}^3 : (x_1, x_2, \theta, u_1, u_2)' \mapsto f(x, u) = (u_1 \cos\theta, u_1 \sin\theta, u_2)'$$

for $\theta \in (-\pi/2, \pi/2)$.

More generally, it is impossible to continuously stabilize any system without drift

$$\dot{x} = u_1 g_1(x) + \ldots + u_m g_m(x) = G(x)u$$

if $m < n$ and $\operatorname{rank}[g_1(0), \ldots, g_m(0)] = m$ (this includes all totally nonholonomic mechanical systems). Indeed, under these conditions, the map $(x, u) \mapsto G(x)u$ cannot contain a neighborhood of zero in its image, when restricted to a small enough neighborhood of zero. Indeed, let us first rearrange the rows of G:

$$G(x) \rightsquigarrow \begin{pmatrix} G_1(x) \\ G_2(x) \end{pmatrix}$$

so that $G_1(x)$ is of size $m \times m$ and is nonsingular for all states x that belong to some neighborhood N of the origin. Then,

$$\begin{pmatrix} 0 \\ a \end{pmatrix} \in \operatorname{Im}[N \times \mathbb{R}^m \to \mathbb{R}^n : (x, u) \mapsto G(x)u] \Rightarrow a = 0$$

(since $G_1(x)u = 0 \Rightarrow u = 0 \Rightarrow G_2(x)u = 0$ too).

If the condition $\text{rank}[g_1(0), \ldots, g_m(0)] = m$ is violated, we cannot conclude a negative result. For instance, the system $\dot{x}_1 = x_1 u$, $\dot{x}_2 = x_2 u$ has $m = 1 < 2 = n$ but it can be stabilized by means of the feedback law $u = -(x_1^2 + x_2^2)$.

Observe that for linear systems, Brockett's condition says that

$$\text{rank}\,[A, B] = n$$

which is the Hautus controllability condition (see e.g. [23], Lemma 3.3.7) at the zero mode.

Idea of the Proof

One may prove Brockett's condition in several ways. A proof based on degree theory is probably easiest, and proceeds as follows (for details see for instance [23], Section 5.9). The basic fact, due to Krasnosel'ski, is that if the system $\dot{x} = F(x) = f(x, k(x))$ has the origin as an asymptotically stable point and F is regular (since k is), then the degree (index) of F with respect to zero is $(-1)^n$, where n is the system dimension. In particular, the degree is also nonzero with respect to points p near enough 0, which means that the equation $F(x) = p$ can be solved for small p, and hence $f(x, u) = p$ can be solved as well. The proof that the degree is $(-1)^n$ follows by exhibiting a homotopy, namely

$$F_t(x^0) = \frac{1}{t}\left[x\left(\frac{t}{1-t}, x^0\right) - x^0\right],$$

between $F_0 = F$ and $F_1(x) = -x$, and noting that the degree of the latter is obviously $(-1)^n$. An alternative (and Brockett's original) proof uses Lyapunov functions. Asymptotic stability implies the existence of a smooth Lyapunov function V for $\dot{x} = F(x) = f(x, k(x))$, so, on the boundary ∂B of a sublevel set $B = \{x \mid V(x) \leq c\}$ we have that F points towards the interior of B, see Figure 6. So for p small, $F(x) - p$ still points to the interior,

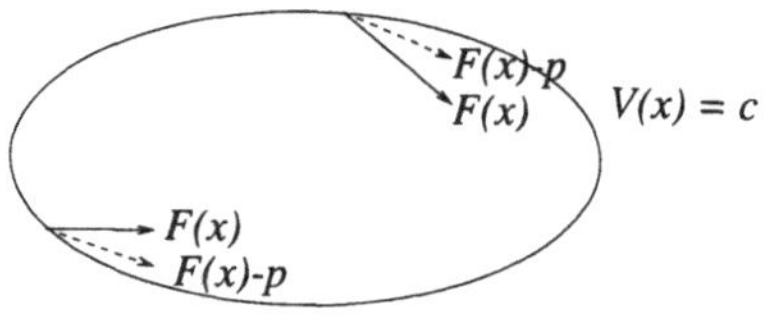

Figure 6: Perturbations of F still point inside B

which means that B is invariant with respect to the perturbed vector field $\dot{x} = F(x) - p$. *Provided that a fixed-point theorem applies to continuous maps* $B \to B$, this implies that $F(x) - p$ must vanish somewhere in B, that is, the equation $F(x) = p$ can be solved. (Because, for each small $h > 0$, the time-h flow ϕ of $F-p$ has a fixed point $x_h \in B$, i.e. $\phi(h, x_h) = x_h$, so picking a convergent subsequence $x_h \to \bar{x}$ gives that $0 = \frac{\phi(h,x_h)-x_h}{h} \to F(\bar{x}) - p$.) A fixed point theorem can indeed be applied, because B is a retract of $\mathbb{R}^n$ (use the flow itself); note that this argument gives a weaker conclusion than the degree condition.

Another Example

There is another nice example of these ideas, which will also be useful later when illustrating Lyapunov techniques (see [1]). It is closely related to the shopping-cart example; in fact, it arises when we control the cart by this procedure: first, we rotate the cart until tangent to, either a circle centered on the x_2-axis and tangent to the x_1-axis (see Figure 7), or the x_1-axis

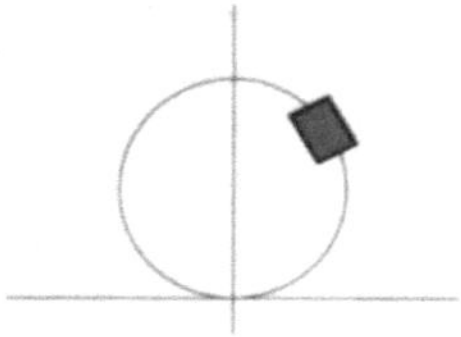

Figure 7: Cart tangent to circle

if we started there; next, we move only with velocities tangential to this circle (steering so as to maintain invariance of the circle). In summary, one obtains a system with state-space $\mathbb{R}^2$, input space $\mathbb{R}$, and equations

$$\dot{x} = g(x)u\,, \text{ where } \quad g(x) = \begin{pmatrix} x_1^2 - x_2^2 \\ 2x_1x_2 \end{pmatrix}.$$

The vector field g and typical orbits of g are shown in Figure 8. In this

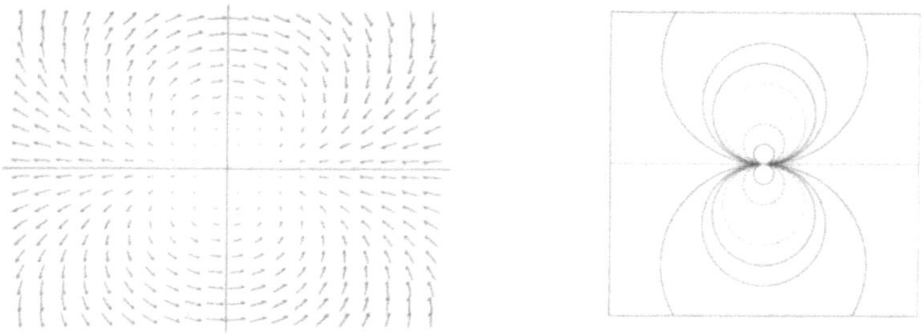

Figure 8: $(x_1^2 - x_2^2)\frac{\partial}{\partial x_1} + 2x_1x_2\frac{\partial}{\partial x_2}$ with typical integral manifolds

system, all motions along the integral curves of g are allowed (clockwise, counterclockwise, or staying at one point). The Brockett condition is satisfied; this is clear if one views the system as a system with complex state space (write $z = x_1 + ix_2$) with equations:

$$\dot{z} = z^2 u\,.$$

The map $(x,u) \mapsto g(x)u$ is $(z,u) \mapsto z^2u$, which is clearly onto any neighborhood of zero, even when restricted to any neighborhood of $z{=}u{=}0$ (take square roots). However, the degree condition fails, as for any continuous feedback one would have $k(x) \neq 0$ for all $x \neq 0$ (otherwise, an equilibrium results), so one obtains degree 2 for $f(x,k(x)) = g(x)k(x)$, and not $(-1)^2 = 1$.

Actually, for this example it is easy to see directly that no regular stabilizing feedback can exist. One proof is by noticing that circles would have to be invariant, so motions restricted to them would result in globally asymptotically stable vector fields in manifolds not diffeomorphic to Euclidean space, contradicting Milnor's theorem. Another proof is even easier: take any path $\gamma : [0,1] \to \mathbb{R}^2$ with $\gamma(0) = (1,0)'$ and $\gamma(1) = (-1,0)'$ and so that $\gamma(r) \neq 0$ for all r. Consider the function $\kappa(r) := k(\gamma(r))$, which is continuous if k was assumed continuous. Since $\kappa(0) < 0$ (because, otherwise, trajectories of $\dot{x} = g(x)k(x)$ starting at $(1,0)'$ cannot converge to zero, since the positive axis is invariant and g points towards the right there) and $\kappa(1) > 0$ (analogous argument), it follows that $\kappa(r_0) = 0$ for some $r_0 \in (0,1)$. Therefore, the point $\bar{x} = \gamma(r_0) \neq 0$ is such that $k(\bar{x}) = 0$ and is therefore an equilibrium point, contradicting the fact that $u = k(x)$ is a stabilizer. We shall return to this example when discussing control-Lyapunov functions.

3.3 Control-Lyapunov Functions

The method of control-Lyapunov functions ("clf's") provides a powerful tool for studying stabilization problems, both as a basis of theoretical developments and as a method for actual feedback design.

Before discussing clf's, let us quickly review the classical concept of Lyapunov functions, through a simple example. Consider first a damped spring-mass system $\ddot{y} + \dot{y} + y = 0$, or, in state-space form with $x_1 = y$ and $x_2 = \dot{y}$, $\dot{x}_1 = x_2$, $\dot{x}_2 = -x_1 - x_2$. One way to verify global asymptotic stability of the equilibrium $x = 0$ is to pick the (Lyapunov) function $V(x_1,x_2) := \frac{3}{2}\,x_1^2 + x_1x_2 + x_2^2$, and observe that $\nabla V(x).f(x) = -\left|x\right|^2 < 0$ if $x \neq 0$, which means that

$$\frac{dV(x(t))}{dt} = -\left|x(t)\right|^2 < 0$$

along all nonzero solutions, and thus the energy-like function V decreases along all trajectories, which, since V is a nondegenerate quadratic form, implies that $x(t)$ decreases, and in fact $x(t) \to 0$. Of course, in this case one could compute solutions explicitly, or simply note that the characteristic equation has all roots with negative real part, but Lyapunov functions are a general technique. (In fact, the classical converse theorems of Massera and Kurzweil show that, whenever a system is GAS, there always exists a smooth Lyapunov function V.)

Now let us modify this example to deal with a control system, and consider a forced (but undamped) harmonic oscillator $\ddot{x} + x = u$, i.e. $\dot{x}_1 = x_2$, $\dot{x}_2 = -x_1 + u$. The damping feedback $u = -x_2$ stabilizes the system, but let us pretend that we do not know that. If we take the same V as before, now the derivatives along trajectories are, using "$\dot{V}(x,u)$" to denote $\nabla V(x).f(x,u)$ and omitting arguments t in $x(t)$ and $u(t)$:

$$\dot{V}(x,u) = -x_1^2 + x_1x_2 + x_2^2 - (x_1 + 2x_2)u\,.$$

This expression is affine in u. Thus, if x is a state such that $x_1 + 2x_2 \neq 0$, then we may pick a control value u (which depends on this current state x) such that $\dot{V} < 0$. On the other hand, if $x_1 + 2x_2 = 0$, then the expression reduces to $\dot{V} = -5x_2^2$ (for any u), which is negative unless x_2 (and hence also $x_1 = -2x_2$) vanishes.

In conclusion, for each $x \neq 0$ there is some u so that $\dot{V}(x,u) < 0$. This is, except for some technicalities to be discussed, the characterizing property of control-Lyapunov functions. For any given compact subset B in $\mathbb{R}^n$, we now pick some compact subset $\mathcal{U}_0 \subset \mathcal{U}$ so that

$$\forall\, x \in B, x \neq 0, \quad \exists\, u \in \mathcal{U}_0 \quad \text{such that} \quad \dot{V}(x,u) < 0\,. \tag{5}$$

In principle, then, we could then stabilize the system, for states in B, by using the steepest descent feedback law:

$$k(x) := \operatorname*{argmin}_{u \in \mathcal{U}_0} \nabla V(x) \cdot f(x,u) \tag{6}$$

("argmin" means "pick any u at which the min is attained"; we restricted $\mathcal{U}$ to be assured that $\dot{V}(x,u)$ attains a minimum). Note that the stabilization problem becomes, in these terms, a set of static nonlinear programming problems: minimize a function of u, for each x. Global stabilization is also possible, by appropriately picking $\mathcal{U}_0$ as a function of the norm of x; later we discuss a precise formulation.

Control-Lyapunov functions, if understood non-technically as the basic paradigm "look for a function $V(x)$ with the properties that $V(x) \approx 0$ if and only if $x \approx 0$, and so that for each $x \neq 0$ it is possible to decrease $V(x)$ by some control action," constitute a very general approach to control (sometimes expressed in a dual fashion, as maximization of some measure of success). They appear in such disparate areas as A.I. game-playing programs (position evaluations), energy arguments for dissipative systems, program termination (Floyd/Dijkstra "variant"), and learning control ("critics" implemented by neural-networks). More relevantly to this paper, the idea underlies much of modern feedback control design, as illustrated for instance by the books [8, 12, 15, 14, 23].

Differentiable clf's: Precise Definition

We say that a continuous function

$$V : \mathbb{R}^n \to \mathbb{R}_{\geq 0}$$

is *positive definite* if $V(x) = 0$ only if $x = 0$, and it is *proper* (or "weakly coercive") if for each $a \geq 0$ the set $\{x \mid V(x) \leq a\}$ is compact, or, equivalently, $V(x) \to \infty$ as $|x| \to \infty$ (radial unboundedness). A property which is equivalent to properness and positive definiteness together is:

$$(\exists\, \underline{\alpha}, \overline{\alpha} \in \mathcal{K}_\infty) \;\; \underline{\alpha}(|x|) \leq V(x) \leq \overline{\alpha}(|x|) \;\; \forall x \in \mathbb{R}^n . \tag{7}$$

A *differentiable control-Lyapunov function* (clf) is a differentiable function $V : \mathbb{R}^n \to \mathbb{R}_{\geq 0}$ which is proper, positive definite, and *infinitesimally decreasing*, meaning that there exists a positive definite continuous function $W : \mathbb{R}^n \to \mathbb{R}_{\geq 0}$, and there is some $\sigma \in \mathcal{N}$, so that

$$\sup_{x \in \mathbb{R}^n} \; \min_{|u| \leq \sigma(|x|)} \; \nabla V(x) \cdot f(x,u) + W(x) \;\leq\; 0 . \tag{8}$$

This is basically the same as condition (5), with $\mathcal{U}_0 =$ the ball of radius $\sigma(|x|)$ picked as a function of x. The main difference is that, instead of saying "$\nabla V(x) \cdot f(x,u) < 0$ for $x \neq 0$" we write $\nabla V(x) \cdot f(x,u) \leq -W(x)$, where W is negative when $x \neq 0$. The two definitions are equivalent, but the "Hamiltonian" version used here is the correct one for the generalizations to be given, to nonsmooth V.

Theorem B (Artstein [1]) *A control-affine system $\dot{x} = g_0(x) + \sum u_i g_i(x)$ admits a differentiable clf if and only if it admits a regular stabilizing feedback.*

The proof of sufficiency is easy: if there is such a k, then the converse Lyapunov theorem, applied to the closed-loop system $F(x) = f(x, k(x))$, provides a smooth V such that

$$L_F V(x) \;=\; \nabla V(x)\, F(x) \;<\; 0 \quad \forall x \neq 0 .$$

This gives that for all nonzero x there is some u (bounded on bounded sets, because k is locally bounded by definition of feedback) so that $\dot{V}(x,u) < 0$; and one can put this in the form (8).

The necessity is more interesting. The original proof in [1] proceeds by a nonconstructive argument involving partitions of unity, but it is also possible to exhibit explicitly a feedback, written as a function

$$k\left(\nabla V(x)\cdot g_0(x)\,,\,\dots\,,\,\nabla V(x)\cdot g_m(x)\right)$$

of the directional derivatives of V along the vector fields defining the system (*universal formulas* for stabilization). Taking for simplicity $m = 1$, one such formula is:

$$k(x) \;:=\; -\,\frac{a(x)+\sqrt{a(x)^2+b(x)^4}}{b(x)} \qquad (0 \text{ if } b=0)$$

where $a(x) := \nabla V(x)\cdot g_0(x)$ and $b(x) := \nabla V(x)\cdot g_1(x)$. (The expression for k is analytic in a, b when $x \neq 0$, because the clf property means that $a(x) < 0$ whenever $b(x) = 0$, see [23] for details.)

Thus, the question of existence of regular feedback, for control-affine systems, reduces to the search for differentiable clf's, and this gives rise to a vast literature dealing with the construction of such V's, see [8, 15, 14, 23] and references therein. Many other theoretical issues are also answered by Artstein's theorem. For example, via Kurzweil's converse theorem one has that the existence of k merely continuous on $\mathbb{R}^n \setminus \{0\}$ suffices for the existence of smooth (infinitely differentiable) V, and from here one may in turn find a k which is smooth on $\mathbb{R}^n \setminus \{0\}$. In addition, one may easily characterize the existence of k continuous at zero as well as regular: this is equivalent to the *small control property*: for each $\varepsilon > 0$ there is some $\delta > 0$ so that $0 < |x| < \delta$ implies that $\min_{|u|\leq\varepsilon} \nabla V(x)\cdot f(x,u) < 0$ (if this property holds, the universal formula automatically provides such a k). We should note that Artstein provided a result valid for general, not necessarily control-affine systems $\dot{x} = f(x,u)$; however, the obtained "feedback" has values in sets of relaxed controls, and is not a feedback law in the classical sense. Later, we discuss a different generalization.

Differentiable clf's will in general not exist, because of obstructions to regular feedback stabilization. This leads us naturally into the twin subjects of discontinuous feedbacks and non-differentiable clf's.

4 Discontinuous Feedback

The previous results and examples show that, in order to develop a satisfactory general theory of stabilization, one in which one proves the implication "asymptotic controllability implies feedback stabilizability," we must allow discontinuous feedback laws $u = k(x)$. But then, a major technical difficulty arises: solutions of the initial-value problem $\dot{x} = f(x,k(x))$, $x(0) = x^0$, interpreted in the classical sense of differentiable functions or even as (absolutely) continuous solutions of the integral equation $x(t) =$

$x^0 + \int_0^t f(x(s), k(x(s)))\, ds$, do not exist in general. The only general theorems apply to systems $\dot{x} = F(x)$ with continuous F. For example, there is no solution to $\dot{x} = -\operatorname{sign} x$, $x(0) = 0$, where $\operatorname{sign} x = -1$ for $x < 0$ and $\operatorname{sign} x = 1$ for $x \geq 0$. So one cannot even pose the stabilization problem in a mathematically consistent sense.

There is, of course, an extensive literature addressing the question of discontinuous feedback laws for control systems and, more generally, differential equations with discontinuous right-hand sides. One of the best-known candidates for the concept of solution of (3) is that of a *Filippov solution* [9, 10], which is defined as the solution of a certain differential inclusion with a multivalued right-hand side which is built from $f(x, k(x))$. Unfortunately, there is no hope of obtaining the implication "asymptotic controllability implies feedback stabilizability" if one interprets solutions of (3) as Filippov solutions. This is a consequence of results in [20, 7], which established that the existence of a discontinuous stabilizing feedback in the Filippov sense implies the Brockett necessary conditions, and, moreover, for systems affine in controls it also implies the existence of regular feedback (which we know is in general impossible).

A different concept of solution originates with the theory of discontinuous positional control developed by Krasovskii and Subbotin in the context of differential games in [13], and it is the basis of the new approach to discontinuous stabilization proposed in [5], to which we now turn.

4.1 Limits of High-Frequency Sampling

By a *sampling schedule* or *partition* $\pi = \{t_i\}_{i \geq 0}$ of $[0, +\infty)$ we mean an infinite sequence

$$0 = t_0 < t_1 < t_2 < \dots$$

with $\lim_{i \to \infty} t_i = \infty$. We call

$$\mathbf{d}(\pi) := \sup_{i \geq 0}(t_{i+1} - t_i)$$

the *diameter* of π. Suppose that k is a given feedback law for system (1). For each π, the *π-trajectory starting from x^0* of system (3) is defined recursively on the intervals $[t_i, t_{i+1})$, $i = 0, 1, \dots$, as follows. On each interval $[t_i, t_{i+1})$, the initial state is measured, the control value $u_i = k(x(t_i))$ is computed, and the constant control $u \equiv u_i$ is applied until time t_{i+1}; the process is then iterated. That is, we start with $x(t_0) = x^0$ and solve recursively

$$\dot{x}(t) = f(x(t), k(x(t_i)))\,,\ t \in [t_i, t_{i+1})\,, \quad i = 0, 1, 2, \dots$$

using as initial value $x(t_i)$ the endpoint of the solution on the preceding interval. The ensuing π-trajectory, which we denote as $x_\pi(\cdot, x^0)$, is defined on some maximal nontrivial interval; it may fail to exist on the entire interval $[0, +\infty)$ due to a blow-up on one of the subintervals $[t_i, t_{i+1})$. We say that it is *well defined* if $x_\pi(t, x^0)$ is defined on all of $[0, +\infty)$.

Definition. The feedback $k : \mathbb{R}^n \to \mathcal{U}$ *stabilizes* the system (1) if there exists a function $\beta \in \mathcal{KL}$ so that the following property holds: For each

$$0 < \varepsilon < K$$

there exists a $\delta = \delta(\varepsilon, K) > 0$ such that, for every sampling schedule π with $\mathbf{d}(\pi) < \delta$, and for each initial state x^0 with $|x^0| \leq K$, the corresponding π-trajectory of (3) is well-defined and satisfies

$$|x_\pi(t, x^0)| \leq \max\{\beta(K, t)\ , \varepsilon\} \quad \forall t \geq 0\,. \tag{9}$$

In particular, we have

$$|x_\pi(t, x^0)| \leq \max\{\beta(|x^0|, t)\ , \varepsilon\} \quad \forall t \geq 0 \tag{10}$$

whenever $0 < \varepsilon < |x^0|$ and $\mathbf{d}(\pi) < \delta(\varepsilon, |x^0|)$ (just take $K := |x^0|$).

Observe that the role of δ is to specify a lower bound on intersampling times. Roughly, one is requiring that

$$t_{i+1} \leq t_i + \theta(|x(t_i)|)$$

for each i, where θ is an appropriate positive function.

Our definition of stabilization is physically meaningful, and is very natural in the context of sampled-data (computer control) systems. It says in essence that a feedback k stabilizes the system if it drives all states asymptotically to the origin and with small overshoot when using *any fast enough sampling schedule*. A high enough sampling frequency is generally required when close to the origin, in order to guarantee small displacements, and also at infinity, so as to preclude large excursions or even blow-ups in finite time. This is the reason for making δ depend on ε and K.

This concept of stabilization can be reinterpreted in various ways. One is as follows. Pick any initial state x^0, and consider any sequence of sampling schedules π_ℓ whose diameters $\mathbf{d}(\pi_\ell)$ converge to zero as $\ell \to \infty$ (for instance, constant sampling rates with $t_i = i/\ell$, $i = 0, 1, 2, \ldots$). Note that the functions $x_\ell := x_{\pi_\ell}(\cdot, x^0)$ remain in a bounded set, namely the ball of radius $\beta(|x^0|, 0)$ (at least for ℓ large enough, for instance, any ℓ so that $\mathbf{d}(\pi_\ell) < \delta(|x^0|/2, |x^0|)$). Because $f(x, k(x))$ is bounded on this ball, these functions are equicontinuous, and (Arzela-Ascoli's Theorem) we may take a subsequence, which we denote again as $\{x_\ell\}$, so that $x_\ell \to x$ as $\ell \to \infty$ (uniformly on compact time intervals) for some absolutely continuous (even Lipschitz) function $x : [0, \infty) \to \mathbb{R}^n$. *We may think of any limit function $x(\cdot)$ that arises in this fashion as a generalized solution of the closed-loop equation (3).* That is, generalized solutions are the limits of trajectories arising from arbitrarily high-frequency sampling when using the feedback law $u = k(x)$. Generalized solutions, for a given initial state x^0, may not be unique – just as may happen with continuous but non-Lipschitz feedback – but there is always existence, and, moreover, for any generalized solution, $|x(t)| \leq \beta(|x^0|, t)$ for all $t \geq 0$. This is precisely the defining estimate for the GAS property. Moreover, if k happens to be regular, then the unique solution of $\dot{x} = f(x, k(x))$ in the classical sense is also the unique generalized solution, so we have a reasonable extension of the concept of solution. (This type of interpretation is somewhat analogous, at least in spirit, to the way in which "relaxed" controls are interpreted in optimal trajectory calculations, namely through high-frequency switching of approximating regular controls.)

Remark. The definition of stabilization was given in [5] in a slightly different form. It was required there that there exist for each $R > 0$ a number $M(R) > 0$, with $\lim_{R \searrow 0} M(R) = 0$, and, for each $0 < r < R$, numbers $\delta_0(r, R) > 0$ and $T(r, R) \geq 0$, such that the following property holds: for each sampling schedule π with $\mathbf{d}(\pi) < \delta_0(r, R)$, each x^0 with $|x^0| \leq R$, and each $t \geq T(r, R)$, it holds that $|x_\pi(t, x^0)| \leq r$, and, in addition, $|x_\pi(t, x^0)| \leq M(R)$ for all $t \geq 0$. This definition is equivalent to the definition that we gave, based on an estimate of the type (9). See Section A.1 for a proof.

4.2 Stabilizing Feedbacks Exist

In the paper [5], the following result was proven by Clarke, Ledyaev, Subbotin, and the author:

Theorem C *The system (1) admits a stabilizing feedback if and only if it is asymptotically controllable.*

Necessity is clear. The sufficiency statement is proved by construction of k, and is based on the following ingredients:

- Existence of a nonsmooth control-Lyapunov function V.
- Regularization on shells of V.
- Pointwise minimization of a Hamiltonian for the regularized V

In order to sketch this construction, we start by quickly reviewing a basic concept from nonsmooth analysis.

Proximal Subgradients

Let V be any continuous function $\mathbb{R}^n \to \mathbb{R}$ (or even, just lower semicontinuous and with extended real values). A *proximal subgradient* of V at the point $x \in \mathbb{R}^n$ is any vector $\zeta \in \mathbb{R}^n$ such that, for some $\sigma > 0$ and some neighborhood $\mathcal{O}$ of x,

$$V(y) \geq V(x) + \zeta \cdot (y - x) - \sigma^2 |y - x|^2 \quad \forall y \in \mathcal{O}.$$

In other words, proximal subgradients are the possible gradients of supporting quadratics at the point x. The set of all proximal subgradients at x is denoted $\partial_P V(x)$. For example (see Figure 9), the function $V(x) = |x|$

Figure 9: $\partial_P |x| (0) = [-1, 1]$ $\qquad$ $\partial_P(-|x|)(0) = \emptyset$

admits any $\zeta \in [-1, 1]$ as a proximal subgradient at $x = 0$ (elsewhere, $\partial_P V(x) = \{\nabla V(x)\}$), while for $V(x) = -|x|$ we have $\partial_P V(0) = \emptyset$ (because there are no possible quadrics that fit inside the graph and touch the corner).

Nonsmooth Control-Lyapunov Functions

A *continuous* (but not necessarily differentiable) $V : \mathbb{R}^n \to \mathbb{R}_{\geq 0}$ is a *control-Lyapunov function* (clf) if it is proper, positive definite, and infinitesimally decreasing in the following generalized sense: there exist a positive definite continuous $W : \mathbb{R}^n \to \mathbb{R}_{\geq 0}$ and a $\sigma \in \mathcal{N}$ so that

$$\sup_{x \in \mathbb{R}^n} \max_{\zeta \in \partial_P V(x)} \min_{|u| \leq \sigma(|x|)} \zeta \cdot f(x, u) + W(x) \leq 0. \tag{11}$$

This is the obvious generalization of the differentiable case in (8); we are still asking that one should be able to make $\nabla V(x) \cdot f(x, u) < 0$ by an appropriate choice of $u = u_x$, for each $x \neq 0$, except that now we replace $\nabla V(x)$ by the proximal subgradient set $\partial_P V(x)$. An equivalent property is to ask that V be a viscosity supersolution of the corresponding Hamilton-Jacobi-Bellman equation.

In the paper [21], the following result was proven by the author:

Theorem D *The system (1) is asymptotically controllable if and only if it admits a continuous clf.*

Not surprisingly, the proof is based on first constructing an appropriate W, and then letting V be the optimal cost (Bellman function) for the problem $\min \int_0^\infty W(x(s))\, ds$. However, some care has to be taken to insure that V is continuous, and the cost has to be adjusted in order to deal with possibly unbounded minimizers. Actually, to be precise, the result as stated here is really a restatement (cf. [26], [5]) of the main theorem given in [21]. See Section A.2 for the details of this reduction.

Regularization

Once V is known to exist, the next step in the construction of a stabilizing feedback is to obtain Lipschitz approximations of V. For this purpose, one considers the Iosida-Moreau inf-convolution of V with a quadratic function:

$$V_\alpha(x) := \inf_{y\in\mathbb{R}^n}\left[V(y)+\frac{1}{2\alpha^2}|y-x|^2\right]$$

where the number $\alpha > 0$ is picked constant on appropriate regions. One has that $V_\alpha(x) \nearrow V(x)$, uniformly on compacts. Since V_α is locally Lipschitz, Rademacher's Theorem insures that V_α is differentiable almost everywhere. The feedback k is then made equal to a pointwise minimizer k_α of the Hamiltonian, at the points of differentiability (compare with (6) for the case of differentiable V):

$$k_\alpha(x) := \operatorname*{argmin}_{u\in\mathcal{U}_0} \nabla V_\alpha(x)\cdot f(x,u)\,,$$

where α and the compact $\mathcal{U}_0 = \mathcal{U}_0(\alpha)$ are chosen constant on certain compacts and this choice is made in between level curves, see Figure 10. The

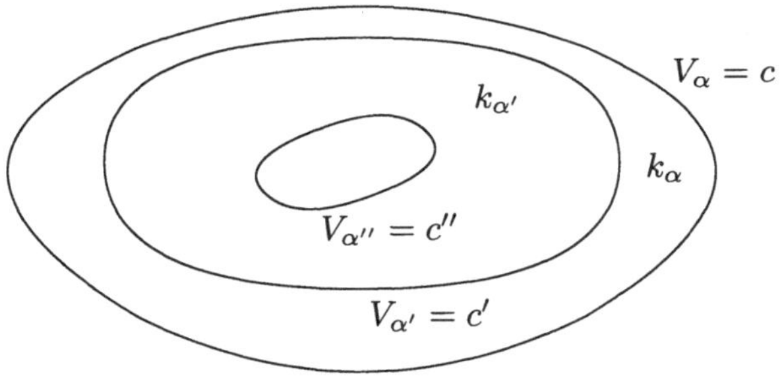

Figure 10: $k = k_\alpha$ on $\{x \mid V_\alpha(x) \le c, V_{\alpha'}(x) > c'\}$

critical fact is that V_α is itself a clf for the original system, at least when restricted to the region where it is needed. More precisely, on each shell of the form

$$C = \{x \in \mathbb{R}^n \mid r \le |x| \le R\}\,,$$

there are positive numbers m and α_0 and a compact subset $\mathcal{U}_0$ such that, for each $0 < \alpha \le \alpha_0$, each $x \in C$, and every $\zeta \in \partial_P V_\alpha(x)$,

$$\min_{u\in\mathcal{U}_0} \zeta\cdot f(x,u) + m \le 0\,.$$

See Section A.3.

Actually, this description is oversimplified, and the proof is a bit more delicate. One must define, on appropriate compact sets

$$k(x) := \underset{u\in\mathcal{U}_0}{\operatorname{argmin}}\, \zeta_\alpha(x)\cdot f(x,u)\,,$$

where $\zeta_\alpha(x)$ is carefully chosen. At points x of nondifferentiability, $\zeta_\alpha(x)$ is not a proximal subgradient of V_α, since $\partial_{\rm P}V_\alpha(x)$ may well be empty. One uses, instead, the fact that $\zeta_\alpha(x)$ happens to be in $\partial_{\rm P}V(x')$ for some $x'\approx x$.

An Example

As a simple example, we consider the system that was obtained from the two-dimensional reduction of the "shopping cart" problem, cf. Figure 8. For this example, continuous stabilization is not possible, and so no differentiable clf's can exist. On the other hand, the system is AC, so one can stabilize it using the techniques just described. A clf for this problem was obtained in [16]:

$$V(x_1,x_2) = \frac{x_1^2+x_2^2}{\sqrt{x_1^2+x_2^2}+|x_1|}$$

(0 if $x_1=x_2=0$) and its level sets are as shown in Figure 11. Note

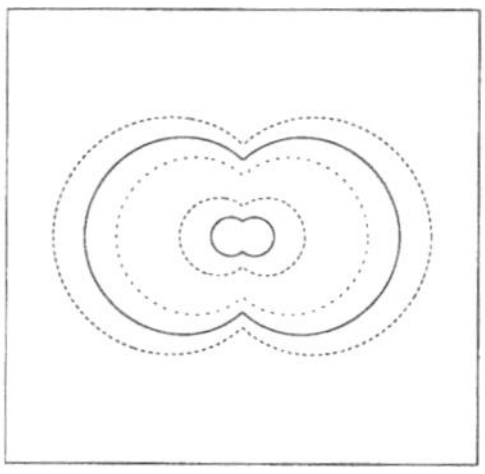

Figure 11: Clf Level Sets for system in Figure 8

that the nonsmoothness happens exactly on the x_2 axis (one has an empty subgradient at those points), and the clf inequality follows from the fact that

$$\inf_u \nabla V(x_1,x_2)\cdot f((x_1,x_2),u) \leq -\frac{1}{2}(x_1^2+x_2^2)$$

at points with $x_1\neq 0$ (the proximal subgradient set is empty otherwise). The stabilizing feedback that results is the obvious one: if to the right of the x_2 axis, move clockwise, if to the left counterclockwise, and make an arbitrary decision (this arbitrariness corresponds to the choice of "$\zeta_\alpha(x)$" in the theory) on the x_2-axis.

5 Sensitivity to Small Measurement Errors

We have seen that every asymptotically controllable system admits a feedback stabilizer k, generally discontinuous, which renders the closed-loop system $\dot{x} = f(x, k(x))$ GAS. On the other hand, one of the main reasons for using feedback is to deal with uncertainty, and one possible source of uncertainty are measurement errors in state estimation. The use of discontinuous feedback means that undesirable behavior –chattering– may arise. In fact, one of the main reasons for the focus on continuous feedback is precisely in order to avoid such behaviors. Thus, we turn now to an analysis of the effect of measurement errors.

Suppose first that k is a continuous function of x. Then, if the error e is small, using the control $u' = k(x+e)$ instead of $u = k(x)$ results in behavior which remains close to the intended one, since $k(x+e) \approx k(x)$; moreover, if $e \ll x$ then stability is preserved. This property of robustness to small errors when k is continuous can be rigorously established by means of a Lyapunov proof, based on the observation that, if V is a Lyapunov function for the closed-loop system, then continuity of $f(x, k(x+e))$ on e means that

$$\nabla V(x) \cdot f(x, k(x+e)) \;\approx\; \nabla V(x) \cdot f(x, k(x)) \;<\; 0\,.$$

Unfortunately, when k is not continuous, this argument breaks down. However, it can be modified so as to avoid invoking continuity of k. Assuming that V is continuously differentiable, one can argue that

$$\nabla V(x) \cdot f(x, k(x+e)) \approx \nabla V(x+e) \cdot f(x, k(x+e)) \;<\; 0$$

(using the Lyapunov property at the point $x+e$ instead of at x). This observation leads to a theorem, formulated below, which says that a discontinuous feedback stabilizer, robust with respect to small observation errors, can be found provided that there is a $\mathcal{C}^1$ clf.

In general, as there are no $\mathcal{C}^1$, but only continuous, clf's, one may not be able to find any feedback law that is robust in this sense. We can see this fact intuitively with an example. Let us take once more the two-dimensional problem illustrated in Figure 8, and let us suppose that we are using the following control law: if to the right of, or exactly on, the x_2 axis, move clockwise, and if to the left move counterclockwise. See Figure 12, which indicates what happens on any circle. The main point that we wish to make is that *this feedback law is extremely sensitive to measurement errors*. Indeed, if the true state x is slightly to the left of the top point, but we mistakenly believe it to be to the right, we use a clockwise motion, in effect bringing the state towards the top, instead of downwards towards the target (the origin); see Figure 13. It is clear that, if we are unlucky enough to consistently make measurement errors that place us on the opposite side

Figure 12: Feedback k on a typical integral manifold

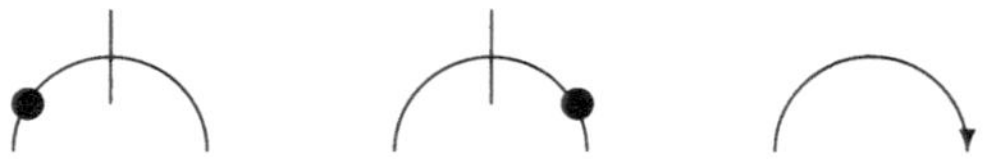

Figure 13: True state; measured state; erroneous motion commanded

of the true state, the result may well be an oscillation (chattering) around the top.

We might say, then, that the discontinuous feedback laws which are guaranteed to always exist by the general theorem in [5] lead to *fussy control*[†] –not to be confused, of course, with "fuzzy" control.

There are many well-known techniques for avoiding chattering, and a very common one is the introduction of deadzones where no action is taken. Indeed, in the above example, one may adopt the following modified control strategy: stay in the chosen direction (even if it might be "wrong") for some minimal time, until we are guaranteed to be far enough from the discontinuity; only after this minimal amount of time, we sample again. At this point, we know for sure on which side of the top we are. (This assumes that we have an upper bound on the magnitude of the error. Also, of course, observation errors when close to the origin will mean that we can only expect "practical" stability, meaning that we cannot be assured of convergence to the origin, but merely of convergence to a neighborhood of the origin whose size depends on the size of the observation errors.)

Such a control strategy is not a pure "continuous time" one, in that a minimum intersample time is required. It can be interpreted, rather, as constructing a *hybrid* (different time scales needed) and *dynamic* (requiring memory) controller. The paper [17] proves a general result showing the possibility of stabilization of every AC system, using an appropriate definition of general hybrid dynamic controllers. The controller given there incorporates an internal model of system. It compares, at appropriate sampling times, the state predicted by the internal model with the –noisy– observations of the state; whenever these differ substantially, a "reseting" is performed in the state of the controller.

Actually, already the feedback constructed in [5], with no modifications needed, can always be used in a manner robust with respect to small ob-

[†]Fussy (adjective): "...requiring ...close attention to details" (Webster).

servation errors, using the idea illustrated with the circle of not sampling again for some minimal period. Roughly speaking, the general idea is as follows.

Suppose that the true current state, let us say at time $t = t_i$, is x, but that the controller uses $u = k(\tilde{x})$, where $\tilde{x} = x + e$, and e is small. Call x' the state that results at the next sampling time, $t = t_{i+1}$. By continuity of solutions on initial conditions, $|x' - \tilde{x}'|$ is also small, where $\tilde{x}'$ is the state that would have resulted from applying the control u if the true state had been $\tilde{x}$. By continuity, it follows that $V_\alpha(x) \approx V_\alpha(\tilde{x})$ and also $V_\alpha(x') \approx V_\alpha(\tilde{x}')$. On the other hand, the construction in [5] provides that $V_\alpha(\tilde{x}') < V_\alpha(\tilde{x}) - d(t_{i+1} - t_i)$, where d is some positive constant (this is valid while we are far from the origin). Hence, if e is sufficiently small compared to the intersample time $t_{i+1} - t_i$, it will necessarily be the case that $V_\alpha(x')$ must also be smaller than $V_\alpha(x)$. See Figure 14. Thinking of

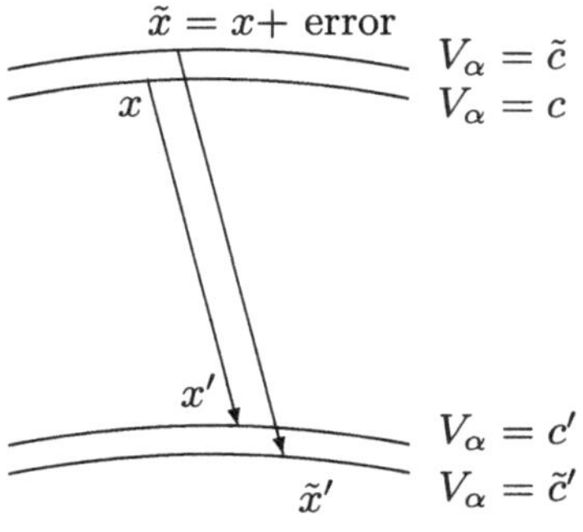

Figure 14: $t_{i+1} \gg t_i \;\Rightarrow\; \tilde{c}' \ll \tilde{c} \;\Rightarrow\; c' < c$

V_α as a Lyapunov function, this means that x' is made "smaller", even if the wrong control $u = k(\tilde{x})$, rather than $k(x)$, is applied.

This discussion may be formalized in several ways. We limit ourselves here to a theorem assuring semiglobal practical stability (i.e., driving all states in a given compact set of initial conditions into a specified neighborhood of zero). For any sampling schedule π, we denote

$$\underline{\mathbf{d}}(\pi) := \inf_{i \geq 0}(t_{i+1} - t_i)\,.$$

If $e : [0, \infty) \to \mathbb{R}^n$ is any function ($e(t)$ is to be thought of as the state estimation error at time t), k is a feedback law, $x^0 \in \mathbb{R}^n$, and π is a sampling schedule, we define the solution of

$$\dot{x} = f(x, k(x + e))\,, \quad x(0) = x^0 \tag{12}$$

as earlier, namely, recursively solving

$$\dot{x}(t) = f(x(t), k(x(t_i) + e(t_i)))$$

with initial condition $x(t_i)$ on the intervals $[t_i, t_{i+1}]$. The feedback stabilizer that was constructed in [5] was defined by patching together feedback laws, denoted there as k_ν (where $\nu = (\alpha, r, R)$ is a triple of positive numbers, with $r < R$). We use these same feedbacks in the statement of the result.

Theorem E *Suppose that the system (1) is asymptotically controllable. Then, there exists a function* $\Gamma \in \mathcal{K}_\infty$ *with the following property. For each* $0 < \varepsilon < K$, *there is a feedback of the type* k_ν, *and there exist positive* $\delta = \delta(\varepsilon, K)$, $\kappa = \kappa(\varepsilon, K)$, *and* $T = T(\varepsilon, K)$, *such that, for each sampling schedule* π *with* $\mathbf{d}(\pi) \leq \delta$, *each* $e : [0, \infty) \to \mathbb{R}^n$ *so that*

$$|e(t)| \leq \kappa \underline{\mathbf{d}}(\pi) \quad \forall t \geq 0,$$

and each x^0 *with* $|x^0| \leq K$, *the solution of the noisy system (12) satisfies*

$$|x(t)| \leq \Gamma(K) \quad \forall t \geq 0$$

and

$$|x(t)| \leq \varepsilon \quad \forall t \geq T.$$

See Section A.4.

5.1 A Necessary Condition

Theorem E insures that stabilization is possible if we sample "just right" (not too slow, so as to preserve stability, but also not too fast, so that observation errors do not cause chattering). It leaves open the theoretical question of precisely under what conditions is it possible to find a state feedback law which is robust with respect to small observation errors and which, on the other hand, is a continuous-time feedback, in the sense of arbitrarily fast sampling. The discussion preceding Theorem E suggests that this objective cannot always be met (e.g., for the circle problem), but that the existence of a $\mathcal{C}^1$ clf might be sufficient for guaranteeing that it can. Indeed, this is what happens, as was proved in the recent paper [18]. We next present the main result from that paper.

We consider systems

$$\dot{x}(t) = f(x(t), k(x(t) + e(t)) + d(t)) \tag{13}$$

in which there are observation errors as well as, now, possible actuator errors $d(\cdot)$. (Actually, robustness to just small actuator errors is not a serious issue; the original paper [5] showed that the feedback laws obtained there stabilize even in the presence of such errors, or even model errors.) We assume that actuator errors $d(\cdot) : [0, \infty) \to \mathcal{U}$ are Lebesgue measurable and locally essentially bounded, and that observation errors $e(\cdot) : [0, \infty) \to \mathbb{R}^n$ are locally bounded.

We wish to define what it means for a feedback k to stabilize (13); roughly, we will ask that, for some $\beta \in \mathcal{KL}$, and for some "tolerance" function $\theta : \mathbb{R}_{\geq 0} \to \mathbb{R}_{\geq 0}$, we have an estimate $|x(t,x^o)| \leq \beta(|x^o|,t)$ provided that $|e(t)| \leq \theta(|x(t)|)$ and $|d(t)| \leq \theta(|x(t)|)$ for all t (small enough errors). However, the definition is somewhat complicated by the need to appropriately choose sampling frequencies.

We define solutions of (13), for each sampling schedule π, in the usual manner, i.e., solving recursively on the intervals $[t_i, t_{i+1})$, $i = 0, 1, \ldots$, the differential equation

$$\dot{x}(t) = f\big(x(t)\,,\, k(x(t_i) + e(t_i)) + d(t)\big) \tag{14}$$

with $x(0) = x^o$. We write $x(t) = x_\pi(t, x^o, d, e)$ for the solution, and say it is *well-defined* if it is defined for all $t \geq 0$.

Definition. The feedback $k : \mathbb{R}^n \to \mathcal{U}$ *stabilizes* the system (13) if there exists a function $\beta \in \mathcal{KL}$ so that the following property holds: For each

$$0 < \varepsilon < K$$

there exist $\delta = \delta(\varepsilon, K) > 0$ and $\eta = \eta(\varepsilon, K)$ such that, for every sampling schedule π with $\mathbf{d}(\pi) < \delta$, each initial state x^o with $|x^o| \leq K$, and each e, d such that $|e(t)| \leq \eta$ for all $t \geq 0$ and $|d(t)| \leq \eta$ for almost all $t \geq 0$, the corresponding π-trajectory of (13) is well-defined and satisfies

$$|x_\pi(t, x^o, d, e)| \leq \max\{\beta(K,t)\ ,\ \varepsilon\} \quad \forall\, t \geq 0\,. \tag{15}$$

In particular, taking $K := |x^o|$, one has that

$$|x_\pi(t, x^o, d, e)| \leq \max\{\beta(|x^o|,t)\ ,\ \varepsilon\} \quad \forall\, t \geq 0$$

whenever $0 < \varepsilon < |x^o|$, $\mathbf{d}(\pi) < \delta(\varepsilon, |x^o|)$, and for all t, $|e(t)| \leq \eta(\varepsilon, |x^o|)$, and $|d(t)| \leq \eta(\varepsilon, |x^o|)$.

The main result in [18] is as follows.

Theorem F *There is a feedback which stabilizes the system (13) if and only if there is a C^1 clf for the unperturbed system (1).*

This result is somewhat analogous to a result obtained, for classical solutions, by Hermes in [11]; see also [10].

It is interesting to note that, as a corollary of Artstein's Theorem, for control-affine systems $\dot{x} = g_0(x) + \sum u_i g_i(x)$ we may conclude that if there is a discontinuous feedback stabilizer that is robust with respect to small noise, then there is also a regular one, and even one that is smooth on $\mathbb{R}^n \setminus \{0\}$.

For non control-affine systems, however, there may exist a discontinuous feedback stabilizer that is robust with respect to small noise, yet there is no regular feedback. For example, consider the following system with $n = 3$ and $m = 1$:

$$\begin{aligned}\dot{x}_1 &= u_2 u_3 \\ \dot{x}_2 &= u_1 u_3 \\ \dot{x}_3 &= u_1 u_2 \,.\end{aligned}$$

Here, there is a $\mathcal{C}^1$ clf, namely the squared norm $(x_1^2 + x_2^2 + x_3^2)$, but there is no possible regular feedback stabilizer, since Brockett's condition fails because points $(0, \neq 0, \neq 0)$ cannot be in the image of $(x, u) \mapsto f(x, u)$. (Because this is a homogeneous system with no drift, Brockett's condition rules out even feedbacks that are not continuous at the origin, see [18] for a remark to that effect.)

The sufficiency part of Theorem F proceeds by taking a pointwise minimization of the Hamiltonian, for a given $\mathcal{C}^1$ clf, i.e. $k(x)$ is defined as any u with $|u| \leq \sigma(|x|)$ which minimizes $\nabla V(x) \cdot f(x, u)$. The necessity part is based on the following technical fact: if the perturbed system can be stabilized, then the differential inclusion

$$\dot{x} \in F(x) := \bigcap_{\varepsilon > 0} \overline{\text{co}}\, f(x, k(x + \varepsilon B))$$

(where B denotes the unit ball in $\mathbb{R}^n$) is strongly asymptotically stable. One may then apply the recent converse Lyapunov theorem of [3] for upper semicontinuous compact convex differential inclusions (which generalized to differential inclusions the theorem from [19] characterizing uniform asymptotic stability of systems with disturbances $\dot{x} = f(x, d)$) to deduce the existence of V.

We can now summarize exactly which implications hold. We write "robust" to mean stabilization of the system subject to observation and actuator noise:

$$\begin{array}{ccccc} \mathcal{C}^1\, V & \Longleftrightarrow & \exists \text{ robust } k & & \\ \Downarrow & & \Downarrow & & \\ \mathcal{C}^0\, V & \Longleftrightarrow & \exists\, k & \Longleftrightarrow & \text{AC} \end{array}$$

Remark. The definition of stabilization of (13) was given in [18] in a slightly different form than here. There, it was required that for each $0 < r < R$ there exist $M = M(R) > 0$ with $\lim_{R \searrow 0} M(R) = 0$, $\delta = \delta(r, R) > 0$, $T = T(r, R) > 0$, and $\eta = \eta(r, R)$, such that, for every partition π with $\mathbf{d}(\pi) < \delta$, each initial state with $|x^0| \leq R$, and each e, d such that $|e(t)| \leq \eta$ for all $t \geq 0$ and $|d(t)| \leq \eta$ for almost all $t \geq 0$, the π-trajectory of $\dot{x} = f(x, k(x + e) + d)$ is defined for all $t \geq 0$ and $|x(t)| \leq r\ \forall\, t \geq T$ and $|x(t)| \leq M(R)\ \forall\, t \geq 0$. This definition is equivalent to the definition that we just gave. The same proof as in Section A.1 applies.

Appendix: Proofs

We fill-in here the proofs of several technical points regarding stabilization.

A.1 Estimates for Stabilization

We prove here that if a feedback k stabilizes in the sense of the definition of stabilization given in [5], then it is also stabilizing in the sense of the definition given here, by means of an estimate of the type (9). (The converse implication is obvious.) Suppose $M(R) > 0$, $\delta_0(r, R) > 0$, and $T(r, R) \geq 0$, are such that $\lim_{R \searrow 0} M(R) = 0$, and whenever $\mathbf{d}(\pi) < \delta_0(r, R)$ and $|x^o| \leq R$, necessarily $|x_\pi(t, x^o)| \leq M(R)$ for all $t \geq 0$ and $|x_\pi(t, x^o)| \leq r$ for all $t \geq T(r, R)$. We first define $\delta = \delta(\varepsilon, R) > 0$, for each $0 < \varepsilon < R$, as follows: pick the smallest positive integer k (necessarily ≥ 2) such that $\frac{R}{k} \leq \varepsilon$, and let

$$\delta(\varepsilon, R) := \min\left\{\delta_0\left(\frac{R}{2}, R\right), \ldots, \delta_0\left(\frac{R}{k}, R\right)\right\}.$$

Next, we define a function

$$\varphi : \mathbb{R}_{\geq 0} \times \mathbb{R}_{\geq 0} \to \mathbb{R}_{\geq 0}$$

as follows. Pick any $R > 0$ and any $t \geq 0$. Let $0 < t_1 < t_2 < \ldots$ be a sequence of real numbers (depending on R) so that $t_i \to \infty$ as $i \to \infty$ and such that

$$t_i \geq T\left(\frac{R}{i+1}, R\right), \quad i = 1, 2, \ldots.$$

Now define $\varphi(R, t) := M(R)$ for $t \in [0, t_1)$ and $\varphi(R, t) := \frac{R}{i+1}$ if $t \in [t_i, t_{i+1})$ for some $i \geq 1$.

We claim now: for each $0 < \varepsilon < R$, each $|x^o| \leq R$, each $t \geq 0$, and each sampling schedule such that $\mathbf{d}(\pi) \leq \delta(\varepsilon, R)$,

$$|x_\pi(t, x^o)| \leq \max\{\varphi(R, t), \varepsilon\}.$$

Pick any such ε, R, x^o, π. Define the sequence $\{t_i\}$ as above, for this R. Let k be the smallest positive integer such that $\frac{R}{k} \leq \varepsilon$. By definition of δ, $\delta(\varepsilon, R) \leq \delta_0(R/j, R)$ for all $j = 2, \ldots, k$. We consider three cases: (i) $t < t_1$, (ii) $t \in [t_1, t_{k-1})$, and (iii) $t \geq t_{k-1}$. In the first case, we know that $|x_\pi(t, x^o)| \leq M(R) = \varphi(R, t)$. In the last case, we know that $|x_\pi(t, x^o)| \leq \varepsilon$, because $\mathbf{d}(\pi) \leq \delta_0(R/k, R)$ implies that $|x_\pi(t, x^o)| \leq R/k \leq \varepsilon$ for all $t \geq T(R/k, R)$, and $t_{k-1} \geq T(R/k, R)$ by definition of the t_i's. In case (ii), we have that there is some $j \in \{2, \ldots, k-1\}$ so that $t \in [t_{j-1}, t_j)$. Since $t_{j-1} \geq T(R/j, R)$ and $\mathbf{d}(\pi) \leq \delta_0(R/j, R)$, $|x_\pi(t, x^o)| \leq R/j = \varphi(R, t)$. The claim is then established.

It only remains to show that there is some function $\beta \in \mathcal{KL}$ so that $\varphi(s, t) \leq \beta(s, t)$ for all s, t. The constructions given in the first section of [19] show that there exists such a β, provided that the following properties hold for φ:

1. For some $\gamma \in \mathcal{K}_\infty$, $\sup_{t>0} \varphi(R,t) \le \gamma(R)$ for all $R > 0$.
2. For each $\varepsilon > 0$ and each $R > 0$ there is some $T(\varepsilon, R)$ such that
$$t \ge T(\varepsilon, R) \;\Rightarrow\; \varphi(R,t) \le \varepsilon\,.$$

The first property is satisfied because M could, without loss of generality, be taken to be of class $\mathcal{K}_\infty$ (if necessary, first use a step function M; then majorize it by a $\mathcal{K}_\infty$ function). The second property holds as well: take without loss of generality $\varepsilon < R$, then pick a positive integer k minimal with $R/k \le \varepsilon$, let the sequence $\{t_i\}$ be as in the construction of φ, and define $T(\varepsilon, R) := t_{k-1}$. Observe that $t \ge t_{k-1}$ implies, by definition of φ, that $\varphi(R,t) \le R/k \le \varepsilon$, as wanted.

A.2 Proximal Form of clf Theorem

We fill-in the details here to show how the proximal subgradient form of the continuous clf existence Theorem D follows from the result in [21]. Before stating the original form of the result, we recall the notion of relaxed control. For each real $s > 0$, let us denote by $\mathcal{U}_s$ the radius-s ball in $\mathcal{U} = \mathbb{R}^m$. A *relaxed* $\mathcal{U}_s$*-valued control* is a measurable map $\omega : I \to \mathbb{P}(\mathcal{U}_s)$, where I is an interval containing zero and $\mathbb{P}(\mathcal{U}_s)$ denotes the set of all Borel probability measures on $\mathcal{U}_s$. Note that ordinary controls can be seen also as relaxed controls, via the natural embedding of $\mathcal{U}_s$ into $\mathbb{P}(\mathcal{U}_s)$ (map any point $u \in \mathcal{U}_s$ into the Dirac delta measure supported at u). Given any $\mu \in \mathbb{P}(\mathcal{U}_s)$, we write $\int_{\mathcal{U}_s} f(x,u)\,d\mu(u)$ simply as $f(x,\mu)$. As with ordinary controls, we also denote by $x(t, x^0, \omega)$ the solution of the initial value problem that obtains from initial state x^0 and relaxed control ω, and we consider the supremum norm $\|\omega\|$, defined as the infimum of the set of s such that $\omega(t) \in \mathbb{P}(\mathcal{U}_s)$ for almost all $t \in I$.

The main result in [21] says that if (and only if) a system is AC, there exist two continuous, positive definite functions $V, W : \mathbb{R}^n \to \mathbb{R}$, with V proper, and a nondecreasing function $\sigma : \mathbb{R}_{\ge 0} \to \mathbb{R}_{\ge 0}$, so that the following property holds: for each $x^0 \in \mathbb{R}^n$ there are a $T > 0$ and a relaxed control $\omega : [0,T) \to \mathbb{P}(\mathcal{U}_{\sigma(|x^0|)})$, so that $x(t) := x(t, x^0, \omega)$ is defined for all $0 \le t < T$ and
$$V(x(t)) - V(x^0) \le -\int_0^t W(x(\tau))\,d\tau \quad \forall t \in [0,T)\,. \tag{16}$$
To simplify notations, let us write $s := \sigma(|x^0|)$. In order to obtain the proximal version of the result, we show that
$$\min_{|u| \le s} \zeta \cdot f(x^0, u) \;\le\; -W(x^0) \tag{17}$$

for all $\zeta \in \partial_P V(x^0)$.

As in [26], we first make an intermediate reduction, showing that

$$\min_{v \in F(x^0,s)} DV(x^0;v) \;\le\; -W(x^0)\,, \tag{18}$$

where $F(x^0,s)$ denotes the (closed) convex hull of $\{f(x^0,u), u \in \mathcal{U}_s\}$, and $DV(x^0;v)$ is the directional subderivate, or contingent epiderivative, of V in the direction of v at x^0, defined as

$$DV(x^0;v) \;:=\; \liminf_{\substack{t \searrow 0 \\ v' \to v}} \;\frac{1}{t}\left[V(x^0+tv') - V(x^0)\right] .$$

(The minimum in (18) is achieved, because the map $v \mapsto DV(x^0;v)$ is lower semicontinuous, see e.g. [4], ex.3.4.1e.) So, let $x(t) = x(t,x^0,\omega)$ be as above, and consider for each $t \in [0,T)$ the vectors

$$r_t \;:=\; \frac{1}{t}\,(x(t)-x^0) \;=\; \frac{1}{t}\int_0^t f(x(\tau),\omega(\tau))\,d\tau \;=\; q_t + p_t\,,$$

with

$$p_t \;:=\; \frac{1}{t}\int_0^t f(x^0,\omega(\tau))\,d\tau\,,$$

where $q_t \to 0$ as $t \searrow 0$ (the existence of such a q_t, for each t, is an easy consequence of the fact that f is locally Lipschitz on x, uniformly on $u \in \mathcal{U}_s$). Moreover, $p_t \in F(x^0,s)$ for all t (because $F(x^0,s)$ is convex, so $f(x^0,\omega(\tau)) \in F(x^0,s)$ for each τ, and then using convexity once more). By compactness, we have that there is some $v \in F(x^0,s)$ and some subsequence $p_{t_j} \to v$ with $t_j \searrow 0$; as $q_t \to 0$, also $v_j := r_{t_j} \to v$. For this v,

$$\begin{aligned} DV(x^0;v) \;&\le\; \liminf_{j\to\infty} \frac{1}{t_j}\left[V(x^0+t_jv_j) - V(x^0)\right] \\ &=\; \liminf_{j\to\infty} \frac{1}{t_j}\left[V(x(t_j)) - V(x^0)\right] \;\le\; -W(x^0) \end{aligned}$$

(using (16)), so (18) indeed holds.

Finally, we show that (18) implies (17). Let $v \in F(x^0,s)$ achieve the minimum, and pick any $\zeta \in \partial_P V(x^0)$. By definition of proximal subgradient, there is some $\mu > 0$ so that, for each $v' \in F(x^0,s)$ and each small $t \ge 0$,

$$\zeta \cdot v' \;\le\; \frac{1}{t}\left[V(x^0+tv') - V(x^0)\right] \;+\; \mu\, t\, |v'|^2$$

so taking limits along any sequence $t \searrow 0$ and $v' \to v$ shows that $\zeta \cdot v \le DV(x^0;v) \le -W(x^0)$. By definition of $F(x^0,s)$, this means that there must exist a $u \in \mathcal{U}_0$ so that also $\zeta \cdot f(x^0,u) \le -W(x^0)$, as desired.

Observe that the proximal condition (17), being linear on the velocities $f(x,u)$, has the great advantage of not requiring convex hulls in order to state, and in that sense is far more elegant than (18).

A.3 V_α is a (Local) clf

We prove here that, on each set $C_{r,R} = \{x \in \mathbb{R}^n \mid r \leq |x| \leq R\}$, there is an $\alpha_0(r,R) > 0$ such that, for each positive $\alpha \leq \alpha_0(r,R)$, the (locally Lipschitz) function V_α behaves like a clf on the set $C_{r,R}$. In order to simplify referencing, we write "([5]n)" to refer to Equation (n) in [5] and do not redefine notations given there. Let $\mathcal{U}_0$ be a compact subset so that $\min_{u\in\mathcal{U}_0} \zeta \cdot f(x,u) \leq -W(x)$ for every $\zeta \in \partial_{\mathrm{P}} V(x)$ and every x in the ball of radius $R + \sqrt{2\beta(R)}$. Let $m_{r,R} := \frac{1}{2}\min\{W(x) \mid x \in C_{r,R}\}$, and let ℓ be so that $|f(x,u) - f(x',u)| \leq \ell\,|x - x'|$ for all $u \in \mathcal{U}_0$ and all x in the ball of radius $R + \sqrt{2\beta(R)}$. (This is almost as in ([5]29), except that, there, ℓ was a Lipschitz constant only with respect to $|x| \leq R$.) Finally, as in [5], $\omega_R(\cdot)$ denotes the modulus of continuity of V on the ball of radius $R + \sqrt{2\beta(R)}$.

Proposition. Let $\alpha \in (0,1]$ satisfy $2\ell\,\omega_R\left(\alpha\sqrt{2\beta(R)}\right) \leq m_{r,R}$. Then, for all $x \in C_{r,R}$ and all $\zeta \in \partial_{\mathrm{P}} V_\alpha(x)$,

$$\min_{u\in\mathcal{U}_0} \zeta \cdot f(x,u) \;\leq\; -m_{r,R}\,.$$

Proof. We start by remarking that, for all $\alpha \in (0,1]$ and all $x \in B_R$,

$$|y_\alpha(x) - x|^2 \;\leq\; 2\alpha^2\,\omega_R\left(\alpha\sqrt{2\beta(R)}\right)\,. \tag{19}$$

This follows from (cf. ([5]19)):

$$\frac{1}{2\alpha^2}\,|y_\alpha(x) - x|^2 \;\leq\; V(x) - V(y_\alpha(x)) \;\leq\; \omega_R\left(|y_\alpha(x) - x|\right)$$

and using that $|y_\alpha(x) - x| \leq \alpha\sqrt{2\beta(R)}$ (cf. Lemma III.3 in [5]) plus the fact that $\omega_R(\cdot)$ is nondecreasing. So

$$|\zeta_\alpha(x)|\cdot|y_\alpha(x) - x| \;=\; \frac{|y_\alpha(x) - x|^2}{\alpha^2} \;\leq\; 2\,\omega_R\left(\alpha\sqrt{2\beta(R)}\right)$$

and hence

$$\zeta_\alpha(x)\cdot[f(y_\alpha(x),u) - f(x,u)] \;\leq\; \ell\,|\zeta_\alpha(x)|\cdot|y_\alpha(x) - x| \;\leq\; 2\ell\,\omega_R\left(\alpha\sqrt{2\beta(R)}\right)$$

for all $u \in \mathcal{U}_0$ and all $x \in B_R$, because $y_\alpha(x) \in B_{R+\sqrt{2\beta(R)}}$ (cf. ([5]20)). Thus,

$$\zeta_\alpha(x)\cdot f(x,u) \;\leq\; \zeta_\alpha(x)\cdot f(y_\alpha(x),u) + 2\ell\,\omega_R\left(\alpha\sqrt{2\beta(R)}\right)\,.$$

For each $x \in C_{r,R}$, we have that $y_\alpha(x) \in B_{R+\sqrt{2\beta(R)}}$ and (cf. ([5]12)) $\zeta_\alpha(x) \in \partial_P V(y_\alpha(x))$, so we can pick a $u_x \in \mathcal{U}_0$ so that $\zeta_\alpha(x) \cdot f(y_\alpha(x), u_x) \leq -W(x) \leq -2m_{r,R}$. Using now the fact that $2\ell\,\omega_R\left(\alpha\sqrt{2\beta(R)}\right) \leq m_{r,R}$, we conclude that

$$\zeta_\alpha(x) \cdot f(x,u) \;\leq\; -m_{r,R}\,.$$

So we only need to prove that $\partial_P V_\alpha(x) \subseteq \{\zeta_\alpha(x)\}$ for all x. This is given in [4], Theorem 1.'5.1, but is easy to prove directly: pick any $\zeta \in \partial_P V_\alpha(x)$; by definition, this means that exists some $\sigma > 0$ such that, for all y near x, $\zeta \cdot (y-x) \leq V_\alpha(y) - V_\alpha(x) + \sigma\,|y-x|^2$, so

$$\zeta \cdot (y-x) \;\leq\; V_\alpha(y) - V_\alpha(x) + \gamma\left(|y-x|\right) \tag{20}$$

for some $\gamma(r) = o(r)$. Adding

$$-\zeta_\alpha(x) \cdot (y-x) \;\leq\; -V_\alpha(y) + V_\alpha(x) + \frac{1}{2\alpha^2}\,|y-x|^2$$

(cf. ([5]13)), we conclude

$$(\zeta - \zeta_\alpha(x)) \cdot (y-x) \;\leq\; o\left(|y-x|\right)\,.$$

Substituting $y = x + h(\zeta - \zeta_\alpha(x))$ and letting $h \searrow 0$ shows that $\zeta = \zeta_\alpha(x)$. (Observe that we have proved more that claimed: Equation (20) is satisfied by any viscosity subgradient ζ; in particular, the gradient of V, if V happens to be differentiable at the point x, must coincide with $\zeta_\alpha(x)$.) ∎

Observe that the feedback constructed in [5] was $k(x) =$ any $u \in \mathcal{U}_0$ minimizing $\zeta_\alpha(x) \cdot f(x,u)$ (where α and $\mathcal{U}_0$ are chosen constant on certain compacts). As V_α is locally Lipschitz, it is differentiable almost everywhere. Thus, the Proposition (see the end of the proof) insures that $\zeta_\alpha(x) = \nabla V_\alpha(x)$ for almost all x. So $k(x) = u$ is, at those points, the pointwise minimizer of the Hamiltonian $\nabla V_\alpha(x) \cdot f(x,u)$ associated to the regularized clf V_α.

A.4 Proof of Theorem E

We will prove the following more precise result. All undefined notations, including the definitions of the functions γ and ρ, can be found in [5].

Theorem G *Pick any $0 < r < R$ so that $2\gamma(r) < \gamma(R)$. Then there exist positive numbers $\alpha, \delta, \kappa, T$ such that, for each partition π with $\mathbf{d}(\pi) \leq \delta$, and each $e : [0,\infty) \to \mathbb{R}^n$ which satisfies $|e(t)| \leq \kappa\,\underline{\mathbf{d}}(\pi)$ for all t, the following property holds: if $x(\cdot)$ satisfies*

$$\dot{x} = f(x, k(x+e))\,, \quad |x(0)| \leq \tfrac{1}{2}\rho(R)\,, \tag{21}$$

where k is the feedback $k_{\alpha,r,R}$, then

$$x(t) \in B_R \quad \forall t \geq 0 \tag{22}$$

and

$$x(t) \in B_r \quad \forall t \geq T\,. \tag{23}$$

Theorem E is a corollary: we first pick any Γ such that $s \leq (1/2)\rho(\Gamma(s))$ for all $s \geq 0$. Then, given ε and K, we can let $R := \Gamma(K)$ (so, $K \leq (1/2)\rho(R)$), we then take any $0 < r < \varepsilon$ such that $2\gamma(r) \leq \gamma(R)$, and apply the above result to $0 < r < R$.

We prove Theorem G through a series of technical steps. Let $0 < r < R$ be given, with $2\gamma(r) < \gamma(R)$. In order to simplify referencing, we write "([5]n)" to refer to Equation (n) in [5].

We start by picking α as any positive number which satisfies ([5]23), ([5]30), and, also, instead of ([5]40), the slightly stronger condition

$$\omega_R\left(\sqrt{2\beta(R)}\alpha\right) < \frac{1}{16}\gamma(r)\,. \tag{24}$$

The function V_α is defined as in ([5]9); because of ([5]13), ([5]11), and ([5]19), the number

$$c := \frac{\sqrt{2\beta(R)}}{\alpha} + \frac{R}{\alpha^2}$$

is a Lipschitz constant for V_α on the set B_R. Without loss of generality, we assume $c \geq 2$. We let $\mathcal{U}_0$ be as in [5], and take the feedback $k = k_{\alpha,r,R}$. The numbers ℓ, m, and Δ are as in ([5]29) and the equation that follows it.

Next, we pick any $\varepsilon_0 > 0$ so that all the following – somewhat redundant – properties hold:

$$B_{\frac{1}{2}\rho(R)} + \varepsilon_0 B \subseteq G_R^\alpha \tag{25}$$

(this is possible because $B_{\rho(R)} \subseteq G_R^\alpha$ by ([5]22)),

$$G_R^\alpha + 2\varepsilon_0 B \subseteq B_R \tag{26}$$

(possible because $G_R^\alpha \subseteq \operatorname{int} B_R$ by ([5]24)),

$$\varepsilon_0 \leq \frac{\gamma(r)}{8c} \leq \frac{\gamma(r)}{16}\,, \tag{27}$$

$$\gamma(r) + c\,\varepsilon_0 < \tfrac{1}{2}\gamma(R) \tag{28}$$

(recall that $2\gamma(r) < \gamma(R)$), and

$$G_r^\alpha + 2\varepsilon_0 B \subseteq G_R^\alpha\,. \tag{29}$$

We let δ_0 be any positive number so that, for every initial state x^0 in the compact set G_R^α, and for each control $u : [0, \delta_0] \to \mathcal{U}_0$, the solution of $\dot{x} = f(x,u)$ with $x(0) = x^\circ$ is defined on the entire interval $[0, \delta_0]$ and satisfies $x(t) \in G_R^\alpha + \varepsilon_0 B$ for all t.

Finally, we pick any $\delta > 0$ which satisfies ([5]33), ([5]41), as well as

$$\delta \le \min\left\{1\,,\, \delta_0\,,\, \frac{2}{\Delta}\varepsilon_0\,,\, \frac{\gamma(r)}{8cm}\right\}, \tag{30}$$

and we let

$$\kappa := \frac{\Delta}{4c}e^{-\ell} \tag{31}$$

and

$$T := \frac{\gamma(R)}{\Delta} \tag{32}$$

(this is twice the value used in ([5]39)).

The main technical fact needed is as follows.

Proposition A.4.1 *Let $0 < r < R$ satisfy $2\gamma(r) < \gamma(R)$. Let $\alpha, \delta, \kappa, T$ be defined as above, and k be the feedback $k_{\alpha,r,R}$. Pick any $\varepsilon > 0$, and consider the following set:*

$$P = P_{r,R,\varepsilon} := \{x \mid x + \varepsilon B \subseteq G_R^\alpha\}\,.$$

Let π be a partition which satisfies

$$\frac{\varepsilon}{\kappa} \le t_{i+1} - t_i \le \delta \quad \forall i = 0, 1, \ldots \tag{33}$$

(that is, $\mathbf{d}(\pi) \le \delta$ and $\varepsilon \le \kappa\,\underline{\mathbf{d}}(\pi)$). Then, for any $e : [0,\infty) \to \mathbb{R}^n$ such that $|e(t)| \le \varepsilon$ for all t, and any $x^\circ \in P$, the solution of $\dot{x} = f(x, k(x+e))$ is defined for all $t \ge 0$, it satisfies (22) and (23), and $x(t_i) \in P$ for all i.

We will prove this via a couple of lemmas, but let us first point out how Theorem G follows from Proposition A.4.1. Suppose given a partition π with $\mathbf{d}(\pi) \le \delta$, an error function e so that $|e(t)| \le \kappa\,\underline{\mathbf{d}}(\pi)$ for all t, and an initial state x° with $|x^\circ| \le \frac{1}{2}\rho(R)$. Let $\varepsilon := \sup_{t\ge0} |e(t)| \le \kappa\,\underline{\mathbf{d}}(\pi)$. Note that

$$\varepsilon \le \kappa\,\underline{\mathbf{d}}(\pi) \le \kappa\,\mathbf{d}(\pi) \le \kappa\,\delta \le \kappa\frac{2}{\Delta}\varepsilon_0 \le \kappa\frac{4c}{\Delta}e^{\ell}\varepsilon_0 = \varepsilon_0\,. \tag{34}$$

Thus, by (25),

$$B_{\frac{1}{2}\rho(R)} \subseteq P\,, \tag{35}$$

so $x^\circ \in P$. Therefore, (22) and (23) hold for the solution of $\dot{x} = f(x, k(x+e))$, as wanted for Theorem G.

We now prove Proposition A.4.1. Observe that (33) implies $\varepsilon \le \kappa\,\underline{\mathbf{d}}(\pi)$, so, arguing as in (34), $\varepsilon \le \varepsilon_0$. It is useful to introduce the following set as well:

$$Q = Q_{r,R,\varepsilon} := G_r^\alpha + \varepsilon B\,.$$

Observe that $Q \subseteq P$, by (29).

We start the proof by establishing an analogue of Lemma IV.2 in [5]:

Lemma A.4.1 *If, for some index i, $x_i := x(t_i) \in P \setminus Q$, then $x(t)$ is defined for all $t \in [t_i, t_{i+1}]$,*

$$x(t) \in B_R \quad \forall t \in [t_i, t_{i+1}], \tag{36}$$

$$V_\alpha(x(t)) \le V_\alpha(x_i) + \varepsilon_0 \quad \forall t \in [t_i, t_{i+1}], \tag{37}$$

and, letting $x_{i+1} := x(t_{i+1})$:

$$x_{i+1} \in P, \tag{38}$$

$$V_\alpha(x_{i+1}) - V_\alpha(x_i) \le -\frac{\Delta}{2}(t_{i+1} - t_i). \tag{39}$$

Proof. By definition of P, we have that $\tilde{x}_i := x_i + e(t_i) \in G_R^\alpha$. Also, $\tilde{x}_i \notin G_r^\alpha$, since otherwise x_i would belong to Q. In particular, $\tilde{x}_i \notin B_{\rho(r)}$. Let $\tilde{x}(\cdot)$ be the solution of $\dot{x} = f(x, k(\tilde{x}_i))$ with $\tilde{x}(t_i) = \tilde{x}_i$ on $[t_i, t_{i+1}]$. By Lemma IV.2 in [5], this solution is well-defined and it holds that $\tilde{x}(t_{i+1}) \in G_R^\alpha$ and $V_\alpha(\tilde{x}(t)) - V_\alpha(\tilde{x}_i) \le -\Delta(t - t_i)$ for all t. As $x_i \in P \subseteq G_R^\alpha$, and $t_{i+1} - t_i \le \mathbf{d}(\pi) \le \delta \le \delta_0$, the definition of δ_0 insures that the solution $x(\cdot)$ of $\dot{x} = f(x, k(\tilde{x}_i))$ with $x(t_i) = x_i$ is indeed well-defined, and it stays in $G_R^\alpha + \varepsilon_0 B \subseteq B_R$ for all t. So, by Gronwall's inequality, we know that

$$|x(t) - \tilde{x}(t)| \le e^{(t-t_i)\ell} |x_i - \tilde{x}_i| \le e^{\delta\ell}\varepsilon$$

for all $t \in [t_i, t_{i+1}]$. Since V_α has Lipschitz constant c on B_R, we have

$$\begin{aligned} &V_\alpha(x(t)) - V_\alpha(x_i) \\ &\quad = V_\alpha(x(t)) - V_\alpha(\tilde{x}(t)) + V_\alpha(\tilde{x}(t)) - V_\alpha(\tilde{x}_i) + V_\alpha(\tilde{x}_i) - V_\alpha(x_i) \\ &\quad \le c e^{\delta\ell}\varepsilon - \Delta(t - t_i) + c\varepsilon \\ &\quad \le \frac{\Delta}{2}(t_{i+1} - t_i) - \Delta(t - t_i) \end{aligned}$$

for all $t \in [t_i, t_{i+1}]$, where we have used that $\delta \le 1$, $\varepsilon \le \kappa\underline{\mathbf{d}}(\pi)$, the definition $\kappa = (\Delta/4c)e^{-\ell}$, and the fact that $\underline{\mathbf{d}}(\pi) \le t_{i+1} - t_i$. In particular, the estimate (39) results at $t = t_{i+1}$, and (37) holds because $(\Delta/2)\delta \le \varepsilon_0$ by (30).

We are only left to prove that $x_{i+1} \in P$. By definition of P, this means that for any given $\eta \in \mathbb{R}^n$ with $|\eta| \le \varepsilon$ it must hold that

$$\eta + x_{i+1} \in G_R^\alpha = \{x \mid V_\alpha(x) \le (1/2)\gamma(R)\}.$$

Pick such an η; then $|\eta + x_{i+1} - \tilde{x}(t_{i+1})| \le \varepsilon + e^{\delta\ell}\varepsilon \le 2e^\ell\varepsilon$, and so, since $\eta + x_{i+1} \in G_R^\alpha + 2\varepsilon_0 B \subseteq B_R$,

$$\begin{aligned} V_\alpha(\eta + x_{i+1}) &\le 2ce^\ell\varepsilon + V_\alpha(\tilde{x}(t_{i+1})) \\ &\le 2ce^\ell\varepsilon + V_\alpha(\tilde{x}_i) - \Delta(t_{i+1} - t_i) \\ &\le V_\alpha(\tilde{x}_i) - \frac{\Delta}{2}(t_{i+1} - t_i) \le \frac{1}{2}\gamma(R) \end{aligned}$$

where we again used $\varepsilon \leq \kappa\, \underline{\mathbf{d}}(\pi)$ as well as the fact that $V_\alpha(\tilde{x}_i) \leq (1/2)\gamma(R)$ (since $\tilde{x}_i \in G_R^\alpha$). ∎

We also need another observation, this one paralleling the proof of Lemma IV.4 in [5].

Lemma A.4.2 *If, for some index i, $x_i := x(t_i) \in Q$, then $x(t)$ is defined for all $t \in [t_i, t_{i+1}]$,*

$$V_\alpha(x(t)) \leq \frac{3}{4}\gamma(r) \quad \forall t \in [t_i, t_{i+1}]\,, \tag{40}$$

and

$$V(x(t)) \leq \frac{7}{8}\gamma(r) \quad \forall t \in [t_i, t_{i+1}]\,. \tag{41}$$

In particular, $x(t) \in B_r$ for all $t \in [t_i, t_{i+1}]$ and $x(t_{i+1}) \in P$.

Proof. The fact that x is defined follows from the choice of δ_0, and we know that $x(t) \in B_R$ for all $t \in [t_i, t_{i+1}]$. So

$$|x(t) - x_i| \leq m\delta \quad \forall t \in [t_i, t_{i+1}]\,. \tag{42}$$

By definition of Q, we may write $x_i = x' + \eta$, for some $x' \in G_r^\alpha$ and some $|\eta| \leq \varepsilon$. Thus, $V_\alpha(x_i) \leq V_\alpha(x') + c\varepsilon_0 \leq \frac{1}{2}\gamma(r) + c\varepsilon_0$ (second inequality by definition of G_r^α). Together with (42), this gives

$$V_\alpha(x(t)) \leq \frac{1}{2}\gamma(r) + c\varepsilon_0 + cm\delta \leq \frac{3}{4}\gamma(r) \quad \forall t \in [t_i, t_{i+1}]$$

(using (27) and (30)). So, using (24) and ([5]21),

$$V(x(t)) \leq V_\alpha(x(t)) + \omega_R\left(\sqrt{2\beta(R)\alpha}\right) \leq \frac{3}{4}\gamma(r) + \frac{1}{16}\gamma(r) < \frac{7}{8}\gamma(r)$$

for all $t \in [t_i, t_{i+1}]$, as wanted. By the definition of γ (see [5]), this means that $x(t) \in B_r$ for all $t \in [t_i, t_{i+1}]$. Finally, if $|\eta| \leq \varepsilon$ then

$$V_\alpha(x(t_{i+1}) + \eta) \leq c\varepsilon + \frac{3}{4}\gamma(r) \leq \frac{1}{2}\gamma(R)$$

(the last inequality by (28)), which means that $x(t_{i+1}) + \eta \in G_R^\alpha$; this implies that $x(t_{i+1}) \in P$. ∎

Back to the proof of Proposition A.4.1, since $x^0 \in P$, Lemmas A.4.1 and A.4.2 guarantee that the solution exists for all t and remains in B_R, and that $x_i := x(t_i) \in P$ for all i.

Moreover, if there is some j so that $x(t_j) \in Q$, then it holds that $V_\alpha(x_i) \leq \frac{3}{4}\gamma(r)$ for all $i > j$. This is because on intervals in which $x_{i-1} \in Q$, we already know that $V_\alpha(x(t_i)) \leq \frac{3}{4}\gamma(r)$, and if instead $x_{i-1} \in P \setminus Q$, then we have $V_\alpha(x_i) < V_\alpha(x_{i-1})$. So, for any such $i > j$, either $x(t) \in B_r$ for all

$t \in [t_i, t_{i+1}]$ (first case) or $V_\alpha(x(t)) \leq \frac{3}{4}\gamma(r) + \varepsilon_0$ for all $t \in [t_i, t_{i+1}]$ (second case). Actually, in this last case we also have

$$\begin{aligned} V(x(t)) &\leq V_\alpha(x(t)) + \omega_R\left(\sqrt{2\beta(R)}\alpha\right) \\ &\leq \frac{3}{4}\gamma(r) + \varepsilon_0 + \omega_R\left(\sqrt{2\beta(R)}\alpha\right) \\ &\leq \frac{3}{4}\gamma(r) + \frac{1}{16}\gamma(r) + \frac{1}{16}\gamma(r) < \gamma(r), \end{aligned}$$

(using (27) and (24)), so, again by definition of γ, also $x(t) \in B_r$ for all $t \in [t_i, t_{i+1}]$. In conclusion, trajectories stay in B_r after the first time that $x(t_j) \in Q$. So we only need to show that there is such a j, with $t_j \leq T$.

Suppose instead that for $i = 0, \ldots, k$ it holds that $x(t_i) \notin Q$, and $t_k > T$. Applying (39) repeatedly,

$$0 \leq V_\alpha(x(t_k)) \leq V_\alpha(x^0) - \frac{\Delta}{2}t_k < \frac{\gamma(R)}{2} - \frac{\Delta}{2}T,$$

(recall that $x^0 \in P$ implies that $x \in G_R^\alpha$), and this contradicts (32). The proof of Proposition A.4.1 is then complete. ∎

This completes the proof. We remark that a more global result is also possible, as follows. We start by picking two sequences $\{r_j, j \in \mathbb{Z}\}$ and $\{R_j, j \in \mathbb{Z}\}$ such that $r_j, R_j \to 0$ as $j \to -\infty$, $r_j, R_j \to \infty$ as $j \to \infty$, and $2R_j \leq \rho(R_{j+1})$, $2\gamma(r_j) < \gamma(R_j)$, and $2r_j \leq \rho(R_{j-1})$ for all j. Next we pick, for each j, positive numbers $\alpha_j, \delta_j, \kappa_j, T_j$ associated as per Proposition A.4.1 to r_j and R_j, and let $k_j := k_{\alpha_j, r_j, R_j}$. (We may assume that values of k_j belong to some fixed $\mathcal{U}_0$ for all $j \leq 0$, and to $\mathcal{U}_j$ for $j > 0$, with all the $\mathcal{U}_j$ compact and forming an increasing sequence.) Since $G_{R_j}^{\alpha_j} \subseteq \operatorname{int} G_{R_{j+1}}^{\alpha_{j+1}}$ for all j (this is proved just as in ([5]48)), there is some sequence of positive numbers $\{\varepsilon_j, j \in \mathbb{Z}\}$ so that $\varepsilon_j < \kappa_j \delta_j$ for all j and also, denoting $P_j := P_{r_j, R_j, \varepsilon_j}$, so that $P_j \subseteq P_{j+1}$ for all j. Note that the choice of the r_j's and R_j's assures that $\bigcup P_j = \mathbb{R}^n \setminus \{0\}$. Since $2r_j \leq \rho(R_{j-1})$ for all j, and using (35), we know that $B_{r_j} \subseteq P_{j-1}$ for all j.

Finally, we define the feedback $k : \mathbb{R}^n \to \mathbb{R}^m$ via:

$$k(x) := k_j(x) \text{ if } x \in P_j \setminus P_{j-1}$$

(and $k(0) = 0$), and let $\varepsilon(x) := \varepsilon_j$, $\mu(x) := \frac{\varepsilon_j}{\kappa_j}$, and $\delta(x) := \delta_j$ for $x \in P_j \setminus P_{j-1}$, for each j (let $\varepsilon_j(0) = 0$ and $0 < m(0) < \delta(0)$ be arbitrary). Observe that, since $\varepsilon_j < \kappa_j \delta_j$ for all j, it holds that $\mu(x) < \delta(x)$ for all x.

Now suppose that x^0 is given, and a partition π and a function $e(\cdot)$ are given so that, recursively along the solution of $\dot{x} = f(x, k(x+e))$,

$$\mu(x(t_i)) \leq t_{i+1} - t_i \leq \delta(x(t_i))$$

and

$$e(t_i) \leq \varepsilon(x(t_i))$$

for all i. Note that, if for some i and j, $x(t_i) \in P_j \setminus P_{j-1}$, then the inequality in (33) holds with $\varepsilon = \varepsilon_j$, $\kappa = \kappa_j$, $\delta = \delta_j$, and this i.

Let $j \in \mathbb{Z}$ be maximal so that $x^0 \in P_j$. Then, it will hold that the solution stays in the bounded set B_{R_j}, and every $x(t_i)$ is again in P_j, until the first sampling time t_i that $x(t_i) \in P_\ell$ for some $\ell < j$. This first time, say t_q, is at most T_j, because $B_{r_j} \subseteq P_{j-1}$. If $x(t_q) = 0$, then the solution stays there forever (since $f(0,0) = 0$). Otherwise, $x(t_q) \in P_\ell \setminus P_{\ell-1}$ for some $\ell < j$, and we may repeat the argument. The conclusion is that the trajectory keeps visiting smaller sets P_ℓ (or it becomes 0 in finite time), with an upper bound $T(i,j)$ (the sum of the corresponding T_ℓ's) on the time required for entering a given P_i, if the initial state x^0 was in a given P_j. Furthermore, given any $0 < r < R$, there are $i < j$ so that $P_i \subseteq B_r \subseteq B_R \subseteq P_j$. Thus, all trajectories in B_R are taken into B_r in a uniform time $T(r,R)$, with bounded overshoot (since trajectories stay in B_{R_j}).

Acknowledgements

I wish to thank Francesca Albertini, Misha Krichman, Daniel Liberzon, Gene Ryan, and Yuan Wang, for many useful comments and suggestions.

References

[1] Artstein, Z., "Stabilization with relaxed controls," *Nonlinear Analysis, Theory, Methods & Applications* **7**(1983): 1163-1173.

[2] Brockett, R.W., "Asymptotic stability and feedback stabilization," in *Differential Geometric Control Theory* (R.W. Brockett, R.S. Millman, and H.J. Sussmann, eds.), Birkhauser, Boston, 1983, pp. 181-191.

[3] Clarke, F.H., Yu.S. Ledyaev, and R.J. Stern, "Asymptotic stability of differential inclusions," submitted.

[4] Clarke, F.H., Yu.S. Ledyaev, R.J. Stern, and P. Wolenski, *Nonsmooth Analysis and Control Theory*, Springer-Verlag, New York, 1998.

[5] Clarke, F.H., Yu. S. Ledyaev, E.D. Sontag, and A.I. Subbotin, "Asymptotic controllability implies feedback stabilization," *IEEE Trans. Automat. Control* **42**(1997): 1394-1407.

[6] Coron, J-M., "Global asymptotic stabilization for controllable systems without drift," *Math of Control, Signals, and Systems* **5**(1992): 295-312.

[7] Coron, J.-M., and L. Rosier, "A relation between continuous time-varying and discontinuous feedback stabilization," *J.Math. Systems, Estimation, and Control* **4**(1994): 67-84.

[8] Freeman, R., and P. V. Kokotović, *Robust Nonlinear Control Design*, Birkhäuser, Boston, 1996.

[9] Filippov, A.F., *Differential Equations with Discontinuous Right-Hand Side*, Nauka, Moscow, 1985 (in Russian), and Kluwer, Dordrecht, 1988 (English translation).

[10] Hájek, O., "Discontinuous differential equations. I, II," *J. Diff. Equations* **32** (1979), 149-170, 171-185.

[11] Hermes, H., "Discontinuous vector fields and feedback control," in *Differential Equations and Dynamical Systems*, Academic Press, New York, 1967, pp. 155-165.

[12] Isidori, A., *Nonlinear Control Systems, Third Edition*, Springer-Verlag, London, 1995.

[13] Krasovskii, N.N., and A.I. Subbotin, *Positional differential games*, Nauka, Moscow, 1974 [in Russian]. French translation *Jeux differentiels*, Editions Mir, Moscou, 1979. Revised English translation *Game-Theoretical Control Problems*, Springer-Verlag, New York, 1988.

[14] Krstić, M., I. Kanellakopoulos, and P. V. Kokotović, *Nonlinear and Adaptive Control Design*, John Wiley & Sons, New York, 1995.

[15] Krstić, M. and H. Deng, *Stabilization of Uncertain Nonlinear Systems*, Springer-Verlag, London, 1998.

[16] Lafferriere, G.A., and E.D. Sontag, "Remarks on control lyapunov functions for discontinuous stabilizing feedback," *Proc. IEEE Conf. Decision and Control, San Antonio*, IEEE Publications, 1993, pp. 306-308.

[17] Ledyaev, Yu.S., and E.D. Sontag, "A remark on robust stabilization of general asymptotically controllable systems," in *Proc. Conf. on Information Sciences and Systems (CISS 97)*, Johns Hopkins, Baltimore, MD, March 1997, pp. 246-251.

[18] Ledyaev, Yu.S., and E.D. Sontag, "A Lyapunov characterization of robust stabilization," *J. Nonlinear Analysis*, to appear. Summarized verion in "Stabilization under measurement noise: Lyapunov characterization," *Proc. American Control Conf.*, Philadelphia, June 1998, pp. 1658-1662.

[19] Lin, Y., E.D. Sontag, and Y. Wang, "A smooth converse Lyapunov theorem for robust stability," *SIAM J. Control and Optimization* **34**(1996): 124-160.

[20] Ryan, E.P., "On Brockett's condition for smooth stabilizability and its necessity in a context of nonsmooth feedback," *SIAM J. Control Optim.* **32**(1994): 1597-1604.

[21] Sontag E.D., "A Lyapunov-like characterization of asymptotic controllability," *SIAM J. Control and Opt.* **21**(1983): 462-471.

[22] Sontag E.D., "Feedback stabilization of nonlinear systems," in *Robust Control of Linear Systems and Nonlinear Control* (M.A. Kaashoek, J.H. van Schuppen, and A.C.M. Ran, eds.) Birkhäuser, Cambridge, MA, 1990, pp. 61-81.

[23] Sontag, E.D., *Mathematical Control Theory, Deterministic Finite Dimensional Systems*, Second Edition, Springer-Verlag, New York, 1998.

[24] Sontag, E.D., "Comments on integral variants of ISS," *Systems and Control Letters* **34**(1998): 93-100.

[25] Sontag, E.D., and H.J. Sussmann, "Remarks on continuous feedback," in *Proc. IEEE Conf. Decision and Control, Albuquerque, Dec. 1980*, IEEE Publications, Piscataway, pp. 916-921.

[26] Sontag, E.D., and H.J. Sussmann, "Nonsmooth Control Lyapunov Functions," in *Proc. IEEE Conf. Decision and Control, New Orleans*, IEEE Publications, 1995, pp. 2799-2805.

Department of Mathematics, Rutgers University, New Brunswick, NJ 08903, USA; Email: `sontag@control.rutgers.edu`

Progress in Systems and Control Theory, Vol. 25

Probabilistic Robustness Analysis and Design of Uncertain Systems

R. Tempo and F. Dabbene

1 Introduction

In the probabilistic approach for robustness analysis and design, one of the objectives is to compute the probability that a control system subject to uncertainty Δ, either real, complex or mixed, restricted to a set $\mathbf{\Delta}$ attains a given performance level γ. Then, it is crucial to derive explicit bounds for the number of samples required to estimate this probability with a certain accuracy and confidence a priori specified. It can be easily shown that the number N of randomly generated samples is independent of the number of blocks of Δ and the size of $\mathbf{\Delta}$. This fact is an immediate consequence of the Law of Large Numbers and is often used in Monte Carlo simulation. In the first part of the paper we discuss several bounds and we show their application to probabilistic robustness analysis. Subsequently, we explain how probabilistic robust design can be performed and we discuss connections with Learning Theory, showing how the problem structure can be taken into account. Current research directions related to sample generation in various sets are finally outlined.

For notational simplicity, in this paper we make use of the same uncertainty description of the standard worst-case robustness methods, which is called the M-Δ configuration and it is now briefly summarized; see e.g. [31] for a more general discussion on this topic. In Figure 1, $M(s)$ represents the deterministic part of the system, which consists of the nominal plant and controller transfer matrices and weighting functions, while Δ is used to describe various perturbations affecting the control system. A very general representation of Δ is given by

$$\Delta = \operatorname{diag}\left[q_1 I_{k_1}, \ldots, q_\ell I_{k_\ell}, \Delta_1, \ldots, \Delta_m\right] \tag{1}$$

with $q_i \in \mathbf{F}, i = 1, \ldots, \ell$ and $\Delta_i \in \mathbf{F}^{k_{\ell+i} \times k_{\ell+i}}, i = 1, \ldots, m$, where $\mathbf{F}$ is either the real or the complex field. The vector $q \doteq [q_1, q_2, \ldots, q_\ell]^T \in \mathbf{F}^\ell$ takes into account real or complex parametric uncertainties affecting the plant, while the matrices $\Delta_i \in \mathbf{F}^{k_{\ell+i} \times k_{\ell+i}}$ are generally introduced to represent high order unmodeled dynamics. The matrix Δ is restricted to a set $\mathbf{\Delta}$ generally described in terms of norm-bounded balls

$$\mathbf{\Delta} = \left\{\operatorname{diag}\left[q_1 I_{k_1}, \ldots, q_\ell I_{k_\ell}, \Delta_1, \ldots, \Delta_m\right] : \|q\|_p \le 1, \|\Delta_i\|_p \le 1, i = 1, \ldots, m\right\}. \tag{2}$$

In the standard μ-theory setting [16] the induced ℓ_2-norm $\|\Delta\|_2 \doteq \sup_{\|x\|_2=1} \|\Delta x\|_2$ is used, even though definitions of μ based on the norm of

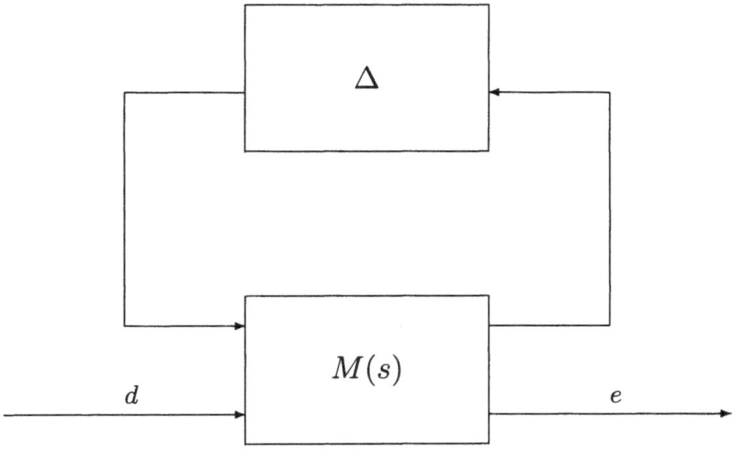

Figure 1: $M - \Delta$ configuration with disturbances d and errors e.

Frobenius have been introduced [20]. The choice of the ℓ_2-norm is clearly equivalent to take $||q||_\infty$ and $||\Delta_i||_2$, $i = 1, \ldots, m$. However, we remark that, depending on the specific problem under attention, different ℓ_p norms may be chosen. For example, in the framework of real parametric uncertainty, q is often bounded by ℓ_∞ norm balls even though sometimes ℓ_2 and ℓ_1 are used; see e.g. [2] and [5].

The classical problem of robustness analysis of the control system shown in Figure 1 is to guarantee that a certain performance requirement is attained for all $\Delta \in \mathbf{\Delta}$. In general, this requirement can be stated in terms of the *performance function*

$$u \doteq u(\Delta)$$

where $u(\Delta)$ is a Lebesgue measurable function of Δ. Without loss of generality in this paper we study a single performance function, but in general the simultaneous attainment of several performance requirements may be considered.

A classical example of performance function deals with the $\mathcal{H}_\infty$ norm of the transfer function $T_{de}(M, \Delta)$ between the disturbances d and the errors e; this transfer function is usually called the upper Linear Fractional Transformation. That is, we set

$$u(\Delta) = ||T_{de}(M, \Delta)||_\infty.$$

Another example of performance function is robust stability of a continuous time MIMO system. To illustrate, we consider the special case when $\Delta = \Delta(q)$ contains only parametric uncertainty q. Assuming a state space description of the system, we write

$$T_{de} = C(sI - A)^{-1}B + D$$

where $A = A(q)$, $B = B(q)$, $C = C(q)$ and $D = D(q)$. Therefore, the performance function $u(\Delta)$ is

$$u(\Delta) = \max(\text{Re } \lambda_1(\Delta), \text{Re } \lambda_2(\Delta), \ldots, \text{Re } \lambda_n(\Delta))$$

where $\lambda_1(\Delta), \ldots, \lambda_n(\Delta)$ are the eigenvalues of $A(q)$.

Further examples of performance requirements, which include sensitivity minimization, disturbance attenuation and tracking, are described in [31]. In general, robustness analysis is equivalent to one of the following two formulations:

Performance Verification Problem: For a given performance level $\gamma > 0$, check whether

$$u(\Delta) \leq \gamma$$

for all $\Delta \in \mathbf{\Delta}$.

Worst-Case Performance Problem: Find $u_{\max}$ such that

$$u_{\max} \doteq \max_{\Delta \in \mathbf{\Delta}} u(\Delta).$$

Focusing on robust design, consider now the configuration of Figure 2, where the transfer matrix $M(s)$ is divided into the plant $P(s)$, which also incorporates weighting functions, and the controller transfer matrix $C(s)$. The problem of robust design can be stated as follows: Find a stabilizing controller $C(s)$ in the set $\mathbf{C}$ of candidate controllers such that the control system attains the desired performance level γ for all $\Delta \in \mathbf{\Delta}$. In this case, we replace the performance function $u(\Delta)$ by

$$u(\Delta, C).$$

The controller C_{opt} achieving the best performance is then the stabilizing controller which guarantees robust performance for the minimum value of γ. That is,

$$C_{opt} \doteq \arg \min_{C \in \mathbf{C}} \max_{\Delta \in \mathbf{\Delta}} u(\Delta, C). \tag{3}$$

One of the drawbacks of this worst-case problem formulation is related to the issue of computational complexity. That is, in several papers [6], [9], [13], [22] and [23] it is shown that various linear control problems, which include robustness analysis and design of control systems subject to general classes of uncertainty, are NP-hard. This drawback, together with the fact that the robustness margins may be too conservative, recently led to the formulation of a probabilistic approach. This approach was also stated in some preliminary papers on Monte Carlo simulation for control system analysis and design; see e.g. [15], [24]. More recent papers studying randomized algorithms and their applications are [10], [11], [19] and [32].

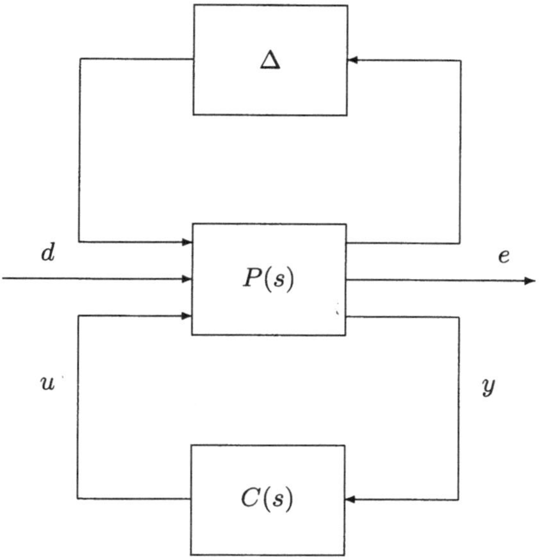

Figure 2: Robust design configuration with inputs u and outputs y.

2 Probabilistic Robustness Analysis

Consider the performance verification problem discussed in the previous section and define the "good" and "bad" sets

$$\begin{aligned} \mathbf{\Delta}_g &\doteq \{\Delta \in \mathbf{\Delta} : u(\Delta) \leq \gamma\}; \\ \mathbf{\Delta}_b &\doteq \{\Delta \in \mathbf{\Delta} : u(\Delta) > \gamma\}. \end{aligned}$$

Clearly, robust performance is attained for all Δ in the set $\mathbf{\Delta}_g$, while it is violated if $\Delta \in \mathbf{\Delta}_b$. The union of $\mathbf{\Delta}_g$ and $\mathbf{\Delta}_b$ obviously coincides with the uncertainty set $\mathbf{\Delta}$. Therefore, the volume

$$\mathrm{vol}(\mathbf{\Delta}_g) \doteq \int_{\mathbf{\Delta}_g} d\Delta$$

can be taken as a measure of the system robustness. We observe that in classical robustness analysis, one of the objectives is to compute the largest value of γ such that performance is guaranteed for all $\Delta \in \mathbf{\Delta}$. In turn, this implies that the set $\mathbf{\Delta}_g$ coincides with $\mathbf{\Delta}$. On the other hand, in a probabilistic setting, we are satisfied if these two sets "approximately" coincide and, in particular, the ratio

$$\frac{\mathrm{vol}(\mathbf{\Delta}_g)}{\mathrm{vol}(\mathbf{\Delta})} \tag{4}$$

is close to one.

Next, we assume that Δ defined in (1) is a random matrix with associated probability density function $f_\Delta(\Delta)$. Formally, for a given set $\mathbf{S}$, we define

$$\text{Prob}\{\Delta \in \mathbf{S}\} = \int_{\mathbf{S}} f_\Delta(\Delta) d\Delta.$$

The probabilistic robustness of the $M - \Delta$ system is therefore stated in terms of the probability that the system satisfies the desired performance. In other words, if Δ is a random matrix, we aim to compute the "weighted" volume of the set $\boldsymbol{\Delta}$ with respect to the density function $f_\Delta(\Delta)$

$$\text{Prob}\{\Delta \in \boldsymbol{\Delta}_g\} = \int_{\boldsymbol{\Delta}_g} f_\Delta(\Delta) d\Delta = \text{Prob}\ \{u(\Delta) \le \gamma\}.$$

Clearly, if the density function $f_\Delta(\Delta)$ is uniform on $\boldsymbol{\Delta}$, the probability $\text{Prob}\{\Delta \in \boldsymbol{\Delta}_g\}$ coincides with the ratio (4) of the two volumes $\text{vol}(\boldsymbol{\Delta}_g)$ and $\text{vol}(\boldsymbol{\Delta})$. We remark that the choice of uniform distribution is common in this setting for its worst-case properties in a certain class of distribution functions [3].

2.1 Performance verification problem

We now study the probabilistic version of the performance verification problem. Given a performance level $\gamma > 0$, we aim to estimate the probability of performance

$$p_\gamma \doteq \text{Prob}\ \{u(\Delta) \le \gamma\}. \tag{5}$$

This probability can be computed by means of a randomized algorithm. To this aim, we generate N independently identically distributed (i.i.d.) samples within $\boldsymbol{\Delta}$

$$\Delta^1, \Delta^2, \ldots, \Delta^N \in \boldsymbol{\Delta}$$

according to the given density function $f_\Delta(\Delta)$. Subsequently, we compute

$$u(\Delta^1), u(\Delta^2), \ldots, u(\Delta^N)$$

and we construct the indicator function

$$I(\Delta^i) = \begin{cases} 1 & \text{if } u(\Delta^i) \le \gamma; \\ 0 & \text{otherwise.} \end{cases} \tag{6}$$

An estimate of p_γ is immediately obtained as

$$\hat{p}_N = \frac{1}{N} \sum_{i=1}^{N} I(\Delta^i)$$

which is equivalent to

$$\hat{p}_N = \frac{N_g}{N}$$

where N_g is the number of samples such that $u(\Delta^i) \leq \gamma$. The estimate $\hat{p}_N$ is usually referred to as the *empirical probability*.

Clearly, for a finite sample size, it is important to know how many samples N are needed to obtain a reliable estimate $\hat{p}_N$ of $p_\gamma = \text{Prob}\{u(\Delta) \leq \gamma\}$. This reliability can be measured in terms of the "closeness" of $\hat{p}_N$ to the probability p_γ. That is,

$$|p_\gamma - \hat{p}_N| = |\text{Prob}\{u(\Delta) \leq \gamma\} - \hat{p}_N| \leq \epsilon$$

for $\epsilon \in (0,1)$. Notice that, since $\hat{p}_N$ is estimated via random sampling, it is a random variable. Therefore, for fixed $\delta \in (0,1)$,

$$\text{Prob}\{|p_\gamma - \hat{p}_N| \leq \epsilon\} \geq 1 - \delta. \tag{7}$$

The problem is then finding the minimal N such that (7) is satisfied for given *accuracy* $\epsilon \in (0,1)$ and *confidence* $\delta \in (0,1)$. An immediate answer to this problem is given by the Bernoulli Law of Large Numbers [4].

Bernoulli Law of Large Numbers: *For any* $\epsilon \in (0,1)$ *and* $\delta \in (0,1)$, *if*

$$N \geq \frac{1}{4\epsilon^2\delta} \tag{8}$$

then

$$\text{Prob}\{|p_\gamma - \hat{p}_N| \leq \epsilon\} \geq 1 - \delta.$$

We observe that the number of samples computed with the Law of Large Numbers is independent of the number of blocks of Δ, the size of $\boldsymbol{\Delta}$ and the density function $f_\Delta(\Delta)$. Unfortunately, the number of samples N may be very large. For example, if $\epsilon = 0.1\%$ and $1 - \delta = 99.9\%$, we obtain $N = 2.5 \cdot 10^8$. We remark, however, that the cost associated with the evaluation of $u(\Delta^i)$ for fixed Δ^i is polynomial-time in many cases. This is true when dealing, for example, with the computation of $\mathcal{H}_\infty$ norms or when stability tests are of concern. Therefore, on the contrary of the worst-case robustness approach (see the discussion in Section 1), we conclude that the total cost to perform probabilistic robustness analysis is polynomial-time.

The Bernoulli bound is an immediate consequence of the Chebyshev inequality. This well-known inequality indicates that a bound on the "spread" of a random variable around its mean value can be computed in terms of the variance of the distribution. More precisely, given a random variable x, for any $\epsilon > 0$, we have

$$\text{Prob}(|x - E(x)| \geq \epsilon) \leq \frac{\sigma^2(x)}{\epsilon^2} \tag{9}$$

where $E(x)$ and $\sigma^2(x)$ are the mean value and the variance of x, respectively. Observe that $\hat{p}_N$ is a random variable binomially distributed with variance $\sigma^2 = p_\gamma(1-p_\gamma)/N$ and mean value p_γ. Substituting in (9), we immediately obtain

$$\text{Prob}\{|p_\gamma - \hat{p}_N| \le \epsilon\} \ge 1 - \frac{p_\gamma(1-p_\gamma)}{N\epsilon^2}.$$

Therefore, the Law of Large Numbers follows immediately if N is taken as in equation (8).

The Chebyshev Inequality can be generalized to obtain the Bienaymé Inequality. For any $\epsilon > 0$ and nonnegative integer m, we have

$$\text{Prob}(|x - E(x)| \ge \epsilon) \le \frac{E(|x - E(x)|^m)}{\epsilon^m}.$$

Clearly, for $m = 2$, we immediately obtain the Chebyshev Inequality. Taking $m = 4$ and proceeding as in the case of the Law of Large Numbers, we obtain the bound

$$N \ge \frac{1}{4\epsilon^2}\sqrt{\frac{3}{\delta}}.$$

which improves upon (8) for $\delta < 1/3$. Taking higher values of m we obtain better bounds. For example, for $m = 6$ we get

$$N \ge \frac{1}{4\epsilon^2}\sqrt[3]{\frac{15}{\delta}}.$$

A more classical bound which improves upon the previous ones and is not based on the Bienaymé inequality, is the Chernoff Bound [12].

Chernoff Bound: *For any $\epsilon \in (0,1)$ and $\delta \in (0,1)$, if*

$$N \ge \frac{\log\frac{2}{\delta}}{2\epsilon^2}$$

then

$$\text{Prob}\{|p_\gamma - \hat{p}_N| \le \epsilon\} \ge 1 - \delta.$$

We remark that the number of samples computed with this bound largely improves upon the bound of Bernoulli. For example, if $\epsilon = 0.1\%$ and $1 - \delta = 99.9\%$, we compute $N = 3.9 \cdot 10^6$. Figure 3 shows a graphical comparison between these bounds.

Finally, we remark that these bounds are *explicit*. That is, given ϵ and δ one can find directly the minimum value of N. On the other hand, when computing the classical upper and lower confidence intervals, the sample

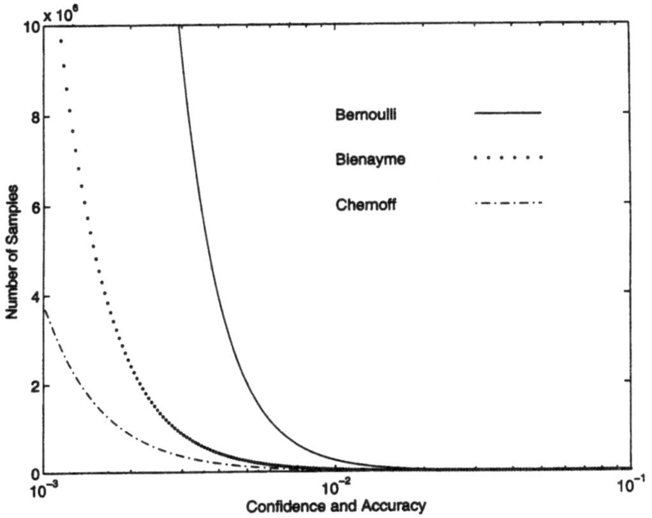

Figure 3: Bernoulli, Bienaymé ($m = 4$) and Chernoff Bounds.

size obtained is not explicit. More precisely, for given $\delta \in (0,1)$, the so-called upper and lower confidence intervals p_L and p_U are such that

$$\mathrm{Prob}\{p_L \leq p_\gamma \leq p_U\} = \delta.$$

The evaluation of this probability requires the solution of the following equations with respect to p_L and p_U

$$\begin{aligned}\sum_{k=0}^{K-1} \binom{N}{k} {p_L}^k (1-p_L)^{N-k} &= 1 - \frac{\delta}{2};\\ \sum_{k=0}^{K} \binom{N}{k} {p_U}^k (1-p_U)^{N-k} &= \frac{\delta}{2}.\end{aligned}$$

Since an explicit solution of these equations is not available, standard tables are generally used; e.g., see [17] and [24].

2.2 Worst-case performance problem

When dealing with the worst-case performance problem, we look for a probabilistic estimate of

$$u_{\max} = \max_{\Delta \in \mathbf{\Delta}} u(\Delta).$$

To this end, we adopt again a random sampling scheme, generating N i.i.d. samples within $\mathbf{\Delta}$

$$\Delta^1, \Delta^2, \ldots, \Delta^N \in \mathbf{\Delta}$$

according to the density function $f_\Delta(\Delta)$ and we compute

$$u(\Delta^1), u(\Delta^2), \ldots, u(\Delta^N).$$

Consequently, we obtain

$$\hat{u}_N \doteq \max_{i=1,2,\ldots,N} u(\Delta^i).$$

For this special formulation of the probabilistic robustness problem a bound on the sample size N is given, for example, in [18] and [26].

<u>**Bound for Worst-Case Performance:**</u> *For any $\epsilon \in (0,1)$ and $\delta \in (0,1)$, if*

$$N \geq \frac{\log \frac{1}{\delta}}{\log \frac{1}{1-\epsilon}}, \tag{10}$$

then

$$\text{Prob}\{\text{Prob}\{u(\Delta) > \hat{u}_N\} \leq \epsilon\} \geq 1 - \delta. \tag{11}$$

In the paper [10] the bound above is shown to be tight if the distribution function $F_\Delta(\Delta)$ of Δ is continuous.

We notice that the worst-case performance problem is a special case of the general empirical probability estimation discussed in the previous subsection. In fact, setting $\gamma = \hat{u}_N$ in (5) we obtain

$$p_\gamma = \text{Prob } \{u(\Delta) \leq \hat{u}_N\} = 1 - \text{Prob } \{u(\Delta) \geq \hat{u}_N\}$$

and

$$\hat{p}_N = \frac{N_g}{N} = 1.$$

Therefore, equation (7) becomes

$$\text{Prob}\{|p_\gamma - \hat{p}_N| \leq \epsilon\} = \text{Prob}\{\text{Prob}\{u(\Delta) \geq \hat{u}_N\} \leq \epsilon\}.$$

Comparing the bound (10) with the Chernoff bound, it can be immediately verified that the number of samples required for the worst-case performance problem is much smaller than that needed for the performance verification problem; see Figure 4. We remark, however, that there is no guarantee that $\hat{u}_N$ is actually close to the maximum $u_{\max}$. Instead, the bound previously discussed only guarantees that the set of points greater than the estimated value has a measure smaller than ϵ and this is true with a probability at

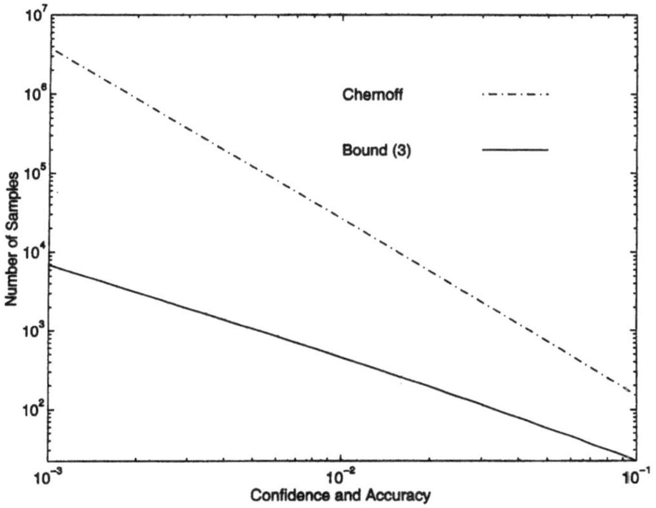

Figure 4: Comparison between Chernoff and Worst-Case Bounds.

least $1-\delta$. In turn, this implies that if the function $u(\Delta)$ is sufficiently smooth, the estimated and actual maximum may be close.

Finally, we remark that the problem

$$\text{Prob}\{|u(\Delta)-\hat{u}_N|\leq\epsilon\}\geq 1-\delta$$

where only one level of probability is considered, requires a sample size that may be an exponential function of the number of blocks in Δ; see [1] for further discussions.

3 Probabilistic Robust Design

In this section, we focus on the probabilistic version of the robust design problem shown in Figure 2 and stated in (3). The description given here is largely based on the paper [29]. We study the following probabilistic design formulation: Given a set of candidate controllers $\mathbf{C}$, find the optimal probabilistic controller $C_{prob}\in\mathbf{C}$ which minimizes the expected value of the performance function $u(\Delta, C)$. That is,

$$C_{prob}\doteq\arg\min_{C\in\mathbf{C}}E[u(\Delta,C)] \tag{12}$$

where

$$E[u(\Delta,C)]\doteq\int_{\mathbf{\Delta}}u(\Delta,C)f_{\Delta}(\Delta)d\Delta. \tag{13}$$

Since the expected value $E[u(\Delta, C)]$ can be interpreted as the weighted volume of $\mathbf{\Delta}$, with weight equal to the performance function $u(\Delta, C)$ and the density function $f_\Delta(\Delta)$, the optimal probabilistic controller C_{prob} guarantees that a certain volume minimization is achieved. For example, if the performance function is defined such that $u(\Delta, C) = 1$ if $\Delta \in \mathbf{\Delta}_b$ and zero otherwise, then $E[u(\Delta, C)]$ coincides with the probability $\text{Prob}\{\Delta \in \mathbf{\Delta}_b\}$. In turn, this probability is minimized when computing the optimal probabilistic controller.

An approximate solution of this problem can be obtained by means of randomized algorithms which are based on sampling both the uncertainty set $\mathbf{\Delta}$ and the controller set $\mathbf{C}$. To this end, we need to solve two separate problems: First, find an estimate of the expected value and, secondly, compute its minimum value. We now study this latter problem.

Estimation of the Minimum: From the definition (13) of expected value, we observe that $E[u(\Delta, C)]$ is a function of C only. Therefore, computing the minimum in (12) via randomization is equivalent to the worst-case performance problem discussed in Subsection 2.2. To show this, consider a density function $f_C(C)$ associated with the controller C. Then, generate M i.i.d. controllers $C^1, C^2, \ldots, C^M \in \mathbf{C}$ according to this density. An estimate of the minimum (12) is given by

$$\min_{i=1,2,\ldots,M} E[u(\Delta, C^i)].$$

Clearly, the sample size given in (10) still holds. That is, given $\epsilon \in (0,1)$ and $\delta \in (0,1)$, if

$$M \geq \frac{\log \frac{1}{\delta}}{\log \frac{1}{1-\epsilon}}$$

then

$$\text{Prob}\{\text{Prob}\{E[u(\Delta, C)] \leq \min_{i=1,2,\ldots,M} E[u(\Delta, C^i)]\} \leq \epsilon\} \geq 1 - \delta. \tag{14}$$

However, for fixed $C^i \in \mathbf{C}$, the exact computation of $E[u(\Delta, C^i)]$ requires the solution of the integral (13) and is very difficult in general. Therefore, we now study its estimation in terms of probability.

Expected Value Estimation: In this case, we follow again a random sampling scheme. That is, we generate N i.i.d. samples within $\mathbf{\Delta}$

$$\Delta^1, \Delta^2, \ldots, \Delta^N \in \mathbf{\Delta}$$

according to the density function $f_\Delta(\Delta)$ and we evaluate the performance function $u(\Delta^1, C^i), \ldots, u(\Delta^N, C^i)$ for a fixed controller $C^i \in \mathbf{C}$. An estimation $\hat{E}_N[u(\Delta, C^i)]$ of the expected value $E[u(\Delta, C^i)]$ is given by

$$\hat{E}_N[u(\Delta, C^i)] \doteq \frac{1}{N}\sum_{k=1}^{N} u(\Delta^k, C^i). \tag{15}$$

This estimate is known as the *empirical mean* and can be simply obtained applying the same technique described in Subsection 2.1 for the computation of the Empirical Probability. The only difference between the two approaches lies in the fact that in (15) the mean value of the performance function is taken instead of the indicator function used in (6). Clearly, the bounds derived in Section 2, and in particular the Chernoff bound, are still valid. We now state this fact precisely.

Chernoff Bound for Empirical Mean: *For fixed controller C^i, for any $\epsilon \in (0,1)$ and $\delta \in (0,1)$, if*

$$N \geq \frac{\log \frac{2}{\delta}}{2\epsilon^2}$$

then

$$\text{Prob}\{|E[u(\Delta, C^i)] - \hat{E}_N[u(\Delta, C^i)]| \leq \epsilon\} \geq 1 - \delta. \tag{16}$$

In order to compute a randomized probabilistic controller, we need to combine the equations (14) and (16). To this end, we notice that if the condition

$$\text{Prob}\{|E[u(\Delta, C^i)] - \hat{E}_N[u(\Delta, C^i)]| \leq \epsilon, i = 1, \ldots, M\} \geq 1 - \delta \tag{17}$$

holds, it follows that

$$\text{Prob}\{|\min_{i=1,\ldots,M} E[u(\Delta, C^i)] - \min_{i=1,\ldots,M} \hat{E}_N[u(\Delta, C^i)]| \leq \epsilon\} \geq 1 - \delta. \tag{18}$$

Next, (14) and (18) imply that

$$\text{Prob}\{\text{Prob}\{E[u(\Delta, C)] \leq \min_{i=1,\ldots,M} \hat{E}_N[u(\Delta, C^i)] - \epsilon\} \leq \epsilon\} \geq (1-\delta)^2 \geq 1-\delta/2. \tag{19}$$

This condition is similar to (14) with some differences. In particular, equation (19) tells us that the estimated minimum $\min_{i=1,\ldots,M} \hat{E}_N[u(\Delta, C^i)]$ is close to the actual one $E[u(\Delta, C)]$ within ϵ in terms of probability. This fact is guaranteed with an accuracy ϵ and a confidence at least $\delta/2$.

We still have to obtain a condition on the number of samples N required to guarantee that (17) holds. To this aim, we notice that equation (17) is equivalent to

$$\text{Prob}\{\exists i : |E[u(\Delta, C^i)] - \hat{E}_N[u(\Delta, C^i)]| > \epsilon\} < \delta. \tag{20}$$

The probability in equation (20) is upper bounded by the sum of the probabilities of the single events

$$\sum_{i=1}^{M} \text{Prob}\{|E[u(\Delta, C^i)] - \hat{E}_N[u(\Delta, C^i)]| > \epsilon\}.$$

Therefore, if

$$\text{Prob}\{|E[u(\Delta, C^i)] - \hat{E}_M[u(\Delta, C^i)]| > \epsilon\} < \frac{\delta}{M}$$

then (20) holds. By repeated application of the Chernoff Bound for Empirical Mean, we immediately obtain that the latter inequality is satisfied if

$$N \geq \frac{\log \frac{2M}{\delta}}{2\epsilon^2}.$$

To conclude, an optimal probabilistic controller has been derived and bounds N and M on the uncertainty and controller samples have been computed. These facts are now summarized.

Randomized Algorithm for Probabilistic Robust Design [29]:
Given the probability density functions $f_C(C)$ and $f_\Delta(\Delta)$, draw M samples $C^i \in \mathbf{C}$ and N samples $\Delta^k \in \mathbf{\Delta}$ according to $f_C(C)$ and $f_\Delta(\Delta)$, respectively. For any $\epsilon \in (0,1)$ and $\delta \in (0,1)$, if

$$M \geq \frac{\log \frac{1}{\delta}}{\log \frac{1}{1-\epsilon}} \quad \text{and} \quad N \geq \frac{\log \frac{2M}{\delta}}{2\epsilon^2}$$

then

$$\text{Prob}\{\text{Prob}\{E[u(\Delta, C)] \leq \min_{i=1,\ldots,M} \hat{E}_N[u(\Delta, C^i)] - \epsilon\} \leq \epsilon\} \geq 1 - \delta/2.$$

The randomized probabilistic controller is given by

$$\hat{C}_{NM} \doteq \arg \min_{i=1,\ldots,M} \frac{1}{N} \sum_{k=1}^{N} u(\Delta^k, C^i).$$

3.1 Connections with Learning Theory

A major drawback of the approach previously described is that the number N of samples drawn in the set $\mathbf{\Delta}$ is a function of the number M of samples taken in the set $\mathbf{C}$. In order to avoid this disadvantage and to compute independent bounds N and M, in [29] an approach based on Learning Theory has been presented. An introductory treatment and an advanced exposition of Learning Theory are given in [30] and [28], respectively. To briefly summarize the application to control system design, consider condition (17)

$$\text{Prob}\{|E[u(\Delta, C^i)] - \hat{E}_N[u(\Delta, C^i)]| \leq \epsilon, i = 1, \ldots, M\} \geq 1 - \delta.$$

This condition can be interpreted as follows: We require that the actual mean $E[u(\Delta, C^i)]$ is sufficiently close to the empirical mean $\hat{E}_N[u(\Delta, C^i)]$ for all controllers C^i, $i = 1, \ldots, M$. In other words, the empirical mean converges for $N \to \infty$ to the actual mean *uniformly* as the function $u(\Delta, C^i)$ varies in the finite set $\mathcal{U}_M \doteq \{u(\Delta, C^i), i = 1, \ldots, M\}$. This convergence condition can be interpreted as a property of the set of functions $\mathcal{U}_M$ and it is known as *Uniform Convergence of Empirical Mean* (UCEM). The results stated in the previous section show that for a finite dimensional set $\mathcal{U}_M$ the UCEM property holds. However, this property may hold also in the case when the class of functions considered is infinite dimensional, if certain conditions are satisfied. In particular, in this case, we require that

$$\text{Prob}\{|E[u(\Delta, C)] - \hat{E}_N[u(\Delta, C)]| \leq \epsilon, \text{for all } u(\Delta, C) \in \mathcal{U}\} \geq 1 - \delta \quad (21)$$

where $\mathcal{U} \doteq \{u(\Delta, C), C \in \mathbf{C}\}$.

Fundamentals results from Learning Theory relate the uniform convergence property to specific features of the set $\mathcal{U}$ [27]. For example, if the class $\mathcal{U}$ is a class of binary-valued functions mapping into the set $\{0, 1\}$, the so-called VC-dimension comes into play. The importance of this parameter is given by the fact that the finiteness of the VC-dimension of the set $\mathcal{U}$ is a sufficient condition for the UCEM property to hold. However, the exact computation of the VC-dimension is not an easy task and in general only an upper bound d can be determined. If the VC-dimension of the set $\mathcal{U}$ or if d is known, then a well-known theorem of Learning Theory (see Theorems 7.2 and 10.2 in [28]) immediately gives the number of samples N needed to compute the empirical mean (15) which is necessary for (21) to hold. This number is given by

$$N \geq \max\left\{\frac{16}{\epsilon^2}\log\frac{4}{\delta}, \frac{32d}{\epsilon^2}\log\frac{32e}{\epsilon^2}\right\},$$

where $\epsilon \in (0, 1)$, $\delta \in (0, 1)$, e is the Euler number and d is an upper bound on the VC-dimension. Since the computation of the upper bound d depends of the specific problem under attention, we say that the *problem structure* is taken into account. An example of the computation of d is given in [29], where robust stabilization of a SISO plant affected by real parametric uncertainty is studied. The SISO controller is parameterized with a *finite* number ℓ of controller parameters and the binary valued performance function is set equal to zero if the controller stabilizes the system and to one otherwise. The obtained bound on the VC-dimension is a function of the Mc Millan degree of the plant and the controller and the number ℓ of controller parameters. Unfortunately, this bound seems quite conservative. Similar considerations hold for the case when the family $\mathcal{U}$ is real valued. In that case the key role is played by the so called P-dimension; see [29].

4 Example

In this section, we give a numerical example of probabilistic robustness analysis. We analyze a spacecraft model originally presented in [25] as an application of robust observers to space flexible structures and we study the loop shaping design proposed in [21]. The model represents the transfer function from the torque applied to the roll axis to the corresponding roll angle of the spacecraft. The state variable model is of the form

$$\begin{aligned} \dot{x} &= Ax + B(u_c + v_d - r_s) \\ y &= Cx \end{aligned}$$

where u_c is the control torque, v_d is a constant disturbance torque, y is the roll angle measurement and r_s is the reference signal provided by the controller. The nominal system matrices given in [25] are

$$\bar{A} = \begin{pmatrix} 0 & 1 & 0 & 0 \\ 0 & 0 & 0 & 0 \\ 0 & 0 & 0 & 1 \\ 0 & 0 & -\omega_n^2 & -2\zeta\omega \end{pmatrix}; \ \bar{B} = \begin{pmatrix} 0 \\ 1.7319 \cdot 10^{-5} \\ 0 \\ 3.7859 \cdot 10^{-4} \end{pmatrix}; \ \bar{C} = (1 \quad 0 \quad 1 \quad 0)$$

where $\omega_n = 1.539$ rad/sec is the natural frequency representing the flexible modes and $\zeta = 0.003$ is the damping. Based upon this data, in [21] the controller described in state space form was designed

$$\begin{aligned} \dot{x}_c &= A_c\, x_c + B_c\, y \\ r_s &= C_c\, x_c \end{aligned}$$

where

$$\begin{aligned} Ac &= \begin{pmatrix} 0 & 10.6591 & -4.3091 & -3.3354 & -7.8802 \\ 0 & -100.5513 & 72.1135 & -73.5563 & -299.0902 \\ 0 & 0.6352 & -0.3807 & 1.0774 & 2.0607 \\ 0 & 0.3630 & -0.6036 & -0.1976 & -0.7720 \\ 0 & 0.7674 & -0.6002 & -0.4013 & -1.6053 \end{pmatrix}; \\ B_c^T &= (0 \quad 384.8650 \quad -1.3706 \quad 0.5944 \quad 0.7300); \\ C_c &= 1.0 \cdot 10^3 (0.02 \quad 5.3296 \quad -2.1545 \quad -1.6677 \quad -3.9401). \end{aligned}$$

The goal of this example is to study probabilistic robustness of the design proposed in [21], when the system is affected by the following uncertainties:

<u>*Real parametric uncertainty:*</u> The natural frequency ω_n and the damping ζ vary around their nominal values

$$\omega_n = 1.539(1 + q_1);$$

$$\zeta = 0.003(1 + q_2),$$

where $\|q\|_\infty \leq \rho$, with $q = [q_1, q_2]^T \in \mathbf{R}^2$.

Real unstructured uncertainty: The state matrix A is assumed to be affected by additive real unstructured uncertainty

$$A = \bar{A} + W_1 \Delta_1$$

where $\Delta_1 \in \mathbf{R}^{4\times 4}$ is such that $\|\Delta_1\|_2 \leq \rho$ and the weight W_1 is equal to 0.5.

Complex uncertainty: To represent high order unmodeled dynamics, the nominal transfer function

$$T(s) = C(sI - A)^{-1}B$$

is affected by an additive complex uncertainty $\Delta_2 \in \mathbf{C}$ such that $\|\Delta_2\|_2 \leq \rho$.

By means of simple algebraic manipulations, this uncertainty structure can be written in the M-Δ configuration. The uncertain block matrix Δ has the form

$$\Delta = \mathrm{diag}\,[q_1 I_3, q_2 I_4, \Delta_1, \Delta_2]$$

and is bounded as $\|\Delta\|_2 \leq \rho$. Next, we consider a robust stability performance function

$$u(\Delta) = \begin{cases} 0 & \text{if the closed loop system is stable;} \\ 1 & \text{otherwise.} \end{cases}$$

Clearly in this case, taking a value of γ in the interval $(0, 1)$, the probability of performance p_γ is simply the probability of robust stability. Therefore we are interested in analyzing the variation of the probability of stability for different values of ρ. In particular, we compute the probability of stability p_γ for 150 grid points of ρ in the interval $[0, 0.15]$, using the described randomized algorithm. For fixed ρ, we evaluate an empirical estimate $\hat{u}_N$ and consequently we obtain an empirical probability $\hat{p}_N$. Requiring an accuracy $\epsilon = 0.02$ and a confidence $\delta = 0.01$, using the Chernoff bound we need at least $N = 7000$ samples to guarantee

$$\mathrm{Prob}\{|p_\gamma - \hat{p}_N| \leq 2\%\} \geq 99\%.$$

The degradation of the probability of robust stability as ρ varies in the interval $[0, 0.15]$ is shown in Figure 5. From this plot we conclude that the system is robustly stable with probability one for $\rho \leq \rho^* \approx 0.7$. For values of ρ in the interval $\rho^* + 10\%\rho^*$, we observe that the probability degradation is less than 1%.

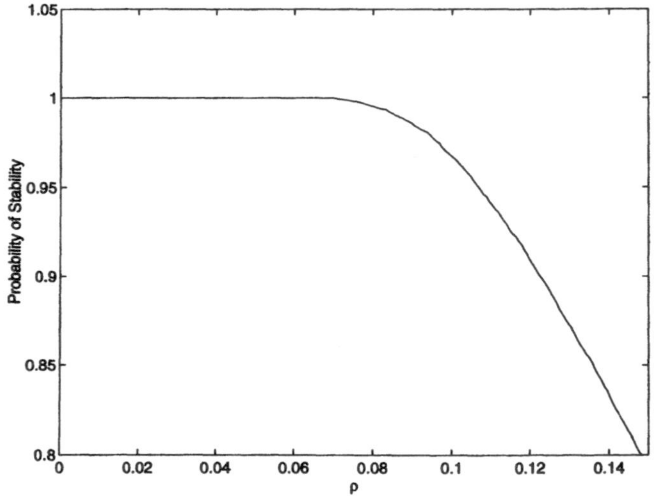

Figure 5: Probabilistic Robustness degradation for $\rho \in [0, 0.15]$

5 Concluding Remarks

In this paper, we described a probabilistic approach for robustness analysis and design of uncertain control systems. A problem which is of interest in this setting is the development of feasible algorithms for sample generation in various sets according to several distributions. For example, given the $M - \Delta$ structure described in the paper and the uncertainty set $\mathbf{\Delta}$, we study the sample generation within $\mathbf{\Delta}$ according to a uniform density function $f_\Delta(\Delta)$. Since $\mathbf{\Delta}$ has a very specific structure, we remark that standard methods available in the Monte Carlo literature (see; e.g. [14]) are indeed not applicable. This problem has a simple solution if Δ consists only of real or complex parametric uncertainties $q_1, q_2, \ldots, q_\ell$ bounded in ℓ_p norms. In [7] an algorithm which is based on the so-called Generalized Gamma Density and generates samples uniformly distributed has been presented. The sample generation problem appears much harder when we are interested in the uniform generation of real or complex matrix samples $\Delta \in \mathbf{F}^{k_\ell \times k_\ell}$ bounded in the induced ℓ_2-norm. In this case, a crucial step is the computation of the probability distribution of the singular values of Δ; see [8].

Acknowledgements: This work was supported by funds of CENS-CNR of Italy. The authors express gratitude to Dr. G. Calafiore for his comments on an earlier version of this paper.

References

[1] Bai, E.-W., R. Tempo and M. Fu (1997). Worst Case Properties of the Uniform Distribution and Randomized Algorithms for Robustness Analysis, *Proceedings of the American Control Conference*, Albuquerque, pp. 861–865; to appear in *Mathematics of Control, Signals, and Systems.*

[2] Barmish, B. R. (1994). *New Tools for Robustness of Linear Systems*, McMillan, New York.

[3] Barmish B. R. and C. M. Lagoa (1997). The Uniform Distribution: A Rigorous Justification for its use in Robustness Analysis, *Mathematics of Control, Signals, and Systems*, vol. 10, pp. 203–222.

[4] Bernoulli, J. (1713). *Ars Conjectandi.*

[5] Bhattacharyya, S. P., H. Chapellat and L. H. Keel (1995). *Robust Control: The Parametric Approach*, Prentice-Hall, Englewood Cliffs.

[6] Blondel V. and J. N. Tsitsiklis (1997). NP-Hardness of Some Linear Control Design Problems, *SIAM Journal of Control and Optimization*, vol. 35, pp. 2118–2127.

[7] Calafiore, G., F. Dabbene and R. Tempo (1998). Uniform Sample Generation of Vectors in ℓ_p Balls for Probabilistic Robustness Analysis; to appear in *Proceedings of the IEEE Conference on Decision and Control*, Tampa.

[8] Calafiore, G., F. Dabbene and R. Tempo (1998). Uniform Sample Generation in Matrix Spaces: the Probability Density Function of the Singular Values, *Technical Report, CENS-CNR 98-3.*

[9] Braatz, R. P., P. M. Young, J. C. Doyle, and M. Morari, (1994). Computational Complexity of μ Calculation, *IEEE Transactions on Automatic Control*, vol. AC-39, pp. 1000–1002.

[10] Chen X. and K. Zhou (1997). A Probabilistic Approach to Robust Control, *Proceedings of the IEEE Conference on Decision and Control*, San Diego, pp. 4894–4895.

[11] Chen X. and K. Zhou (1997). On the Probabilistic Characterization of Model Uncertainty and Robustness, *Proceedings of the IEEE Conference on Decision and Control*, San Diego, pp. 3816–3821.

[12] Chernoff, H. (1952). A Measure of Asymptotic Efficiency for Test of Hypothesis Based on the Sum of Observations, *Annals of Mathematical Statistics*, vol. 23, pp. 493–507.

[13] Coxson, G. E. and C. L. DeMarco (1994). The Computational Complexity of Approximating the Minimal Perturbation Scaling to Achieve Instability in an Interval Matrix, *Mathematics of Control, Signal and Systems*, vol. 7, pp. 279–291.

[14] Devroye, L. (1986). *Non-Uniform Random Variate Generation*, Springer-Verlag, New York.

[15] Djavdan, P., H. J. A. F. Tulleken, M. H. Voetter, H. B. Verbruggen and G. J. Olsder (1989). Probabilistic Robust Controller Design, *Proceedings of the IEEE Conference on Decision and Control*, Tampa, pp. 2164–2172.

[16] Doyle, J. (1982). Analysis of Feedback Systems with Structured Uncertainty, *IEE Proceedings*, vol. 129, part D, pp. 242–250.

[17] Fukunaga, K. (1972). *Introduction to Statistical Pattern Recognition*, Academic Press, New York.

[18] Khargonekar, P. P. and A. Tikku (1996). Randomized Algorithms for Robust Control Analysis Have Polynomial Time Complexity, *Proceedings of the Conference on Decision and Control*, Kobe, Japan, pp. 3470–3475.

[19] Khargonekar, P. and A. Yoon (1997). Computational Experiments in Robust Stability Analysis, *Proceedings of the Conference on Decision and Control*, San Diego, pp. 3260–3264.

[20] Khatri, S. H. and P. A. Parrilo (1998). Spherical μ, *Proceedings of the American Control Conference*, Philadelphia, pp. 2314–2318.

[21] McFarlane, D. C. and K. Glover (1990). *Robust Controller Design Using Normalized Coprime Factor Plant Descriptions*, Springer-Verlag, pp. 132–142.

[22] Nemirovskii, A. (1993). Several NP-Hard Problems Arising in Robust Stability Analysis, *Mathematics of Control, Signals and Systems*, vol. 6, pp. 99–105.

[23] Polijak S. and J. Rohn (1993). Checking Robust Non-Singularity is NP-Hard, *Mathematics of Control, Signals and Systems*, vol. 6, pp. 1–9.

[24] Ray, L. R. and R. F. Stengel (1993). A Monte Carlo Approach to the Analysis of Control System Robustness, *Automatica*, vol. 29, pp. 229–236.

[25] Salehi, S. (1985). Application of Adaptive Observers to the Control of Flexible Spacecraft, *IFAC Symposium on Control in Space*, Toulouse, France.

[26] Tempo, R., E. W. Bai and F. Dabbene (1997). Probabilistic Robustness Analysis: Explicit Bounds for the Minimum Number of Samples, *Systems and Control Letters*, vol. 30, pp. 237–242.

[27] Vapnik V. N. and A. Ya Chervonenkis (1971). On the Uniform Convergence of Relative Frequencies to Their Probabilities, *Theory of Probability and its Applications*, vol. 16, pp. 264–280.

[28] Vidyasagar, M. (1997). *A Theory of Learning and Generalization*, Springer-Verlag, London.

[29] Vidyasagar, M. (1997). Statistical Learning Theory: An Introduction and Applications to Randomized Algorithms, *Proceedings of the European Control Conference*, Brussels, Belgium, pp. 161–189.

[30] Sontag, E. D. (1998). VC Dimension of Neural Networks; to appear in *Neural Networks and Machine Learning*, Springer-Verlag, London.

[31] Zhou, K., J. C. Doyle and K. Glover (1996). *Robust and Optimal Control*, Prentice-Hall, Upper Saddle River.

[32] Zhu, X., Y. Huang and J. Doyle (1996). Soft vs. Hard Bounds in Probabilistic Robustness Analysis, *Proceedings of the Conference on Decision and Control*, Kobe, Japan, pp. 3412–3417.

CENS-CNR, Politecnico di Torino, Corso Duca degli Abruzzi 24, 10129 Torino, Italy
tempo@polito.it, dabbene@polito.it

Progress in Systems and Control Theory, Vol. 25

An Approach to Observer Design

Maria Elena Valcher [1], Jan C. Willems

1 Introduction

In the last decade the behavioral point of view [8, 11, 12] has received an increasingly broader acceptance as an approach for modeling dynamical systems, and now is generally viewed as a cogent framework for system analysis. One of the reasons of its success has to be looked for in the fact that it does not start with the input/output point of view for describing how a system interacts with its environment, but focuses on the set of system trajectories, the *behavior*, and hence on the mathematical model describing the relations among all system variables.

By assuming this point of view, important aspects of the classical system theory have been translated and solved, thus leading to interesting results, which are powerful generalizations of well-known theorems obtained within the input/output or state space contexts. In particular, quite recently the control problem has been posed in the behavioral setting [13], where it can be naturally viewed as a problem of systems interconnection. Although several issues have already been analyzed in some detail, the important question of estimating some system variables, not available for measurements, from others, which are measured, has not been treated, yet.

The synthesis of an observer of the state for a (linear time-invariant) state space system has been the object of a considerable interest in classical system theory [1, 7]. The original theory of state observers was concerned with the problem of reconstructing (or estimating) the state from the corresponding inputs and outputs. This problem has been later generalized in various ways, and in the last years there has been a great deal of research aiming at state observers in the presence of unknown inputs (disturbances) [2, 3, 9].

In this contribution we will investigate the observer problem for linear time-invariant (continuous-time) dynamical systems, that are described in behavioral terms by means of a set of differential equations. More precisely, we will consider a dynamical system $\Sigma = (\mathbb{R}, \mathbb{R}^{\mathtt{w}_1+\mathtt{w}_2}, \mathfrak{B})$, whose trajectories $(\mathbf{w}_1, \mathbf{w}_2)$ satisfy some set of differential equations

$$R_2(\frac{d}{dt})\mathbf{w}_2 = R_1(\frac{d}{dt})\mathbf{w}_1,$$

and we will assume that $\mathbf{w}_1$ can be measured, while $\mathbf{w}_2$ is unknown. The

[1] Part of the results achieved in this paper were obtained during the author's stay in Groningen, which was financially supported by the ESF Cosy Projecty.

natural goal is that of designing an estimator of $\mathbf{w}_2$ based on the knowledge of $\mathbf{w}_1$, such that its estimation error goes to zero asymptotically.
To reach this goal, we shall first introduce the notions of observability and detectability of $\mathbf{w}_2$ from $\mathbf{w}_1$, and then provide necessary and sufficient conditions for the existence of an (asymptotic) observer. This then leads to a complete parametrization of all possible observers, in terms of the polynomial matrices involved in the system description.

Section 4 deals with the problem of determining under what conditions it is possible to obtain (asymptotic) observers with a proper transfer matrix. As for classical state space models, the estimate $\hat{\mathbf{w}}_2$ produced by the observer can be used, together with the measured variable $\mathbf{w}_1$, to control the whole plant, thus obtaining an "observer-based controller" for the original plant. As this paper is a preliminary report on the argument, all proofs will be skipped. For the details, we refer the interested reader to [10].

In the following, integers that refer to dimensions of linear spaces and/or sizes of matrices are always denoted in typewriter fonts. For instance, $\mathbb{R}^{\mathtt{w}_i}$ denotes the linear space of real column vectors with $\mathtt{w}_i$ components, $\mathbb{R}^{\mathtt{r}\times\mathtt{w}_i}$ is the space of real $\mathtt{r}\times\mathtt{w}_i$ matrices, $I_{\mathtt{w}_i}$ the identity matrix of size $\mathtt{w}_i$. We also make the following convention: vectors $\mathbf{w}_i, \mathbf{d}$ and ℓ are elements of $\mathbb{R}^{\mathtt{w}_i}$, $\mathbb{R}^{\mathtt{d}}$ and $\mathbb{R}^{\mathtt{l}}$. However, in keeping with the usual notation, when dealing with state space systems we will assume that the state vector $\mathbf{x}$ is $\mathtt{n}$-dimensional, the output vector $\mathbf{y}$ is $\mathtt{p}$-dimensional and the input vector $\mathbf{u}$ is $\mathtt{m}$-dimensional.

2 Observability and detectability

Consider a dynamical system $\Sigma = (\mathbb{R}, \mathbb{R}^{\mathtt{w}_1+\mathtt{w}_2}, \mathfrak{B})$, with trajectories $(\mathbf{w}_1, \mathbf{w}_2)$, whose behavior $\mathfrak{B}$ is specified by the set of differential equations

$$R_2(\frac{d}{dt})\mathbf{w}_2 = R_1(\frac{d}{dt})\mathbf{w}_1, \tag{2.1}$$

with $R_1 \in \mathbb{R}[\xi]^{\mathtt{r}\times\mathtt{w}_1}$ and $R_2 \in \mathbb{R}[\xi]^{\mathtt{r}\times\mathtt{w}_2}$ polynomial matrices, and $\mathtt{w}_i := \dim\mathbf{w}_i$, $i = 1, 2$. In the sequel, we will assume that the trajectories $(\mathbf{w}_1, \mathbf{w}_2)$ belong to $\mathfrak{L}^{loc}(\mathbb{R}, \mathbb{R}^{\mathtt{w}_1+\mathtt{w}_2})$, the space of all measurable functions f from $\mathbb{R}$ to $\mathbb{R}^{\mathtt{w}_1+\mathtt{w}_2}$ for which the integral $\int_{t_1}^{t_2} \|f(t)\|\, dt$ is finite for all t_1 and t_2. The set of trajectories $(\mathbf{w}_1, \mathbf{w}_2)$ satisfying (2.1) will be denoted, for short, by $\ker\left([R_2(\frac{d}{dt}) \mid -R_1(\frac{d}{dt})]\right)$.

If we assume that $\mathbf{w}_1$ can be exactly measured, while $\mathbf{w}_2$ is completely unknown, it is natural to search for necessary and sufficient conditions for the existence of an estimator of $\mathbf{w}_2$, based on the knowledge of $\mathbf{w}_1$, whose estimation error goes to zero asymptotically. The first step toward this end is that of introducing the notions of observability [12] and detectability.

Definition 2.1 *Consider a dynamical system* $\Sigma = (\mathbb{R}, \mathbb{R}^{\mathtt{w}_1+\mathtt{w}_2}, \mathfrak{B})$*, whose behavior* $\mathfrak{B}$ *is described as follows*

$$\mathfrak{B} := \left\{ (\mathbf{w}_1, \mathbf{w}_2) \in \left(\mathbb{R}^{\mathtt{w}_1+\mathtt{w}_2}\right)^{\mathbb{R}} : R_2(\frac{d}{dt})\mathbf{w}_2 = R_1(\frac{d}{dt})\mathbf{w}_1 \right\}, \tag{2.2}$$

with $\mathtt{w}_i := \dim \mathbf{w}_i$, $i = 1, 2$. *We say that* $\mathbf{w}_2$ *is*

- observable *from* $\mathbf{w}_1$*, if* $(\mathbf{w}_1, \mathbf{w}_2), (\mathbf{w}_1, \bar{\mathbf{w}}_2) \in \mathfrak{B}$ *implies* $\mathbf{w}_2 = \bar{\mathbf{w}}_2$*;*
- detectable *from* $\mathbf{w}_1$*, if* $(\mathbf{w}_1, \mathbf{w}_2), (\mathbf{w}_1, \bar{\mathbf{w}}_2) \in \mathfrak{B}$ *implies*

$$\mathbf{w}_2(t) - \bar{\mathbf{w}}_2(t) \xrightarrow[t \to +\infty]{} 0.$$

Proposition 2.1 *Consider the dynamical system* $\Sigma = (\mathbb{R}, \mathbb{R}^{\mathtt{w}_1+\mathtt{w}_2}, \mathfrak{B})$ *described by (2.2), with* $R_1 \in \mathbb{R}[\xi]^{\mathtt{r}\times\mathtt{w}_1}$ *and* $R_2 \in \mathbb{R}[\xi]^{\mathtt{r}\times\mathtt{w}_2}$*. Then*

i) $\mathbf{w}_2$ *is observable from* $\mathbf{w}_1$ *if and only if* R_2 *is a right prime matrix, or, equivalently, if and only if* $R_2(\lambda)$ *has full column rank for all* $\lambda \in \mathbb{C}$*;*

ii) $\mathbf{w}_2$ *is detectable from* $\mathbf{w}_1$ *if and only if* R_2 *is of full column rank and the g.c.d. of its maximal order minors is Hurwitz, or, equivalently, if and only if* $R_2(\lambda)$ *has full column rank for all* $\lambda \in \mathbb{C}^+ := \{\lambda \in \mathbb{C} : \mathfrak{Re}(\lambda) \geq 0\}$.

The previous definitions of observability and detectability are consistent with those for state space models. Actually, given an ($\mathtt{n}$-dimensional) state space model, with $\mathtt{m}$ inputs and $\mathtt{p}$ outputs, i.e.

$$\begin{aligned} \frac{d\mathbf{x}}{dt} &= F\mathbf{x}(t) + G\mathbf{u}(t), \\ \mathbf{y}(t) &= H\mathbf{x}(t) + J\mathbf{u}(t), \end{aligned} \qquad t \geq 0, \tag{2.3}$$

the set of its trajectories is equivalently described, in behavioral terms, as the set $\mathfrak{B}$ of all sequences $(\mathbf{x}, \mathbf{u}, \mathbf{y})$ satisfying

$$\begin{bmatrix} \frac{d}{dt}I_{\mathtt{n}} - F \\ H \end{bmatrix} \mathbf{x} = \begin{bmatrix} G & 0 \\ -J & I_{\mathtt{p}} \end{bmatrix} \begin{bmatrix} \mathbf{u} \\ \mathbf{y} \end{bmatrix}. \tag{2.4}$$

By the previous proposition, $\mathbf{x}$ is observable from $(\mathbf{u}, \mathbf{y})$ if and only if

$$\operatorname{rank} \begin{bmatrix} \lambda I_{\mathtt{n}} - F \\ H \end{bmatrix} = \mathtt{n}, \qquad \forall\, \lambda \in \mathbb{C},$$

and is detectable from $(\mathbf{u}, \mathbf{y})$ if and only if

$$\operatorname{rank} \begin{bmatrix} \lambda I_{\mathtt{n}} - F \\ H \end{bmatrix} = \mathtt{n}, \qquad \forall\, \lambda \in \mathbb{C}^+.$$

These represent the well-known observability and detectability PBH tests for state space models.

3 Asymptotic observers design

Consider the dynamical system described by (2.2), with $\mathbf{w}_1$ the measured variable and $\mathbf{w}_2$ the to-be-estimated one. The problem we will now address is that of introducing a sound definition of "observer". As a first requirement, an observer of $\mathbf{w}_2$ from $\mathbf{w}_1$ for system Σ should "accept" every sequence $\mathbf{w}_1$, which is part of a behavior trajectory $(\mathbf{w}_1, \mathbf{w}_2)$, and correspondingly produce some (in general, not unique) estimated trajectory $\hat{\mathbf{w}}_2$. This amounts to saying that an observer of Σ should not introduce additional constraints on the $\mathbf{w}_1$ components of the system trajectories. We refer to such a dynamical system as to an "acceptor" of the signal $\mathbf{w}_1$ for Σ. As a further requirement, it is reasonable to assume that the output of an observer is consistent when tracking $\mathbf{w}_2$, by this meaning that when the trajectories $\hat{\mathbf{w}}_2$ and $\mathbf{w}_2$ coincide for a sufficiently long time, for instance in $(-\infty, 0]$, then they coincide all over the time. Therefore, an observer for Σ is a system that, corresponding to every $(\mathbf{w}_1, \mathbf{w}_2)$ in $\mathfrak{B}$, produces an estimate $\hat{\mathbf{w}}_2$ of the trajectory $\mathbf{w}_2$, and does not loose track of the correct trajectory once it has followed it over a sufficiently long time. Such an observer is said to be asymptotic if the estimate $\hat{\mathbf{w}}_2$ it provides represents a good asymptotic estimate of $\mathbf{w}_2$, namely the sequence $\mathbf{w}_2(t) - \hat{\mathbf{w}}_2(t)$ goes to zero as t goes to $+\infty$. An asymptotic observer for Σ which produces an estimate $\hat{\mathbf{w}}_2$ of $\mathbf{w}_2$ which coincides with $\mathbf{w}_2$ at each time instant is an exact observer. These notions are formalized in the following definitions.

Definition 3.1 *Consider the dynamical system* $\Sigma = (\mathbb{R}, \mathbb{R}^{\mathtt{w}_1+\mathtt{w}_2}, \mathfrak{B})$, *whose behavior* $\mathfrak{B}$ *is described by (2.2). The set of differential equations*

$$Q(\frac{d}{dt})\hat{\mathbf{w}}_2 = P(\frac{d}{dt})\mathbf{w}_1 \tag{3.5}$$

with P *and* Q *polynomial matrices of suitable dimensions, is said to describe*

- *an* acceptor of $\mathbf{w}_1$ for Σ *if for every* $(\mathbf{w}_1, \mathbf{w}_2) \in \mathfrak{B}$ *there exists* $\hat{\mathbf{w}}_2$ *such that* $(\mathbf{w}_1, \hat{\mathbf{w}}_2)$ *satisfies (3.5);*

an acceptor (3.5) is

- *an* observer of $\mathbf{w}_2$ from $\mathbf{w}_1$ for Σ *if whenever* $(\mathbf{w}_1, \mathbf{w}_2)$ *is in* $\mathfrak{B}$, *and* $(\mathbf{w}_1, \hat{\mathbf{w}}_2)$ *satisfies (3.5) with* $\hat{\mathbf{w}}_2(t) = \mathbf{w}_2(t)$ *for* $t \in (-\infty, 0]$, *then* $\hat{\mathbf{w}}_2(t) = \mathbf{w}_2(t)$ *for all* $t \in \mathbb{R}$.

The observer of $\mathbf{w}_2$ *from* $\mathbf{w}_1$ *for* Σ, *(3.5), is said to be*

- *an* asymptotic observer *if for every* $(\mathbf{w}_1, \mathbf{w}_2)$ *in* $\mathfrak{B}$, *and* $(\mathbf{w}_1, \hat{\mathbf{w}}_2)$ *satisfying (3.5), we have* $\lim_{t\to+\infty} \mathbf{w}_2(t) - \hat{\mathbf{w}}_2(t) = 0$, *and*
- *an* exact observer *if for every* $(\mathbf{w}_1, \mathbf{w}_2)$ *in* $\mathfrak{B}$ *and* $(\mathbf{w}_1, \hat{\mathbf{w}}_2)$ *satisfying (3.5), we have* $\mathbf{w}_2 = \hat{\mathbf{w}}_2$.

In the sequel, as $\mathbf{w}_1$ will always represent the measured variable and $\mathbf{w}_2$ the unmeasurable one, we will refer to the acceptors of $\mathbf{w}_1$ for Σ and to the observers of $\mathbf{w}_2$ from $\mathbf{w}_1$ for Σ simply as acceptors and observers for Σ.

Given an acceptor, described by (3.5), its behavior $\hat{\mathfrak{B}}$ is the set of all solutions $(\mathbf{w}_1, \hat{\mathbf{w}}_2)$ of the differential equation equation (3.5), and, by definition, it satisfies the following condition

$$\mathcal{P}_1\mathfrak{B} := \{\mathbf{w}_1 : \exists(\mathbf{w}_1, \mathbf{w}_2) \in \mathfrak{B}\} \subseteq \{\mathbf{w}_1 : \exists(\mathbf{w}_1, \hat{\mathbf{w}}_2) \in \hat{\mathfrak{B}}\} =: \mathcal{P}_1\hat{\mathfrak{B}}.$$

Among all the trajectories of $\hat{\mathfrak{B}}$, however, we will be interested only in those that are produced corresponding to the trajectories of $\mathfrak{B}$, namely in the set

$$\{(\mathbf{w}_1, \hat{\mathbf{w}}_2) \in \hat{\mathfrak{B}} : \mathbf{w}_1 \in \mathcal{P}_1\mathfrak{B}\}.$$

So, by assuming this point of view, it seems reasonable to regard as *equivalent* two acceptors, in particular two observers, for the same system Σ, not if their behaviors $\hat{\mathfrak{B}}_1$ and $\hat{\mathfrak{B}}_2$ coincide, but if their behaviors satisfy the following condition

$$\{(\mathbf{w}_1, \hat{\mathbf{w}}_2) \in \hat{\mathfrak{B}}_1 : \mathbf{w}_1 \in \mathcal{P}_1\mathfrak{B}\} = \{(\mathbf{w}_1, \hat{\mathbf{w}}_2) \in \hat{\mathfrak{B}}_2 : \mathbf{w}_1 \in \mathcal{P}_1\mathfrak{B}\}.$$

For an observer described by (3.5), the difference variable $\mathbf{e} := \mathbf{w}_2 - \hat{\mathbf{w}}_2$ represents, of course, the *estimation error.* So, the previous definitions can be paraphrased by saying that an observer is asymptotic (exact) if the set of its estimation error trajectories constitutes a behavior, denoted by $\mathfrak{B}_e$, which is autonomous and stable (the zero autonomous behavior). Since autonomous behaviors can always be represented as kernels of nonsingular square matrices [8], this amounts to saying that there exists some Hurwitz (unimodular) matrix $\Delta \in \mathbb{R}[\xi]^{\mathtt{w}_2 \times \mathtt{w}_2}$, i.e. a nonsingular matrix whose determinant is a Hurwitz polynomial (a nonzero constant term), such that $\mathfrak{B}_e = \ker\Delta$. The characteristic polynomial of the behavior $\mathfrak{B}_e$, namely $\det \Delta$, will be called the *error-dynamics characteristic polynomial* (see, also, [13]).

Of course, an acceptor for Σ always exists: one can choose, for instance, Σ itself. So, the existence of an acceptor is not an issue. Necessary and sufficient conditions for the existence of (asymptotic or exact) observers, instead, are given in the following proposition.

Proposition 3.1 *Consider a dynamical system* $\Sigma = (\mathbb{R}, \mathbb{R}^{\mathtt{w}_1+\mathtt{w}_2}, \mathfrak{B})$, *whose behavior* $\mathfrak{B}$ *is described by (2.2).*

(i) A necessary and sufficient condition for the existence of an observer for Σ *is that* R_2 *in (2.2) is a full column rank polynomial matrix;*

(ii) a necessary and sufficient condition for the existence of an asymptotic observer for Σ *is that* $\mathbf{w}_2$ *is detectable from* $\mathbf{w}_1$*;*

(iii) a necessary and sufficient condition for the existence of an exact observer for Σ *is that* $\mathbf{w}_2$ *is observable from* $\mathbf{w}_1$*.*

As our main interest in the paper is in asymptotic observers, from now on we will assume that $\mathbf{w}_2$ is detectable from $\mathbf{w}_1$. As a consequence, it is easy to see that the behavior $\mathfrak{B}$ can be equivalently described by means of a set of equations of the following type

$$D_2(\frac{d}{dt})\mathbf{w}_2 = N_1(\frac{d}{dt})\mathbf{w}_1 \tag{3.6}$$

$$0 = D_1(\frac{d}{dt})\mathbf{w}_1, \tag{3.7}$$

where D_2 is nonsingular Hurwitz. Also, it entails no loss of generality assuming that D_1 is of full row rank $\mathtt{d}_1$. In order to obtain a complete parametrization of the (asymptotic/exact) observers of Σ, we need the following technical lemma, stating that given any acceptor for Σ (in particular, an observer), it is possible to obtain an equivalent one, (i.e., producing the same set of trajectories $(\mathbf{w}_1, \hat{\mathbf{w}}_2)$ for every $\mathbf{w}_1$ in $\mathcal{P}_1\mathfrak{B}$), for which matrix Q is of full row rank.

Lemma 3.1 *If $Q(\frac{d}{dt})\hat{\mathbf{w}}_2 = P(\frac{d}{dt})\mathbf{w}_1$ is an acceptor (in particular, an observer) for Σ, there exists an equivalent acceptor (observer) $\bar{Q}(\frac{d}{dt})\hat{\mathbf{w}}_2 = \bar{P}(\frac{d}{dt})\mathbf{w}_1$ with $\bar{Q}$ of full row rank.*

Under the hypothesis that the matrix Q appearing in the observer equation is of full row rank, we can obtain deeper insights into the algebraic properties of the polynomial matrices P and Q involved in the observer description, and explicitly relate them to the matrices D_2, D_1 and N_1.

Theorem 3.1 *Consider a plant Σ whose behavior $\mathfrak{B}$ is described by (3.6) - (3.7), with D_2 Hurwitz and D_1 of full row rank $\mathtt{d}_1$. If P and Q are polynomial matrices, with Q of full row rank, then*

$$Q(\frac{d}{dt})\hat{\mathbf{w}}_2 = P(\frac{d}{dt})\mathbf{w}_1,$$

is an (asymptotic) observer for Σ if and only if there exists a nonsingular (Hurwitz) matrix $Y \in \mathbb{R}[\xi]^{\mathtt{w}_2 \times \mathtt{w}_2}$ and a polynomial matrix $X \in \mathbb{R}[\xi]^{\mathtt{w}_2 \times \mathtt{d}_1}$ such that

$$[\,Q(\xi) \quad -P(\xi)\,] = [\,Y(\xi) \quad X(\xi)\,] \begin{bmatrix} D_2(\xi) & -N_1(\xi) \\ 0 & -D_1(\xi) \end{bmatrix}. \tag{3.8}$$

Moreover, the set $\mathfrak{B}_e$ of all possible error trajectories coincides with ker $(Q(\frac{d}{dt}))$, which amounts to saying that we can assume $\Delta = Q$.

Remark 3.1 *For the problem analysis it has been useful to adopt the behavior description (3.6) - (3.7). If we assume, however, that the behavior* $\mathfrak{B}$ *is described as in (2.2), with* $\mathbf{w}_2$ *detectable from* $\mathbf{w}_1$*, the asymptotic observers for* Σ *are those and those only described by (3.5) with* Q *and* P *polynomial matrices,* Q *nonsingular Hurwitz, satisfying*

$$[\, Q(\xi) \quad -P(\xi)\,] = T(\xi)\,[\, R_2(\xi) \quad R_1(\xi)\,] \tag{3.9}$$

for some polynomial matrix T.

Also, by assuming that the matrix Q *appearing in the observer description is of full row rank, and hence nonsingular square, we have obtained a complete parametrization of all possible (asymptotic) observers. Loosening this constraint, indeed, would only produce a wider set of representations, not necessarily full row rank, for the same observers.*

As a further result, we are now interested in analyzing what performances can be achieved from the asymptotic observers in terms of error dynamics. These performances can be evaluated in terms of error-dynamics characteristic polynomials, as analyzed in the following corollary.

Corollary 3.1 *Consider a dynamical system whose behavior* $\mathfrak{B}$ *is described by (3.6) - (3.7), with* D_2 *Hurwitz and* D_1 *of full row rank. Then*

i) for every (asymptotic) observer for Σ*, the error-dynamics characteristic polynomial* $\det \Delta$ *is a (Hurwitz) polynomial satisfying the divisibility condition* $\det D_2 \mid \det \Delta$ *(i.e.* $\det D_2$ *divides* $\det \Delta$*);*

ii) for every (Hurwitz) polynomial $\delta \in \mathbb{R}[\xi]$ *with* $\det D_2 \mid \delta$*, there exists an (asymptotic) observer whose error-dynamics characteristic polynomial coincides with* δ.

As in the previous section, it is interesting to see how the above general results specialize to the case of state space models. Indeed, if we consider the behavior description (2.4) of a state space model, and assume that (H, F) is a detectable pair, we can apply the previous reasonings and express all asymptotic state estimators as

$$Q(\frac{d}{dt})\hat{\mathbf{x}} = P_u(\frac{d}{dt})\mathbf{u} + P_y(\frac{d}{dt})\mathbf{y}, \tag{3.10}$$

with (Q, P_u, P_y) a triple of polynomial matrices satisfying the following constraints:

i) Q is nonsingular Hurwitz;

ii) $[\, Q(\xi) \quad -P_u(\xi) \quad -P_y(\xi)\,] = [\, Y(\xi) \quad X(\xi)\,] \begin{bmatrix} \xi I_n - F & -G & 0 \\ H & J & -I_p \end{bmatrix},$

for suitable polynomial matrices Y and X.

It is easily seen that among all possible asymptotic observers, there is, in particular, Luenberger (full-order feedback) state observer. Indeed, if L is any $\mathtt{n} \times \mathtt{p}$ real matrix such that $F + LH$ is asymptotically stable, it is sufficient to assume $Y(\xi) = I_{\mathtt{n}}$ and $X(\xi) = -L$ in (ii), thus obtaining the state observer

$$\left(\frac{d}{dt} I_{\mathtt{n}} - F - LH\right) \hat{\mathbf{x}} = G\mathbf{u} - L\mathbf{y}, \tag{3.11}$$

which satisfies condition (i), and whose estimation error dynamics matrix Δ coincides with $\xi I_{\mathtt{n}} - F - LH$. Notice that such an observer is endowed with a strictly proper rational transfer matrix $\hat{W}(\xi) := (\xi I_{\mathtt{n}} - F - LH)^{-1} [G \quad -L]$.

4 Proper asymptotic observers

Theorem 3.1 provides us with a useful parametrization of the observers for a system described by (3.6) - (3.7). Indeed, all the observers for Σ can be described by means of the differential equation

$$(YD_2)(\frac{d}{dt})\ \hat{\mathbf{w}}_2 = (YN_1 + XD_1)(\frac{d}{dt})\ \mathbf{w}_1, \tag{4.12}$$

with X and Y polynomial matrices of suitable dimensions, and the additional constraint that Y is Hurwitz if the observer is asymptotic.

This parametrization can be fruitfully exploited to investigate further relevant issues, in particular, that of determining the existence of (strictly) proper asymptotic observers, endowed with a (strictly) proper transfer matrix. Clearly, this amounts to searching for conditions guaranteeing that there exists a matrix pair (Y, X), with Y Hurwitz, such that

$$\hat{W}(\xi) := [Y(\xi)D_2(\xi)]^{-1}[Y(\xi)N_1(\xi) + X(\xi)D_1(\xi)]$$

is (strictly) proper rational.

As shown in the following proposition, autonomous behaviors described as in (3.6) - (3.7) always admit strictly proper asymptotic observers.

Proposition 4.1 *Let Σ be a dynamical system, whose behavior $\mathfrak{B}$ is described by(3.6) - (3.7), with D_2 Hurwitz and D_1 having full row rank $\mathtt{d}_1$. If $\mathfrak{B}$ is autonomous or, equivalently, D_1 is a nonsingular square matrix, then there exists a strictly proper asymptotic estimator of $\mathbf{w}_2$ from $\mathbf{w}_1$.*

The general problem, when $\mathfrak{B}$ is an arbitrary (not necessarily autonomous) behavior, is a little more involved. In order to solve it, we refer to the original behavior description and assume, without loss of generality, that $\mathfrak{B}$ is described by the differential equation

$$R_2(\frac{d}{dt})\mathbf{w}_2 = R_1(\frac{d}{dt})\mathbf{w}_1, \tag{4.13}$$

with $[\, R_2 \quad -R_1 \,] \in \mathbb{R}[\xi]^{\mathtt{r}\times(\mathtt{w}_2+\mathtt{w}_1)}$ a row reduced matrix [4] with row degrees $h_1, h_2, \ldots, h_{\mathtt{r}}$, $\mathtt{r} = \mathtt{w}_1 + \mathtt{d}_1$. Of course, $\mathbf{w}_2$ is assumed to be detectable from $\mathbf{w}_1$. The first step toward the solution is given by the following lemma, where conditions for the existence of (strictly) proper observers for Σ, not necessarily asymptotic, are provided.

Lemma 4.1 *Consider a dynamical system Σ with behavior $\mathfrak{B}$ described by (4.13) and $\mathbf{w}_2$ detectable from $\mathbf{w}_1$. If $[\, R_{2hr} \quad -R_{1hr} \,]$ denotes the leading row coefficient matrix [4] of $[\, R_2 \quad -R_1 \,]$, then*

i) a necessary and sufficient condition for the existence of an observer for Σ with proper transfer matrix $\hat{W}$ is that R_{2hr} has full column rank $\mathtt{w}_2$;

ii) a necessary and sufficient condition for the existence of an observer for Σ with strictly proper transfer matrix $\hat{W}$ is that there exists $S \in \mathbb{R}^{\mathtt{w}_2\times(\mathtt{w}_2+\mathtt{d}_1)}$ such that $S\,[\, R_{2hr} \quad -R_{1hr} \,] = [\, I_{\mathtt{w}_2} \quad 0 \,]$.

Remark 4.1 *The condition that R_{2hr} has full column rank can be given the following interpretation. Consider the system Σ. Following the theory developed in [12], the system variables $\mathbf{w}^T = [\mathbf{w}_1^T \ \mathbf{w}_2^T]$ can always be partitioned into inputs and outputs. This can be done, however, in many ways. The full column rank condition on R_{2hr} ensures that a possible partition is the one described in the following picture, where $\mathbf{w}_1$ and $\mathbf{w}_2$ have been partitioned as $\mathbf{w}_1 = \begin{bmatrix} \mathbf{w}_{1,1} \\ \mathbf{w}_{1,2} \end{bmatrix}$ and $\mathbf{w}_2 = \begin{bmatrix} \mathbf{w}_{2,1} \\ \mathbf{w}_{2,2} \end{bmatrix}$, and Σ_1 and Σ_2 have both proper transfer matrices.*

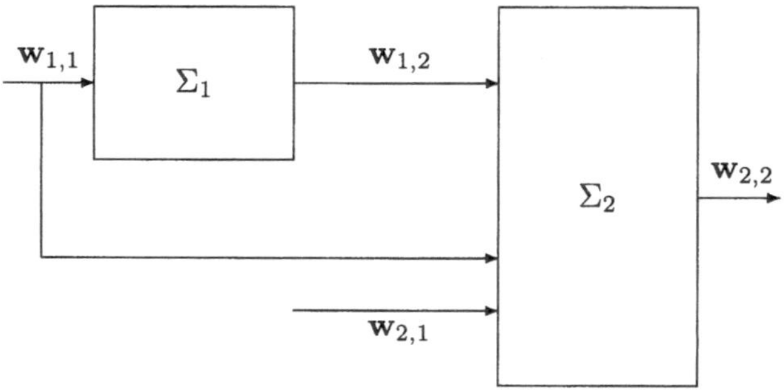

Fig. 1

Theorem 4.1 *Consider a dynamical system Σ with behavior $\mathfrak{B}$ described by (4.13) and $\mathbf{w}_2$ detectable from $\mathbf{w}_1$. If there exists a proper observer for Σ then there exists a proper asymptotic one.*

Remark 4.2 *As a consequence of the above proposition and lemma, once we reduce the matrix $[\,R_2 \quad -R_1\,]$ involved in the behavior description to row reduced form, the existence of a proper asymptotic observer is immediately checked by simply verifying that R_{2hr} has full column rank. When so, by resorting the constructive procedure given in the proof of Theorem 4.1 [10], one can explicitly construct such an observer.*

Consider a state space model (2.4), with (H, F) a detectable pair. Also in this context one can look for (strictly) proper asymptotic state estimators, namely observers described by

$$Q(\frac{d}{dt})\hat{\mathbf{x}} = P_u(\frac{d}{dt})\mathbf{u} + P_y(\frac{d}{dt})\mathbf{y}, \tag{4.14}$$

with

i) Q nonsingular Hurwitz;

ii) $[\,Q(\xi) \quad -P_u(\xi) \quad -P_y(\xi)\,] = [\,Y(\xi) \quad X(\xi)\,] \begin{bmatrix} \xi I_n - F & -G & 0 \\ H & J & -I_p \end{bmatrix},$

for suitable polynomial matrices Y and X;

iii) $Q^{-1}[P_u \,|P_y]$ (strictly) proper rational.

As previously noticed, the Luenberger observer is endowed with a strictly proper rational transfer matrix. So, the existence of a proper state observer is not an issue. It can be interesting, instead, to determine what matrices Q, and hence Δ, possibly describe the estimation error dynamics. The first step toward the solution is given by the following lemma, which proves that in order to fulfill condition (iii) above it is sufficient that $Q^{-1}P_y$ is proper rational.

Lemma 4.2 *Consider an asymptotic state observer described as in (4.14), with matrices Q, P_u and P_y satisfying conditions (i) and (ii). Such an observer is (strictly) proper, i.e. fulfills (iii), if and only if $Q^{-1}P_u$ is (strictly) proper.*

As an immediate consequence of the above lemma, the search for proper asymptotic state observers is equivalent to the problem of determining proper asymptotic state observers for the autonomous system

$$\begin{bmatrix} \frac{d}{dt}I_{\mathrm{n}} - F \\ H \end{bmatrix} \mathbf{x} = \begin{bmatrix} 0 \\ I_{\mathrm{p}} \end{bmatrix} \mathbf{y}. \tag{4.15}$$

This result is rather intuitive, since it expresses the fact that the forced state evolution could be easily removed without affecting the solution of the proper observer problem. Furthermore, it allows us to reduce ourselves to the special case of autonomous behaviors, previously analyzed, thus getting the following result.

Proposition 4.2 *Let $D_\ell^{-1}N_\ell$ be a left coprime MFD of $H(\xi I_{\mathrm{n}} - F)^{-1}$, with D_ℓ row-reduced with row indices $h_1, h_2, \ldots, h_{\mathrm{p}}$. For every polynomial pair $(\bar{Y}, \bar{X})$ such that $Q(\xi) := \bar{Y}(\xi)(\xi I_{\mathrm{n}} - F) + \bar{X}(\xi)H$ is row reduced with row degrees lower bounded by $\max_i h_i - 1$ ($\max_i h_i$), there exists a new pair (Y, X) such that $Y(\xi)(\xi I_{\mathrm{n}} - F) + X(\xi)H = Q(\xi)$ and*

$$Q(\frac{d}{dt})\hat{\mathbf{x}} = X(\frac{d}{dt})\mathbf{y}$$

is a proper (strictly proper) state observer for (4.15), and hence

$$Q(\frac{d}{dt})\hat{\mathbf{x}} = [YG - XJ](\frac{d}{dt})\mathbf{u} + X(\frac{d}{dt})\mathbf{y},$$

is a proper (strictly proper) state observer for the state model (2.4).

This result admits a rather interesting interpretation. As the row indices $h_1, h_2, \ldots, h_{\mathrm{p}}$ are the well-known *observability indices* of the pair (H, F) [4, 5, 6], the previous proposition states that it is always possible to obtain a state observer (4.14) where Q, and hence the error dynamics matrix Δ (see Theorem 3.1), is row reduced with row degrees lower bounded by the maximum of the observability indices. So, these indices somehow provide a constraint on the minimal complexity the asymptotic state observers possibly exhibit. This situation strictly reminds of an analogous one for the classical output feedback compensator, where, instead, the reachability indices are involved [5, 6].

The characterization of strictly proper state observers is much simpler than the characterization of general proper state observers, as shown by the following proposition.

Proposition 4.3 *Let*

$$Q(\frac{d}{dt})\hat{\mathbf{x}} = P_u(\frac{d}{dt})\mathbf{u} + P_y(\frac{d}{dt})\mathbf{y},$$

be an asymptotic state observer, whose matrices Q, P_u and P_y satisfy conditions (i) and (ii) for some (uniquely determined) matrix pair (Y, X). Such an observer is strictly proper if and only if $Y^{-1}X$ is proper.

5 The control problem

The control problem that will be considered is that of designing a suitable device (*controller*), modeled as a dynamical system Σ_c, that can be applied to the plant Σ thus producing a resulting system with desired properties. As recently emphasized in [13], the control problem is naturally stated as an interconnection problem, and the behavioral framework is very convenient for this. So, one has to look for some system $\Sigma_c = (\mathbb{R}, \mathbb{R}^{\mathtt{w}_1+\mathtt{w}_2}, \mathfrak{B}_c)$ that, once connected to Σ as shown in Fig. 2 below, results in a *controlled system*

$$\Sigma \wedge \Sigma_c := (\mathbb{Z}, \mathbb{F}^q, \mathfrak{B} \cap \mathfrak{B}_c), \tag{5.16}$$

with a desired behavior.

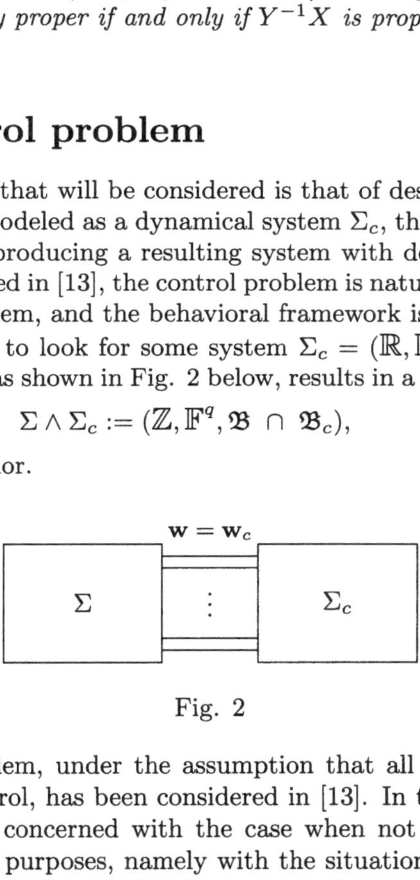

Fig. 2

The control problem, under the assumption that all system variables are available for control, has been considered in [13]. In this contribution, instead, we shall be concerned with the case when not all variables are accessible for control purposes, namely with the situation when the set of system variables $\mathbf{w}$ can be partitioned into two subvectors $\mathbf{w}^T = [\mathbf{w}_1^T \ \mathbf{w}_2^T]$ of which only $\mathbf{w}_1$ is available for control. Such a controller, that operates by restricting the set of admissible trajectories for the variable $\mathbf{w}_1$, will be called a $\mathbf{w}_1$-*based controller.*

In order to investigate what possibilities are offered by a controller of this kind, we start by assuming (without loss of generality) that the plant behavior is described by the set of differential equations:

$$D_2(\frac{d}{dt})\mathbf{w}_2 = N_1(\frac{d}{dt})\mathbf{w}_1 \tag{5.17}$$

$$0 = D_1(\frac{d}{dt})\mathbf{w}_1, \tag{5.18}$$

with D_2 and D_1 both of full row rank, and consider a $\mathbf{w}_1$-based controller defined by the following representation

$$0 = C_1(\frac{d}{dt})\mathbf{w}_1, \tag{5.19}$$

where C_1 is a polynomial matrix (see Fig. 3).

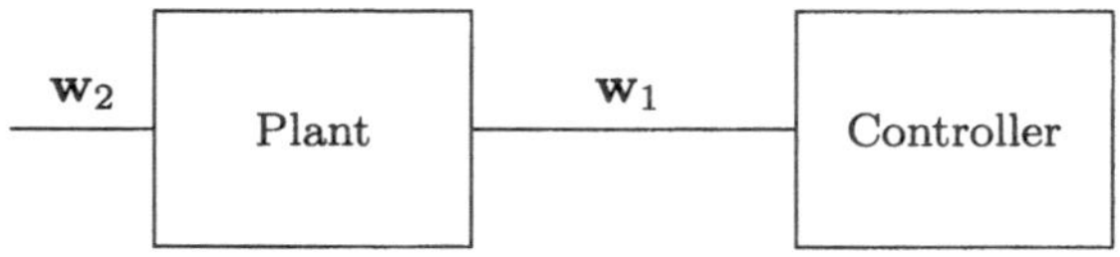

Fig. 3

As usual, the main goal one wants to achieve is that of making the resulting controlled system autonomous and possibly stable (sometimes with a preassigned characteristic polynomial). In this case, the $\mathbf{w}_1$-based controller is said to be *stabilizing*. The possibility of obtaining these properties are strictly related to the properties of Σ, as shown in the following theorem.

Theorem 5.1 *Consider the dynamical system Σ, whose behavior $\mathfrak{B}$ satisfies the differential equations (5.17) - (5.18). A necessary and sufficient condition for the existence of a $\mathbf{w}_1$-based controller that makes the resulting controlled system autonomous is that there exists an observer for Σ, i.e. D_2 is nonsingular square. When so,*

i) if Σ is autonomous, namely if D_1 is also nonsingular square, then every $\mathbf{w}_1$-based controller makes the controlled system autonomous with its characteristic polynomial χ_{res} satisfying

$$\det D_2 \mid \chi_{\text{res}} \mid \det D_2 \det D_1, \tag{5.20}$$

and, conversely, for every polynomial χ_{res} satisfying (5.20), there exists a $\mathbf{w}_1$-based controller such that the resulting autonomous system has characteristic polynomial χ_{res};

ii) if Σ is not autonomous (rank $D_1 =: \mathtt{d}_1 < \mathtt{w}_1$), then for every $\mathbf{w}_1$-based controller that makes the controlled system autonomous, the characteristic polynomial χ_{res} satisfies

$$\det D_2 \mid \chi_{\text{res}}, \tag{5.21}$$

and, conversely, for every polynomial χ_{res} satisfying (5.21), there exists a $\mathbf{w}_1$-based controller that makes the resulting system autonomous with characteristic polynomial χ_{res};

Consequently,

iii) there exists a stabilizing $\mathbf{w}_1$*-based controller if and only if* $\mathbf{w}_2$ *is detectable from* $\mathbf{w}_1$*, and*

iv) there exists a $\mathbf{w}_1$*-based controller that makes the resulting connected system autonomous with an arbitrarily chosen characteristic polynomial if and only if* $\mathbf{w}_2$ *is observable from* $\mathbf{w}_1$*.*

As we have just seen, the possibility of achieving certain results by means of a $\mathbf{w}_1$-based controller depends, indeed, on how much information about the "missing variable", $\mathbf{w}_2$, can be deduced from $\mathbf{w}_1$. In particular, the possibility of stabilizing Σ by constraining only $\mathbf{w}_1$ depends on the fact that the information about $\mathbf{w}_2$ is "asymptotically correct", namely that $\mathbf{w}_2$ is detectable from $\mathbf{w}_1$. If so, a reasonable solution could be to use certainty equivalence, i.e. to exploit an asymptotic estimate $\hat{\mathbf{w}}_2$ of $\mathbf{w}_2$, obtained by means of a suitable observer, and to design a controller Σ_c which makes use of the pair $(\mathbf{w}_1, \hat{\mathbf{w}}_2)$ as if it was $(\mathbf{w}_1, \mathbf{w}_2)$.

The situation just described represents the generalization of the analogous one for state space models, and we will call a controller with this structure, connected to the original plant as shown in Fig. 4, below, an *observer based controller.* Our interest, as usual, is in observer based controllers that make the resulting connected system autonomous and stable, and hence are *stabilizing.*

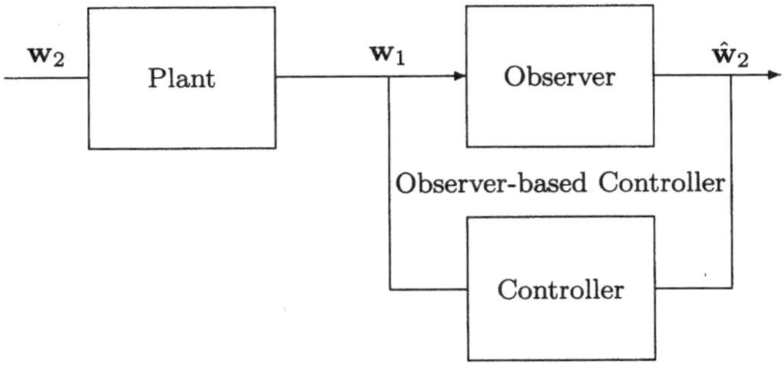

Fig. 4

Once again, we assume that the system behavior $\mathfrak{B}$ is described by (5.17) - (5.18), with D_2 Hurwitz and D_1 of full row rank, and introduce an asymptotic observer for Σ

$$Q(\frac{d}{dt})\hat{\mathbf{w}}_2 = P(\frac{d}{dt})\mathbf{w}_1 \tag{5.22}$$

with Q and P satifying condition (3.8) for some polynomial matrices X and Y, with Y Hurwitz. If we introduce a controller whose behavior $\mathfrak{B}_c$ is described by the following set of differential equations

$$C_2(\frac{d}{dt})\hat{\mathbf{w}}_2 = C_1(\frac{d}{dt})\mathbf{w}_1, \tag{5.23}$$

then the behavior of the whole connected system in Fig. 4, $\Sigma_{\text{res}} = (\mathbb{R}, \mathbb{R}^{\mathtt{w}_1+2\mathtt{w}_2}, \mathfrak{B}_{\text{res}})$, is described by

$$\begin{bmatrix} D_2 & 0 & -N_1 \\ 0 & 0 & -D_1 \\ 0 & Q & -P \\ 0 & C_2 & -C_1 \end{bmatrix} (\frac{d}{dt}) \begin{bmatrix} \mathbf{w}_2 \\ \hat{\mathbf{w}}_2 \\ \mathbf{w}_1 \end{bmatrix} = 0. \tag{5.24}$$

As $\mathfrak{B}_{\text{res}}$ can be expressed as the kernel of the polynomial matrix

$$M(\xi) := \begin{bmatrix} D_2(\xi) & 0 & -N_1(\xi) \\ 0 & 0 & -D_1(\xi) \\ 0 & Q(\xi) & -P(\xi) \\ 0 & C_2(\xi) & -C_1(\xi) \end{bmatrix},$$

in order to make it autonomous, we have to choose C_1 and C_2 so that M is of full column rank. If this is the case, $\mathfrak{B}_{\text{res}}$ is stable if and only if the g.c.d. of the maximal order minors of M, the characteristic polynomial χ_{res} of $\mathfrak{B}_{\text{res}}$, is Hurwitz.

Theorem 5.2 *Consider the controlled system described in Fig. 4 and set* $\mathtt{d}_1 := \text{rank} D_1$. *If Σ is autonomous, namely $\mathtt{d}_1$ coincides with $\mathtt{w}_1$, then*

i) for every controller (5.23) the resulting controlled system Σ_{res} is autonomous with its characteristic polynomial χ_{res} satisfying

$$\det D_2 \mid \chi_{\text{res}} \mid \det D_2 \det Q \det D_1, \tag{5.25}$$

and, conversely,

ii) for every polynomial χ_{res} which satisfies (5.25), there exists a controller (5.23) such that the resulting autonomous system has characteristic polynomial χ_{res}.

Otherwise, if $\mathtt{d}_1 < \mathtt{w}_1$, then

iii) for every controller (5.23) which makes the whole system $\Sigma_{\rm res}$ autonomous, the characteristic polynomial $\chi_{\rm res}$ is a multiple of $\det D_2$*, and*

iv) for every polynomial $\chi_{\rm res}$ which is multiple of $\det D_2$ *there exists a controller (5.23) such that the resulting system is autonomous with characteristic polynomial $\chi_{\rm res}$.*

Necessary and sufficient conditions for the existence of stabilizing observer-based controllers are immediately derived, as a straightforward consequence of the previous result.

Corollary 5.1 *Given a plant Σ, whose behavior $\mathfrak{B}$ is described by (5.17) - (5.18), with D_2 nonsingular square and D_1 of full row rank, a necessary and sufficient condition for the existence of a stabilizing observer-based controller is that $\mathbf{w}_2$ is detectable from $\mathbf{w}_1$, namely D_2 is Hurwitz.*

Remark 5.1 *It is worthwhile noticing that the gap existing between the results achieved, in terms of characteristic polynomials, by an observer based controller and by a $\mathbf{w}_1$-based controller in case of detectability is only apparent, and due to the fact that the characteristic polynomial we refer to in the case of an observer based controller takes into account also the observer dynamics. If we considered just the behavior of the plant Σ, namely the projection of $\mathfrak{B}_{\rm res}$ onto the variables $\mathbf{w}_1$ and $\mathbf{w}_2$, we would obtain the same results, as indeed, observer based controllers and $\mathbf{w}_1$-based controllers, under the detectability assumption, are completely equivalent.*

Also, notice that the development of observer-based controllers can of course be combined with the notions of singular and regular ($\mathbf{w}$-based) controllers extensively discussed in [13]. However, we have preferred not to enter into these ramifications here.

References

[1] Bongiorno J.J. and Youla D.C. (1968). On observers in multivariable control systems. *Int. J. Contr.*, **8**, 221–243.

[2] Chang S.K. and Hsu P.L. (1994). On the application of the {1}-inverse to the design of general structured unknown input observers. *Int. J. Syst. Sci.*, **25**, 2167–2186.

[3] Hautus M.L.J. (1983). Strong detectability and observers. *Linear Alg. Appl.*, **50**, 353–368.

[4] Kailath T. (1980). *Linear Systems.* Prentice Hall, Inc..

[5] Kučera V. (1979). *Discrete linear control: the polynomial equation approach.* J. Wiley & Sons, New York.

[6] Kučera V. (1991). *Analysis and design of discrete linear control systems.* Academia, Praga.

[7] Luenberger D.G. (1966). Observers for multivariable systems. *IEEE Trans. Aut.Contr.*, **AC-11**, 190–197.

[8] Polderman J.W. and Willems J.C. (1997). *Introduction to Mathematical Systems Theory: A behavioral approach.* Springer-Verlag.

[9] Tse E. (1973). Observer-estimators for discrete-time systems. *IEEE Trans. Aut. Contr.*, **AC-18**, 10–16.

[10] Valcher M.E. and Willems J.C. (1998). Observer synthesis in the behavioral approach. Submitted for pubblication to *IEEE Trans. on Automatic Control.*

[11] Willems J.C. (1988). Models for dynamics. *Dynamics reported*, **2**, 171–269.

[12] Willems J.C. (1991). Paradigms and puzzles in the theory of dynamical systems. *IEEE Trans. Aut.Contr.*, **AC-36**, 259–294.

[13] Willems J.C. (1997). On interconnections, control, and feedback. *IEEE Trans. Aut. Contr.*, **AC-42**, 326–339.

Dip. di Elettronica ed Informatica, Univ. di Padova, via Gradenigo 6a, 35131 Padova, Italy

Research Institute for Mathematics and Computing Science, University of Groningen, P.O. Box 800, 9700 AV Groningen, The Netherlands

Progress in Systems and Control Theory, Vol. 25

Group Codes and Behaviors

G. David Forney, Jr.

1 Introduction

The close connections between convolutional coding theory and linear system theory are well known. A convolutional encoder is simply a discrete-time linear time-invariant (LTI) system over a *finite* field, always finite-dimensional, in general multivariable. These connections were exploited in the 1970's to develop an algebraic structure theory of convolutional codes [5], the key elements of which turned out later to be useful in linear system theory [6, 14].

More recently, Willems and his school [25, 26] have developed a behavioral approach to system theory, in which all properties of a system are deduced from its behavior (the set of all possible trajectories of the system). Behavioral system theory appears to have two major branches. The foundations of the field are developed in set-theoretic language [25, Secs. 1-3]. Most of the behavioral literature, however, focusses on LTI systems, *cf.* [25, Sec. 4]. The most important property of linear behavioral systems is that they always have well-defined, essentially unique minimal state realizations, unlike general behavioral systems.

The behavioral approach is highly congenial to coding theorists, since a code is nothing more than a discrete-time, discrete-alphabet behavior. Moreover, in coding theory the code is the fundamental object of study; finding an encoder to generate it is secondary. A binary convolutional code, for example, is simply an LTI behavior over the binary field $\mathbb{F}_2$; it has many possible encoders (even minimal encoders).

Recent developments in coding theory have focussed attention on broader classes of systems than LTI systems. In particular, there has been great interest in:

- Minimal state realizations (trellises) of linear block codes [23], which are necessarily time-varying.
- Group codes; in particular, geometrically uniform group codes for Euclidean-space coding (*e.g.,* lattices) [7].

These developments have motivated fundamental research into the properties of group codes (behaviors), possibly time-varying [9]. It appears from this work that some of the most important properties of LTI systems depend only on their group property, not on linearity or time-invariance. In particular, group codes (behaviors) have well-defined, essentially unique minimal state realizations.

Our intent in this paper is to outline briefly the main properties of group behaviors for a system-theoretic audience. In Section 2, we introduce group codes and behaviors and minimal state realizations thereof, using the fundamental State Space Theorem. In Section 3, we discuss controllability and observability properties of group behaviors. In Section 4, we show that even duality is well defined for group behaviors, at least over finite abelian groups (or, more generally, over locally compact abelian groups). We discuss a number of fundamental dynamical dualities.

Very recently, the hottest topic in coding theory has been "codes defined on graphs," which include "tail-biting" codes and capacity-approaching codes such as turbo codes and low-density parity-check codes [1, 16]. Such codes may be regarded as behaviors defined on unconventional, unordered time axes. In Section 5, we give a brief introduction to this exciting topic.

2 Minimal Realizations of Group Behaviors

In this section we define group codes, or equivalently discrete-time group behaviors, and show that they have well-defined, essentially unique minimal state realizations. This key property follows from the fundamental State Space Theorem.

Our basic reference for behavioral system theory is [25]; see also [26]. Our basic reference for group codes is [9].

2.1 Group Codes and Behaviors

Group codes and discrete-time group behaviors are essentially identical concepts. With a group code, however, the symbol alphabet is usually finite, and the ordering of the time axis may be immaterial.

In behavioral system theory, a (discrete-time) *system* is a triple $\Sigma = (I, \mathcal{A}, \mathcal{B})$, where

- the *index set* (time axis) I is a discrete, ordered set, which without loss of generality may be identified with a subinterval of the integers, $I \subseteq \mathbb{Z}$;

- the *symbol sequence space* $\mathcal{A} = \bigotimes_{k \in I} A_k$ is the Cartesian product of a collection of *symbol alphabets* $A_k, k \in I$;

- the *behavior* $\mathcal{B}$ is a subset $\mathcal{B} \subseteq \mathcal{A}$.

The behavior is the set of sequences (trajectories) $\mathbf{a} \in \mathcal{A}$ that may actually occur.

A *group system* is a system $\Sigma = (I, \mathcal{A}, \mathcal{B})$ that has a componentwise group property; *i.e.*,

- the symbol sequence space $\mathcal{A} = \bigotimes_{k \in I} A_k$ is the *direct product* of a collection of *symbol groups* $A_k, k \in I$;
- the behavior $\mathcal{B}$ is a *subgroup* of $\mathcal{A}$.

(In group theory a group system is called a subdirect product.)

Most behavioral literature focusses on linear time-invariant (LTI) systems. A (strictly) *linear time-invariant system* is a group system in which:

- the time axis is the set of integers, $I = \mathbb{Z}$;
- the symbol groups A_k are all equal to a common vector space V over a field $\mathbb{F}$, so the symbol sequence space is the infinite direct product $\mathcal{A} = V^{\mathbb{Z}}$, a vector space over $\mathbb{F}$;
- the behavior $\mathcal{B}$ is a *subspace* of $\mathcal{A}$;
- $\mathcal{B}$ is invariant under time shifts; *i.e.*, $D\mathcal{B} = \mathcal{B}$, where D is the *unit delay operator*.

A *group* (resp. *linear*) *code* is no more nor less than a group (resp. linear) behavior. However, in coding the symbol groups A_k are usually finite and always discrete; *e.g.*, in the LTI case, V is usually a finite-dimensional vector space over a finite field $\mathbb{F}$.

A time-invariant convolutional code is an LTI behavior over a finite field $\mathbb{F}$. Consequently there are close connections between convolutional coding theory and LTI system theory.

For a block code, however, the time axis I is a *finite* subset of the integers, so a block code is necessarily time-varying. Moreover, two block codes are usually regarded as equivalent if one is a permutation of the other; *i.e.*, the ordering of the time axis is immaterial.

2.2 State Realizations

A (conventional) state realization of a behavior $\mathcal{B} \subseteq \mathcal{A}$ is defined as follows.

For simplicity, assume that the time axis is $I = \mathbb{Z}$. For each $k \in \mathbb{Z}$, define a *state space* S_k and a *local behavior* $\mathcal{B}_k \subseteq S_k \times A_k \times S_{k+1}$. The *state sequence space* is then $\mathcal{S} = \bigotimes_k S_k$, the *configuration space* is $\mathcal{A}^+ = \mathcal{A} \times \mathcal{S}$, and the *global behavior* $\mathcal{B}^+$ is the set of all configurations $(\mathbf{a}, \mathbf{s}) \in \mathcal{A}^+$ such that $(s_k, a_k, s_{k+1}) \in \mathcal{B}_k$ for all $k \in \mathbb{Z}$. The realization is a *state realization* of $\mathcal{B}$ if the projection $\mathcal{B}^+_{|\mathcal{A}}$ of the global behavior $\mathcal{B}^+$ onto the symbol sequence space $\mathcal{A}$ is equal to $\mathcal{B}$.

The realization is a *group state realization* of a group behavior $\mathcal{B}$ if each state space S_k is a group and each local behavior $\mathcal{B}_k$ is a subgroup of the direct product $S_k \times A_k \times S_{k+1}$.

(Later we will discuss finite-support (and Laurent) sequence spaces $\mathcal{A}$. In such cases $\mathcal{S}$ and $\mathcal{A}^+$ must be similarly restricted, and thus $\mathcal{B} \subseteq \mathcal{A}$ and $\mathcal{B}^+ \subseteq \mathcal{A}^+$ as well.)

A state realization is said to be *minimal* if for all k the state space S_k is the "smallest" possible in any state realization of a given behavior $\mathcal{B}$. This is a strong notion of minimality, and non-group behaviors need not have a minimal state realization in this sense.

Willems [25] shows that a behavior $\mathcal{B}$ has a well-defined, essentially unique (up to relabeling of state spaces) minimal state realization if and only if:

- If two pasts have one future in common, then they have all futures in common; or, equivalently,
- If two futures have one past in common, then they have all pasts in common.

2.3 The State Space Theorem

We now show that any group behavior has a well-defined, essentially unique minimal group state realization. This important property is a corollary of the State Space Theorem.

A state space arises from a cut of the time axis I at time k, which partitions I into two disjoint subsets, the *past* $k^- = \{i \in I|\ i < k\}$ and the *future* $k^+ = \{i \in I|\ i \geq k\}$. We will also consider more general two-way partitions of I into disjoint subsets $\{J, K\}$, where $K = I - J$.

Given a group behavior $\mathcal{B}$ defined on I and a subset $J \subseteq I$, we define:

- The *projection* $\mathcal{B}_{|J} = \{\mathbf{a}_{|J}|\ \mathbf{a} \in \mathcal{B}\}$;
- The *cross-section* $\mathcal{B}_J = \{\mathbf{a}_{|J}|\ \mathbf{a} \in \mathcal{B}, \mathbf{a}_{|K} = \mathbf{e}\}$,

where $K = I - J$ and $\mathbf{e}$ denotes the identity sequence. It is immediate that $\mathcal{B}_{|J}$ and $\mathcal{B}_J$ are groups, and that $\mathcal{B}_J$ is a normal subgroup of $\mathcal{B}_{|J}$.

The following easy but extremely important result for group systems was shown in [9]; for linear systems, the essence of this result is in [25].

Theorem 2.1 (**State Space Theorem** *[25, 9]) Given a group behavior $\mathcal{B}$ defined on a time axis I, let $\{J, K\}$ be any two-way partition of I. Then the following three quotient groups are isomorphic:*

1. *the* two-sided state space $\mathcal{B}/(\mathcal{B}_J \times \mathcal{B}_K)$;
2. *the* J-induced state space $\mathcal{B}_{|J}/\mathcal{B}_J$;
3. *the* K-induced state space $\mathcal{B}_{|K}/\mathcal{B}_K$.

The proof is elementary. (Indeed, this is the first theorem about subdirect products in [12].)

The *state space* induced by a two-way partition of I into $\{J, K\}$ will be taken to be any group S_J that is isomorphic to any of the three state spaces of the State Space Theorem. If the two-way partition is $\{k^-, k^+\}$, then we will denote the corresponding state space by S_k.

The *state map* σ_J will be defined by abuse of notation as any of the three natural maps from $\mathcal{B}$ to $\mathcal{B}/(\mathcal{B}_J \times \mathcal{B}_K)$, from $\mathcal{B}_{|J}$ to $\mathcal{B}_{|J}/\mathcal{B}_J$, or from $\mathcal{B}_{|K}$ to $\mathcal{B}_{|K}/\mathcal{B}_K$, followed by the isomorphism from the appropriate quotient group to S_J. The state map is thus a homomorphism. Again, if the two-way partition is $\{k^-, k^+\}$, we will denote the corresponding state map by σ_k.

If $(\mathbf{a}_{|k^-}, \mathbf{a}_{|k^+})$ is a sequence in $\mathcal{B}$, then we say that $\mathbf{a}_{|k^+}$ is a future of the past $\mathbf{a}_{|k^-}$, and *vice versa*. It follows from the proof of the State Space Theorem that $\mathbf{a}_{|k^+}$ is a future of $\mathbf{a}_{|k^-}$ if and only if

$$\sigma_k(\mathbf{a}_{|k^-}) = \sigma_k(\mathbf{a}_{|k^+}).$$

Therefore $\mathcal{B}$ has a well-defined, essentially unique minimal state realization with state spaces S_k.

Indeed, the *canonical state realization*, defined as the set of all sequences

$$\mathcal{B}^+ = \{(\mathbf{a}, \mathbf{s}) |\ \mathbf{a} \in \mathcal{B}, s_k = \sigma_k(\mathbf{a})\ \forall k\},$$

is such a minimal realization. Moreover, the canonical state realization is a group realization, since the state spaces are all groups and the local behaviors

$$\mathcal{B}_k = \{(\sigma_k(\mathbf{a}), a_k, \sigma_{k+1}(\mathbf{a})) |\ \mathbf{a} \in \mathcal{B}\}$$

are all homomorphic images of $\mathcal{B}$.

$\mathcal{B}^+$ is minimal if and $s_{k+1} \neq e$; $\mathbf{g}$.*sequence* $\in \mathcal{B}$. The canonical realization must be

The projection $\mathcal{B}^+_{|\mathcal{S}}$ of the canonical realization $\mathcal{B}^+$ onto the state sequence space $\mathcal{S} = \bigotimes_k S_k$ is called the *state behavior* $\sigma(\mathcal{B})$ of $\mathcal{B}$. The state behavior is a homomorphic image of $\mathcal{B}$, and thus is itself a group behavior.

Finally, there is a well-defined *input group* at time k, namely

$$U_k = \{a_k \in \mathcal{A}_k |\ \exists\{e, a_k, s_{k+1}\} \in \mathcal{B}_k\}.$$

By the group property, the set $U(s_k)$ of possible symbols a_k starting from any state s_k is a coset of U_k. Using these facts, the canonical realization may be extended to a minimal input/output (I/O) realization of $\mathcal{B}$ [9].

The minimal realization of a group behavior $\mathcal{B}$ of course depends on the time axis I. If the time axis is permuted, then the minimal realization (in particular, the sizes of the state spaces) can change dramatically. The topic of "trellis complexity of linear block codes" addresses the problem of finding the minimal state complexity of realizations of linear block codes over all time axis permutations (see [23] for a recent review).

2.4 Notes on the State Space Theorem

We pause to make a few remarks about the State Space Theorem.

We have already noted that it applies to *any* two-way partition $\{J, K\}$ of the time axis I. For example, for the trivial two-way partition $\{\emptyset, I\}$, the state space $S_\emptyset = S_I$ is trivial (isomorphic to $\{e\}$).

Time-invariance is not required.

The groups need not be abelian. However, there are certain simplifications in the abelian case. In particular, we find a "fourth state space":

Theorem 2.2 (**Reciprocal State Space Theorem** *[10]*) *Given a group behavior $\mathcal{B}$ defined on a time axis I, let $\{J, K\}$ be any two-way partition of I, and let S_J be the state space induced by this partition. If and only if S_J is abelian, then $\mathcal{B}$ is a normal subgroup of $\mathcal{B}_{|J} \times \mathcal{B}_{|K}$, and the* reciprocal state space *$(\mathcal{B}_{|J} \times \mathcal{B}_{|K})/\mathcal{B}$ is isomorphic to S_J.*

Note however that even when S_J is nonabelian, $\mathcal{B}$ is a subgroup of $\mathcal{B}_{|J} \times \mathcal{B}_{|K}$, and the set of left (or right) cosets of $\mathcal{B}$ in $\mathcal{B}_{|J} \times \mathcal{B}_{|K}$ is in bijection with S_J. Whether or not the nonabelian case is fundamentally different or just more cumbersome remains an open question. Recent work by Fagnani and Zampieri [4] suggests that the nonabelian case may not be fundamentally different.

3 Controllability and Observability

The State Space Theorem arises from a two-way partition of the time axis. Controllability and observability properties arise from three-way partitions of the time axis. For group codes and behaviors, controllability may be identified with generatability and observability with checkability (completeness). Later, we will see that controllability and observability are duals.

3.1 Controllability and Generatability

In Willems' behavioral system theory, controllability is defined as a property of a behavior rather than of a realization, as in classical system theory. The following natural definition expresses controllability in terms of the ability to connect any trajectory up to time m with any trajectory beyond time n by an appropriate connection during the interval $[m, n)$:

A system $\Sigma = (I, \mathcal{A}, \mathcal{B})$ is *$[m, n)$-controllable* if for any $\mathbf{a}, \mathbf{a}' \in \mathcal{B}$, there exists a trajectory $\mathbf{a}'' \in \mathcal{B}$ such that $\mathbf{a}''_{|m^-} = \mathbf{a}_{|m^-}$ and $\mathbf{a}''_{|n^+} = \mathbf{a}'_{|n^+}$.

Equivalently, this property may be expressed as follows:

Definition 3.1 *A system Σ is $[m, n)$-controllable if*

$$\mathcal{B}_{|I-[m,n)} = \mathcal{B}_{|m^-} \times \mathcal{B}_{|n^+}.$$

This notion of controllability is set-theoretic and may be extended to any three-way partition of I, not just $\{m^-, [m,n), n^+\}$.

In the group case, an alternative notion that we call *generatability* turns out to be equivalent to controllability. Recall that a cross-section $\mathcal{B}_J$ is the set of all trajectories in $\mathcal{B}$ whose support is contained in J.

A group system $\Sigma = (I, \mathcal{A}, \mathcal{B})$ is *$[m,n)$-generatable* if every $\mathbf{a} \in \mathcal{B}$ can be expressed as the combination under the group operation of $\mathcal{B}$ of some $\mathbf{a}' \in \mathcal{B}_{n^-}$ and some $\mathbf{a}'' \in \mathcal{B}_{m^+}$. In other words, using "+" to denote the group operation of $\mathcal{B}$:

Definition 3.2 *A system Σ is $[m,n)$-generatable if*

$$\mathcal{B} = \mathcal{B}_{n^-} + \mathcal{B}_{m^+}.$$

It is then easy to show:

Theorem 3.1 *A group system Σ is $[m,n)$-controllable if and only if it is $[m,n)$-generatable.*

A system is called *L-controllable* if it is $[m, m+L)$-controllable for all m and *strongly controllable* if it is L-controllable for some integer L; the *controller memory* of a strongly controllable system is the least such L. It follows from this theorem that a group system is L-controllable if and only if $\mathcal{B}$ is generated by the finite sequences in $\mathcal{B}$ whose support length is not greater than $L+1$.

Note that if we regard a partitioned time axis $I = \{m^-, [m,n), n^+\}$ as a finite time axis of length three, then a system is $[m,n)$-controllable on I if and only if it is 1-controllable on the equivalent length-3 time axis.

3.2 Observability and Checkability

Willems has been reluctant to define observability as a property of a behavior dual to controllability. At least for discrete-time group behaviors, we see no problem in making such a definition. The following definition [17] corresponds to what Willems calls "finite memory":

A system $\Sigma = (I, \mathcal{A}, \mathcal{B})$ is *$[m,n)$-observable* if for any $\mathbf{a}, \mathbf{a}' \in \mathcal{B}$ such that $\mathbf{a}_{|[m,n)} = \mathbf{a}'_{|[m,n)}$, the concatenation of $\mathbf{a}_{|m^-}$, $\mathbf{a}_{|[m,n)} = \mathbf{a}'_{|[m,n)}$ and $\mathbf{a}'_{|n^+}$ is in $\mathcal{B}$.

Again, this notion of observability is set-theoretic, and may be extended to any three-way partition of I.

In the group case, it suffices to consider trajectories such that $\mathbf{a}_{|[m,n)} = \mathbf{e}$, which leads to the following equivalent formulation:

Definition 3.3 *A group system Σ is $[m,n)$-observable if*

$$\mathcal{B}_{I-[m,n)} = \mathcal{B}_{m^-} \times \mathcal{B}_{n^+}.$$

An alternative notion that we call *checkability* (Willems: "completeness") turns out to be equivalent to observability.

A system $\Sigma = (I, \mathcal{A}, \mathcal{B})$ is $[m,n)$-*checkable* if every $\mathbf{a} \in \mathcal{A}$ such that $\mathbf{a}_{|n^-} \in \mathcal{B}_{|n^-}$ and $\mathbf{a}_{|m^+} \in \mathcal{B}_{|m^+}$ is actually in $\mathcal{B}$. In other words:

Definition 3.4 *A system Σ is $[m,n)$-checkable if*

$$\mathcal{B} = \{\mathbf{a} \in \mathcal{A} |\ \mathbf{a}_{|n^-} \in \mathcal{B}_{|n^-}, \mathbf{a}_{|m^+} \in \mathcal{B}_{|m^+}\}.$$

Checkability is also a set-theoretic property. In the group case, it is easy to show:

Theorem 3.2 *A group system Σ is $[m,n)$-observable if and only if it is $[m,n)$-checkable.*

A system is called *L-observable* if it is $[m, m+L)$-observable for all m and *strongly observable* if it is L-observable for some integer L; the *observer memory* of a strongly controllable system is the least such L. It follows from this theorem that a complete group system [25] is L-observable if and only if it is L-checkable; *i.e.*, if and only if every sequence $\mathbf{a} \in \mathcal{A}$ that satisfies all checks corresponding to a sliding window of length $L+1$, $\mathbf{a}_{|[m,m+L]} \in \mathcal{B}_{|[m,m+L]}$ for all m, is actually in $\mathcal{B}$. (Willems: "L-complete" = "memory L + complete.")

Note that if we regard a partitioned time axis $I = \{m^-, [m,n), n^+\}$ as a finite time axis of length three, then a system is $[m,n)$-observable on I if and only if it is 1-observable on the equivalent length-3 time axis.

4 Duality

It is well known that duality is an important tool in the study of linear systems. It is less well known that, like the system-theoretic properties already discussed, duality is fundamentally a property of topological groups. In this section we recapitulate some basic properties of Pontryagin duality and their system-theoretic consequences. (For LTI systems, Pontryagin duality reduces to the ordinary notions of duality.)

4.1 Pontryagin Duality

Pontryagin duality theory applies to locally compact abelian (LCA) groups such as real Euclidean n-space $\mathbb{R}^n$, and underlies all Fourier (harmonic) analysis [21, 22, 13]. For simplicity, we focus here on finite abelian groups, whose topology is naturally discrete.

There are two basic dualities in Pontryagin duality theory: character group duality and orthogonal subgroup duality.

The *character group* $G^\wedge$ of an LCA group G is the group under composition of all continuous homomorphisms $h: G \to \mathcal{C}$ from G into the complex unit circle $\mathcal{C}$. The fundamental Pontryagin duality theorem is that $G^\wedge$ is LCA and the character group of $G^\wedge$ is $G^{\wedge\wedge} = G$. If one defines a "pairing" $\langle\ ,\ \rangle: G^\wedge \times G \to \mathcal{C}$ by $\langle h, g\rangle = h(g)$, then the characters $g: G^\wedge \to \mathcal{C}$ of $G^\wedge$ are precisely the set defined by $g(h) = \langle h, g\rangle$.

If G is isomorphic to a finite cyclic group $\mathbb{Z}_m$, then its character group $G^\wedge$ is isomorphic to $\mathbb{Z}_m$, and the pairing $\langle h, g\rangle$ may be written explicitly as

$$\langle h, g\rangle = \omega^{gh},$$

where $\omega = e^{2\pi i/m}$ is a complex mth root of unity, both G and $G^\wedge$ are identified with $\mathbb{Z}_m$, and the product gh is taken in $\mathbb{Z}_m$.

If G is any finite abelian group, then it is isomorphic to a direct product $G_1 \times \cdots \times G_n$ of a finite collection of cyclic groups $\{G_i, 1 \le i \le n\}$, and the elements of G may correspondingly be written as $\mathbf{g} = (g_1, \ldots, g_n)$. Its character group is the direct product $G^\wedge = G_1{}^\wedge \times \cdots \times G_n{}^\wedge$ of the corresponding character groups $\{G_i{}^\wedge, 1 \le i \le n\}$, each character may be written as $\mathbf{h} = (h_1, \ldots, h_n)$, and the pairing $\langle \mathbf{h}, \mathbf{g}\rangle$ may be written as

$$\langle \mathbf{h}, \mathbf{g}\rangle = \prod_i \langle h_i, g_i\rangle.$$

The character group $G^\wedge$ of a finite abelian group G is thus isomorphic to G.

If G and $G^\wedge$ are dual character groups, then $g \in G$ and $h \in G^\wedge$ are said to be *orthogonal* if $\langle h, g\rangle = 1$ (the identity of $\mathcal{C}$). If S is a subgroup of G, then the *orthogonal subgroup* (annihilator) of S is the set $S^\perp$ of all $h \in G^\wedge$ that are orthogonal to all $g \in S$. The second basic Pontryagin duality theorem is that $S^\perp$ is a *closed* subgroup of $G^\wedge$, and $S^{\perp\perp}$ is the closure $\bar{S}$ of S in G. Thus S is a closed subgroup of G (*i.e.*, a topological subgroup) if and only if $S^{\perp\perp} = S$. Moreover, S is completely characterized by its orthogonal subgroup $S^\perp$ if and only if S is closed in G.

Finally, a third elementary duality result is that if S and $S^\perp$ are orthogonal subgroups of G and $G^\wedge$, respectively, then $S^\perp$ may be identified with (and thus is isomorphic to) the character group of the quotient group G/S. More generally, if $T \subseteq S \subseteq G$, then $S^\perp \subseteq T^\perp \subseteq G^\wedge$, and $T^\perp/S^\perp$ may be identified with the character group $(S/T)^\wedge$ of S/T. The latter result reduces to the former if $S = G$, because $G^\perp = \{0\}$ and *vice versa.*

4.2 Sequence Space Duality

A group behavior $\mathcal{B}$ was defined earlier as a subgroup of a direct product sequence space $\mathcal{A}$. In the present topological context, it is natural to insist that $\mathcal{B}$ be closed in $\mathcal{A}$, and to define the *dual behavior* as the orthogonal subgroup $\mathcal{B}^\perp$ in the *dual sequence space* $\mathcal{A}^\wedge$, the character group of $\mathcal{A}$. We therefore need to understand the topology of a direct product sequence space $\mathcal{A}$ and the properties of its character group $\mathcal{A}^\wedge$.

A direct product sequence space is defined as the direct product $\mathcal{A} = \bigotimes_{k \in I} A_k$ of a collection of symbol groups $\{A_k, k \in I\}$. If the component groups are finite abelian, then $\mathcal{A}$ is abelian, but if the time axis I is infinite, then $\mathcal{A}$ is not in general finite. Moreover, the topology of $\mathcal{A}$ is not in general discrete. If the component groups are discrete, then the natural topology of $\mathcal{A}$ is the topology of pointwise convergence (two sequences are "close" if they agree over a large central interval).

An infinite direct product of LCA groups is not necessarily LCA (although an infinite direct product of compact groups is necessarily compact). Nonetheless, it has been shown [15] that:

Theorem 4.1 *The character group of a direct product $\bigotimes_{k \in I} G_k$ of LCA groups G_k is the direct sum $\bigoplus_{k \in I} G_k\hat{}$ of their character groups $G_k\hat{}$.*

A *direct sum* $\mathcal{A}_f = \bigoplus_{k \in I} A_k$ is defined as in group theory as the set of all sequences $\mathbf{a} \in \mathcal{A} = \bigotimes_{k \in I} A_k$ that have finite support (*i.e.*, only finitely many nonzero components). If all A_k are discrete, then $\mathcal{A}_f$ naturally has the discrete topology. If I is finite, then $\mathcal{A}_f = \mathcal{A}$, but if I is infinite, then in general $\mathcal{A}_f$ is a nonclosed proper subgroup of $\mathcal{A}$, and its *completion* (closure in $\mathcal{A}$) is $(\mathcal{A}_f)^c = \mathcal{A}$.

If each symbol group A_k is finite abelian, then $A_k\hat{}$ may be identified with A_k. The character group of the direct product sequence space $\mathcal{A} = \bigotimes_{k \in I} A_k$ is then the direct sum sequence space $\mathcal{A}\hat{} = \mathcal{A}_f = \bigoplus_{k \in I} A_k$. The pairing between two sequences $\mathbf{a} \in \mathcal{A}$ and $\mathbf{a}' \in \mathcal{A}_f$ is the componentwise product

$$\langle \mathbf{a}, \mathbf{a}' \rangle = \prod\nolimits_k \langle a_k, a_k{}' \rangle.$$

The product is well defined because only finitely many pairings $\langle a_k, a_k{}' \rangle$ are unequal to 1. (Sometimes the direction of time is reversed in the dual sequence space— *i.e.*, $\langle a_k, a_k{}' \rangle$ is replaced by $\langle a_k, a_{-k}{}' \rangle$— in order to harmonize this componentwise definition with the convolutional definition of inner product usually used for LTI systems.)

When the symbol groups are finite (discrete and compact) and I is infinite, $\mathcal{A}$ is compact but not discrete, while $\mathcal{A}\hat{} = \mathcal{A}_f$ is discrete but not compact. This illustrates the general proposition that the character group of a compact abelian group is discrete and *vice versa* [22].

In behavioral system theory, sequence spaces are usually taken to be Cartesian products, and behaviors are usually taken to be "complete;" *i.e.*, closed under the topology of pointwise convergence. Then it is natural to focus on group behaviors that are closed subgroups of such "complete" (direct product) sequence spaces. However, it then follows that the dual sequence space must be a "finite-support" (direct sum) sequence space, so the dual (orthogonal) behavior must be a group of finite-support sequences. Thus any theory of linear or group behaviors that addresses duality must embrace finite-support as well as complete sequence spaces and behaviors.

(In the literature of LTI behaviors it has been recognized that the dual of a complete behavior is a "Laurent-polynomial" (finite-support) behavior, but the group-theoretic foundation of this observation has not been widely noted; see however [20].)

For example, the *infinite repetition code* C over a group G is the set of all sequences $\mathbf{g} = (\ldots, g, g, g, \ldots) \in G^{\mathbb{Z}}$, a closed ("complete") subgroup of the complete sequence space $G^{\mathbb{Z}}$. The dual (orthogonal) code $C^{\perp}$ then must be defined in the dual finite-support sequence space $((G\hat{})^{\mathbb{Z}})_f$, where $G\hat{}$ is the character group of G. Indeed, $C^{\perp}$ is easily seen to be the *infinite zero-sum code* over $G\hat{}$, namely the set of all finite-support sequences over $G\hat{}$ whose component sum is zero:

$$C^{\perp} = \{\mathbf{h} \in ((G\hat{})^{\mathbb{Z}})_f | \sum_k h_k = 0\}.$$

The component sum $\sum_k h_k$ is well defined because $\mathbf{h}$ has finite support. Note that $C^{\perp}$ is not "complete" (closed in $(G\hat{})^{\mathbb{Z}}$); its completion $(C^{\perp})^c$ is the entire complete sequence space $(G\hat{})^{\mathbb{Z}}$.

By the orthogonal subgroup duality theorem, a subgroup $\mathcal{B}$ of a complete sequence space $\mathcal{A}$ is closed ("complete") if and only if it is the orthogonal subgroup to its orthogonal subgroup $\mathcal{B}^{\perp}$ in the dual finite-support sequence space $\mathcal{A}\hat{}$. Thus $\mathcal{B}$ is complete if and only if it is orthogonal to a set of finite-support "check" sequences that generate $\mathcal{B}^{\perp}$. For group behaviors, this amounts to the usual definition of completeness in behavioral system theory: a behavior $\mathcal{B}$ is complete if every sequence in $\mathcal{A}$ that looks like a sequence in $\mathcal{B}$ during every finite interval is actually in $\mathcal{B}$.

Discrete groups can be treated without reference to any topological properties. It may therefore be easiest to characterize a complete, compact behavior $\mathcal{B}$ by the properties of its discrete, finite-support dual $\mathcal{B}^{\perp}$. On the other hand, compactness is also useful.

A more symmetric approach is obtained if one considers "Laurent" (*e.g.*, in the LTI case, Laurent power series) sequence spaces and behaviors. A Laurent sequence space may be defined as the direct product of a finite (direct sum) sequence space on the past and a complete (direct product) sequence space on the future, where the location of the cut between the past and the future in I may be arbitrary. The dual sequence space is then an anti-Laurent space, the direct product of a direct sum on the future with a direct sum on the past— or, if the direction of time is reversed in the dual space, then it is a Laurent sequence space. However, certain behaviors (*e.g.*, repetition or zero-sum codes) cannot be "Laurentized."

4.3 State Space Duality

We now discuss some fundamental dynamical dualities between the properties of dual behaviors $\mathcal{B}$ and $\mathcal{B}^{\perp}$ in dual sequence spaces.

The most fundamental duality is between projections and cross-sections, and is a simple consequence of the componentwise definition of pairings:

Theorem 4.2 *If $\mathcal{B}$ and $\mathcal{B}^{\perp}$ are dual behaviors defined on a time axis I and $J \subseteq I$, then $\mathcal{B}_{|J}$ and $(\mathcal{B}^{\perp})_J$ are dual behaviors defined on J.*

In the language of coding theory, the dual of a punctured group code is the corresponding shortened code of its dual group code. For example, given any finite $J \subset \mathbb{Z}$, the dual of a finite (punctured) repetition code $C_{|J}$ defined on J is the finite (shortened) zero-sum code $(C^{\perp})_J$ defined on J.

By quotient group duality, we immediately have the following dualities for the quotient groups that define the state spaces of $\mathcal{B}$ and $\mathcal{B}^{\perp}$:

Theorem 4.3 *If $\mathcal{B}$ and $\mathcal{B}^{\perp}$ are dual behaviors defined on a time axis I and $J \subseteq I$, then*

- *The J-induced state space $\mathcal{B}^{\perp}{}_{|J}/\mathcal{B}^{\perp}{}_J$ of $\mathcal{B}^{\perp}$ may be identified with the character group of the J-induced state space $\mathcal{B}_{|J}/\mathcal{B}_J$ of $\mathcal{B}$;*
- *The two-sided state space $\mathcal{B}^{\perp}/((\mathcal{B}^{\perp})_J \times (\mathcal{B}^{\perp})_{I-J})$ of $\mathcal{B}^{\perp}$ may be identified with the character group of the reciprocal two-sided state space $(\mathcal{B}_{|J} \times \mathcal{B}_{|I-J})/\mathcal{B}$ of $\mathcal{B}$.*

Briefly, every state space of $\mathcal{B}^{\perp}$ is isomorphic to the character group of the corresponding state space of $\mathcal{B}$.

If all state spaces of $\mathcal{B}$ are finite abelian groups, then Theorem 4.3 says that every state space of $\mathcal{B}^{\perp}$ is isomorphic to the corresponding state space of $\mathcal{B}$. (Similarly, if all state spaces of $\mathcal{B}$ are vector spaces, then every state space of $\mathcal{B}^{\perp}$ is a vector space of the same dimension as the corresponding state space of $\mathcal{B}$.) It follows that a minimal state realization of $\mathcal{B}^{\perp}$ has the same set of state complexities as a minimal state realization of $\mathcal{B}$. This simple result is probably the most important dynamical duality between $\mathcal{B}$ and $\mathcal{B}^{\perp}$.

4.4 Duality of Controllability and Observability

Similarly, we can now show that controllability and observability are duals:

Theorem 4.4 *If $\mathcal{B}$ and $\mathcal{B}^{\perp}$ are dual behaviors defined on a time axis I, then*

- *$\mathcal{B}^{\perp}$ is $[m, n)$-observable if and only if $\mathcal{B}$ is $[m, n)$-controllable;*
- *$\mathcal{B}^{\perp}$ is $[m, n)$-checkable if and only if $\mathcal{B}$ is $[m, n)$-generatable.*

Each of these propositions follows directly from projection/cross-section duality (Theorem 4.2), although of course they are equivalent to each other by Theorem 3.1 or 3.2.

It follows that the observer memory of $\mathcal{B}^\perp$ is equal to the controller memory of $\mathcal{B}$, and *vice versa.* This is probably the second most important dynamical duality between $\mathcal{B}$ and $\mathcal{B}^\perp$.

4.5 Weak Controllability and Observability

In view of the connection between controllability and generatabilty, it is natural to define a system to be *weakly* (wide-sense) *controllable* if its behavior $\mathcal{B}$ is generated by its finite-support sequences. Here "generated" is used in both an algebraic and topological sense; *i.e.,* $\mathcal{B}$ must be the closure $\bar{\mathcal{B}}_f$ of its subgroup $\mathcal{B}_f$. (A similar definition was introduced previously by Fagnani [3].)

If the behavior $\mathcal{B}$ is a finite-support behavior, then it is automatically weakly controllable, so this definition is redundant. However, if $\mathcal{B}$ is a complete behavior, then it is weakly controllable if and only if $\mathcal{B} = (\mathcal{B}_f)^c$. For example, an infinite repetition code C is not weakly controllable, since C_f consists of the single sequence $\mathbf{0}$, whose completion is only $(C_f)^c = \{\mathbf{0}\}$, a proper subset of C.

Similarly, it is natural to define a system to be *weakly* (wide-sense) *observable* if its behavior $\mathcal{B}$ is the dual of a behavior $\mathcal{B}^\perp$ that is generated by finite-support sequences; in other words, if $\mathcal{B}$ can be characterized by a set of finite-interval checks. The relation $\mathcal{B} = \mathcal{B}^{\perp\perp}$ ensures that $\mathcal{B}$ is closed.

If the behavior $\mathcal{B}$ is a complete behavior, then it is automatically weakly observable, so this definition is redundant. However, if $\mathcal{B}$ is a finite-support behavior, then it is weakly observable if and only if $\mathcal{B} = (\mathcal{B}^c)_f$. For example, an infinite zero-sum code $C^\perp$ over $G\hat{}$ is not weakly observable, since $(C^\perp)^c$ is the complete sequence space $(G\hat{})^{\mathbb{Z}}$, whose finite subset is $((G\hat{})^{\mathbb{Z}})_f$, a proper superset of $C^\perp$.

By definition, $\mathcal{B}$ is weakly observable if and only if $\mathcal{B}^\perp$ is weakly controllable, and *vice versa.*

It is easy to see that strong controllability implies weak controllability, and likewise for observability. Conversely, Fagnani [3] has shown that if $\mathcal{B}$ is complete, compact and time-invariant, then weak controllability implies strong controllability (finite controller memory). Similarly, if $\mathcal{B}$ is finite-support, discrete and time-invariant, then weak observability implies strong observability (finite observer memory).

A system is "well-behaved" if it is both controllable and observable. In this case it does not matter much whether we characterize the system as complete or finite-support (or Laurent); the complete version of its behavior is just the completion $(\mathcal{B}_f)^c$ of the finite-support version $\mathcal{B}_f$, which in turn is the finite-support subset $(\mathcal{B}^c)_f$ of the complete version $\mathcal{B}^c$.

Convolutional codes are usually required to be both controllable and observable. The dual of a controllable and observable system is evidently controllable and observable.

property of having an kernel representation. image representation automatically has an observable. The behavioral system theory finite-support behaviors; from

Note that these definitions of weak controllability and observability do not depend on the ordering of the time axis I. They are therefore applicable to two-dimensional systems, for example, or to some of the generalized realizations to be discussed in the next section.

5 Generalized State Realizations

In this final section, we briefly describe some recent work on a class of generalized state realizations and their graphical representations [8]. This work is motivated by the intense current interest in "codes defined on graphs," a class of codes that includes "tail-biting" codes, low-density parity-check (LDPC) codes, and turbo codes. Codes defined on graphs with cycles can be decoded with iterative decoding algorithms that often achieve near-optimum performance, with complexity much less than that of optimal trellis-based decoding algorithms [1, 16].

5.1 Definitions

Let $\Sigma = (I_{\mathcal{A}}, \mathcal{A} = \bigotimes_{k \in I_{\mathcal{A}}} A_k, \mathcal{B})$ be a behavioral system defined on an unordered, discrete time axis $I_{\mathcal{A}}$. A *generalized state realization* of $\mathcal{B}$ is then defined as follows.

Define a collection $\{S_j, j \in I_{\mathcal{S}}\}$ of *state spaces* S_j and a collection $\{B_i, i \in I_{\mathcal{B}}\}$ of *local behaviors* B_i, where $I_{\mathcal{S}}$ and $I_{\mathcal{B}}$ are discrete index sets. Each local behavior is defined as a subset

$$B_i \subseteq \bigotimes_{k \in I_{\mathcal{A}}(i)} A_k \times \bigotimes_{j \in I_{\mathcal{S}}(i)} S_j,$$

where $I_{\mathcal{A}}(i) \subseteq I_{\mathcal{A}}$ and $I_{\mathcal{S}}(i) \subseteq I_{\mathcal{S}}$ are the subsets of the corresponding index sets that are *involved* in the local behavior B_i. We require that each symbol alphabet A_k be involved in precisely one local behavior, and each state space S_j be involved in precisely two local behaviors.

Note that a conventional state realization satisfies these requirements (except possibly for dummy unary state spaces at the beginning and end of finite-time-axis realizations).

The *global behavior* defined by these local behaviors is the set $\mathcal{B}^+$ of all configurations $(\mathbf{a}, \mathbf{s}) \in \mathcal{A} \times \mathcal{S}$ such that $(\mathbf{a}_{|I_{\mathcal{A}}(i)}, \mathbf{s}_{|I_{\mathcal{S}}(i)}) \in \mathcal{B}_i$ for all $i \in I_{\mathcal{B}}$, where $\mathcal{A} = \bigotimes_{k \in I_{\mathcal{A}}} A_k$ and $\mathcal{S} = \bigotimes_{j \in I_{\mathcal{S}}} S_j$ are the *symbol* and *state configuration spaces*, respectively. The realization is a *generalized state realization* of $\mathcal{B}$ if the projection $\mathcal{B}^+_{|\mathcal{A}}$ of the global behavior $\mathcal{B}^+$ onto the symbol configuration space $\mathcal{A}$ is equal to $\mathcal{B}$.

The realization is a *generalized group state realization* of a group behavior $\mathcal{B}$ if each state space S_j is a group and each local behavior $\mathcal{B}_i$ is a subgroup of the direct product $\bigotimes_{k \in I_{\mathcal{A}}(i)} A_k \times \bigotimes_{j \in I_S(i)} S_j$.

A convenient method of expressing a global behavior in terms of local behaviors is by means of indicator functions. In general, if S is a set with elements $s \in S$, then the *indicator function* of S will be defined as the function $\delta(s \in S)$ that is equal to 1 if $s \in S$ and 0 otherwise. From the above definition, the indicator function of the global behavior $\mathcal{B}^+$ may be written as the product of the indicator functions of the local behaviors:

$$\delta((\mathbf{a}, \mathbf{s}) \in \mathcal{B}^+) = \prod_{i \in I_{\mathcal{B}}} \delta((\mathbf{a}_{|I_{\mathcal{A}}(i)}, \mathbf{s}_{|I_S(i)}) \in \mathcal{B}_i). \tag{5.1}$$

A generalized state realization has a nice graphical representation, as follows. A *graph with leaves* is defined by a set of vertices, a set of ordinary edges, each incident on two vertices, and a set of *leaf edges*, each incident on only one vertex. The *graph* of a generalized state realization will be defined as the graph with leaves in which

- the vertices represent the local behaviors B_i;
- the ordinary edges represent the state spaces S_j, where an edge S_j is incident on the two vertices representing the two local behaviors in which S_j is involved; and
- the leaf edges represent the symbol alphabets A_k, where a leaf edge A_k is incident on the single vertex representing the unique local behavior in which A_k is involved.

A generalized state realization may be characterized by the properties of its graph, such as connectedness or cycle-freedom. Notice for example that the product (5.1) *factors* into the product of two independent equations if and only if the graph is *disconnected*.

5.2 Examples

For example, a *tail-biting realization* is defined on a finite circular time axis $I_{\mathcal{A}} = I_S = I_{\mathcal{B}} = \mathbb{Z}_m$ of *length* m. As with conventional state realizations, each local behavior $B_k \subseteq S_k \times A_k \times S_{k+1}, k \in \mathbb{Z}_m$, is defined in terms of the "current state" S_k, "current symbol" A_k, and "next state" S_{k+1}, where the subscript $k+1$ is evaluated in $\mathbb{Z}_m$ (*i.e.*, "mod m"). Thus each state space is involved in precisely two local behaviors, and each symbol alphabet is involved in precisely one local behavior. The graph of a tail-biting realization has m vertices arranged in a circle, each vertex connected to two neighbors by an ordinary (state) edge, and each also bearing one leaf (symbol) edge.

For another example, a *parity-check code* C is the set of all sequences $\mathbf{a} = \{a_k, 1 \leq k \leq n\}$ of n binary symbols $a_k \in \mathbb{Z}_2$ that satisfy $H\mathbf{a} = \mathbf{0}$, where H is an $r \times n$ binary parity-check matrix. In other words, code sequences $\mathbf{a}$ satisfy r parity-check equations of the form

$$\sum_{k \in I_r(i)} a_k = 0, 1 \leq i \leq r,$$

where $I_r(i)$ is the set of column indices of the 1's in the ith row of H. Similarly, the set of row indices of the 1's in the kth column of H will be denoted by $I_c(k)$.

Each parity-check equation defines a local behavior B_i. Such a set of local behaviors does not however define a generalized state realization, because each symbol a_k is in general involved in multiple parity checks. To fix this problem, define a set of binary *replicas* $\{s_{ik}, 1 \leq i \leq r, k \in I_r(i)\}$ and a set of $n + r$ constraints (local behaviors) as follows:

1. For $1 \leq k \leq n$, all replicas $s_{ik}, i \in I_c(k)$, must equal a_k;
2. For $1 \leq i \leq r$, $\sum_{k \in I_r(i)} s_{ik} = 0$.

Now each replica (state) variable S_{ik} is involved in precisely two of these constraints, one of Type 1 and one of Type 2, and each symbol variable A_k is involved in precisely one constraint (of Type 1).

Thus we have constructed a generalized state realization that evidently represents the same code C. All state variables are binary, and the number of state variables is the number of nonzero elements in the parity-check matrix H of C. For this reason it is particularly interesting to consider *low-density parity-check codes*, defined as parity-check codes with a *sparse* parity-check matrix H.

By a similar introduction of replicas, any code represented by a "factor graph" [16] may be represented by a generalized state realization.

5.3 Cut Sets and Conditioning

A *cut set* of a connected graph is a minimal set of edges such that removal of that set partitions the graph into two disconnected subgraphs. A graph is *cycle-free* if every edge is by itself a cut set.

Fixing a variable in Equation (5.1) (*i.e.*, conditioning on a value of that variable) effectively removes that variable from the equation, and removes the corresponding edge in the corresponding graph [16]. Fixing a set of variables corresponding to a cut set partitions the graph into two disconnected subgraphs, and correspondingly factors (5.1) into two equations dependent on disjoint subsets of symbol and state variables. Thus cut sets in the graph correspond to instances of conditional independence.

From this observation follows the important cut set lower bound [24, 2]:

Theorem 5.1 *(***Cut Set Lower Bound***) Let I be the symbol index set of a group behavior $\mathcal{B}$, let S be a cut set in the graph of a generalized state realization of $\mathcal{B}$, and let J and $I - J$ be the two disjoint subsets of symbol indices in the graph partition induced by removal of S. Then the product of the alphabet sizes of the set of variables corresponding to the set of edges in S is lowerbounded by the size of the J-induced state space of $\mathcal{B}$.*

The idea is that the set of variables corresponding to the set of edges in the cut set S form a "super-state," whose alphabet size is lowerbounded by the State Space Theorem.

In a cycle-free generalized state realization, each state space is a cut set. A canonical minimal realization of a group behavior can then be constructed in a manner similar to the construction of a conventional canonical state realization, where each state space is chosen to be the canonical state space that is suggested by Theorem 5.1.

By using generalized state realizations with cycles, however, dramatic reductions in state complexity can be obtained. Cut sets then may comprise multiple state variables, and the minimum complexity implied by the Cut Set Lower Bound may be distributed across these variables.

For example, with a tail-biting realization, every cut set comprises two state variables; consequently, the minimum state complexity of both such variables can be as small as the square root of the value given by the Cut Set Lower Bound. It has been shown, for example, that there exists a tail-biting realization of the $(24, 12, 8)$ Golay code in which the size of each state space is only 16, whereas in any comparable conventional state realization at least one state space must have as many as 256 states [2].

The cut set idea leads to the *sum-product algorithm* [16], an efficient iterative algorithm for optimal "decoding" of cycle-free generalized state realizations. All operations in this algorithm are local, so it may be applied also to realizations with cycles, even though there is no guarantee of optimality (or even convergence) in this case. This algorithm is widely used to decode capacity-approaching turbo codes and LDPC codes [16].

5.4 Dual Generalized State Realizations

We now aim to define a dual realization of a generalized state realization of a group behavior $\mathcal{B}$ such that the dual realization necessarily generates the dual behavior $\mathcal{B}^{\perp}$.

It is natural to define the symbol groups of the dual realization as the character groups $\{A_k\hat{}, k \in I_{\mathcal{A}}\}$ of the primal symbol groups A_k, the state spaces as the character groups $\{S_j\hat{}, j \in I_{\mathcal{S}}\}$ of the primal state spaces S_j, and the local behaviors as the orthogonal subgroups $\{(B_i)^{\perp}, i \in I_{\mathcal{B}}\}$ of the primal local behaviors B_i in the dual subconfiguration spaces $\bigotimes_{k \in I_{\mathcal{A}}(i)} A_k\hat{} \times \bigotimes_{j \in I_{\mathcal{S}}(i)} S_j\hat{}$. Of course if all symbol and state groups are finite abelian, then the same symbol and state groups may be used.

We use a trick that was introduced by Mittelholzer [19] to construct dual trellises (conventional state realizations): invert the sign of each dual state space $S_j\hat{}$ in precisely one of its two appearances in dual local behaviors $(B_i)^\perp$. With this trick, we have:

Theorem 5.2 *If $\mathcal{B}$ and $\mathcal{B}'$ are the behaviors generated by dual generalized group state realizations as defined above, then $\mathcal{B}$ and $\mathcal{B}'$ are orthogonal.*

The proof involves calculating the product of all local pairings between two valid configurations $(\mathbf{a}, \mathbf{s}) \in \mathcal{A} \times \mathcal{S}$ and $(\mathbf{a}', \mathbf{s}') \in \mathcal{A}\hat{} \times \mathcal{S}\hat{}$; because all state pairings $\langle s_j, s'_j \rangle$ appear precisely twice with opposite signs, they all cancel, leaving the product of symbol pairings $\langle a_k, a'_k \rangle$ equal to 1. Thus Theorem 5.2 depends essentially on the topological constraints that we have imposed on generalized state realizations.

Orthogonality does not necessarily imply duality. We have proved that $\mathcal{B}$ and $\mathcal{B}'$ are actually duals ($\mathcal{B}' = \mathcal{B}^\perp$) under mild additional conditions:

- There is a one-to-one correspondence between $\mathcal{B}^+$ and $\mathcal{B}$;
- All state spaces S_k are trim; *i.e.*, $(\mathcal{B}^+)_{|S_k} = S_k$.

We continue to investigate refinements of these conditions.

An immediate consequence is that if a (finite abelian) group behavior $\mathcal{B}$ has a generalized state realization that meets these conditions, then its dual $\mathcal{B}^\perp$ has a generalized state realization with the same graph topology and state spaces of the same size. This is a sweeping generalization of the corresponding result for minimal conventional state realizations. Note that this corollary holds for any graph topology, cycle-free or not.

For example, consider the generalized state realization of a parity-check code given in Section 5.2. The dual realization uses the same symbol groups A_k and state spaces (replicas) S_{ik}, but replaces the primal constraints by the following dual constraints:

1. For $1 \le k \le n$, $a_k = \sum_{i \in I_c(k)} s_{ik}$;
2. For $1 \le i \le r$, all s_{ik} must be equal, $k \in I_r(i)$.

It may be seen that these constraints define a binary linear code C' with r "information bits" corresponding to the common values of the states s_{ik} with first index i, $1 \le i \le r$, and n "output bits" a_k equal to the sum of the information bits $s_{ik}, i \in I_c(k)$. Thus C' has a generator matrix G equal to the parity-check matrix H of C (defined by the sets $I_r(i)$ or $I_c(k)$), so by the usual duality theory of binary linear codes $C' = C^\perp$, as desired. In this way any binary linear code with generator matrix G has a generalized state realization with a number of binary state spaces equal to the number of nonzero elements in G.

Conclusion

The purpose of this paper has been to introduce some recent developments in coding theory to a system-theoretic audience, both because they seem interesting for behavioral system theory, and also because we hope to attract some system theorists to work in coding theory. We also hope to have convinced system theorists that certain basic questions about linear systems are best addressed at the group-theoretic level.

Acknowledgments

I am grateful to H.-A. Loeliger, J. Rosenthal, M. D. Trott, J. C. Willems and S. Zampieri for their comments on this paper.

References

[1] Aji S. M. and R. J. McEliece (1998). The generalized distributive law. *IEEE Trans. Inform. Theory*, to appear.

[2] Calderbank A. R., G. D. Forney, Jr. and A. Vardy (1998). Minimal tail-biting trellises: The Golay code and more. *IEEE Trans. Inform. Theory*, to appear.

[3] Fagnani F. (1997). Shifts on compact and discrete Lie groups: Algebraic-topological invariants and classification problems. *Adv. Math.* **127**, 283–306.

[4] Fagnani F. and S. Zampieri (1998). Minimal syndrome formers for group codes. *IEEE Trans. Inform. Theory*, to appear.

[5] Forney G. D., Jr. (1970). Convolutional codes I: Algebraic structure. *IEEE Trans. Inform. Theory* **16**, 720–738.

[6] Forney G. D., Jr. (1975). Minimal bases of rational vector spaces, with applications to multivariable linear systems. *SIAM J. Control* **13**, 493–520.

[7] Forney G. D., Jr. (1991). Geometrically uniform codes. *IEEE Trans. Inform. Theory* **36**, 1241–1260.

[8] Forney G. D., Jr. (1998). Generalized state realizations and graphical representations of codes and dual codes. In preparation.

[9] Forney G. D., Jr. and M. D. Trott (1993). The dynamics of group codes: State spaces, trellis diagrams and canonical encoders. *IEEE Trans. Inform. Theory* **39**, 1491–1513.

[10] Forney G. D., Jr. and M. D. Trott (1995). Controllability, observability and duality in behavioral group systems. *Proc. 34th Conf. Dec. Ctrl.*, New Orleans, **3**, 3259–3264.

[11] Forney G. D., Jr. and M. D. Trott (1998). The dynamics of group codes: Dual abelian group codes and systems. In preparation.

[12] Hall M., Jr. (1959). *The Theory of Groups*, MacMillan.

[13] Hewitt E. and K. A. Ross (1979). *Abstract Harmonic Analysis I* (second ed.), Springer.

[14] Kailath T. (1980). *Linear Systems*, Prentice-Hall.

[15] Kaplan S. (1948). Extension of the Pontrjagin duality I: Infinite products. *Duke Math. J.* **15**, 649–658.

[16] Kschischang F., B. J. Frey and H.-A. Loeliger (1998). Factor graphs and the sum-product algorithm. *IEEE Trans. Inform. Theory*, to appear.

[17] Loeliger H.-A., G. D. Forney Jr., T. Mittelholzer and M. D. Trott (1994). Minimality and observability in group systems. *Linear Alg. Appl.* **205–206**, 937–963.

[18] Loeliger H.-A. and T. Mittelholzer (1996). Convolutional codes over groups. *IEEE Trans. Inform. Theory* **42**, 1660–1686.

[19] Mittelholzer T. (1995). Convolutional codes over groups: A pragmatic approach. *Proc. 33d Allerton Conf. Commun. Control Comput.*, Allerton IL, 380–381.

[20] Oberst U. (1990). Multidimensional constant linear systems. *Acta Appl. Math.* **20**, 1–175.

[21] Pontryagin L. (1946). *Topological Groups*, Princeton U. Press.

[22] Rudin W. (1990). *Fourier Analysis on Groups*, Wiley.

[23] Vardy A. (1998). Trellis structure of codes, in *Handbook of Coding Theory*, V. Pless and W. C. Huffman eds., Elsevier.

[24] Wiberg N., H.-A. Loeliger and R. Kötter (1995). Codes and iterative decoding on general graphs. *Euro. Trans. Telecommun.* **6**, 513–526.

[25] Willems J. C. (1989). Models for dynamics, in *Dynamics Reported* **2**, U. Kirchgraber and H. O. Walther eds., Wiley, 171–269.

[26] Willems J. C. (1991). Paradigms and puzzles in the theory of dynamical systems. *IEEE Trans. Automat. Contr.* **42**, 259–294.

Laboratory of Information and Decision Systems, MIT, Cambridge MA 02139, USA

Progress in Systems and Control Theory, Vol. 25

The Berlekamp-Massey algorithm, error-correction, keystreams and modeling

Margreet Kuijper[1]

1 Introduction

The areas of system theory and coding theory are more or less equally "young", having their origins around the fifties. Throughout their history, there have been observations on the existence of connections between the two areas. One of the first of these observations was concerned with the Berlekamp-Massey algorithm, derived in [5, 22] for the purpose of decoding BCH/Reed-Solomon codes. Indeed, following upon Massey's exposition in [22], Sain pointed out in [26] that what the Berlekamp-Massey algorithm solves is "a version of the widely conceived engineering black box problem". More specifically, the problem formulation is the following:

Problem Statement 1.1 *Let $a_0, a_1, \ldots, a_N$ be a given sequence of numbers in a field $\mathbb{F}$. Find polynomials $f(s)$ and $g(s)$ for which*

$$\frac{g(s)}{f(s)} = a_0 + a_1 s^{-1} + a_2 s^{-2} + \cdots + a_N s^{-N} + \phi(s) s^{-(N+1)},$$

with ϕ proper rational and $f(s)$ of minimal degree.

This is indeed a black box problem: knowing only its partial response $a_0, a_1, \ldots, a_N$ to an impulse input, we seek to model the black box that relates input and output by constructing a transfer function $g(s)/f(s)$ of minimal McMillan degree.

In the system-theoretic literature, the above problem is known as the *minimal partial realization problem* (scalar case), a terminology that can already be found in Kalman's early paper [15]. It is a classical system-theoretic problem that has given rise to a long sequence of papers providing various solutions for it (e.g. [1, 2, 3, 7, 11, 15, 16, 26]). The majority of these solutions is based on a Hankel approach in which constant matrices are the key players. Exceptions are e.g. the early paper [7] and the recent work [3]—both are based on a polynomial approach; the latter is set in a behavioral framework. In this semi-tutorial paper we present the connection with the applications in coding theory for which the Berlekamp-Massey algorithm was originally designed, as well as the applications in cryptography for which it is currently mostly used. We give an overview of behavioral results on modeling that are relevant to this material and present

[1]Supported by the Australian Research Council

the general multivariable algorithm of [19]. We believe that the preceding scalar-case paper [17] by Jan C. Willems and the author is an example of a fruitful interaction between two different areas—through a close study of the Berlekamp-Massey algorithm it was possible to fill in the details of an iterative behavioral modeling procedure, whereas at the same time the behavioral formulation enabled an explanation of the workings of the Berlekamp-Massey algorithm as well as an extension to the multivariable case. An application of the multivariable algorithm to the minimal partial realization of periodically time-varying systems can be found in [21].

The above problem can be phrased a bit differently in terms of a linear recurrence relation. Let us first define this important notion. A sequence $a_1, a_2, \ldots, a_N$ of numbers in a field $\mathbb{F}$ is said to satisfy a *linear recurrence relation* $(f_1, \ldots, f_L)$ of length L if

$$a_{j+L} + f_1 a_{j+L-1} + \cdots + f_L a_j = 0 \quad \text{for } j = 1, \ldots, N - L.$$

The smaller L, the more linear structure underlying $a_1, a_2, \ldots, a_N$ is captured; a linear recurrence relation of smallest possible length is called a *shortest linear recurrence relation* for $a_1, a_2, \ldots, a_N$. What do linear recurrence relations have to do with the above Problem Statement 1.1 of minimal partial realization? It can be easily seen that the denominator polynomial $f(s) := s^L + f_1 s^{L-1} + \cdots + f_{L-1}s + f_L$ of a partial realization serves as a linear recurrence relation for $a_1, a_2, \ldots, a_N$; indeed, this follows from the fact that the numerator polynomial $g(s) = f(s)(a_0 + a_1 s + \cdots + a_N s^N)$ has degree $\leq L$. The above Problem Statement 1.1 can thus be phrased alternatively as

Problem Statement 1.2 *Let $a_1, a_2, \ldots, a_N$ be a given sequence of numbers in a field $\mathbb{F}$. Find a shortest linear recurrence relation $(f_1, \cdots, f_L)$ for $a_1, a_2, \ldots, a_N$.*

The Berlekamp-Massey algorithm solves exactly this problem; it was designed ([5, 22]) for the purpose of solving the key step in the error correction of BCH/Reed-Solomon codes. These are block codes that have found many applications, ranging from Compact Discs to deep space communication. In section 2 we give a detailed explanation as to why this key step amounts to finding a shortest linear recurrence relation for a finite sequence of numbers from a finite field.

In most present applications involving Reed-Solomon codes, preference is given to the implementation of a combination of codes with low error correcting capability over the implementation of one code with high error correcting capability. As a result, trivial ad hoc decoding procedures are feasible and implemented in e.g. CD players. Although for BCH codes research on the application of the Berlekamp-Massey algorithm for decoding purposes is very much active today (see subsection 2.2), the dominant application of the algorithm has shifted to the area of cryptography. Cryptographic devices use sequences (*keystreams*) $a_1, a_2, \ldots, a_N$ that are generated deterministically by a finite state machine. One method to "attack" such a keystream with the aim of decryption is to use the Berlekamp-Massey algorithm: from a few initial data, knowledge of the entire sequence is obtained in the form of an underlying shortest linear recurrence relation. The length of that shortest linear recurrence relation is called the *linear complexity* of the sequence. From the point of view of the cryptographic designer it is important to employ sequences that have a high degree of unpredictability. This involves requirements not only on the complexity of the sequence (should be high) but also on the local linear complexities of all the subsequences beginning with a_1. The vector of local linear complexities is called the *linear complexity profile* of the sequence. More details are presented in subsection 2.3. Apart from serving as a cryptanalyst's attack method on a keystream, the Berlekamp-Massey algorithm is used as a device for calculating the linear complexity profile of a sequence, thereby serving as a tool for assessment of predictability of a sequence. In fact, because of its iterative nature, the algorithm is ideally suited for a linear complexity profile calculation.

2 Error-correction and keystreams

In this section, we have a closer look at the applications in the areas of coding theory and cryptography. We aim at a system theoretic formulation and seek to clarify the connection with system theory in this way. Let us first turn to the area in which the Berlekamp-Massey algorithm originated and give a brief outline of Reed-Solomon codes and subsequently BCH codes. We do not assume any prior knowledge on the part of the reader and introduce only those concepts that are absolutely necessary for the purposes of this paper. Since we will derive a result that is relevant to decoding BCH codes in subsection 4.2, it is our aim in this section to provide the reader with a real understanding of the basic principles.

2.1 Reed-Solomon codes

In 1960, a paper was published by I.S. Reed and G. Solomon which presented a new class of error-correcting codes. The codes proved to be an important part of the telecommunications revolution of this century. Reed-Solomon codes, as they were subsequently called, have found many applications, from compact discs to deep space communication.

A Reed-Solomon code is a linear block code for which the parity check matrix H has a structure which can perhaps be best described as "doubly

Vandermonde":

$$H = \begin{bmatrix} 1 & \gamma_1 & \cdots & \gamma_1^{n-1} \\ 1 & \gamma_2 & \cdots & \gamma_2^{n-1} \\ \vdots & \vdots & \vdots & \\ 1 & \gamma_N & \cdots & \gamma_N^{n-1} \end{bmatrix}, \tag{1}$$

where the γ_i's ($i = 1, \ldots, N$) are *consecutive* elements of a finite field $\mathbb{F}$. This means that, writing $\mathbb{F}$ in terms of one of its primitive elements α as $\mathbb{F} = \{0, 1, \alpha, \ldots, \alpha^{n-2}\}$, we require γ_i to be of the form $\gamma_i = \alpha^{b+i}$ for some integer b and for $i = 1, \ldots, N$. Let us, for the sake of simplicity, restrict ourselves to the case $b = -1$. We then have

$$H = \begin{bmatrix} 1 & 1 & \cdots & 1 \\ 1 & \alpha & \cdots & \alpha^{(n-1)} \\ 1 & \alpha^2 & \cdots & \alpha^{2(n-1)} \\ \vdots & \vdots & \vdots & \\ 1 & \alpha^{N-1} & \cdots & \alpha^{(N-1)(n-1)} \end{bmatrix}.$$

The corresponding Reed-Solomon code is defined as the linear subspace

$$C = \{\mathbf{c} \in \mathbb{F}^n \mid H\mathbf{c} = 0\}.$$

If a codeword $\mathbf{c}$ is corrupted by errors, resulting in $\mathbf{c} + e$, then the *syndrome vector* a, defined as

$$a := H(\mathbf{c} + e) = He$$

yields information about the error locations and error values. The aim of error correction is to find a unique vector e with a minimum number of nonzero components such that

$$a = He. \tag{2}$$

Note that every N columns in H are independent so that codewords in C should have at least $N + 1$ nonzero entries. In fact, the *minimum distance* of a Reed-Solomon code (i.e. the minimum number of entries on which two codewords differ) equals $N + 1$ so that uniqueness of e is guaranteed if the number of errors is smaller than $(N + 1)/2$: the *random error-correcting capability* of a Reed-Solomon code equals $(N + 1)/2$.

A system-theoretic reformulation of the error-correction problem is the following:

Problem Statement 2.1 *Let $a_1, a_2, \ldots, a_N$ be a given sequence of numbers in a field $\mathbb{F}$, corresponding to the components of a syndrome vector. Find a minimal state-space realization*

$$\begin{aligned} x(k+1) &= Ax(k) + Bu(k) \\ y(k) &= Cx(k) \end{aligned}$$

with B of size $L \times 1$ and the matrices A and C in the form

$$A = \begin{bmatrix} \lambda_1 & & \\ & \ddots & \\ & & \lambda_L \end{bmatrix}, \quad C = \begin{bmatrix} 1 & \cdots & 1 \end{bmatrix},$$

(with λ_i's in $\mathbb{F}$) such that an impulse input $u(0) = 1, u(1) = 0, \cdots$ produces the output response $y(0) = 0, y(1) = a_1, \ldots, y(N) = a_N$.

Indeed, the above realization gives rise to the equation

$$\begin{bmatrix} a_1 \\ a_2 \\ \vdots \\ a_N \end{bmatrix} = \begin{bmatrix} C \\ CA \\ \vdots \\ CA^{N-1} \end{bmatrix} B$$
$$= \begin{bmatrix} 1 & 1 & \cdots & 1 \\ \lambda_1 & \lambda_2 & \cdots & \lambda_L \\ \lambda_1^2 & \lambda_2^2 & \cdots & \lambda_L^2 \\ \vdots & \vdots & \vdots & \vdots \\ \lambda_1^{N-1} & \lambda_2^{N-1} & \cdots & \lambda_L^{N-1} \end{bmatrix} B. \quad (3)$$

If the solution (A, B, C) is minimal and unique (up to similarity), then (3) uniquely solves (2): the error locations are determined by the λ_i's ($i = 1, \cdots, L$) and the error values are given by the entries of B, respectively. Note that L equals the number of errors that occurred. The above problem statement differs from Problem Statement 1.1 in the kind of representation that is aimed for. In Problem Statement 1.1 it is a transfer function of minimal McMillan degree; in Problem Statement 2.1 it is a state space description of minimal size in diagonal form. The relationship between the two representations is of course that the zeros of the polynomial $f(s)$ of Problem Statement 1.1 are the diagonal elements $\lambda_1, \ldots, \lambda_L$ of A. Thus error-correction for Reed-Solomon codes comes down to the computation of a unique shortest linear recurrence relation $(f_1, f_2, \ldots, f_L)$ for the components of a syndrome vector, such that the polynomial $f(s) = s^L + f_1 s^{L-1} + \ldots + f_L$ has distinct zeros in $\mathbb{F}$. In the coding theoretic literature the polynomial $f(s) = s^L + f_1 s^{L-1} + \ldots + f_L$ is called the *error locator polynomial*: its zeros determine the error locations. For an *arbitrary* sequence $a_1, a_2, \ldots, a_N$ (i.e. not necessarily a syndrome vector) it is well-known that there exists a unique shortest linear recurrence relation for $a_1, a_2, \ldots, a_N$ of length δ if and only if $\delta < (N+1)/2$. Thus, in the situation that $a_1, a_2, \ldots, a_N$ are components of a syndrome vector, a unique solution of Problem Statement 2.1 is guaranteed if $L < (N+1)/2$. This corresponds with the above-mentioned random error-correcting capability of a Reed-Solomon code.

Thus we have found that error correction for Reed-Solomon codes can be achieved by SISO minimal partial realization.

In practical binary applications the finite field $\mathbb{F}$ in which the symbols of a Reed-Solomon code take their value, is a binary extension field, denoted as $GF(2^M)$ (the *Galois field* with 2^M elements). Its elements are M-tuples of binary numbers. Thus one corrupted symbol can stand for a string of corrupted bits and it is for this reason that Reed-Solomon codes are known for their "burst-error correcting capability". This in contrast to BCH codes, the subject of the next subsection.

2.2 BCH codes

Introduced by A. Hocquenghem in 1959 as a generalization of the elementary Hamming codes, BCH codes were simultaneously published in a work by R.C. Bose and D.K. Ray-Chaudhuri and this explains their name. Although discovered independently from Reed-Solomon codes, it was later realized that the two types of codes are closely related. In fact, BCH codes and Reed-Solomon codes differ only in the type of field in which the codeword components take their value–the codeword components of a BCH code are required to be from a base field whereas those of a Reed-Solomon code take their value in an extension field. Let us clarify this further and restrict ourselves to the binary case for the sake of simplicity. A BCH code is then defined in exactly the same way as a Reed-Solomon code except that the code word symbols are required to be *binary* instead of from $GF(2^M)$. Thus a binary BCH code is defined as

$$C = \{\boldsymbol{c} \in \{GF(2)\}^n \mid H\boldsymbol{c} = 0\}$$

with H given as in (1)

$$H = \begin{bmatrix} 1 & \gamma_1 & \cdots & \gamma_1^{n-1} \\ 1 & \gamma_2 & \cdots & \gamma_2^{n-1} \\ \vdots & \vdots & \vdots & \\ 1 & \gamma_N & \cdots & \gamma_N^{n-1} \end{bmatrix}.$$

Again the γ_i's $(i = 1, \ldots, N)$ are required to be *consecutive* elements of an extension field $GF(2^M)$, say α^{b+1}, α^{b+2}, $\ldots$, α^{b+N} for some integer b.

The decoding of BCH codes is identical to the decoding of Reed-Solomon codes (note, however, that for binary BCH codes error values need not be calculated since they are necessarily equal to 1). Again, error-correction can be achieved if the number of errors $L < (N+1)/2$. In contrast to a Reed-Solomon code, however, a BCH code may have a minimum distance that exceeds $N+1$ and thus an increased random error-correcting capability: the so-called "actual minimum distance" of the code may exceed its so-called "designed minimum distance" of $N+1$. Due to the fact that the codeword components are binary instead of from the extension field $GF(2^M)$ we have that an equality of the type

$$\begin{bmatrix} 1 & \gamma_i & \cdots & \gamma_i^{n-1} \end{bmatrix} \begin{bmatrix} c_0 \\ \vdots \\ c_{n-1} \end{bmatrix} = 0$$

implies

$$\begin{bmatrix} 1 & \gamma_i^2 & \cdots & \gamma_i^{2(n-1)} \end{bmatrix} \begin{bmatrix} c_0 \\ \vdots \\ c_{n-1} \end{bmatrix} = 0.$$

As a result, there exists a set of $\tilde{\gamma}_i$'s, containing our previously specified set of consecutive elements $\{\alpha^{b+1}, \alpha^{b+2}, \ldots, \alpha^{b+N}\}$, such that the code is equivalently described by the parity check matrix

$$\tilde{H} = \begin{bmatrix} 1 & \tilde{\gamma}_1 & \cdots & \tilde{\gamma}_1^{n-1} \\ 1 & \tilde{\gamma}_2 & \cdots & \tilde{\gamma}_2^{n-1} \\ \vdots & \vdots & \vdots & \\ 1 & \tilde{\gamma}_{\tilde{N}} & \cdots & \tilde{\gamma}_{\tilde{N}}^{n-1} \end{bmatrix}.$$

There are many examples of BCH codes where more than one string of N consecutive elements are among the $\tilde{\gamma}_i$'s. Recent research [8] points out that the occurrence of, say, m strings of N consecutive zeros can be used (under certain conditions) to perform error correction beyond the designed error-correcting capability.

Let us have a closer look and reformulate this in system-theoretic terms; denote the m strings of N consecutive zeros by $\alpha^{b_i+1}, \ldots, \alpha^{b_i+N}$ and the corresponding syndrome components by $a_1^i, \ldots, a_N^i$ $(i = 1, \ldots, m)$.

Error correction amounts to finding a unique vector e with a minimum number of nonzero components such that

$$\begin{bmatrix} a_1^1 \\ \vdots \\ a_N^1 \\ a_1^2 \\ \vdots \\ a_N^2 \\ \vdots \\ a_1^m \\ \vdots \\ a_N^m \end{bmatrix} = \begin{bmatrix} 1 & \alpha^{b_1+1} & \ldots & \alpha^{(b_1+1)(n-1)} \\ \vdots & \vdots & \vdots & \\ 1 & \alpha^{b_1+N} & \ldots & \alpha^{(b_1+N)(n-1)} \\ 1 & \alpha^{b_2+1} & \ldots & \alpha^{(b_2+1)(n-1)} \\ \vdots & \vdots & \vdots & \\ 1 & \alpha^{b_2+N} & \ldots & \alpha^{(b_2+N)(n-1)} \\ \vdots & \vdots & \vdots & \\ 1 & \alpha^{b_m+1} & \ldots & \alpha^{(b_m+1)(n-1)} \\ \vdots & \vdots & \vdots & \\ 1 & \alpha^{b_m+N} & \ldots & \alpha^{(b_m+N)(n-1)} \end{bmatrix} e. \tag{4}$$

This problem can be reformulated in system-theoretic terms as:

Problem Statement 2.2 *Let $a_1^i, \ldots, a_N^i$ $(i = 1, \ldots, m)$ be m given sequences of numbers in a field $\mathbb{F}$, corresponding to the syndrome components. Find a minimal state-space realization*

$$\begin{aligned} x(k+1) &= Ax(k) + Bu(k) \\ y(k) &= Cx(k) \end{aligned}$$

with B of size $L \times m$ and the matrices A and C in canonical form

$$A = \begin{bmatrix} \lambda_1 & & \\ & \ddots & \\ & & \lambda_L \end{bmatrix}, \quad C = \begin{bmatrix} 1 & \cdots & 1 \end{bmatrix}$$

(with λ_i's in $\mathbb{F}$) such that an impulse at the i'th input $u_i(0) = 1, u_i(1) = 0, \cdots$ produces the output response $y(1) = a_1^i, \ldots, y(N) = a_N^i$ $(i = 1, \ldots, m)$.

Indeed, the above realization gives rise to the equation

$$\begin{aligned} \begin{bmatrix} a_1^1 & \cdots & a_1^m \\ \vdots & \vdots & \vdots \\ a_N^1 & \cdots & a_N^m \end{bmatrix} &= \begin{bmatrix} C \\ CA \\ \vdots \\ CA^{N-1} \end{bmatrix} B \\ &= \begin{bmatrix} 1 & 1 & \cdots & 1 \\ \lambda_1 & \lambda_2 & \cdots & \lambda_L \\ \lambda_1^2 & \lambda_2^2 & \cdots & \lambda_L^2 \\ \vdots & \vdots & \vdots & \vdots \\ \lambda_1^{N-1} & \lambda_2^{N-1} & \cdots & \lambda_L^{N-1} \end{bmatrix} B, \end{aligned}$$

which is equivalent to (4). As in the case of error-correction for Reed-Solomon codes, it follows that the L error locations are given by the λ_i's $(i = 1, \cdots, L)$ if the solution is minimal and unique. The error correction problem can alternatively be formulated as the problem of finding a unique shortest *common* linear recurrence relation $(f_1, f_2, \ldots, f_L)$ for the sequences $a_1^i, \ldots, a_N^i$ $(i = 1, \cdots, m)$, such that the polynomial $f(s) = s^L + f_1 s^{L-1} + \ldots + f_L$ has distinct zeros in $\mathbb{F}$. In [8] a "generalized Berlekamp-Massey algorithm" is presented that produces a shortest common linear recurrence relation for arbitrary sequences $a_1^i, \ldots, a_N^i$ $(i = 1, \cdots, m)$. It has been shown by Roos in [23] that, under certain assumptions on the b_i's, a lower bound ("Roos bound") for the minimum distance d_{min} is given by

$$N + m \leq d_{min}.$$

Subsequently it has been shown in [8] that, under somewhat stronger assuptions on the b_i's, the generalized Berlekamp-Massey algorithm is guaranteed to produce the error locator polynomial if $L < (N+m)/2$. Thus, for certain

BCH codes, error correction up to $(N+m)/2$ can be achieved by minimal partial realization with multiple input and single output (MISO).

In subsection 4.2 we derive the generalized Berlekamp-Massey algorithm of [8] as a special case of an algorithm for behavioral modeling of partial data. This brings about a reformulation of the algorithm in which not only L but also m other integers are updated at every step. These integers are the major players in a necessary and sufficient condition for uniqueness of a shortest common linear recurrence relation for arbitrary sequences of numbers $a_1^i, \ldots, a_N^i$ $(i = 1, \cdots, m)$. In the situation that these numbers are syndrome components, this gives rise to a new sufficient condition for error-correction of BCH codes beyond the designed error-correcting capability. The relationship of this condition with the above-mentioned Roos bound is still a subject of further investigation.

2.3 Keystreams

As stated in the introduction, the Berlekamp-Massey algorithm nowadays finds its dominant application in the area of cryptography. The so-called *keystreams* $a_1, a_2, \ldots, a_N$ that are used in cryptographic devices are generated deterministically by a finite state machine and are therefore periodic. The length of the generating linear recurrence relation is called the *linear complexity* of the sequence. As it calculates this linear recurrence relation, the Berlekamp-Massey algorithm can be used to attack such a keystream for the purpose of decryption.

Typically, the cryptographic designer wants a binary sequence to have a long period in which the values resemble the outcomes of a fair coin tossing experiment. The concept of a *pseudo-noise sequence*, introduced by Golomb [10], incorporates many of the desired characteristics. A pseudo-noise sequence is defined as a binary sequence that is generated by a linear recurrence relation of, say, length L for which the corresponding polynomial is irreducible and primitive. Such a sequence has uniform statistical properties and a period that is maximal $(= 2^L - 1)$ over all sequences of linear complexity L. However, despite the fact that a pseudo-noise sequence has many of the statistical properties of a fair coin tossing experiment, it is highly predictable. This is due to the fact that its linear complexity is relatively small compared to the period of the sequence: only $2L$ initial bits are needed to predict the whole sequence.

Alternative criteria were therefore defined by Rueppel in [24]. First of all he requires not only the period but also the linear complexity of the sequence to be high— typically close to the period. His further requirements apply to the local linear complexities of all the subsequences beginning with a_1. Denoting the linear complexity of the subsequence $a_1, \ldots, a_k$ by $L(k)$, the *linear complexity profile* of $a_1, a_2, \ldots, a_N$ is defined as the vector of local linear complexities $(L(1), L(2), \ldots, L(N))$. Rueppel [24] shows that the expected value of $L(N)$ of a sequence $a_1, a_2, \ldots, a_N$, coming from N independent and uniformly distributed binary random variables, is approximately $N/2$ for large N. He then argues that it is therefore reasonable to require that, as a function of N, the linear complexity profile approximately follows a line of slope $1/2$. Because of its iterative nature, the Berlekamp-Massey algorithm is an ideal tool for calculating the linear complexity profile.

When $\mathbb{F}$ is an infinite field, the algebraic-geometric concept of "genericity" makes sense. Here a property of a sequence $\in \mathbb{F}^N$ is called "generic" if the set of sequences $\in \mathbb{F}^N$ that do not have the property is an algebraic variety ($\neq \mathbb{F}^N$). In subsection 3.2 we show that the linear complexity of a sequence $a_1, a_2, \ldots, a_N$ can be interpreted as one of the observability indices of a system of order $N+1$ with two outputs and no inputs. This provides an alternative way to explain the wellknown result that, generically,

$$L(N) = \lfloor \frac{N+1}{2} \rfloor.$$

3 Iterative modeling of a behavior–the algorithm

All problem statements that were presented in the above, center around one issue: the modeling of a given impulse response in a minimal way. Let us now have a look at a behavioral formulation of this problem and see how this gives rise to algorithms in a very natural fashion. It turns out that both the Berlekamp-Massey algorithm and its extension, the generalized Berlekamp-Massey algorithm, find their place in this theory. In fact, the theory allows for a fully multivariable algorithm which is a further extension of the generalized Berlekamp-Massey algorithm.

The modeling of truncated trajectories into minimal behaviors has been described in [3]. Before presenting this view in full detail, we first give a brief outline of the relevant part of the behavioral theory.

3.1 Quick overview of behavioral modeling

In the behavioral approach [29]–[31], a system is essentially defined as a set of trajectories. We will be concerned with linear shift-invariant behaviors on the time-set $\mathbb{Z}_-$ of the form $\mathcal{B} = \ker R(\sigma)$, where R is a polynomial $g \times q$-matrix and σ is the forward shift operator:

$$\sigma((\ldots, w_2, w_1, w_0)) := (\ldots, w_2, w_1).$$

The behavior $\mathcal{B}$ consists of trajectories $\boldsymbol{w} : \mathbb{Z}_- \mapsto \mathbb{R}^q$, for which

$$R(\sigma)\boldsymbol{w} = 0. \tag{5}$$

The representation (5) is called a *kernel representation* of $\mathcal{B}$.

Let us repeat some notions from [29], and start with the following lemma (see e.g. [18, Th. 3.9] for a detailed proof).

Lemma 3.1 *Let* $R_1 \in \mathbb{F}^{g_1 \times q}[s]$ *and* $R_2 \in \mathbb{F}^{g_2 \times q}[s]$. *Then*

$$ker\ R_1(\sigma) \subset\ ker\ R_2(\sigma)$$

if and only if there exists a polynomial matrix $F \in \mathbb{F}^{g_2 \times g_1}[s]$ *such that*

$$R_2 = FR_1.$$

It is a corollary of the above lemma that polynomial matrices R_1 and R_2 of full row rank represent the same behavior if and only if there exists a unimodular matrix U (i.e., a polynomial matrix with constant nonzero determinant) such that $R_1 = UR_2$.

As a measure of complexity of a model we introduce the *order* $n(\mathcal{B})$ of a behavior. It is defined as the minimum value of the sum of the row degrees of R, where the minimum is taken over all possible full row rank kernel representations (5) of $\mathcal{B}$. This minimum is attained exactly when R is "row reduced".

Definition 3.1 *Let* $R \in \mathbb{R}^{g \times q}[s]$ *have full row rank. Define* $R_d \in \mathbb{R}^{g \times q}$ *as the* leading row coefficient matrix *of* R, *i.e., the constant matrix that consists of the coefficients of the highest degree terms in each row of* R. *Define* R *to be* row reduced *if* R_d *has full row rank.*

When a matrix R is not row reduced, a unimodular matrix U can be found such that UR is row reduced. A procedure is given in [33, p. 27], see also [18, p. 24] where it is shown that not only the sum of the minimal row degrees is an invariant of a behavior, but also the minimal row degrees themselves are invariants of a behavior–they are the *observability indices* of $\mathcal{B}$.

Let us next assume that we have a *data set* $\mathbf{D} = \{\boldsymbol{b}_0, \boldsymbol{b}_1, \ldots, \boldsymbol{b}_N\}$ where $\boldsymbol{b}_i \in (\mathbb{R}^q)^{\mathbb{Z}_-}$ are observed trajectories ($i = 0, 1, \ldots, N$). A behavior $\mathcal{B}$ is called an *unfalsified model* for $\mathbf{D}$ if $\mathbf{D} \subseteq \mathcal{B}$. A model $\mathcal{B}_1$ is called *more powerful* than a model $\mathcal{B}_2$ if $\mathcal{B}_1 \subsetneq \mathcal{B}_2$. A model $\mathcal{B}^*$ is called the *most powerful unfalsified model (MPUM)* for $\mathbf{D}$, if $\mathcal{B}^*$ is unfalsified for $\mathbf{D}$ and $\mathbf{D} \subseteq \mathcal{B} \implies \mathcal{B}^* \subseteq \mathcal{B}$. It has been shown in [30] that a unique MPUM $\mathcal{B}^*$ exists for $\mathbf{D}$. However, note that a kernel representation (5) of $\mathcal{B}^*$ is far from unique.

We are now ready to present the procedure of [31, p. 289], which provides a framework for the iterative construction of a kernel representation of the MPUM for $\mathbf{D} = \{\boldsymbol{b}_0, \boldsymbol{b}_1, \ldots, \boldsymbol{b}_N\}$. It can be easily understood from Lemma 3.1.

Procedure 3.1 *([31]) Initially define*

$$R_{-1} := I \quad \textit{(where I is the identity matrix).}$$

Proceed iteratively as follows for $k = 0, 1, \ldots, N$. *Define, after receiving* $\{\boldsymbol{b}_0, \boldsymbol{b}_1, \ldots, \boldsymbol{b}_k\}$, *the* k-th error trajectory $\boldsymbol{e}_k$ *as*

$$\boldsymbol{e}_k := R_{k-1}(\sigma)\boldsymbol{b}_k.$$

Compute a kernel representation $V_k(\sigma)\boldsymbol{w} = 0$ *of the MPUM for* $\{\boldsymbol{e}_k\}$. *Then define*

$$R_k := V_k R_{k-1}.$$

Theorem 3.1 *([31]) For* $k = 0, 1, \ldots, N$, *the kernel representation*

$$R_k(\sigma)\boldsymbol{w} = 0,$$

with R_k *defined in Procedure 3.1, represents the MPUM for* $\{\boldsymbol{b}_0, \boldsymbol{b}_1, \ldots, \boldsymbol{b}_k\}$.

With the above procedure we only need to be able to compute a MPUM representation for a *single* trajectory (the error trajectory) in order to derive an MPUM representation for a *finite set* of trajectories. In order to deal with the requirements on minimality, at each step k ($k = 0, 1, \ldots, N$), we can achieve row reducedness of the representation $R_k(\sigma)\boldsymbol{w} = 0$ by choosing V_k in such a way that $V_k R_{k-1}$ remains row reduced. For our specific application of modeling of a truncated impulse response, in employing the above procedure, we can force the error trajectory to be of an extremely simple form. As a result, the choice of update matrix is restricted. The crucial part of the design then consists of the specific choice of the update matrix at each step that guarantees not only that row reducedness is preserved but also that the error trajectory at the next step is again in the desired simple form. This is a preliminary rough outline of our solution which will be detailed in the next section—let us first reformulate the original problem in behavioral terms.

3.2 Reformulation in terms of behaviors

In this subsection we go back to our original problem of modeling an impulse response in a minimal way and give a summary of the ideas of [3] in which this problem is formulated in a behavioral setting. For the sake of clarity we first deal with the SISO case and Problem Statement 1.1. For this, we need to consider linear shift-invariant behaviors on the time-set $\mathbb{Z}$ with values in $\mathbb{R}^2$ of the form

$$\mathcal{B} = \ker R(\sigma^*),$$

where R is a polynomial $g \times 2$-matrix and σ^* is the backward shift operator:

$$\sigma^* \boldsymbol{w}(t) := \boldsymbol{w}(t+1).$$

Writing the truncated behavior $\mathcal{B}_N$ as

$$\mathcal{B}_N = \{w : (-\infty, N] \mapsto \mathbb{R}^2 \mid \exists \tilde{w} \in \mathcal{B} : w = \tilde{w}_{|(-\infty,N]}\},$$

we can formulate Problem Statement 1.1 in behavioral terms as

Problem Statement 3.1 *Let $a_0, a_1, \ldots, a_N$ be a given sequence of numbers in a field $\mathbb{F}$. Find polynomials $f(s)$ and $g(s)$ with $f(s)$ of minimal degree, such that the representation*

$$f(\sigma^*)y = g(\sigma^*)u$$

represents a behavior $\mathcal{B}$, defined on $\mathbb{Z}$, for which $\mathcal{B}_N$ contains the trajectory

$$\left(\ldots, \begin{bmatrix} 0 \\ 0 \end{bmatrix}, \begin{bmatrix} a_0 \\ 1 \end{bmatrix}, \begin{bmatrix} a_1 \\ 0 \end{bmatrix}, \ldots, \begin{bmatrix} a_N \\ 0 \end{bmatrix}\right).$$

Of course we could just as well have formulated this in terms of the forward shift operator σ.

Problem Statement 3.2 *Let $a_0, a_1, \ldots, a_N$ be a given sequence of numbers in a field $\mathbb{F}$. Find polynomials $c(s)$ and $p(s)$ with $c(0) \neq 0$ and deg $[c(s) \quad p(s)] :=$ max $\{$ deg $c(s)$, deg $p(s)\}$ minimal, such that the representation*

$$c(\sigma)y = p(\sigma)u$$

represents a behavior $\mathcal{B}$, defined on $\mathbb{Z}$, for which $\mathcal{B}_N$ contains the trajectory

$$\left(\ldots, \begin{bmatrix} 0 \\ 0 \end{bmatrix}, \begin{bmatrix} a_0 \\ 1 \end{bmatrix}, \begin{bmatrix} a_1 \\ 0 \end{bmatrix}, \ldots, \begin{bmatrix} a_N \\ 0 \end{bmatrix}\right).$$

The relationship between the pair $(f(s), h(s))$ of Problem Statement 3.1 and the above pair $(c(s), p(s))$ is of course a reciprocal one:

$$[c(s) \quad p(s)] := [f(s) \quad h(s)]^r.$$

(Here the *reciprocal* $P^r(s)$ of a polynomial matrix $P(s) = P_n s^n + P_{n-1}s^{n-1} + \cdots + P_1 s + P_0$ ($P_n \neq 0$) is defined by $P^r(s) := P_n + P_{n-1}s + \cdots + P_1 s^{n-1} + P_0 s^n$.)

But now the problem can just as well be formulated on the time set $\mathbb{Z}_-$ instead of $\mathbb{Z}$! This is the crucial step towards an elegant solution based on behavioral modeling. The problem can be equivalently formulated on $\mathbb{Z}_-$ as follows:

Problem Statement 3.3 *Let $a_0, a_1, \ldots, a_N$ be a given sequence of numbers in a field $\mathbb{F}$. Find polynomials $c(s)$ and $p(s)$ with $c(0) \neq 0$ and deg $[c(s) \quad p(s)] :=$ max $\{$ deg $c(s)$, deg $p(s)\}$ minimal, such that the representation*

$$c(\sigma)y = p(\sigma)u$$

represents a behavior $\mathcal{B}$, defined on $\mathbb{Z}_-$, that contains the trajectory

$$\boldsymbol{b}_N = \left(\ldots, \begin{bmatrix} 0 \\ 0 \end{bmatrix}, \begin{bmatrix} a_0 \\ 1 \end{bmatrix}, \begin{bmatrix} a_1 \\ 0 \end{bmatrix}, \ldots, \begin{bmatrix} a_N \\ 0 \end{bmatrix}\right).$$

The advantage of modeling on $\mathbb{Z}_-$ is that it allows us to employ the theory of finite-dimensional behavioral modeling of the previous subsection: the MPUM for $\{\boldsymbol{b}_N\}$ is simply the linear span of $\boldsymbol{b}_N$ and its shifts $\sigma\boldsymbol{b}_N$, $\sigma^2\boldsymbol{b}_N$ and $\sigma^N\boldsymbol{b}_N$. Its order is $N+1$ and a representation is easily given:

$$\begin{bmatrix} 1 & -(a_0 + a_1\sigma + \cdots + a_N\sigma^N) \\ 0 & \sigma^{N+1} \end{bmatrix} \boldsymbol{w} = 0. \tag{6}$$

Any other kernel representation can be obtained by left multiplication by a unimodular matrix–in particular, the matrix in (6) can be brought into row reduced form by left multiplication by a unimodular matrix. The sought after polynomials $c(s)$ and $p(s)$ then make up one of the rows of the row reduced representation. More specifically ([4, Sect. IX-A] and [17, Th. 7]), if the second row vanishes at $s = 0$, then the first row of the row reduced representation necessarily equals a solution $[c(s) \quad -p(s)]$. Note that its row degree is an observability index of $\mathcal{B}$.

We are now ready to fill in more details of the general iterative outline of Procedure 3.1: we need to choose the update matrices V_k such that, at every step, a row reduced matrix R_k emerges whose second row vanishes at $s = 0$. This will be done in the next section. It turns out that the resulting algorithm is precisely the Berlekamp-Massey algorithm! (Compare in particular Berlekamp's original four-polynomial-formulation, see also [6]). Thus, the connection between coding theory and system theory, that was alluded to in section 1, is made fully explicit.

Remark 3.1 *An alternative* noniterative *solution method consists of constructing a unimodular polynomial matrix U such that left multiplication of the matrix in (6) by U results in a row reduced matrix. For this, the procedure in [33] can be used. In this specific case, the procedure essentially coincides with the euclidean algorithm, applied to the polynomials s^{N+1} and $a_0 + a_1 s + \cdots + a_N s^N$, see [20] for more details. The euclidean algorithm is a wellknown solution method in the coding theoretic literature [25].*

For the multivariable case, the reformulation of the problem in behavioral terms is a straightforward generalization of the SISO case: in the $p \times m$ case we need to model m vector impulse responses of the form

$$\boldsymbol{b}_i = \left(\ldots, \begin{bmatrix} 0_p \\ 0_m \end{bmatrix}, \begin{bmatrix} A_0^i \\ e_i \end{bmatrix}, \begin{bmatrix} A_1^i \\ 0_m \end{bmatrix}, \ldots, \begin{bmatrix} A_N^i \\ 0_m \end{bmatrix}\right) \quad (i = 1, \ldots, m).$$

Here the A_j^i's are vectors in $\mathbb{F}^p$, 0_m denotes the zero vector of length m and e_i is the i-th unit vector of length m. The problem statement generalizes to:

Problem Statement 3.4 *Let A_0^i, $A_1^i, \ldots, A_N^i$ $(i = 1, \ldots, m)$ be m given sequences of vectors in $\mathbb{F}^p$. Find polynomial matrices $C(s)$ and $P(s)$ of* *sizes $p \times p$ and $p \times m$, respectively, with $C(0)$ nonsingular and the sum of the row degrees of $[C(s) \quad P(s)]$ minimal, such that the representation*

$$C(\sigma)\boldsymbol{y} = P(\sigma)\boldsymbol{u}$$

represents a behavior $\mathcal{B}$, defined on $\mathbb{Z}_-$, that contains the trajectories $\boldsymbol{b}_i$ $(i = 1, \ldots, m)$ defined above.

Completely analogous to the SISO case, a solution can be obtained by constructing a row reduced MPUM representation for $\mathcal{B}$, involving a $(p+m) \times (p+m)$ polynomial matrix. It follows from [4, Sect. IX-A] that, if the last m rows vanish at $s = 0$, then the first p rows of such a representation give a solution $[C(s) \quad -P(s)]$.

Thus the multivariable problem stated above calls for an algorithm with analogous properties as in the SISO case. Again the general iterative outline of Procedure 3.1 can be used. Completely analogous to the SISO case, we need to choose the update matrices V_k such that, at every step, a row reduced matrix R_k emerges that has its last m rows vanish at $s = 0$. The design of these update matrices turns out to be nontrivial, see [19]. We will present the multivariable algorithm of [19] in subsection 4.1 below.

4 The algorithm

In this section we present the general algorithm of [19] for multiple inputs and multiple outputs. The algorithm is designed along the lines of the behavioral modeling of the previous section. For $p = m = 1$ the algorithm comes down to the scalar version of [17], which coincides with the Berlekamp-Massey algorithm. For $m = 1$ the algorithm essentially coincides with the (SIMO) algorithm of [32] for vector time series; for multiple inputs and multiple outputs, a sequential application of the algorithm of [32] differs from the algorithm below since the latter processes the data in an interleaved way.

After presenting the general MIMO algorithm in the next subsection, we concentrate on the multiple-input-single-output (MISO) case in subsection 4.2. Although the MISO algorithm is a special case of the general MIMO algorithm, we will spell it out in detail. We recall a uniqueness result and show its relevance to the decoding of BCH codes beyond their designed error-correcting capability.

4.1 MIMO case: the general algorithm

The modeling of a finite set of vector impulse responses or, equivalently, the minimal partial realization of a finite sequence of matrices $A_0, A_1, \ldots, A_N$, is solved by the following algorithm. Define A_j^i as the i'th column of A_j $(i = 1, \ldots, m)$. The algorithm below processes the columns $A_0^1, A_0^2, \ldots, A_0^m$,

$A_1^1, \ldots, A_k^i$ iteratively. At each step it produces a $(p+m) \times (p+m)$ polynomial matrix R_{km+i} and updates its row degree vector L_{km+i}, denoted as

$$L_{km+i} = \begin{bmatrix} L_{km+i}(1) \\ \vdots \\ L_{km+i}(p+m) \end{bmatrix}.$$

In the formulation of the algorithm, e_j denotes the jth unit vector of length $p+m$ and $\Pi_{(j,\ell)}$ is the permutation, operating on vectors of length $p+m$, that interchanges the jth and ℓth elements.

Algorithm 4.1 *([19])*
Denote $C_j := \begin{bmatrix} I_p & 0 \end{bmatrix} R_j \begin{bmatrix} I_p \\ 0 \end{bmatrix}$.

Initially define $R_0 := I_{p+m}$ *and* $L_0 := 0$ *and proceed iteratively as follows. Define, after receiving* $A_0^1, A_0^2, \ldots, A_0^m, A_1^1, \ldots, A_k^i$, *the vector* Δ_{km+i} *as the coefficient-vector of* s^k *in* $C_{km+i-1}(s)(A_0^i + A_1^i s + \cdots + A_k^i s^k)$. *Denote*

$$\Delta_{km+i} = \begin{bmatrix} \Delta_{km+i}(1) \\ \vdots \\ \Delta_{km+i}(p) \end{bmatrix}.$$

Let j_* *be the smallest integer for which* $L_{km+i-1}(j_*)$ *is minimal among*

$$\{L_{km+i-1}(j) \mid \Delta_{km+i}(j) \neq 0,\ j = 1, \ldots, p\}.$$

Compute the matrix R_{km+i} *as*

$$R_{km+i} := V_{km+i} R_{km+i-1},$$

where V_{km+i} *is a* $(p+m) \times (p+m)$ *matrix, that is defined as follows.*

–If $\Delta_{km+i} = 0$ *or* $L_{km+i-1}(j_*) > L_{km+i-1}(p+i)$,

then define V_{km+i} *as the identity matrix except for the following entries*

- *for* $j = 1, \ldots, p$, *define the* $(j, p+i)$*-entry as* $-\Delta_{km+i}(j)$
- *define the* $(p+i, p+i)$*-entry as* s.

Update L_{km+i} *as*

$$L_{km+i} := L_{km+i-1} + e_{p+i}.$$

–Otherwise,

define V_{km+i} *as the identity matrix except for the following entries*

- *for* $j = 1, \ldots, p$ *and* $j \neq j_*$, *define the* (j, j_*)*-entry as* $-\Delta_{km+i}(j)/\Delta_{km+i}(j_*)$.
- *define the* $(j_*, p+i)$*-entry as* $-\Delta_{km+i}(j_*)$.
- *define the* $(p+i, j_*)$*-entry as* $s/\Delta_{km+i}(j_*)$ *and the* $(p+i, p+i)$*-entry as zero.*

Update L_{km+i} *as*

$$L_{km+i} := \Pi_{(j_*,p+i)}(L_{km+i-1}) + e_{p+i}.$$

Theorem 4.1 *([19]) Let* $A_0, A_1, \ldots, A_N$ *be a finite sequence of* $p \times m$ *constant matrices. Let the above algorithm operate on* $A_0, A_1, \ldots, A_N$*; define* R *as the* $(p+m) \times (p+m)$ *polynomial matrix* $R_{(N+1)m}$ *that is finally produced by the algorithm. Let* $L(1), \ldots, L(p+m)$ *be its row degrees. Define polynomial matrices* C *and* P *by*

$$[C \quad -P] := [\ I_p \quad 0\ ]\, R.$$

Define $[E \quad -H]$ *as the row reciprocal of* $[C \quad -P]$. *Then* E *is row reduced and* $T := E^{-1}H$ *is a partial realization for* $A_0, A_1, \ldots, A_N$ *that has minimal McMillan degree* $L(1) + \cdots + L(p)$.

Furthermore, a parametrization of all minimal partial realizations is obtained from

$$[C \quad -P] := [\ I_p \quad Q\ ]\, R,$$

where Q *is a* $p \times m$ *polynomial matrix, such that the set of row degrees of* $[I_p \quad Q]\, R$ *equals* $\{L(1), \ldots, L(p)\}$.

In particular, T *is unique if and only if*

$$L(r) < L(p+j) \text{ for all } \quad r \in \{1, \ldots, p\} \text{ and } j \in \{1, \ldots, m\}$$

4.2 MISO case: $p = 1$; a result on decoding BCH codes beyond their error correcting capability

In the case of multiple inputs and single output ($p = 1$), the above algorithm processes m sequences of numbers $a_0^i, \ldots, a_N^i$ ($i = 1, \ldots, m$) and produces a shortest common linear recurrence relation for these sequences. In this case the algorithm is essentially the same as Feng & Tzeng's algorithm of [8], although its formulation is different. The main difference is that the above matrix-based formulation involves an explicit update at each step of not only the error locator polynomial but also $m^2 + 2m$ other polynomials. The degrees of all of these $(m+1)^2$ polynomials can be employed to derive a uniqueness result that is relevant to the decoding of BCH codes beyond the designed error-correcting capability. Note that for $m = 1$ the algorithm below equals the Berlekamp-Massey algorithm (compare in particular its formulation in [6]).

Algorithm 4.2

Denote $c_j := \begin{bmatrix} 1 & 0_m{}^T \end{bmatrix} R_j \begin{bmatrix} 1 \\ 0_m \end{bmatrix}$.

Initially define $R_0 := I_{m+1}$ *and* $L_0 := 0$.

Proceed iteratively as follows.
Define, after receiving $a_0^1, a_0^2, \ldots, a_0^m, a_1^1, \ldots, a_k^i$, *the number* Δ_{km+i} *as the coefficient of* s^k *in* $c_{km+i-1}(s)(a_0^i + a_1^i s + \cdots + a_k^i s^k)$.

Compute the matrix R_{km+i} *as*

$$R_{km+i} := V_{km+i} R_{km+i-1},$$

where V_{km+i} *is a* $(m+1) \times (m+1)$ *matrix, that is defined as the identity matrix except for its* $(1, i+1)$*-entry which is defined as* $-\Delta_{km+i}$ *and the following entries:*

-If $\Delta_{km+i} = 0$ *or* $L_{km+i-1}(1) > L_{km+i-1}(1+i)$,

define the $(i+1, i+1)$*-entry as* s.
Update L_{km+i} *as*

$$L_{km+i} := L_{km+i-1} + e_{i+1}.$$

-Otherwise,

define the $(i+1, 1)$*-entry as* s/Δ_{km+i} *and the* $(i+1, i+1)$*-entry as zero.*
Update L_{km+i} *as*

$$L_{km+i} := \Pi_{(1,i+1)}(L_{km+i-1}) + e_{i+1}.$$

In [8] it is proven that for certain BCH codes the algorithm achieves error correction if the number of errors is smaller than $(N+m)/2$. As a corollary of Theorem 4.1, we can conclude the following result.

Theorem 4.2 *Let* $a_0^i, \ldots, a_N^i$ $(i = 1, \ldots, m)$ *be* m *sequences, corresponding to* m *syndrome vectors. Let the above algorithm operate on* $a_0^i, \ldots, a_N^i$ $(i = 1, \ldots, m)$; *define* R *as the* $(m+1) \times (m+1)$ *polynomial matrix* $R_{(N+1)m}$ *that is finally produced by the algorithm. Let* $L(1), \ldots, L(m+1)$ *be its row degrees. Let the polynomial* $c(s) := 1 + c_1 s + \ldots + c_{L(1)} s^{L(1)}$ *be the left upper element of* $R_{(N+1)m}$. *Then the reciprocal* $s^{L(1)} + c_1 s^{L(1)-1} + \cdots + c_{L(1)}$ *is the error locator polynomial if it has distinct zeros in* $\mathbb{F}$ *and*

$$L(1) < L(i) \quad \text{for } i = 2, \ldots, m+1 \tag{7}$$

It is a topic of further research to prove directly that the conditions in [8] imply (7) and use the above result to possibly derive more general conditions under which the generalized Berlekamp-Massey algorithm produces the error locator polynomial.

Conclusions

Since its derivation in 1968, the Berlekamp-Massey algorithm has been widely studied (e.g. [9, 12, 13, 28]). As pointed out in [14], Massey's observation that the algorithm computes a shortest-length linear feedback shift register has opened the door to connections with many problems. In this paper we concentrated on the connection with system theory. We first gave a tutorial account on the algorithm's error-correction and cryptographic applications and its connection with the classical system-theoretic problem of minimal partial realization. We then recalled the extension [19] of the algorithm to the multivariable case (calculating a "shortest linear recurrence relation" for a sequence of matrices) and showed its relevance to BCH error correction beyond the designed error correcting capability.

The presented material is an example of how the interplay between two areas can give rise to a different view on well established topics that leads to new results that have a relevance to both areas.

Acknowledgement

I thank Serdar Boztas for helpful discussions. I also thank Joachim Rosenthal for his valuable comments.

References

[1] Anderson, B.M., F.M. Brasch jr. and P.V. Lopresti (1975). The sequential construction of minimal partial realizations from finite input-output data. *SIAM J. Control* **13**, 552-571.

[2] Antoulas, A.C. (1986). On recursiveness and related topics in linear systems. *IEEE Trans. Aut. Control* **31**, 1121-1135.

[3] Antoulas, A.C. (1994). Recursive modeling of discrete-time time series, in *Linear Algebra for Control Theory*, P. Van Dooren and B. Wyman eds., Springer-Verlag, IMA **62**, 1-20.

[4] Antoulas, A.C. and J.C. Willems (1993). A behavioral approach to linear exact modeling. *IEEE Trans. Aut. Control* **38**, 1776-1802.

[5] Berlekamp, E.R. (1968). *Algebraic Coding Theory*, New York, McGraw-Hill.

[6] Blahut, R.E. (1983). *Theory and Practice of Error Control Codes*, Addison-Wesley.

[7] Dickinson, B.W., M. Morf and T. Kailath (1974). A minimal realization algorithm for matrix sequences. *IEEE Trans. Aut. Control* **19**, 31-38.

[8] Feng, G-L. and K.K. Tzeng (1991). A generalization of the Berlekamp-Massey algorithm for multisequence shift-register synthesis with applications to decoding cyclic codes. *IEEE Trans. Info. Theory* **37**, 1274-1287.

[9] Fitzpatrick, P. and G.H. Norton (1995). The Berlekamp-Massey algorithm and linear recurring sequences over a factorial domain. *Applicable Algebra in Engineering, Communication and Computing* **6**, 309-323.

[10] Golomb, S.W. (1967). *Shift register sequences*, Holden-Day, San Francisco.

[11] Gragg, W.B. and A. Lindquist (1983). On the partial realization problem. *Lin. Alg. Appl.* **50**, 277-319.

[12] Imamura, K. and W. Yoshida (1987). A simple derivation of the Berlekamp-Massey algorithm and some applications. *IEEE Trans. Info. Theory* **33**, 146-150.

[13] Jonckheere, E. and C. Ma (1989). A simple Hankel interpretation of the Berlekamp-Massey algorithm. *Lin. Alg. Appl.* **125**, 65-76.

[14] Kailath, T. (1994). Encounters with the Berlekamp-Massey algorithm, in *Communications and Cryptography; two sides of one tapestry (1994 Massey Symposium)*, R. Blahut et al. eds., Kluwer, 209-220.

[15] Kalman, R.E. (1971). On minimal partial realizations of a linear input/output map, in *Aspects of Network and System Theory*, R.E. Kalman and N. DeClaris eds., Holt, Rinehart and Winston, New York, 385-408.

[16] Kalman, R.E. (1979). On partial realizations, transfer functions, and canonical forms. *Acta Polytechnica Scandinavica, Math. Comp. Sc. Series*, **31**, 9-32.

[17] Kuijper, M. and J.C. Willems (1997). On constructing a shortest linear recurrence relation. *IEEE Trans. Aut. Control* **42**, 1554-1558.

[18] Kuijper, M. (1994). *First-Order Representations of Linear Systems.* Series on "Systems and Control: Foundations and Applications", Birkhäuser, Boston.

[19] Kuijper, M. (1997). An algorithm for constructing a minimal partial realization in the multivariable case. *Systems & Control Letters* **31**, 225-233.

[20] Kuijper, M. (1997). Partial realization and the euclidean algorithm, to appear in *IEEE Trans. Aut. Control.*

[21] Kuijper, M. (1997). How to construct a periodically time-varying system of minimal lag: an algorithm based on twisting, to appear in *Automatica.*

[22] Massey, J.L. (1969). Shift-register synthesis and BCH decoding. *IEEE Trans. Info. Theory* **15**, 122-127.

[23] Roos, C. (1983). A new lower bound for the minimum distance of a cyclic code. *IEEE Trans. Info. Theory* **29**, 330-332.

[24] Rueppel, R.A. (1986). *Analysis and Design of Stream Ciphers*, Springer-Verlag.

[25] Sugiyama, Y., Kasahara, M., Hirasawa, S. and T. Namekawa (1975). A method for solving key equation for decoding Goppa codes. *Information and Control* **27**, 87-99.

[26] Sain, M.K. (1975). Minimal torsion spaces and the partial input/output problem. *Information and Control* **29**, 103-124.

[27] Van Barel, M. and A. Bultheel (1989). A canonical matrix continued fraction solution of the minimal (partial) realization problem. *Lin. Alg. Appl.* **122/123/124**, 973-1002.

[28] L.R. Welch and R.A. Scholtz (1979). Continued fractions and Berlekamp's algorithm. *IEEE Trans. Info. Theory* **25**, 19-27.

[29] Willems, J.C. (1986). From time series to linear system. Part I: Finite-dimensional linear time invariant systems. *Automatica* **22**, 561-580.

[30] Willems, J.C. (1986). From time series to linear system. Part II: Exact modeling. *Automatica* **22**, 675-694.

[31] Willems, J.C. (1991). Paradigms and puzzles in the theory of dynamical systems. *IEEE Trans. Aut. Control* **36**, 259-294.

[32] Willems, J.C. (1997). Fitting data sequences to linear systems. in *Systems and Control in the Twenty-First Century*, C.I. Byrnes, B.N. Datta, C.F. Martin and D.S. Gilliam eds., Birkhäuser, Boston, (invited papers presented at the 12th MTNS, 1996, St. Louis), 405-416.

[33] Wolovich, W.A. (1974). *Linear Multivariable Systems*, Springer Verlag, New York.

Department of Electrical and Electronic Engineering, University of Melbourne, Parkville, Victoria 3052, Australia
E-mail: `m.kuijper@ee.mu.oz.au`

Progress in Systems and Control Theory, Vol. 25

An Algebraic Decoding Algorithm for Convolutional Codes

Joachim Rosenthal[1]

1 Introduction

The class of convolutional codes generalizes the class of linear block codes in a natural way. The construction of convolutional codes which have a large free distance and which come with an efficient decoding algorithm is a major task. Contrary to the situation of linear block codes there exists only very few algebraic construction of convolutional codes.

It is the purpose of this article to introduce a new iterative algebraic decoding algorithm which is capable of decoding convolutional codes which have a certain underlying algebraic structure. The algorithm exploits the algebraic structure of the convolutional code and it achieves its best performance if some naturally associated block codes can be efficiently decoded in an algebraic manner.

In order to achieve this goal we will work with a classical state space description of a so called systematic encoder. Using this description we will derive a general procedure which will allow one to extend known decoding algorithms for block codes (like e.g. the Berlekamp Massey algorithm) to convolutional codes.

In the coding literature there exist several decoding algorithms for convolutional codes. Maybe the most prominent one is the Viterbi decoding algorithm which applies the principle of dynamic programming to compute the transmitted message sequence. It was shown by Forney [6] that this algorithm computes the message sequence in a maximum likelihood fashion. The disadvantage of this algorithm is that it becomes computationally infeasible if the degree of the encoder is larger than 20. On the side of the Viterbi algorithm there are several sub-optimal algorithms and we would like to mention Massey's threshold decoding algorithm [9], the sequential decoding algorithm and the feedback decoding algorithm [7, 8, 12]. More recently there has been a significant interest in some iterative decoding algorithms in connection with the decoding of low density parity check codes and other codes defined on general graphs and we refer to [17, 20].

The iterative decoding algorithm which we will present in this paper seems to be different from above ideas. Indeed the algorithm iteratively computes the state vector x_t inside the trellis diagram (see [7, 8]) by making use of the algebraic structure of the convolutional code.

[1]Supported in part by NSF grant DMS-96-10389.

Once a state x_τ has been correctly computed we will show how to compute in an algebraic manner a new state vector $x_{\tau+\Theta}$, where Θ is a positive integer which depends on the underlying code. Once $x_{\tau+\Theta}$ is computed all code words between time τ and time $\tau+\Theta$ are generally computed through an algebraic decoding scheme.

Similarly to the known algebraic decoding algorithms for block codes it is required that the convolutional code has a certain algebraic structure. In this way the algorithm cannot be applied efficiently to arbitrary convolutional codes. On the other hand if the convolutional code is of Reed Solomon type (see [15]) or of BCH type (see [16, 21]) then the algorithm is capable of decoding convolutional codes in situations where the Viterbi decoding algorithm would not be feasible because of complexity considerations.

The paper is structured as follows: In the next section we summarize some basic notions for block codes and convolutional codes. Emphasis will be on the state space representation for a systematic encoder. In Section 3 we first provide exact conditions on the convolutional code and on the transmitted error pattern which guarantee that the iterative decoding algorithm as presented in Section 4 does compute the transmitted message. In Section 5 we address issues of complexity and we describe two variations where we expect the algorithm to perform very efficiently. In Section 6 we will show that the Berlekamp Massey algorithm or any of its recent improvements (see e.g. [3]) can be invoked to iteratively decode the Reed Solomon and BCH type convolutional codes as presented in [15, 16, 21].

2 Convolutional Codes and their State Space Description

In this section we will provide a short tutorial on block codes and convolutional codes. More details on our state space approach are given in [11, 13, 15, 16, 21]. Comprehensive textbooks on convolutional codes are [7, 8, 12].

Let $\mathbb{F} = \mathbb{F}_q$ be the Galois field of q elements. If

$$\varphi: \mathbb{F}^k \longrightarrow \mathbb{F}^n$$

is a monomorphism we say that $\mathcal{C} := \operatorname{im}(\varphi) \subset \mathbb{F}^n$ is a linear block code and φ is an encoder. Let G be an $n \times k$ matrix representing the linear map φ. The encoding process given by φ is then described by

$$m \longmapsto v = Gm.$$

One says G is a generator matrix for the code $\mathcal{C}$, $m \in \mathbb{F}^k$ is a message vector and $v \in \mathbb{F}^n$ is a code vector. If S is a $k \times k$ invertible matrix then G and

$\tilde{G} := GS$ generate the same code and we say that G and $\tilde{G}$ are equivalent encoders.

We say G is a *systematic encoder* if G has the particular form $\begin{pmatrix} Y \\ I_k \end{pmatrix}$, where I_k is the $k \times k$ identity matrix and Y is a matrix of size $(n-k) \times k$. A systematic encoder has the property that k message symbols $m \in \mathbb{F}^k$ will be transmitted (in an 'unencoded manner') together with $(n-k)$ *parity check symbols* $y = Ym \in \mathbb{F}^{n-k}$.

If the transmitted data has more than k symbols it will be necessary to break down the data into several message blocks. Let $m_0, \ldots, m_\gamma$ be $\gamma + 1$ blocks of messages to be transmitted. If one introduces the polynomial vectors

$$m(z) := \sum_{i=0}^{\gamma} m_i z^i \in \mathbb{F}^k[z] \quad \text{and} \quad v(z) := \sum_{i=0}^{\gamma} v_i z^i \in \mathbb{F}^n[z]$$

then the total encoding scheme

$$m_i \longmapsto v_i = G m_i, \; i = 0, \ldots, \gamma$$

can be compactly described through the module homomorphism

$$\hat{\varphi} : \; \mathbb{F}^k[z] \longrightarrow \mathbb{F}^n[z], \quad m(z) \longmapsto v(z) = Gm(z).$$

It was the idea of Elias [2] to replace in above encoding scheme the generator matrix G with a generator matrix $G(z)$ and to allow in this way general module homomorphisms as encoding schemes.

Using this point of view we define a convolutional code $\mathcal{C}$ as a $\mathbb{F}[z]$ sub-module of $\mathbb{F}^n[z]$. If $V(z)$ is a $k \times k$ unimodular matrix then the encoders $G(z)$ and $\tilde{G}(z) := G(z)V(z)$ define the same convolutional code and we will say that $G(z)$ and $\tilde{G}(z)$ are equivalent encoders. Without loss of generality we can therefore assume that $G(z)$ is in column proper form with column degree $\nu_1 \geq \cdots \geq \nu_k$.

The integer $\delta := \nu_1 + \cdots + \nu_k$ is an invariant of the code (module) $\mathcal{C} \subset \mathbb{F}^n[z]$. We call δ the degree (or complexity) of the code $\mathcal{C}$. Convolutional codes of degree $\delta = 0$ correspond in this way to linear block codes.

Remark 2.1 In the coding literature [5, 7, 12] a convolutional code is often defined as a linear subspace of $\mathcal{F}^n$, where $\mathcal{F} := \mathbb{F}((z))$ is the field of formal Laurent series. If one takes this approach the column degrees are no more invariants of the code and the degree δ is hence also not an invariant of the code but rather a property of the particular encoder. This is one reason why we consider the presented module theoretic approach as appealing. In the same time there seems to exist no practical necessity to have a framework for messages of infinite length.

Remark 2.2 A module theoretic approach to convolutional codes was introduced by Fornasini and Valcher [4, 18] in the context of two dimensional codes. If one works with multidimensional codes then it becomes very difficult if one works with the field of formal Laurent series $\mathbb{F}((z_1, \ldots, z_m))$. In [4, 18] Fornasini and Valcher define a code as a submodule of R^n, where $R = \mathbb{F}[z_1, z_2, z_1^{-1}, z_2^{-1}]$ is the ring of Laurent polynomials in the variables z_1, z_2. In this way codes become dual to complete linear behaviors defined on $\mathbb{Z} \times \mathbb{Z}$. Weiner [19] defines a convolutional code as a submodule of R^n, where R is the polynomial ring $R = \mathbb{F}[z_1, \ldots, z_m]$. In this framework codes are dual to linear complete behaviors defined on $\mathbb{N}^m$.

Since submodules of $\mathbb{F}^n[z]$, i.e. convolutional codes, are dual to linear complete behaviors they have natural state space descriptions. In the sequel we follow [15, 16] and explain this relation.

Partition the generator matrix $G(z)$ into $G(z) = \begin{pmatrix} Y(z) \\ U(z) \end{pmatrix}$, where $U(z)$ is of size $k \times k$ and $Y(z)$ is of size $(n-k) \times k$. For simplicity assume that $\deg \det U(z) = \delta$, the degree of the encoder $G(z)$. Let $X(z)$ be a basis matrix of size ν (compare with [15, 16]). Then one has the result:

Lemma 2.3 *There exist matrices $A \in \mathbb{F}^{\delta \times \delta}$, $B \in \mathbb{F}^{\delta \times k}$, $C \in \mathbb{F}^{(n-k) \times \delta}$, and $D \in \mathbb{F}^{(n-k) \times k}$ such that*

$$\ker \begin{pmatrix} zI - A & 0 & -B \\ -C & I & -D \end{pmatrix} = \operatorname{im} \begin{pmatrix} X(z) \\ Y(z) \\ U(z) \end{pmatrix}. \tag{2.1}$$

In particular $v(z) = \begin{pmatrix} y(z) \\ u(z) \end{pmatrix} \in \mathbb{F}^n[z]$ is a code word if and only if there is a polynomial vector $x(z) \in \mathbb{F}^\delta[z]$ with

$$\begin{pmatrix} zI - A & 0 & -B \\ -C & I & -D \end{pmatrix} \begin{pmatrix} x(z) \\ y(z) \\ u(z) \end{pmatrix} = 0. \tag{2.2}$$

Remark 2.4 One immediately verifies that the matrices A, B, C, D form a realization for the transfer function $Y(z)U(z)^{-1}$, i.e. one has the relation $Y(z)U(z)^{-1} = C(zI - A)^{-1}B + D$. If the high order coefficient matrix of $U(z)$ is the identity matrix then it is possible to compute the matrices A, B, C, D 'by inspection' [14].

It is possible to give (2.2) a dynamical interpretation. For this let

$$x(z) = x_0 z^\gamma + x_1 z^{\gamma-1} + \ldots + x_\gamma;\ x_t \in \mathbb{F}^\delta, t = 0, \ldots, \gamma,$$

$$u(z) = u_0 z^\gamma + u_1 z^{\gamma-1} + \ldots + u_\gamma;\ u_t \in \mathbb{F}^k, t = 0, \ldots, \gamma,$$

and let

$$y(z) = y_0 z^\gamma + y_1 z^{\gamma-1} + \ldots + y_\gamma;\ y_t \in \mathbb{F}^{n-k}, t = 0, \ldots, \gamma.$$

Then one verifies that (2.2) is equivalent with:

$$\begin{aligned} x_{t+1} &= Ax_t + Bu_t, \\ y_t &= Cx_t + Du_t, \\ v_t &= \begin{pmatrix} y_t \\ u_t \end{pmatrix}, \ x_0 = 0,\ x_{\gamma+1} = 0. \end{aligned} \tag{2.3}$$

In these equations the sequence of vectors u_t represents the *message vectors*, the sequence of y_t represents the *parity vectors* and the sequence of v_t represents the set of *code vectors.* The equations in (2.3) define the state space realization of the systematic convolutional encoder $\begin{pmatrix} Y(z)U(z)^{-1} \\ I_k \end{pmatrix}$.

Remark 2.5 In the coding literature [5, 10, 11] one often finds a state space description, where the message words m_i are the inputs and the code words v_i are the outputs. Such an A, B, C, D representation is related to the generator matrix $G(z)$ via the relation $G(z^{-1}) = C(zI - A)^{-1}B + D$. The state space realization (2.3) is different.

Assume that the encoder is at state x_τ. Using (2.3) we immediately derive an algebraic dependence between the message vectors u_t and the parity check vectors y_t (compare with [15, 16, 21]):

Proposition 2.6 (Local Description of Trajectories) *Let $\tau, \gamma \in \mathbb{Z}_+$ be positive integers with $\tau < \gamma$. Assume that the encoder is at state x_τ at time $t = \tau$. Then any code sequence $\left\{ \binom{y_t}{u_t} \right\}_{t \geq 0}$ governed by the dynamical system* (2.3) *must satisfy:*

$$\begin{pmatrix} y_\tau \\ y_{\tau+1} \\ \vdots \\ \vdots \\ y_\gamma \end{pmatrix} = \begin{pmatrix} C \\ CA \\ \vdots \\ \vdots \\ CA^{\gamma-\tau} \end{pmatrix} x_\tau + \begin{pmatrix} D & 0 & \cdots & & 0 \\ CB & D & \ddots & & \vdots \\ CAB & CB & \ddots & \ddots & \\ \vdots & & \ddots & \ddots & 0 \\ CA^{\gamma-\tau-1}B & CA^{\gamma-\tau-2}B & \cdots & CB & D \end{pmatrix} \begin{pmatrix} u_\tau \\ u_{\tau+1} \\ \vdots \\ \vdots \\ u_\gamma \end{pmatrix}.$$

Moreover the evolution of the state vector x_t is given over time as:

$$x_t = A^{t-\tau}x_\tau + \left(A^{t-\tau-1}B \ \dots \ B\right)\begin{pmatrix} u_\tau \\ \vdots \\ u_{t-1} \end{pmatrix}; \ t = \tau+1, \tau+2, \dots, \gamma+1. \tag{2.4}$$

In this paper we will mainly use the code description of Proposition 2.6 to arrive at an iterative decoding algorithm of the convolutional code. As it turns out it is possible to construct the matrices A, B, C in a way which will allow one to use iteratively known decoding algorithms for block codes to arrive at the decoding of the convolutional code.

3 Basic Assumptions and Main Results

The decoding task is as follows: Assume a sequence of code words $\{v_t\}_{t\geq 0} = \left\{\binom{y_t}{u_t}\right\}_{t\geq 0}$ was sent and the sequence

$$\{\hat{v}_t\}_{t\geq 0} = \left\{\begin{pmatrix} \hat{y}_t \\ \hat{u}_t \end{pmatrix}\right\}_{t\geq 0}$$

has been received. The decoding problem then asks for the minimization of the error

$$\text{error} := \min_{\{v_t\}\in C} \sum_{t=0}^{\infty} \text{dist}\,(v_t, \hat{v}_t) = \min \left(\sum_{t=0}^{\infty} (\text{dist}\,(u_t, \hat{u}_t) + \text{dist}\,(y_t, \hat{y}_t)) \right), \tag{3.1}$$

where 'dist' does denote the usual Hamming distance between two vectors, i.e. dist $(v, \hat{v})$ is equal to the number of coordinates where v and $\hat{v}$ differ. If no transmission error did occur then $\{\hat{v}_t\}_{t\geq 0}$ is a valid trajectory and the error value in (3.1) is zero. If $\{\hat{v}_t\}_{t\geq 0}$ is not a valid trajectory then the decoding task asks for the computation of the 'nearest trajectory' with respect to the Hamming metric. The decoding problem has therefore the characteristic of a discrete tracking problem. The difficulty lies in the fact that the Hamming metric is not induced by a positive quadratic form and it is therefore not possible to apply standard techniques from LQ theory immediately.

Remark 3.1 If the transmission is done over the 'Gaussian channel' then the natural metric on $\mathbb{F}^n$ is not the Hamming metric but rather a metric which is induced by the Euclidean metric through some modulation scheme. The received signals are in this situation some points in Euclidean space and the decoding task asks for the minimization of the error (3.1) which can be any positive real number. Even in this situation standard methods used in the study of the linear quadratic regulator problem cannot be applied. The basic difficulty comes this time from the fact that the set of code trajectories is $\mathbb{F}$ linear but not $\mathbb{R}$ linear. In either case it seems that decoding of general convolutional codes is in terms of computational complexity a 'hard' problem.

In the sequel we will work with the Hamming metric and we will approach the decoding task by combining ideas used in the decoding of linear block codes and systems theoretic descriptions such as the one given in Proposition 2.6. Let

$$\left\{\begin{pmatrix} f_t \\ e_t \end{pmatrix}\right\}_{t\geq 0} := \left\{\begin{pmatrix} \hat{y}_t - y_t \\ \hat{u}_t - u_t \end{pmatrix}\right\}_{t\geq 0} \tag{3.2}$$

be the sequence of errors.

Assumption 3.2 Consider a convolutional code $\mathcal{C}$ described by the matrices A, B, C, D having sizes $\delta \times \delta$, $\delta \times k$, $(n-k) \times \delta$ and $(n-k) \times k$ respectively and let $T > \Theta$ be integers satisfying:

1. A is invertible, the matrix

$$\begin{pmatrix} B & AB & \dots & A^{T-1}B \end{pmatrix} \tag{3.3}$$

has full row rank δ and its rows form the parity check matrix of a block code of distance at least d_1.

2. The matrix

$$\begin{pmatrix} C \\ CA \\ \vdots \\ CA^{\Theta-1} \end{pmatrix} \tag{3.4}$$

has full column rank δ and its columns define the generator matrix of a block code of distance d_2.

Assumption 3.2 implies that (A, B) forms a controllable pair and (A, C) forms an observable pair, in particular (2.3) forms a minimal state space representation of a non-catastrophic encoder (see [16] for details). The integer Θ appearing in (3.4) is necessarily larger than the observability index of the matrix pair (A, C). The observability index describes the maximal number of consecutive zero code words $v_t = \binom{y_t}{u_t}$ starting from a nonzero state. In the coding literature [5] this number appears as the solution for the zero run problem.

Conditions (3.3) and (3.4) are stronger than the simple controllability and observability requirement and they imply that valid code trajectories have necessarily certain distance properties. The following lemma makes this precise.

Lemma 3.3 *Assume the matrices A, B, C and the integers T, Θ satisfy the conditions of Assumption 3.2. Assume that*

$$\left\{\begin{pmatrix} y_t \\ u_t \end{pmatrix}\right\}_{t\geq 0} \text{ and } \left\{\begin{pmatrix} \tilde{y}_t \\ \tilde{u}_t \end{pmatrix}\right\}_{t\geq 0}$$

are two set of codewords both satisfying (2.3). Let $\{x_t\}_{t\geq 0}$ and $\{\tilde{x}_t\}_{t\geq 0}$ be the corresponding set of state vectors. If there is a $\tau \geq 0$ with

$$x_\tau = \tilde{x}_\tau \text{ and } x_{\tau+1} \neq \tilde{x}_{\tau+1}$$

then for any γ satisfying $\tau + T > \gamma \geq \tau$ one has that

$$\sum_{t=\tau}^{\gamma} (\text{dist}\,(u_t, \tilde{u}_t) + \text{dist}\,(y_t, \tilde{y}_t)) \geq \min\left(d_1, \left\lfloor \frac{\gamma - \tau}{\Theta} \right\rfloor + 1\right). \tag{3.5}$$

Proof: The proof is established by induction over the integer $\eta := \lfloor \frac{\gamma-\tau}{\Theta} \rfloor$. Since $x_{\tau+1} \neq \tilde{x}_{\tau+1}$ it follows that $u_\tau \neq \tilde{u}_\tau$ and the result is therefore true for $\eta = 0$. Assume now that the result has already been proved for $\eta = k$ and let $\gamma = k\Theta + \tau$. By induction hypothesis we can assume that $\sum_{t=\tau}^{\gamma} (\text{dist}\,(u_t, \tilde{u}_t) + \text{dist}\,(y_t, \tilde{y}_t)) \geq \min(d_1, k+1)$. If $x_{\gamma+1} = \tilde{x}_{\gamma+1}$ then necessarily one has

$$(B\ AB\ \ldots\ A^{\gamma-\tau}B) \begin{pmatrix} u_\gamma - \tilde{u}_\gamma \\ \vdots \\ u_\tau - \tilde{u}_\tau \end{pmatrix} = 0. \tag{3.6}$$

By Assumption 3.2 it follows that $\sum_{t=\tau}^{\gamma} \text{dist}\,(u_t, \tilde{u}_t) \geq d_1$. In this situation the proof would be complete. If $x_{\gamma+1} \neq \tilde{x}_{\gamma+1}$ then either

$$\sum_{t=\gamma+1}^{\gamma+\Theta} \text{dist}\,(u_t, \tilde{u}_t) \geq 1$$

(in this case the induction step would be complete as well) or alternatively $u_t = \tilde{u}_t$ for $t = \gamma+1, \ldots, \gamma+\Theta$. In the latter situation it follows from Proposition 2.6 and from the second condition of Assumption 3.2 that $\sum_{t=\gamma+1}^{\gamma+\Theta} \text{dist}\,(y_t, \tilde{y}_t) \geq d_2 \geq 1$. In either case we did show the claim for $\gamma + \Theta = (k+1)\Theta + \tau$, i.e. for $\eta = k+1$. □

The main result of this section is formulated in the following Theorem. The result shows that under certain assumptions on the weight distribution of the errors $\{e_t, f_t\}$ it is possible to decode the received message uniquely. The proof of this theorem will be established in the next section through an explicit iterative decoding algorithm.

Theorem 3.4 *Let A, B, C, D be matrices satisfying the conditions of Assumption 3.2. Consider a received message*

$$\left\{\begin{pmatrix} \hat{y}_t \\ \hat{u}_t \end{pmatrix}\right\}_{t\geq 0} = \left\{\begin{pmatrix} y_t + f_t \\ u_t + e_t \end{pmatrix}\right\}_{t\geq 0}. \tag{3.7}$$

Assume that for any $\tau \geq 0$ the error sequence

$$\begin{pmatrix} f_\tau \\ e_\tau \end{pmatrix}, \ldots, \begin{pmatrix} f_{\tau+T-1} \\ e_{\tau+T-1} \end{pmatrix}$$

has weight at most

$$\lambda := \min\left(\left\lfloor \frac{d_1 - 1}{2} \right\rfloor, \left\lfloor \frac{T}{2\Theta} \right\rfloor\right).$$

In this situation it is possible to uniquely compute the transmitted sequence $\left\{\binom{y_t}{u_t}\right\}_{t\geq 0}$.

Remark 3.5 The major task in the decoding procedure will be the decoding of the block codes defined in (3.3) and (3.4). If the matrices A, B, C are chosen in a way which allows one to decode the block codes defined in (3.3) and (3.4) through an efficient algebraic decoding algorithm then one is led to an efficient algebraic decoding algorithm of the associated convolutional code.

Remark 3.6 If $\lfloor \frac{d_1-1}{2} \rfloor = \lambda$ it follows from [16, Theorem 3.1] that the free distance of the convolutional code defined by the matrices A, B, C, D is at least d_1. Theorem 3.4 essentially states that decoding is possible if no more than 'half the free distance' errors occur in any time interval of length T.

4 The Decoding Algorithm

The presented decoding algorithm is an iterative algorithm. We hence will assume that the received message has already been correctly decoded up to time τ. In other words we will assume that the code words

$$\begin{pmatrix} y_0 \\ u_0 \end{pmatrix}, \begin{pmatrix} y_1 \\ u_1 \end{pmatrix}, \ldots, \begin{pmatrix} y_{\tau-1} \\ u_{\tau-1} \end{pmatrix}$$

have been computed correctly and that the state x_τ is known. Under these conditions we will show how to decode at least another Θ code vectors.

We start with some preliminary remarks: Assume for a moment that the message vectors

$$\hat{u}_{\tau+T-\Theta+1}, \hat{u}_{\tau+T-\Theta+2}, \ldots, \hat{u}_{\tau+T} \tag{4.1}$$

have been correctly received. Assume also that the error sequence

$$f_{\tau+T-\Theta+1}, f_{\tau+T-\Theta+2}, \ldots, f_{\tau+T} \tag{4.2}$$

has weight at most $\lfloor \frac{d_2-1}{2} \rfloor$ where d_2 is the distance of the block code introduced in Assumption 3.2. From Proposition 2.6 it follows that

$$\begin{pmatrix} y_{\tau+T-\Theta+1} \\ y_{\tau+T-\Theta+2} \\ \vdots \\ \vdots \\ y_{\tau+T} \end{pmatrix} - \begin{pmatrix} D & 0 & \cdots & & 0 \\ CB & D & \ddots & & \vdots \\ CAB & CB & \ddots & \ddots & \\ \vdots & & \ddots & \ddots & 0 \\ CA^{\Theta-2}B & CA^{\Theta-3}B & \cdots & CB & D \end{pmatrix} \begin{pmatrix} u_{\tau+T-\Theta+1} \\ u_{\tau+T-\Theta+2} \\ \vdots \\ \vdots \\ u_{\tau+T} \end{pmatrix}$$

$$= \begin{pmatrix} C \\ CA \\ \vdots \\ \vdots \\ CA^{\Theta-1} \end{pmatrix} x_{\tau+T-\Theta+1}. \tag{4.3}$$

In particular the left hand side of the previous equation is in the column space of the block code generated by the columns of

$$\begin{pmatrix} C \\ CA \\ \vdots \\ CA^{\Theta-1} \end{pmatrix}. \tag{4.4}$$

Since this last code has distance d_2 it is possible to both compute the errors appearing in (4.2) and the state vector $x_{\tau+T-\Theta+1}$. Since the state vector $x_{\tau+T-\Theta+1}$ is also equal to

$$x_{\tau+T-\Theta+1} = A^{T-\Theta+1}x_\tau + \left(A^{T-\Theta}B \ \ldots \ B\right) \begin{pmatrix} u_\tau \\ \vdots \\ u_{\tau+T-\Theta} \end{pmatrix} \tag{4.5}$$

it is possible to compute the error sequence $e_\tau, \ldots, e_{\tau+T-\Theta}$ from the syndrome vector

$$\left(A^{T-\Theta}B \ \ldots \ B\right) \begin{pmatrix} \hat{u}_\tau \\ \vdots \\ \hat{u}_{\tau+T-\Theta} \end{pmatrix} - x_{\tau+T-\Theta+1} - A^{T-\Theta+1}x_\tau. \tag{4.6}$$

Once the error sequence $e_\tau, \ldots, e_{\tau+T-\Theta}$ has been computed we can compute the sequence of states $x_{\tau+1}, \ldots, x_{\tau+T-\Theta+1}$ and the sequence of parity vectors $y_\tau, \ldots, y_{\tau+T-\Theta}$ using the defining equation (2.3). Because of the assumptions the error sequence

$$\begin{pmatrix} f_\tau = y_\tau - \tilde{y}_\tau \\ e_\tau = u_\tau - \tilde{u}_\tau \end{pmatrix}, \ldots, \begin{pmatrix} f_{\tau+T-\Theta} = y_{\tau+T-\Theta} - \tilde{y}_{\tau+T-\Theta} \\ e_{\tau+T-\Theta} = u_{\tau+T-\Theta} - \tilde{u}_{\tau+T-\Theta} \end{pmatrix} \tag{4.7}$$

must have weight at most λ.

After these preliminary remarks we explain the algorithm.

In a first step we attempt to compute the state vector $x_{\tau+T-\Theta+1}$ from identity (4.3) and the error sequence (4.7) from identities (4.6) and (2.3). Several things might happen then.

A) It is possible that we cannot compute the state vector $x_{\tau+T-\Theta+1}$ from identity (4.3).

B) It is possible that we did compute a state vector $x_{\tau+T-\Theta+1}$ and the resulting error sequence has weight

$$\sum_{t=\tau}^{\tau+T-\Theta} (\mathrm{wt}(f_t) + \mathrm{wt}(e_t)) > \lambda. \tag{4.8}$$

C) It is possible that we compute a state vector $x_{\tau+T-\Theta+1}$ and the weight of the error sequence appearing in (4.8) is less than or equal to λ.

In the sequel we will show that after some possible iterations we will always end up with the situation C).

In situations A) and B) we can conclude that either sequence (4.1) was wrong or the weight of the sequence appearing in (4.2) is larger than $\lfloor \frac{d_2-1}{2} \rfloor$. In both these cases we will attempt to compute the state vector $x_{\tau+T-2\Theta+1}$ using the new sequence of message vectors

$$\hat{u}_{\tau+T-2\Theta+1}, \hat{u}_{\tau+T-2\Theta+2}, \ldots, \hat{u}_{\tau+T-\Theta} \tag{4.9}$$

and parity check vectors

$$\hat{y}_{\tau+T-2\Theta+1}, \hat{y}_{\tau+T-2\Theta+2}, \ldots, \hat{y}_{\tau+T-\Theta}. \tag{4.10}$$

Since there were mistakes in the last Θ code words, i.e. the weight of the sequences appearing in (4.1) and (4.2) were nonzero we can assume that there were at most $\lambda - 1$ errors among the received sequence

$$\begin{pmatrix} \hat{y}_\tau \\ \hat{u}_\tau \end{pmatrix}, \ldots, \begin{pmatrix} \hat{y}_{\tau+T-\Theta} \\ \hat{u}_{\tau+T-\Theta} \end{pmatrix}.$$

Again if we cannot compute either $x_{\tau+T-2\Theta+1}$ or if the weight

$$\sum_{t=\tau}^{\tau+T-2\Theta} (\mathrm{wt}(f_t) + \mathrm{wt}(e_t)) > \lambda - 1$$

we conclude that either sequence (4.9) was wrong or there were more than $\lfloor \frac{d_2-1}{2} \rfloor$ mistakes in the sequence appearing in (4.10).

Proceeding in this way iteratively we will find after h iterations that the state vector $x_{\tau+T-h\Theta+1}$ can be computed from the data

$$\hat{u}_{\tau+T-h\Theta+1}, \hat{u}_{\tau+T-h\Theta+2}, \ldots, \hat{u}_{\tau+T-(h-1)\Theta}$$

and

$$\hat{y}_{\tau+T-h\Theta+1}, \hat{y}_{\tau+T-h\Theta+2}, \ldots, \hat{y}_{\tau+T-(h-1)\Theta}$$

and in addition we have that the weight

$$\sum_{t=\tau}^{\tau+T-h\Theta} (\mathrm{wt}(f_t) + \mathrm{wt}(e_t)) \leq \lambda - h + 1.$$

In other words we did arrive at situation C) after h iterations. In general it would be wrong to assume that the state $x_{\tau+T-h\Theta}$ is a correct state. However Lemma 3.3 and the assumption on the error pattern as formulated in Theorem 3.4 will guarantee that the state $x_{\tau+\Theta}$ is correctly computed. Indeed if the computed state $x_{\tau+\Theta}$ would be different from the true state $x_{\tau+\Theta}$ then the computed code sequence

$$\begin{pmatrix} y_\tau \\ u_\tau \end{pmatrix}, \ldots, \begin{pmatrix} y_{\tau+T} \\ u_{\tau+T} \end{pmatrix}$$

would be in distance more than $\min\left(d_1, \lfloor \frac{T}{\Theta} \rfloor + 1\right)$ apart from the true code sequence. This is not possible since we assumed that at most λ errors did occur. The computed state $x_{\tau+\Theta}$ has therefore to be correct.

Under the given assumptions we also conclude that

$$\begin{pmatrix} y_\tau \\ u_\tau \end{pmatrix}, \ldots, \begin{pmatrix} y_{\tau+\Theta-1} \\ u_{\tau+\Theta-1} \end{pmatrix}.$$

has been decoded correctly. In this way Θ additional time units were decoded.

The algorithm proceeds now again from the beginning by replacing the initial state x_τ with the state $x_{\tau+\Theta}$.

Remark 4.1 Crucial for the algorithm was the computation of a new state vector $x_{\tau+\Theta}$. Due to the fact that we have been working with a systematic encoder one has a certain asymmetry between the assumption on the error patterns among the input sequence $\{u_t\}$ and the output sequence $\{y_t\}$. In [1] a new method for computing the state vector $x_{\tau+\Theta}$ was announced for codes having rate $1/n$. This method seems to have advantages over the one presented here and it will be addressed in upcoming research.

In the next section we will show that under certain probabilistic assumptions one can speed up the algorithm considerably.

5 Complexity Considerations and Variations of the Algorithm

It is clear from the described algorithm that the block codes described in Assumption 3.2 have to be decoded over and over again. It is hence desirable to construct (A, B, C) matrices where these codes have both good distance properties and come with efficient decoding algorithms.

Even in these cases up to $\lfloor \frac{T}{\Theta} \rfloor$ vectors have to be decoded by the two block codes described in Assumption 3.2 to decode in the worst case just Θ time units. The algorithm takes into consideration many unlikely events and it 'cautiously' assumes that $x_{\tau+\Theta}$ is correct despite the fact that there is already a preliminary estimate for $x_{\tau+T-h\Theta-1}$.

We see in principle two ways how to speed up the algorithm considerably:

Variation 1: One way which will guarantee that the algorithm performs much quicker is to assume that the number of errors λ which are permitted in every time interval of length T is less than the number specified in Theorem 3.4. For this assume e.g. that for any $\tau \geq 0$ the error sequence

$$\begin{pmatrix} f_\tau \\ e_\tau \end{pmatrix}, \ldots, \begin{pmatrix} f_{\tau+T-1} \\ e_{\tau+T-1} \end{pmatrix}$$

has weight at most

$$\tilde{\lambda} := \min \left(\left\lfloor \frac{d_1 - 1}{2} \right\rfloor, \left\lfloor \frac{T}{4\Theta} \right\rfloor \right).$$

In contrast with Theorem 3.4 we assume that at most half the errors do occur over any time period T if $\lfloor \frac{T}{2\Theta} \rfloor < \lfloor \frac{d_1-1}{2} \rfloor$. In this situation one verifies with the help of Lemma 3.3 that not only the state vector $x_{\tau+\Theta}$ is correct but that even the state vector $x_{\tau+\frac{T}{2}}$ has been correctly computed. In this variation up to $\lfloor \frac{T}{2\Theta} \rfloor$ vectors have to be decoded by the two block codes described in Assumption 3.2 in order to decode $\frac{T}{2}$ time units.

Variation 2: Assume that the distance d_2 of the block code

$$\begin{pmatrix} C \\ CA \\ \vdots \\ CA^{\Theta-1} \end{pmatrix}$$

is relatively large, i.e. $d_2 >> 1$. Under certain probabilistic assumptions it is then possible to proceed directly with the state vector $x_{\tau+T-h\Theta-1}$ after h iterations of the algorithm. This works particularly well if one rejects the computation of the state vector $x_{\tau+T-j\Theta-1}$ in the jth iteration of the algorithm as soon as the weight of the error sequence

$$f_{\tau+T-(j+1)\Theta+1}, f_{\tau+T-j\Theta+2}, \ldots, f_{\tau+T-j\Theta} \tag{5.1}$$

is more than a certain fraction of d_2. By doing this it becomes highly unlikely that a wrong state vector $x_{\tau+T-h\Theta-1}$ was computed. Indeed the computation of a wrong state vector $x_{\tau+T-h\Theta-1}$ where the weight of the parity check sequence appearing in (5.1) was low can only happen if in the local description of the code as provided in Proposition 2.6 the errors among the message vectors and parity check vectors do cancel in a very particular way. In this way one can assume with high probability that after h iterations we actually will find a correct state vector $x_{\tau+T-h\Theta-1}$. The probability will depend on the fraction of d_2 which one uses for rejecting the error sequence appearing in (5.1).

During the decoding process of

$$\begin{pmatrix} \hat{y}_\tau \\ \hat{u}_\tau \end{pmatrix}, \begin{pmatrix} \hat{y}_{\tau+1} \\ \hat{u}_{\tau+1} \end{pmatrix}, \ldots, \begin{pmatrix} \hat{y}_{\tau+T-(h+1)\Theta-2} \\ \hat{u}_{\tau+T-(h+1)\Theta-2} \end{pmatrix}$$

one has one more verification that $x_{\tau+T-h\Theta-1}$ was actually correctly computed.

We conclude the section with some complexity estimates for the second variation of the algorithm. For this assume that in average the error sequence

$$\begin{pmatrix} f_\tau \\ e_\tau \end{pmatrix}, \ldots, \begin{pmatrix} f_{\tau+\Theta-1} \\ e_{\tau+\Theta-1} \end{pmatrix}$$

has weight one. In other words we do assume that over Θ time units the expected number of errors is one. In average there are therefore at most $h = 2$ iterations needed until it is possible to compute the new state vector $x_{\tau+T-h\Theta-1}$. But this means that for the decoding of T message vectors we need to decode in average up to two block codes of rate $\frac{\delta}{(n-k)\Theta}$ having the form (3.4) and in average one block code of rate $\frac{\delta}{kT}$ having the form (3.3).

The described decoding algorithm seems to be particularly well suited in situations where once in a while there is a burst error and where in the remaining time the transmission is error free.

In the next section we explain a situation where both the block codes appearing in (3.3) and in (3.4) are Reed Solomon type block codes.

6 Decoding of Reed Solomon and BCH type Convolutional Codes

In order that Assumption 3.2 is satisfied it will be necessary that the parity check matrix appearing in (3.3) has some good distance properties. Moreover the presented algorithm works best if the associated block code can be decoded efficiently.

In [15, 16, 21] convolutional codes were presented where the block code defined by (3.3) is a BCH code. In the sequel we will illustrate the decoding algorithm using these codes. For the sake of presentation we will illustrate the algorithm only for Reed Solomon convolutional codes as presented in [15] and we leave the extension to the general BCH situation to the reader.

Let α be a primitive of the field $\mathbb{F}_q$ and assume that $n > k$ are positive integers with $k \geq n-k$. Consider the matrices

$$A := \begin{pmatrix} \alpha^k & 0 & \cdots & 0 \\ 0 & \alpha^{2k} & \ddots & \vdots \\ \vdots & \ddots & \ddots & 0 \\ 0 & \cdots & 0 & \alpha^{\delta k} \end{pmatrix},$$

$$B := \begin{pmatrix} 1 & \alpha & \alpha^2 & \cdots & \alpha^{k-1} \\ 1 & \alpha^2 & \alpha^4 & \cdots & \alpha^{2(k-1)} \\ \vdots & \vdots & \vdots & & \vdots \\ 1 & \alpha^\delta & \alpha^{2\delta} & \cdots & \alpha^{\delta(k-1)} \end{pmatrix},$$

$$C := \begin{pmatrix} 1 & 1 & \cdots & 1 \\ \alpha & \alpha^2 & \cdots & \alpha^\delta \\ \alpha^2 & \alpha^4 & \cdots & \alpha^{2\delta} \\ \vdots & \vdots & & \vdots \\ \alpha^{n-k-1} & \alpha^{2(n-k-1)} & \cdots & \alpha^{\delta(n-k-1)} \end{pmatrix},$$

$$D := \begin{pmatrix} 1 & 1 & \cdots & 1 \\ \alpha & \alpha^2 & \cdots & \alpha^k \\ \vdots & \vdots & & \vdots \\ \alpha^{(n-k-1)} & \alpha^{2(n-k-1)} & \cdots & \alpha^{k(n-k-1)} \end{pmatrix}.$$

It has been shown in [15]:

Theorem 6.1 *Assume $|\mathbb{F}_q| = q > \delta k \left\lceil \frac{\delta}{n-k} \right\rceil$. Then the convolutional code $\mathcal{C}$ defined by the matrices A, B, C, D represents an observable, rate k/n convolutional code with degree δ and free distance*

$$d_f(\mathcal{C}) \geq \delta + 1. \tag{6.1}$$

Since the free distance of this code is at least $\delta+1$ it should be possible to decode up to $\lfloor \frac{\delta}{2} \rfloor$ errors. Actually we will show (compare with Remark 3.6) that the decoding algorithm presented in Section 4 is capable of decoding the received message if at most $\lfloor \frac{\delta}{2} \rfloor$ errors do occur in any time interval of length T where T is defined as

$$T := \delta \left\lceil \frac{\delta}{n-k} \right\rceil . \tag{6.2}$$

In order to apply Theorem 3.4 we define

$$\Theta := \left\lceil \frac{\delta}{n-k} \right\rceil . \tag{6.3}$$

Because of the assumed number of field elements we have $q > kT$. The block code described in (3.3) defines therefore a Reed Solomon code of distance $\delta + 1$. This code is in particular a maximum distance separable (MDS) code.

The block code appearing in (3.4) describes a MDS block code as well, although the distance is small since we did choose the smallest possible value for Θ. Assumption 3.2 is therefore in place with $d_1 = \delta + 1$ and $d_2 \geq 1$. Finally note that the number λ appearing in Theorem 3.4 is equal to

$$\lambda = \left\lfloor \frac{d_1 - 1}{2} \right\rfloor = \left\lfloor \frac{T}{2\Theta} \right\rfloor = \left\lfloor \frac{\delta}{2} \right\rfloor .$$

According to Theorem 3.4 it is possible to decode a message word if at most λ mistakes did occur over any time interval of length T. The decoding algorithm of Section 4 does therefore fully apply.

In order to illustrate the second variation of the algorithm as presented in Section 5 let

$$\tilde{T} := 2T = 2\delta \left\lceil \frac{\delta}{n-k} \right\rceil \quad \text{and} \quad \tilde{\Theta} := 2\Theta = 2 \left\lceil \frac{\delta}{n-k} \right\rceil . \tag{6.4}$$

Assume that the field size $q > k\tilde{T} = 2kT$. With these choices the block code described in (3.4) has distance at least δ. If in the iteration of the decoding algorithm we require that the weight of the computed error sequence described in (5.1) is e.g. at most $\frac{1}{10}\delta$ then the likelihood that an accepted state vector $x_{\tau+T-h\Theta-1}$ is actually a correct state is very high. (The balls centered around the code words and having radius $\frac{1}{10}\delta$ are a small fraction inside the total configuration space).

Conclusions

We presented an iterative decoding algorithm for convolutional codes whose performance mainly depends on the availability of good algorithms to decode the block codes appearing in (3.3) and (3.4). If these block codes are of Reed Solomon type (as described in [15]) or of BCH type (as described in [16, 21]) then the major decoding task can be accomplished by iteratively applying e.g. the Berlekamp Massey algorithm.

Acknowledgments.
I am grateful to Brian Allen, Dan Costello, Dave Forney, Heide Glüsing-Lüerssen, Hans Schumacher, Roxana Smarandache, Paul Weiner and Eric York for helpful discussions in connection with the presented research.

References

[1] B. M. ALLEN AND J. ROSENTHAL (1998). Parity-Check decoding of convolutional codes whose systems parameters have desirable algebraic properties, to appear in *Proceedings of the 1998 IEEE International Symposium on Information Theory*, 307, Boston, MA.

[2] P. ELIAS (1955). Coding for Noisy Channels. *IRE Conv. Rec.* 4, 37–46.

[3] P. FITZPATRICK (1995). On the Key Equation. *IEEE Trans. Inform. Theory* IT-41, No. 5, 1290–1302.

[4] E. FORNASINI AND M.E. VALCHER (1994). Algebraic Aspects of 2D Convolutional Codes. *IEEE Trans. Inform. Theory* IT-40, No. 4, 1068–1082.

[5] G. D. FORNEY (1973). Structural Analysis of Convolutional Codes via Dual Codes. *IEEE Trans. Inform. Theory* IT-19, No. 5, 512–518.

[6] G. D. FORNEY (1974). Convolutional Codes II: Maximum Likelihood Decoding. *Inform. Control* 25, 222–266.

[7] R. JOHANNESSON AND K. SH. ZIGANGIROV (1999). *Fundamentals of Convolutional Coding.* IEEE Press, New York.

[8] S. LIN AND D. COSTELLO (1983). *Error Control Coding: Fundamentals and Applications.* Prentice-Hall, Englewood Cliffs, NJ.

[9] J. L. MASSEY (1963). *Threshold decoding.* MIT Press, Cambridge, Massachusetts.

[10] J. L. MASSEY AND M. K. SAIN (1967). Codes, Automata, and Continuous Systems: Explicit Interconnections. *IEEE Trans. Automat. Contr.* AC-12, No. 6, 644–650.

[11] R.J. MCELIECE. The Algebraic Theory of Convolutional Codes. In *Handbook of Coding Theory*, R. Brualdi, W.C. Huffman and V. Pless (eds.). Elsevier Science Publishers, Amsterdam, The Netherlands, 1998, to appear.

[12] PH. PIRET (1988). *Convolutional Codes, an Algebraic Approach.* MIT Press, Cambridge, MA.

[13] J. ROSENTHAL (1997). Some Interesting Problems in Systems Theory which are of Fundamental Importance in Coding Theory, in *Proc. of the 36th IEEE Conference on Decision and Control*, 4574–4579, San Diego, California.

[14] J. ROSENTHAL AND J. M. SCHUMACHER (1997). Realization by Inspection. *IEEE Trans. Automat. Contr.* AC-42, No. 9, 1257–1263.

[15] J. ROSENTHAL, J. M. SCHUMACHER AND E.V. YORK (1996). On Behaviors and Convolutional Codes. *IEEE Trans. Inform. Theory* 42, No. 6, 1881–1891.

[16] J. ROSENTHAL AND E.V. YORK (Oct. 1997). BCH Convolutional Codes. , Tech. rep., University of Notre Dame, Dept. of Mathematics, Preprint # 271. Available at http://www.nd.edu/~rosen/preprints.html.

[17] R. M. TANNER (1983). A recursive approach to low complexity codes. *IEEE Trans. Inform. Theory* 27, No. 5, 533–547.

[18] M.E. VALCHER AND E. FORNASINI (1994). On 2D Finite Support Convolutional Codes: an Algebraic Approach. *Multidim. Sys. and Sign. Proc.* 5, 231–243.

[19] P. WEINER (1998). *Multidimensional Convolutional Codes.* Ph.D. thesis, University of Notre Dame, Available at http://www.nd.edu/~rosen/preprints.html.

[20] N. WIBERG, H.A. LOELIGER AND R. KOETTER (1995). Codes and Iterative Decoding on General Graphs. *European Trans. on Telecommunications* 6, No. 5, 513–525.

[21] E.V. YORK (1997). *Algebraic Description and Construction of Error Correcting Codes, a Systems Theory Point of View.* Ph.D. thesis, University of Notre Dame, Available at http://www.nd.edu/~rosen/preprints.html.

Department of Mathematics, University of Notre Dame, Notre Dame, Indiana 46556, USA.
e-mail: Rosenthal.1@nd.edu, *URL:* http://www.nd.edu/~rosen/

Progress in Systems and Control Theory, Vol. 25

Introduction to Mathematical aspects of computer vision

J. Malik and P. Perona

Introduction

The goal of the minisymposium *Mathematical Aspects of Computer Vision* was to introduce the audience of MTNS98 to areas of research and open problems in computer vision that are mathematical, or, more generally, theoretical in nature. A further goal was to promote cross-fertilization between system and control and vision researchers.

We provide here a context by outlining the nature of the 'vision problem.' We then introduce the topics covered by the symposium speakers. We conclude with a bibliography for further reading.

Brief taxonomy of vision

Machine vision is an applied science whose objective is, similarly to human vision, to extract from images information about the world. Vision tasks may be organized into four broad categories:

1. *Reconstruction.* From a set of images, or from an image sequence, build the 3D model of an environment, determining spatial layout by finding the locations and poses of surfaces and estimating surface color, reflectance and texture properties. Reconstruct the motion of the camera through an unknown environment.

2. *Control.* Visually guided control of vehicle navigation and manipulation are typical examples. Locomotion tasks include satellite docking, navigating a robot around obstacles or controlling the speed and direction of a car driving down a freeway. Reaching, grasping and insertion operations are examples of manipulation.

3. *Grouping and tracking.* Grouping is the association of image pixels, into regions corresponding to single objects or parts of objects. Tracking is matching these groups from one time frame to the next. Grouping is used in the segmentation of different kinds of tissues in an ultrasound image or in traffic monitoring to distinguish and track individual vehicles.

4. *Recognition.* Determining the class of particular objects that have been imaged (this is a face), as well as recognize specific instances such

as faces of particular individuals (this is Gandhi's face). Identifying gaits, expressions and gestures from time sequences. Estimating the physical properties of a surface (the blade of that knife looks smooth and cold).

Machine vision researchers come in two flavors: engineers and scientists. The first aim at building artificial vision systems that can augment and/or substitute people in tasks such as quality control, inspection, piloting vehicles, reading handwritten forms. The second aim at understanding the physical, geometrical and computational foundations that underlie both human and machine vision.

In order to appreciate the nature of the theoretical and practical issues that face vision researchers, one must first understand the nature of the input and output data and the degrees of freedom of the solutions. Images are large matrices: two-dimensional for still-frames, three-dimensional for time sequences and tomographic data. Each entry (pixel) in the matrix is either a scalar or a small vector encoding color. Typical image (image sequence) sizes are 1 Mbyte to 1Gbyte. The desired output of a visual system is different in nature: either a symbol (e.g. Gandhi), or a proposition (e.g. a pedestrian is crossing the road in front of me), or a control signal (e.g. the amount of steering that takes my car by the pedestrian without hitting her). An additional degree of complexity is given by the fact that the control interacts with the signal: as I drive my car my eyes move tracking potential obstacles. More subtly, the areas of my visual field that I process may change as I shift my attention.

How can we convert an image, a large matrix of numbers, into a small set of logical/geometrical descriptors and control signals? One may start by considering the opposite problem: given a physical scene how does its image come about? This is the domain of computer graphics. Given a set of surfaces that have an assigned shape and are made of specified materials and are lit in a certain way calculate the image that a camera positioned in a certain spot will capture. Computer graphics is highly successful and it relies on a detailed knowledge of the geometry, optics and physics of how light, striking a textured surfaces that is made of a given material and has a certain orientation in space, bounces off towards a viewer forming an image. Vision may be thought of as a large inverse problem, inverse graphics. In this very general formulation, the problem is probably insoluble. What we need to examine is the recovery of information adequate to support the various forms of visually guided behavior.

In general vision tasks are under constrained: the image of a Greek column impinging on our retina is both consistent with the 3D marble object (a cylinder whose surface has constant reflectance properties and which is lit from a well-defined direction) and with the 2D photograph of the object (the photograph is a flat surface with variable reflectance, whose direction of lighting is quite unimportant). Similarly the image of an ellipse is consistent both with an elliptical object that is parallel to our retina, and of a circle that is slanted. Most often an infinite family of worlds is consistent with an image. The image of a line is consistent with any planar curve seen edge-on. Yet, we seem to be able to extract useful and unambiguous information from our visual system. Clearly, there must be some other source of information which allows us to choose the 'best' solution amongst many possibilities. This information is the a-priori knowledge of how the world is: piecewise-smooth, with low curvatures having higher likelihood than high curvatures. Also the 'generic viewpoint' assumption, which says that it is highly unlikely that we see a flat object edge-on.

Vision, therefore, could be thought of in a Bayesian setting. Given images and prior probability densities, recover posterior densities of appropriate scene parameters. What is the structure of this map? Unfortunately the problem is so complex and nonlinear that it is in practice impossible to synthesize this 'vision map' from the specification of a task, e.g. vision-based navigation or recognition. Vision researchers parcel the problem into manageable sub-problems in order to solve it:

1. *Early vision.* The image is filtered in order to extract simple structure at many scales of resolution and orientations. At this stage one may extract local texture descriptors, local image motion, significant brightness gradients, local curvature, local stereoscopic disparity.

2. *Grouping and segmentation.* Image locations that share common properties are grouped into regions. The boundaries and 2D shape and velocity of these regions are calculated and represented explicitly.

3. *Mid-level vision.* The regions are organized into depth and into larger groupings corresponding to objects. Hypotheses are formed as to occlusion, 3D shape and object structure.

4. *Extraction of geometry and motion.* The global 3D structure, pose and motion of the objects in the scene is computed.

5. *Recognition.* Surfaces, objects and behaviors in the scene are associated to classes.

These modules are interconnected in a feedback fashion so that both bottom-up and top-down processing are possible. Elements of signal processing, applied probability and statistics, geometry, optics, linear algebra, functional analysis, numerical analysis and computer science are brought to bear on the problem.

Contributions to the symposium

Early vision : J. Koenderink sets the stage by motivating the fact that structure in images comes at multiple scales of resolution [21, 18]. G.

Sapiro and P. Perona [32, 33] present mechanisms, based on nonlinear diffusion-like partial differential equations, for analyzing brightness and orientation images at multiple scales of resolution. See also [2, 43, 30, 40, 6, 7, 3, 35, 42, 41, 24, 28, 31, 23, 29, 10, 26].

Grouping : J. Malik argues that grouping is a fundamental step for all vision tasks and presents a technique for grouping that is based on graph spectral factorization.

Structure recovery : Faugeras discusses the geometry underlying structure from motion and presents animations demonstrating the complete reconstruction of an indoors scene from a small set of images. Soatto analyzes the problem of recovering depth from a number of images that are taken from the same position under different focal settings of the lens. Cipolla [34, 16, 8] explores the role of the projections of the boundaries of curved surfaces in estimating the surfaces' shape. Mallat discusses how surface orientation may be recovered from texture using wavelet decompositions. See also [12, 4, 14, 5, 11, 1, 36, 9, 27, 13, 15, 22, 21, 25, 20, 19].

Tracking : Blake presents his *condensation* algorithm [17]. The aim is tracking when the underlying probability density is not Gaussian, and therefore the standard Kalman filter cannot be used. On the same topic see also [39, 38, 37]

References

[1] Y. Amit, D. Geman, and K. Wilder. Joint induction of shape features and tree classifiers. *IEEE Transactions on Pattern Analysis and Machine Intellegence*, 19(11):1300–1305, 1997.

[2] S. Angenent, G. Sapiro, and A. Tannenbaum. On the affine heat equation for non-convex curves. *J. American Math. Soc.*, 11(3):601–634, 1998.

[3] M.J.. Black, G. Sapiro, D.H. Marimont, and D. Heeger. Robust anisotropic diffusion. *IEEE Trans. on Image Processing*, 7(3):421–432, 1998.

[4] A. Blake, B. Bascle, M. Isard, and J. MacCormick. Statistical models of visual shape and motion. *Phil. Trans. of the Royal Soc. of London Series A-Math. Phys. and Engr. Sciences* , 356(1740):1283–1301, 1998.

[5] E. Calabi, P.J. Olver, C. Shakiban, A. Tannenbaum, and S. Haker. Differential and numerically invariant signature curves applied to object recognition. *Internat. J. of Computer Vision* , 26(2):107–135, 1998.

[6] V. Caselles, J.M. Morel, G. Sapiro, and A. Tannenbaum. Introduction to the special issue on partial differential equations and geometry-driven diffusion in image processing and analysis. *IEEE Trans. on Image Processing*, 7(3):269–273, 1998.

[7] V. Caselles, J.M. Morel, and C. Sbert. An axiomatic approach to image interpolation. *IEEE Trans. on Image Processing*, 7(3):376–386, 1998.

[8] R. Cipolla. The visual motion of curves and surfaces. *Phil. Trans. of the Royal Soc. of London Series A-Math. Phys. and Engr. Sciences* , 356(1740):1103–1118, 1998.

[9] R. Cipolla, G. Fletcher, and P. Giblin. Following cusps. *Internat. J. of Computer Vision* , 23(2):115–129, 1997.

[10] I. DAUBECHIES, S. Mallat, and A.S. WILLSKY. Special issue on wavelet transforms and multiresolution signal analysis - introduction. *IEEE Trans. on Info. Theory* , 38(2):529–531, 1992.

[11] O. Faugeras and R. Keriven. Variational principles, surface evolution, pde's, level set methods, and the stereo problem. *IEEE Trans. on Image Processing*, 7(3):336–344, 1998.

[12] O. Faugeras and T. Papadopoulo. Grassmann-cayley algebra for modelling systems of cameras and the algebraic equations of the manifold of trifocal tensors. *Phil. Trans. of the Royal Soc. of London Series A-Math. Phys. and Engr. Sciences* , 356(1740):1123–1150, 1998.

[13] O. Faugeras and L. Robert. What can two images tell us about a third one? *Internat. J. of Computer Vision* , 18(1):5–19, 1996.

[14] O. Faugeras, L. Robert, S. Laveau, G. Csurka, C. Zeller, C. Gauclin, and I. Zoghlami. 3-d reconstruction of urban scenes from image sequences. *Computer Vision and Image Understanding* , 69(3):292–309, 1998.

[15] D. Geman and B. Jedynak. An active testing model for tracking roads in satellite images. *IEEE Transactions on Pattern Analysis and Machine Intellegence*, 18(1):1–14, 1996.

[16] P. Giblin. Apparent contours: an outline. *Phil. Trans. of the Royal Soc. of London Series A-Math. Phys. and Engr. Sciences* , 356(1740):1087–1102, 1998.

[17] M. Isard and A. Blake. Condensation - conditional density propagation for visual tracking. *Internat. J. of Computer Vision* , 29(1):5–28, 1998.

[18] J.J. Koenderink. The structure of images. *Biological Cybernetics*, 50(5):363–370, 1984.

[19] J.J. Koenderink. Optic flow. *Vision Research* , 26(1):161–179, 1986.

[20] J.J. Koenderink and A.J. Vandoorn. Facts on optic flow. *Biological Cybernetics*, 56(4):247–254, 1987.

[21] J.J. Koenderink and A.J. Vandoorn. Generic neighborhood operators. *IEEE Transactions on Pattern Analysis and Machine Intellegence*, 14(6):597–605, 1992.

[22] J.J. Koenderink and A.J. Vandoorn. 2-plus-one-dimensional differential geometry. *Pattern Recognition Letters* , 15(5):439–443, 1994.

[23] G. Koepfler , C. Lopez, and J.M. Morel. A multiscale algorithm for image segmentation by variational method. *SIAM Journal on Numerical Analysis*, 31(1):282–299, 1994.

[24] P. Kube and P. Perona. Scale-space properties of quadratic feature detectors. *IEEE Transactions on Pattern Analysis and Machine Intellegence*, 18(10):987–999, 1996.

[25] J. MALIK. Interpreting line drawings of curved objects. *Internat. J. of Computer Vision* , 1(1):73–103, 1987.

[26] J. Malik and P. Perona. Preattentive texture-discrimination with early vision mechanisms. *J. of the Optical Soc. of America A-Optical Image Science and Vision*, 7(5):923–932, 1990.

[27] J. Malik and R. Rosenholtz. Computing local surface orientation and shape from texture for curved surfaces. *Internat. J. of Computer Vision* , 23(2):149–168, 1997.

[28] S. Mallat. Wavelets for a vision. *Proceddings of IEEE*, 84(4):604–614, 1996.

[29] S. Mallat and S. Zhong. Characterization of signals from multiscale edges. *IEEE Transactions on Pattern Analysis and Machine Intellegence*, 14(7):710–732, 1992.

[30] J.M. Morel. The mumford-shah conjecture in image processing. *Asterisque*, (241):221–242, 1997.

[31] P. Perona. Deformable kernels for early vision. *IEEE Transactions on Pattern Analysis and Machine Intellegence*, 17(5):488–499, 1995.

[32] P. Perona. Orientation diffusions. *IEEE Trans. on Image Processing*, 7(3):457–467, 1998.

[33] P. Perona and J. Malik. Scale-space and edge-detection using anisotropic diffusion. *IEEE Transactions on Pattern Analysis and Machine Intellegence*, 12(7):629–639, 1990.

[34] J. Sato and R. Cipolla. Quasi-invariant parameterisations and matching of curves in images. *Internat. J. of Computer Vision* , 28(2):117–136, 1998.

[35] K. Siddiqi, Y.B. Lauziere, A. Tannenbaum, and S.W. Zucker. Area and length minimizing flows for shape segmentation. *IEEE Trans. on Image Processing*, 7(3):433–443, 1998.

[36] S. Soatto. 3-d structure from visual motion: Modeling, representation and observability. *Automatica*, 33(7):1287–1312, 1997.

[37] S. Soatto, R. Frezza, and P. Perona. Motion estimation via dynamic vision. *IEEE Transactions on Automatic Control*, 41(3):393–413, 1996.

[38] S. Soatto and P. Perona. Recursive 3-d visual motion estimation using subspace constraints. *Internat. J. of Computer Vision* , 22(3):235–259, 1997.

[39] S. Soatto and P. Perona. Reducing "structure from motion": A general framework for dynamic vision part 1: Modeling. *IEEE Transactions on Pattern Analysis and Machine Intellegence*, 20(9):933–942, 1998.

[40] P.C. Teo, G. Sapiro, and B.A. Wandell. Creating connected representations of cortical gray matter for functional mri visualization. *IEEE Transactions on Medical Imaging*, 16(6):852–863, 1997.

[41] Y.L. You, W.Y. Xu, A. Tannenbaum, and M. Kaveh. Behavioral analysis of anisotropic diffusion in image processing. *IEEE Trans. on Image Processing*, 5(11):1539–1553, 1996.

[42] S.C. Zhu and D. Mumford. Prior learning and gibbs reaction-diffusion. *IEEE Transactions on Pattern Analysis and Machine Intellegence*, 19(11):1236–1250, 1997.

[43] S.C. Zhu, Y.N. Wu, and D. Mumford. Filters, random fields and maximum entropy (frame): Towards a unified theory for texture modeling. *Internat. J. of Computer Vision* , 27(2):107–126, 1998.

[44] Faugeras, O. (1993). *Three-Dimensional Computer Vision: A Geometric Viewpoint.* Cambridge: MIT Press.

[45] Horn, B. K. P. and Brooks, M.J.(1989). *Shape from Shading.* Cambridge: MIT Press.

[46] Isard, M.and Blake, A. (1996). Contour tracking by stochastic propagation of conditional density. In *Proceedings of Fourth European Conference on Computer Vision. ECCV '96*, Cambridge, UK. Edited by: Buxton, B.; Cipolla, R. Berlin, Germany: Springer-Verlag, 1996. p. 343-356 vol.1.

[47] Haralick, R. M. and L.G. Shapiro. (1992). Computer and Robot Vision, Volumes I and II. Reading, MA: Addison-Wesley.

[48] Horn, B. K. P. (1986). *Robot Vision.* Cambridge: MIT Press.

[49] Koenderink, J. J. (1990). *Solid Shape.* Cambridge: MIT Press.

[50] Marr, D. (1982). *Vision.* San Francisco: Freeman.

[51] Nalwa, V.S. (1993). *A Guided Tour of Computer Vision.* Reading, MA: Addison Wesley.

[52] E. Trucco, E. and Verri, A.(1998). *Introductory Techniques for 3-D Computer Vision*, NJ:Prentice-Hall, 1998.

[53] Ullman, S. (1996). *High-level Vision: Object Recognition and Visual Cognition.* Cambridge: MIT Press.

[54] Shi, J. and Malik, J (1997). Normalized cuts and image segmentation. In *Proceedings of the 1997 IEEE Computer Society Conference on Computer Vision and Pattern Recognition*, San Juan, Puerto Rico. pp. 731-737.

University of California at Berkeley. Email: malik@cs.berkeley.edu

California Institute of Technology and Università di Padova.
Email: perona@caltech.edu

Progress in Systems and Control Theory, Vol. 25

The Structure and Motion of Surfaces

R. Cipolla and P.R.S. Mendonça

1 Introduction

For smooth curved surfaces the dominant image feature is the *apparent contour* or outline. This is the projection of the *contour generator* – the locus of points on the surface which separate visible and occluded parts. The contour generator is dependent of the local surface geometry and the viewpoint. Each viewpoint will generate a different contour generator. This paper addresses the problem of recovering the 3D shape and motion of curves and surfaces from image sequences of apparent contours.

For *known* viewer motion the visible surfaces can then be reconstructed by exploiting a spatio-temporal parametrization of the apparent contours and contour generators under viewer motion. A natural parametrization exploits the contour generators and the epipolar geometry between successive viewpoints. The *epipolar parametrization* (Cipolla & Blake 1992) leads to simplified expressions for the recovery of depth and surface curvatures from image velocities and accelerations and known viewer motion.

The parametrization is, however, degenerate when the apparent contour is singular since the ray is tangent to the contour generator (Koenderink & Van Doorn 1976) and at *frontier points* (Giblin & Weiss 1994) when the epipolar plane is a tangent plane to the surface. At these isolated points the epipolar parametrization can no longer be used to recover the local surface geometry. This paper reviews the epipolar parametrization and shows how the degenerate cases can be used to recover surface geometry and *unknown* viewer motion from apparent contours of curved surfaces. The general, affine and special case of circular motion are considered. Practical implementations are outlined.

Structure and motion from image sequences of point features has attracted considerable attention and a large number of algorithms exist to recover both the spatial configuration of the points and the motion compatible with the views. A key component of these algorithms is the recovery of the *epipolar geometry* between distinct views (Luong & Faugeras 1996). The structure and motion problem for curves and curved surfaces is more challenging. For curved surfaces the dominant image feature is the *apparent contour* which is the projection of the curve on the surface (*contour generator*) dividing visible and occluded parts. The contour generator is dependent on viewpoint and local surface geometry (via tangency and conjugacy constraints) and each viewpoint will generate a different contour generator. The image curves are therefore projections of different space curves and there is no correspondence between points on the curves in the two images.

The family of contour generators generated under continuous viewer motion can be used to represent the visible surface. Giblin & Weiss (1987) and Cipolla & Blake (1992) have shown how the spatio-temporal analysis of deforming image apparent contours or outlines enables computation of local surface curvature along the corresponding contour generator on the surface. To perform the analysis, however, a spatio-temporal parametrization of image-curve motion is needed, but is under-constrained . The *epipolar* parametrization is most naturally matched to the recovery of surface curvature. In this parametrization (for both the spatio–temporal image and the surface), *correspondence* between points on successive snapshots of an apparent contour and contour generator is set up by matching along epipolar lines and epipolar planes respectively. The parametrization leads to simplified expressions for the recovery of depth and surface curvature from image velocities and accelerations and known viewer motion.

A degeneracy of the epipolar parametrization occurs when contour generators from subsequent viewpoints intersect to form an envelope. This occurs when the epipolar plane is also a tangent plane to the surface. These isolated surface points are called *frontier points* (Giblin & Weiss 1994). The surface can not be reconstructed at these points by the epipolar parametrization since the contour generator is locally stationary. However the frontier points correspond to real, fixed feature points on the surface which are visible in two views. They can be used to recover the viewer motion.

This paper addresses the problem of recovering the 3D shape and motion of curves and surfaces from image sequences of apparent contours. As with point features, the epipolar geometry plays an important role in both the recovery of the motion and in the reconstruction of the surface.

2 Reconstruction Under Known Viewer Motion

2.1 Viewing Geometry

Consider a smooth surface M. For each vantage point, $\mathbf{c}$, the sets of points, $\mathbf{r}$, on the surface for which the visual ray is tangent to M can be defined. This is called the *contour generator*, Γ, and is the set of points $\mathbf{r}$ for which

$$(\mathbf{r} - \mathbf{c}).\mathbf{n} = 0 \qquad (2.1)$$

where $\mathbf{n}$ is the unit normal to the surface at $\mathbf{r}$. The contour generator is usually (but not always, see section 3) a smooth curve on the surface separating the visible from the occluded parts and can be parametrized using say s as a parameter.

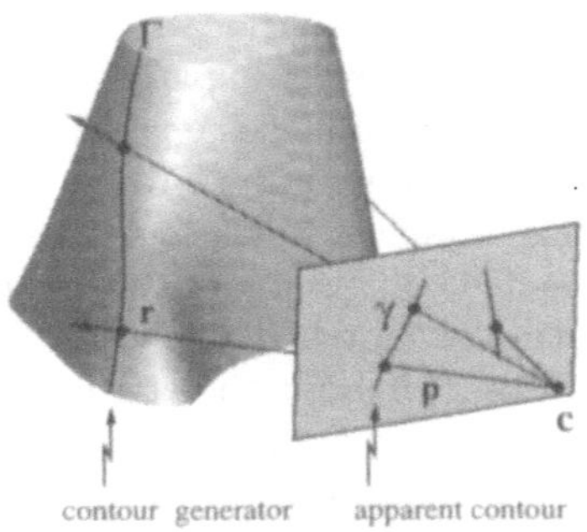

Figure 1: Viewing geometry and parametrization of the surface. For each viewpoint, $\mathbf{c}$, the family of rays which are tangent to the surface define the contour generator, $\mathbf{\Gamma}$. The image of the contour generator is called the apparent contour, γ. A surface point, $\mathbf{r}$, has position $\mathbf{c} + \lambda\mathbf{p}$ where λ is the distance from the viewing centre along a ray with direction given by the unit vector $\mathbf{p}$.

The image, γ, of the contour generator, $\mathbf{\Gamma}$, is called the apparent contour or outline and is the intersection of the set of rays which are tangent to the surface and the imaging surface. Without loss of generality and for mathematical simplicity, we consider perspective projection onto the unit sphere. An apparent contour point, $\mathbf{p}$, (a unit vector specifying the direction of the visual ray) satisfies:

$$\mathbf{r} = \mathbf{c} + \lambda\mathbf{p} \tag{2.2}$$

$$\mathbf{p}.\mathbf{n} = 0 \tag{2.3}$$

where λ is the distance along the ray to the surface point, $\mathbf{r}$, from the position of the centre of projection $\mathbf{c}$. It is natural to attempt a parametrization of M which is 'compatible' with the motion of the camera centre, in the sense that contour generators are parameter curves. This is the basic constraint of the spatio-temporal parametrization, that, imposed over (2.2) and (2.3) yields

$$\mathbf{r}(s,t) = \mathbf{c}(t) + \lambda(s,t)\mathbf{p}(s,t) \tag{2.4}$$

$$\mathbf{p}(s,t).\mathbf{n}(s,t) = 0. \tag{2.5}$$

However, the spatio-temporal parametrization of the apparent contours, $\mathbf{p}(s,t)$, and the surface, $\mathbf{r}(s,t)$, is not unique. The choice of the t-parameter curves, $\mathbf{p}(s_0,t)$ and $\mathbf{r}(s_0,t)$, for fixed s_0, is under-constrained.

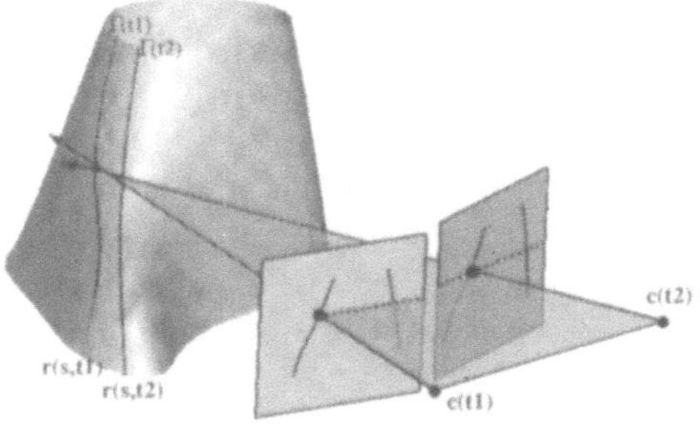

Figure 2: Epipolar parametrization. The surface is parametrized locally by the contour generators from successive viewpoints and *epipolar* curves defined by the intersection of the pencil of epipolar planes and the surface.

2.2 Epipolar parametrization

A natural choice of parametrization is the *epipolar* parametrization. In this parametrization the *correspondence* between points on successive snapshots of an apparent contour and contour generator are set up by matching along epipolar planes. Namely the corresponding ray in the next viewpoint (in an infinitesimal sense), is chosen so that it lies in the epipolar plane defined by the viewer's translational motion and the ray in the first viewpoint.

The epipolar parametrization is defined by (Cipolla & Blake 1992):

$$\mathbf{r}_t \wedge \mathbf{p} = 0 \tag{2.6}$$

and leads to the following *epipolar matching* condition

$$[\mathbf{p}_t, \mathbf{c}_t, \mathbf{p}] = 0 \tag{2.7}$$

such that the t-parameter curves are defined to lie instantaneously in the epipolar plane defined by the ray and direction of translation. This parametrization leads to simplified expressions for the recovery of depth and surface curvature. By simple manipulation of equations (2.4) to (2.7) and their spatial and temporal derivatives (denoted below by subscripts s and t) it is easy to show that the local surface geometry (e. g. depth, Gaussian curvature and surface normal) can be recovered from spatio-temporal derivatives (up to second order) of the apparent contours and the *known* viewer motion (Cipolla & Blake 1992) as follows:

1. The orientation of the surface normal, $\mathbf{n}$, can be recovered from a single view from the vector product of the ray direction, $\mathbf{p}$ and the

tangent to the apparent contour, $\mathbf{p}_s$:

$$\mathbf{n} = \frac{\mathbf{p} \wedge \mathbf{p}_s}{|\mathbf{p} \wedge \mathbf{p}_s|}. \tag{2.8}$$

2. If the contour generator is smooth at $\mathbf{r}$ then its tangent is in a *conjugate* direction (with respect to the second fundamental form) to the visual ray $\mathbf{p}$ and:

$$\mathbf{p}.\mathbf{n}_s = 0 \tag{2.9}$$

3. Under viewer motion and the epipolar parametrization, a given point on a contour generator, $\mathbf{r}$, will *slip* over the surface with velocity given by $\mathbf{r}_t$ and which depends on the distance, λ, and surface curvature (normal curvature in direction of the ray):

$$\mathbf{r}_t = -\left(\frac{\mathbf{c}_t.\mathbf{n}}{\lambda\kappa^t}\right)\mathbf{p} \tag{2.10}$$

   Note that the velocity is inversely proportional to the surface curvature and is zero in the limiting case of viewing a space curve or crease. The latter can be simply treated as apparent contours with infinite curvature along the ray direction.

4. Depth (distance along the ray, λ) can be computed from the deformation of the apparent contour, ($\mathbf{p}_t$), under known viewer motion (translational velocity $\mathbf{c}_t$):

$$\lambda = -\frac{\mathbf{c}_t.\mathbf{n}}{\mathbf{p}_t.\mathbf{n}} \tag{2.11}$$

5. The Gaussian curvature at a point on the apparent contour, K, can be recovered from the depth, λ, the normal curvature κ^t along the line of sight and the *geodesic* curvature of the apparent contour, κ^p:

$$K = \frac{\kappa^p\kappa^t}{\lambda} \tag{2.12}$$

   Since the normal section in the direction of the ray must always be convex at a point on the apparent contour, the sign of the Gaussian curvature is determined by the sign of the curvature of the apparent contour. Convexities, concavities and inflections of an apparent contour correspond to elliptic, hyperbolic and parabolic surface points respectively (Koenderink 1984).

Figure 3 illustrates the epipolar parametrization and the reconstruction of a strip of surface at the contour generator. Details of the camera calibration and the detection and tracking of the apparent contours with B-spline snakes can be found in Cipolla & Blake (1992).

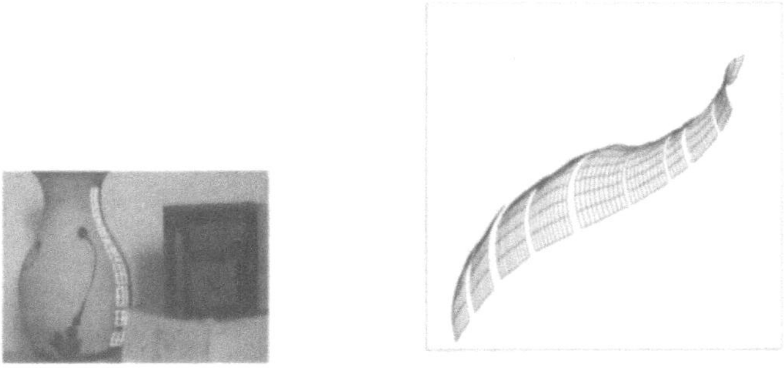

Figure 3: Recovery of surface geometry. The surface is recovered as a family of s-parameter curves – the contour generators – and t-parameter curves – portions of the *osculating* circles measured in each epipolar plane. The strip is shown projected into the image of the scene from a different viewpoint and after extrapolation (Cipolla & Blake 1992).

3 The frontier and epipolar tangencies

The degeneracy case of the epipolar parametrization for epipolar planes (spanned by the direction of translation and the ray) which coincide with tangent planes to the surface will occur at a finite set of points on the surface where the surface normal $\mathbf{n}$ is perpendicular to the direction of translation:

$$\mathbf{c}_t.\mathbf{n} = 0. \tag{3.1}$$

This condition implies that the contour generator is locally stationary ($\mathbf{r}_t = 0$). In fact (see Figure 4) consecutive contour generators will intersect at points where the epipolar plane is tangent to the surface.

The points of contact on the surface are called *frontier points* because for continuous motion the locus of intersections of consecutive contour generators in an infinitesimal sense define a curve on the surface which represents the boundary of the visible region swept out by the contour generators under viewer motion.

For larger discrete motions the contour generators defined by the discrete viewpoints also intersect at points on the surface where the epipolar plane is tangent to the surface. This is easily seen if we consider the motion to be linear. $\mathbf{c}_t$ is then a *constant* vector, and the frontier point on the surface at time t satisfies the frontier condition at subsequent times. The frontier degenerates to a point on the surface. In the discrete case the frontier points are defined by the condition

$$\Delta\mathbf{c}.\mathbf{n} = 0 \tag{3.2}$$

where $\Delta\mathbf{c} = \mathbf{c}(t_2) - \mathbf{c}(t_1)$ and $\mathbf{n}$ is the surface normal at the point in which in the two contour generators for each viewpoint intersect.

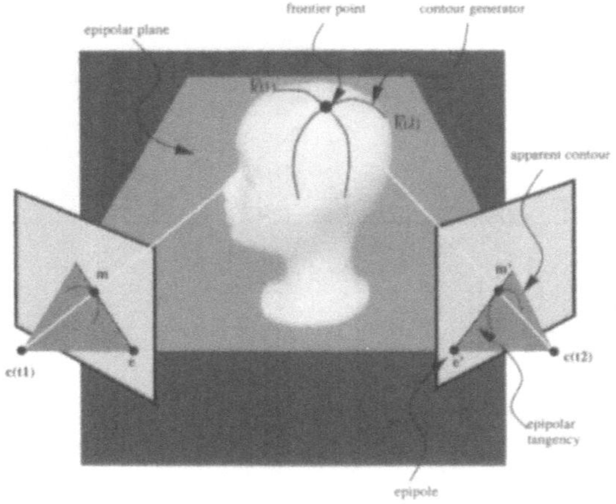

Figure 4: A frontier point appears the intersection of two consecutive contour generators and is visible in both views. The frontier point projects to a point on the apparent contour which is an epipolar tangency point.

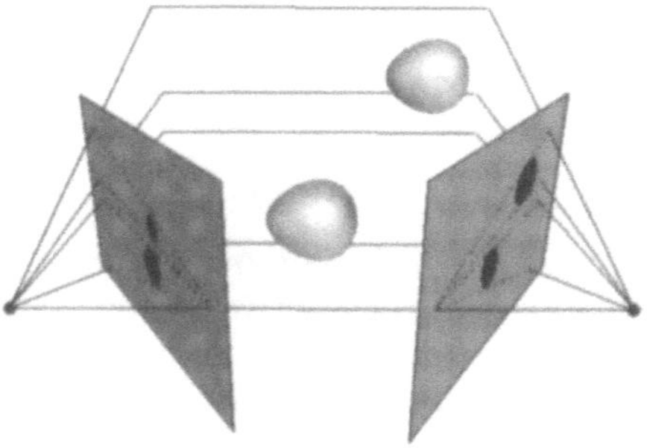

Figure 5: Epipolar geometry and epipolar tangencies under arbitrary motion

The frontier point projects to a point on the apparent contour in both views such that its tangent passes through the epipole. It is an epipolar tangency point since the tangent plane is also the epipolar plane.

The surface curvature can not be recovered by the epipolar parametrization at these points since the contour generator is locally stationary. However frontier points correspond to real, fixed feature points on the surface which are visible in both views, once detected they can be used to provide a constraint on viewer motion. In fact they can be used in the same way as points in the recovery of the epipolar geometry via the epipolar constraint.

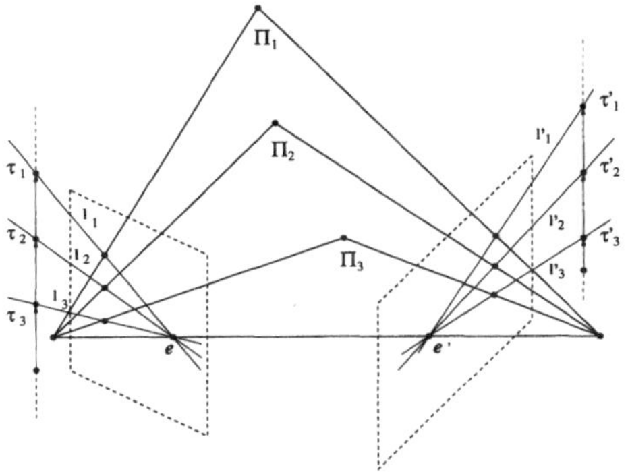

Figure 6: The epipolar geometry of an uncalibrated stereo pair of images is completely specified by the image positions of the epipoles and 3 pairs of corresponding epipolar lines. The projective parameters τ and τ' represent the intersection of the epipolar line and the line at infinity. The directions in the two views are related by a homography.

4 Recovery of Viewer Motion

4.1 Parametrisation of the Fundamental Matrix

The epipolar geometry between two uncalibrated views is completely determined by 7 independent parameters: the position of the epipoles in the two views $(u_e, v_e, 1)^T$ and $(u'_e, v'_e, 1)^T$ and the 3 parameters of the homography relating the pencil of epipolar lines in view 1 to those in view 2

$$\tau' = -\frac{h_2\tau + h_1}{h_4\tau + h3} \tag{4.1}$$

where τ and τ' represent the directions of a pair of corresponding epipolar lines in the first and second images respectively. The transformation is fixed by three pairs of epipolar line correspondences (Luong & Faugeras 1996).

The epipolar geometry can be conveniently specified by the Fundamental Matrix, F, (3×3 matrix defined up to an arbitrary scale and of rank two) such that the image co-ordinates (projective representation) of a pair of corresponding points, $\mathbf{m}$ and $\mathbf{m}'$, must satisfy the epipolar constraint:

$$\mathbf{m}'^T\mathbf{F}\mathbf{m} = 0 \tag{4.2}$$

and where the left and right epipoles ($\mathbf{e}$ and $\mathbf{e}'$) are given by the null space of F and F^T respectively.

This gives the following parametrization of the fundamental matrix:

$$\mathbf{F} = \begin{pmatrix} h_1 & h_2 & -u_e h_1 - v_e h_2 \\ h_3 & h_4 & -u_e h_3 - v_e h_4 \\ -u'_e h_1 - v'eh_3 & -u'_e h_2 - v'eh_4 & u_e u'_e h_1 + v_e u'_e h_2 + u_e v'eh_3 + v_e v'eh_4 \end{pmatrix}$$

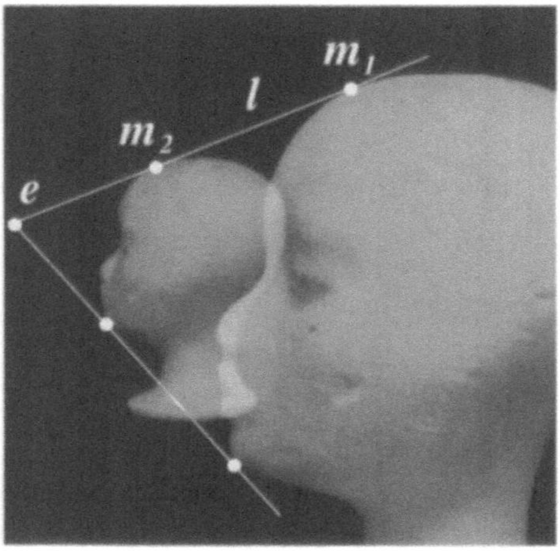

Figure 7: Under pure translation the frontier/epipolar tangency point moves along the epipolar line since the position of the epipole and the direction of the epipolar lines do not change. From a minimum of two bi-tangents of the apparent contour in two views it is possible to recover the epipole, $\mathbf{e}$.

4.2 The Pure Translation Case

Under pure translation the epipolar geometry is completely determined by the position of the epipole in a single view. The position of the epipole is the same in both views if the intrinsic parameters do not change and the epipolar lines have the same directions and are *auto-epipolar*. The bitangents at two consecutive apparent contours are epipolar tangencies and hence the projection of frontier points. The intersection of at least two distinct tangencies (epipolar lines) determines the position of the epipole. See Figure 7.

The solution is no longer trivial in the case of arbitrary motion with rotation. There is in fact no closed form solution since the epipoles are needed to define the epipolar tangency points (and frontier points) and these are needed to determine the epipoles.

4.3 The General Motion Case

The solution proceeds as a search and optimization problem to find the position of the epipoles in both views such that the epipolar tangencies in the first view are related to the set of epipolar tangencies in the second view by a one-dimensional homography (Cipolla, Åström & Giblin 1995).

A suitable cost function is needed. A geometric criterion (distance) is used in the estimation of the fundamental matrix from point correspon-

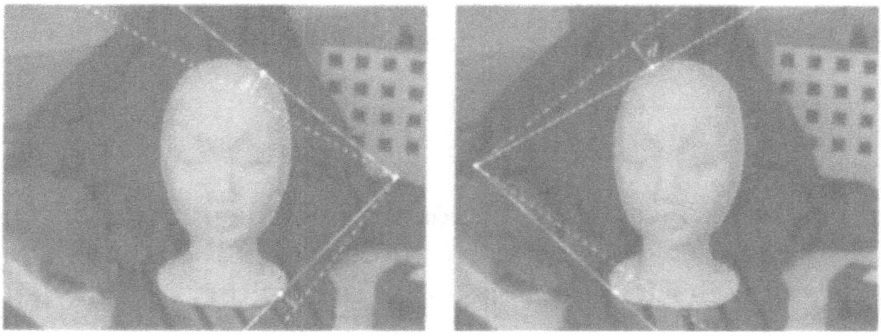

Figure 8: Illustration of the cost function to be minimised in the motion estimation algorithm. From the initial guess of the epipoles the homography is determined, and epipolar tangencies are transfered from one image to the other. The length d is the distance from a tangency point in the first image and an epipolar line obtained by the transfer of an epipolar tangency from the second image. The distance d' is found in the same way, interchanging the roles of the images. The cost function is then the sum $\sum_i(d_i^2 + d_i'^2)$ for each matching pair i of putative epipolar tangencies.

dences and can also be used in the case of curves. The geometric distance is computed as the sum over all tangency points of the square of the distance between the image point and the corresponding epipolar line from the tangency point in the other view, as shown in in Figure 8

The key to a successful implementation is to ensure that the search space is reduced and that the optimization begins from a good starting point using approximate knowledge of the camera motion or point correspondences.

The solution proceeds as follows:

1. Start with an initial guess or estimate of the epipoles in both views.

2. Compute the epipolar tangencies, $\mathbf{m}_i(\mathbf{e})$ and $\mathbf{m'}_i(\mathbf{e}')$, in both views respectively. These are points on the apparent contours with tangents passing through the epipole.

3. Estimate the elements of the homography between the pencil of tangencies in both views. This can be done linearly by minimising

$$\sum_i (h_4\tau_i\tau_i' + h_3\tau_i' - h_2\tau_i - h_1)^2 \tag{4.3}$$

   by least squares over all pairs of correspondences (τ and τ').

4. The fundamental matrix is now given by the parametrization above and the distance criterion, i. e. sum of squared distances between tangency point and corresponding epipolar line, can be computed as

below:

$$C = \sum_i \left(\frac{1}{(\mathbf{Fm}_i)_1^2 + (\mathbf{Fm}_i)_2^2} + \frac{1}{(\mathbf{F}^T\mathbf{m}'_i)_1^2 + (\mathbf{F}^T\mathbf{m}'_i)_2^2} \right) (\mathbf{m}'^T_i \mathbf{Fm}_i)^2.$$

5. Minimise the distance by the conjugate gradient method. The search space is restricted to the four co-ordinates of the epipoles only. This requires the first-order partial derivatives of the cost function (4)with respect to the co-ordinates of the epipoles which can be computed analytically but are more conveniently estimated by numerical techniques.

At each iteration of the algorithm, steps 1 to 4 are repeated, and the positions of the epipoles are refined. The search is stopped when the root-mean-square distance converges to a minimum (usually less than 0.1 pixels). It is of course not guaranteed to find a unique solution.

A number of the experiments were carried out with simulated data (with noise) and known motion (Figure 9). The apparent contours were automatically extracted from the sequence by fitting B-splines (Cham & Cipolla 1996). 5-10 iterations each for 4 different initial guesses for the position of the epipole were sufficient to find the correct solution to within an root-mean-square error of 0.1 pixel per tangency point.

Figure 11 shows an example with real data whose apparent contours are detected and automatically tracked using B-splines snakes. A solution is found very quickly which minimises the geometric distances but as with all structure from motion algorithms, a limited field of view and small variation in depths result in a solution which is sensitive to image localisation errors.

4.3.1 The Affine Case

When the field of view is narrow or the depth variation are small compared with the distance from the camera to the scene, the epipoles will be far from the image centre, and the epipolar lines will be approximately parallel. This viewing geometry suggests the use of an *affine camera model* (Mundy & Zisserman 1992) and *affine epipolar geometry* (Shapiro, Mundy & Zisserman 1995), that assumes that the epipoles will be at infinity, and reduces the degrees of freedom of the fundamental matrix, which will then take the form:

$$\mathbf{F} = \begin{bmatrix} 0 & 0 & c \\ 0 & 0 & d \\ a & b & e \end{bmatrix}. \tag{4.4}$$

There are two circumstances when the *affine fundamental matrix* may be used. The first is when the affine model can be used to describe the cameras. If the principal point of a camera is (u_0, v_0), the variation of depth in the scene is ΔZ and the mean distance of the features of the scene to the camera is Z_{mean}, the difference of the image of a point taken from a projective camera (u, v) and its image at the affine camera, (u_a, v_a) is given by

$$u - u_a = (u - u_0)\Delta Z/Z; \tag{4.5}$$
$$v - v_a = (v - v_0)\Delta Z/Z. \tag{4.6}$$

When the field of view is narrow, the terms $u - u_0$ and $v - v_0$ will be small. In this case, or when the depth variation of the scene is much smaller than its mean depth, e. g. $\Delta Z/Z < 0.1$, the error due to the affine approximation is negligible.

Other favorable situations for the use of the affine fundamental matrix are when the motion is restricted to translation orthogonal to the optical ray and cyclorotation. In this case the affine fundamental matrix can be used even if the affine camera model is inadequate. It is important to notice that a rotation by a small angle around a distant axis is a good approximation for such motion.

As scale factors are not important, the affine fundamental matrix has only four degrees of freedom, and can be linearly computed from 4 point correspondences. Each epipole, being at infinity, is described by a single parameter, corresponding to the direction in the image plane. This observation suggests another parametrization for the fundamental matrix, where the directions of the epipoles are made explicit. If ϕ and ϕ' are the directions of the epipoles in the first and second images, the affine fundamental matrix can be expressed as

$$\mathbf{F} = \begin{bmatrix} 0 & 0 & \alpha' \sin\phi' \\ 0 & 0 & -\alpha' \cos\phi' \\ -\alpha \sin\phi & \alpha\cos\phi & \sqrt{1 - \alpha^2 - \alpha'^2} \end{bmatrix} \tag{4.7}$$

where the parameters α and α' are related to the distances between epipolar lines on each image.

The geometric interpretation of the parameters α and α' can be seen in figure 11. It is easy to show that they are proportional to the distance between epipolar lines, or, in the notation of figure 11,

$$\begin{bmatrix} \alpha \\ \alpha' \end{bmatrix} = \frac{\begin{bmatrix} d_1' - d_2' \\ d_2 - d_1 \end{bmatrix}}{\sqrt{(d_2 d_1' + d_1 d_2')^2 + (d_1' - d_2')^2 + (d_1 - d_2)^2}}. \tag{4.8}$$

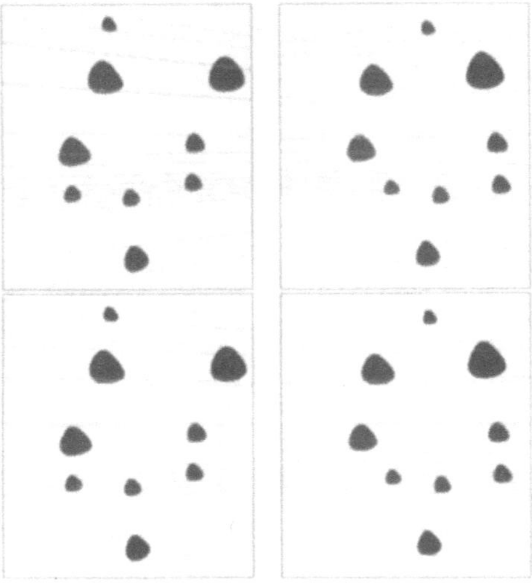

Figure 9: Starting point for optimization (above). An initial guess of the position of the epipoles is used to determine epipolar tangencies in both views and the homography relating the epipolar lines. For each tangency point the corresponding epipolar line is drawn in the other view. The distances between epipolar lines and tangency points are used to search for the correct positions of the epipoles. Convergence to local minimum after 5 iterations (below). The epipolar lines are tangent to apparent contours in both views.

Figure 10: Local minimum obtained by iterative scheme to estimate the epipolar geometry from 8 epipolar tangencies.

In the affine case the epipolar tangencies will be parallel lines, with directions given by the corresponding epipole, and, as in the projective case, the epipolar tangencies will touch the apparent contours at corresponding points. Since the number of degrees of freedom of the affine fundamental matrix is 4, this will also be the number of epipolar tangencies necessary for its computation.

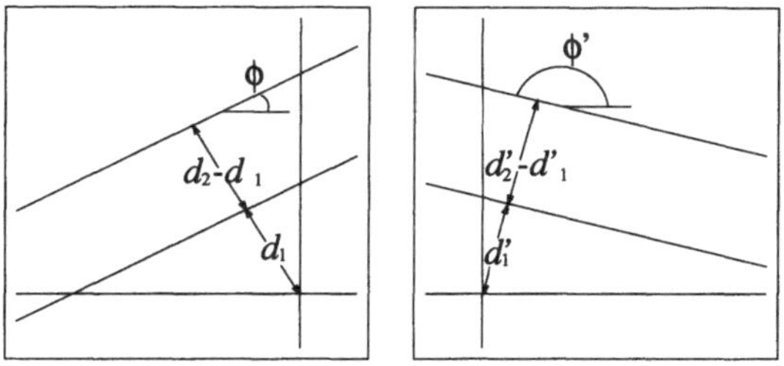

Figure 11: Geometric interpretation of the parametrization of the affine fundamental matrix presented in (4.7). The directions of the epipoles on each image are given by the angles ϕ and ϕ', and the parameters α and α' are proportional to the differences of distances $d_2' - d_1'$ and $d_2 - d_1$, respectively (see (4.8)).

Thus, the algorithm for computation of the epipolar geometry from apparent contours in the affine case is described as follows:

1. Start with initial estimates for the directions of both epipoles.
2. Determine correspondences $\mathbf{m}_i$ and $\mathbf{m}_i'$ from epipolar tangencies consistent with the directions of the epipoles.
3. Compute the affine fundamental matrix from the epipoles and the correspondences. This must be done by using the parametrization given in (4.7).
4. Minimise the the sum of geometric distances from the tangent points on the contours to the corresponding epipolar lines. The search is restricted to the two directions of the epipoles, and the cost function is the same as given by (4).

Experimental Results. The algorithm was tested in the images shown in figure 10, with the directions of the epipoles initialised at 0°. The resultant epipolar lines are shown in figure 12. There is some discrepancy between the results found by the general algorithm and the ones found by the algorithm for the affine case. The solution found by the general algorithm may be a local minima, but, since perspective effects seem to be fairly strong, the affine model may not be suitable for that particular situation. Nevertheless, the bottom epipolar line may give a clue that the affine approximation is indeed a good estimate. If the angles of the epipolar lines had opposite signals in respect the horizontal axis, the points put under correspondence by the epipolar geometry would be at opposite places in the bottom of the sculpture, which is, obviously, an incorrect solution. So the angles must have the same sign. Moreover, the actual values found were such that correspondent tangent points are located at the very bottom of the sculpture, which is consistent with a rotation motion around the sculpture.

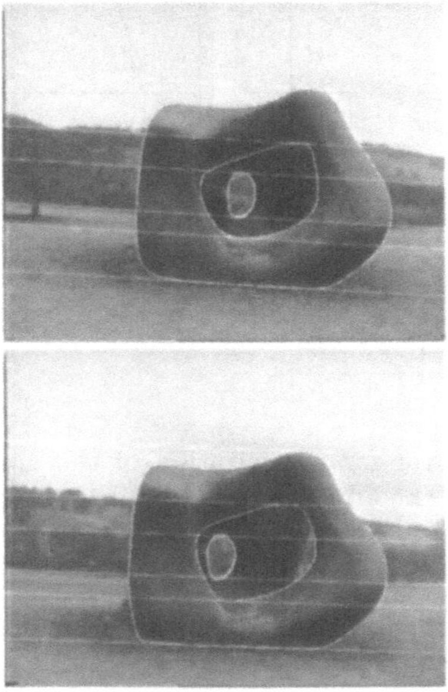

Figure 12: Estimated affine epipolar geometry from the apparent contours of the Moore sequence. The result is not in agreement with the one found by the general method, shown in figure 10.

Consider now three images taken from a camera undergoing constant motion. In this case, the epipolar geometry of adjacent viewpoints will also be fixed, and if two epipolar tangencies are available, a pair of estimates for the directions of the epipoles will produce, for images showing two frontier points, 4 correspondences, or 2 correspondences for each pair of images. A typical example of such configuration is *circular motion* – when a single camera views an object undergoing a constant rotation about a fixed but arbitrary axis. This idea was explored in the experiments that follow.

The algorithm previously described was implemented for the images shown in figure 13. Convergence was achieved after 4 to 5 iterations of both Davidon-Fletcher-Powell or Levenberg-Marquardt optimization methods. To evaluate the results, a calibration grid substituted the dummy and the same rotation was applied (see figure 14). The epipolar lines on the right image correctly match the points correspondent to the ones marked on left, showing the quality of the estimated epipolar geometry.

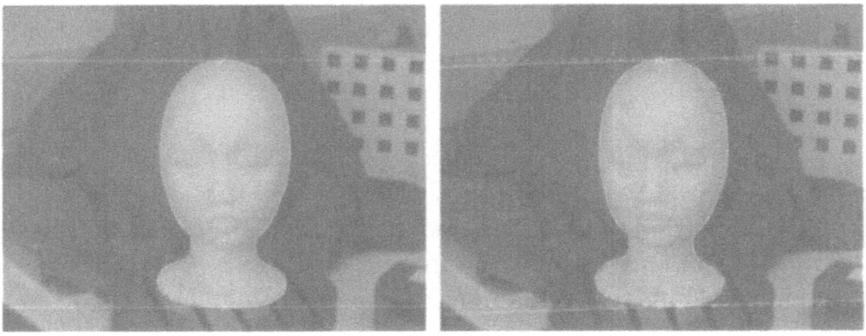

a) b)

c)

Figure 13: Images and contours used in the computation of the affine fundamental matrix. The rotation angle between successive snapshots is 5°. The continuous lines and × marks are corresponding epipolar lines and epipolar tangencies from images a) and b), and the dashed lines and circles are corresponding epipolar lines and epipolar tangencies from images b) and c), after convergence of the algorithm.

4.4 The Circular Motion Case

There are two main image features in a stereo pair of images obtained by a camera viewing an object undergoing rotation: the image of the rotation axis and the horizon line (Armstrong, Zisserman & Hartley 1996). If the camera intrinsic parameters are kept fixed during the rotation, the projection of the rotation axis will be a line fixed pointwise in the pair of images. The horizon line is the projection of the rotation plane, i. e., the plane orthogonal to the rotation axis that contains the camera centres. Since the rotation plane contains the camera centres, it also contains the line that passes through the camera centres. Thus, the epipoles must lie on the horizon line.

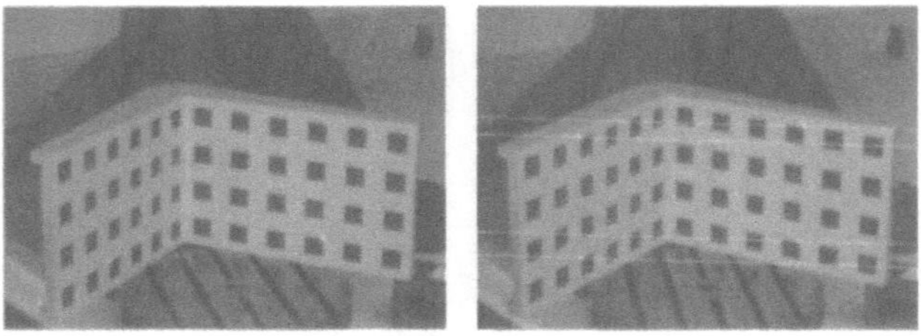

Figure 14: Evaluation of the estimated affine epipolar geometry. The fundamental matrix estimated from the apparent contours of the rotating dummy was tested in a calibration grid under the same motion. Four pairs of matching points are marked in the images. The epipolar lines corresponding to the points on the top were ploted in the bottom image, and the distance between the lines and points is always less than one pixel, showing the accuracy of the estimation.

4.4.1 Determination of the Epipolar Geometry

The determination of the epipoles and the image of the rotation axis, henceforth called *rotation line*, is enough to fix the epipolar geometry. Since the rotation line is fixed pointwise, the epipoles and three points in the rotation line determine an homography of the epipolar lines from one image to the other. This homography, together with the epipoles, fixes the fundamental matrix and thus the epipolar geometry of the system. So, in the circular motion case the fundamental matrix has 6 degrees of freedom: 4 for the coordinates of the two epipoles plus 2 for the equation of the rotation line.

The relation between the epipoles, the rotation line and the homography is represented in figure 15. If the images are rotated in such way that the horizon line is horizontal, the homography has a simpler expression, given by

$$\tau' = \frac{\tau}{k - \tau(1+k)\cot\theta}, \tag{4.9}$$

where $k = a/b$, θ is the angle between the rotation line and the horizon line and τ' and τ are the tangents of the angles ψ' and ψ as shown in figure 15. The parameters a and b are the distance from e and e' to the intersection of the rotation and horizon lines, as shown in figure 15. Hence, it is possible to parametrize the fundamental matrix in the case of rotation motion with only 6 parameters: 4 for the coordinates of the epipoles, that fix the horizon line, 1 for the intersection of the rotation line with the horizon line and 1 for the angle between the rotation line and the horizon line.

The planes intersecting the camera centres and tangent to the surface being viewed by the cameras define epipolar lines in both images. If these epipolar lines are superimposed in a single image, they must intersect in a point at the rotation line. So, assuming that the epipoles are known, two of

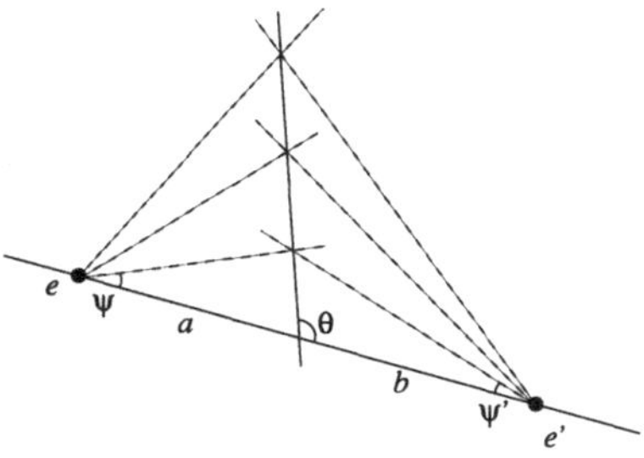

Figure 15: The homography shown in (4.9) relates the tangents τ and τ' of the angles ψ and ψ'.

such tangent planes would fix the rotation line and thus the homography. The position of the epipoles is controlled by four degrees of freedom (the coordinates of the epipoles), and four more tangent planes would provide enough information to determine the epipolar geometry of the pair of cameras, in the case of rotation. If more tangent planes cannot be determined but another image from a third camera, whose position relative to the second camera is the same as the relative position of the second camera to the first one, the new tangency points and the second and third camera centres will determine another pair of tangent planes and correspondent epipolar lines between images two and three. Proceeding with this, at the fourth image one would have came up with the six necessary tangent planes, and the epipolar geometry would be fixed. It is necessary that adjacent cameras be related by the same rotation and the intrinsic parameters are kept constant so the epipoles will not move. Intermediate solutions would also be possible, e. g., three tangencies and three images with adjacent cameras related by the same rotation. In general we must have

$$n(i-1) \geq 6, \tag{4.10}$$

where n is the number of tangencies in each image and i is the number of images in order to find the epipolar geometry from apparent contours in the case of rotation.

We can now summarise the algorithm for estimation of the epipolar geometry in the circular motion case:

- Initialise the epipoles at random positions.
- Superimpose adjacent images and determine epipolar tangencies.
- Fit by least-squares a line to the set of points generated by the intersection of each pair of correspondent epipolar lines. The sum of the square distance of each point to the fitted line is the cost function

which will be then minimised through a search for the correct position for the epipoles.

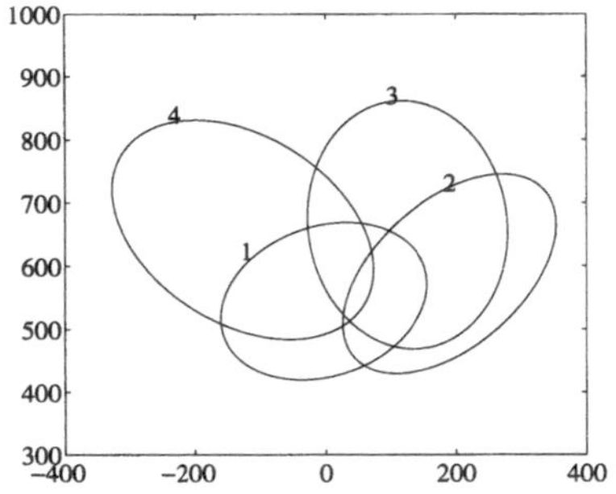

Figure 16: Superposition of the four apparent contours of the ellipsoid used for testing the algorithm.

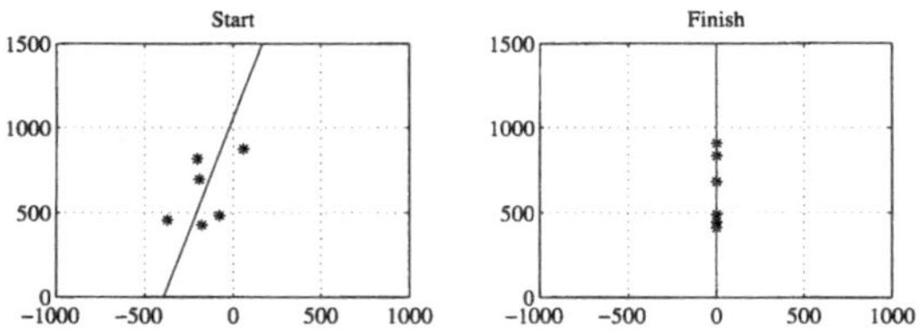

Figure 17: Intersection of epipolar lines at first and last iteration of the algorithm. The fitted line is the estimate of the rotation line.

Preliminary Experiments. In this first experiment with synthetic data, 4 images of a single ellipsoid (figure 16) were taken at successive positions related by a rotation of $2\pi/5$rad around a fixed axis. The epipoles were initialised at coordinates $e_0 = (4000, 200)$ and $e'_0 = (-3000, 400)$, far from the true values $e = (2752.8, 0)$ and $e' = (-2752.8, 0)$. The epipolar tangencies passing through the epipoles were computed, and for each pair of adjacent cameras, their intersection was computed. A line was fitted to the 6 points so produced, and the sum of the square distance in pixels of each point to the fitted line is the cost function to be minimised. The gradient of this function with respect to the coordinates of the epipoles was estimated numerically, and the Davidon-Fletcher-Powell or Levenberg-Marquardt methods were used to optimise the position of the epipoles. In figure 17 the initial and final set of intersection points is presented, together with the correspondent estimate for the rotation line.

In figure 18 the initial and final epipolar configuration are shown. Each picture presents the superposition of two adjacent images of the ellipsoid and the correspondent tangent lines for each epipole and contour.

The final values found for the coordinates of the epipoles were $e = (2752.9, 0.0)$ and $e' = (2752.8, 0.0)$, very close, by any criteria, of the true values. The experiment was reproduced with noise levels up to one pixel, and the measure Em (Luong & Faugeras 1996) for the relative error in the position of the epipoles, given by

$$Em = \frac{1}{2}\left(\frac{\|e - e_0\|}{\min(\|e\|, \|e_0\|)} + \frac{\|e' - e'_0\|}{\min(\|e'\|, \|e'_0\|)}\right), \tag{4.11}$$

was always under 0.2.

Other experiments, with different values of the baseline angle, β, were also done, and, *for large baseline angle*, e. g. $\beta > \pi/10$, the results were similar. For smaller values of β, though, the sensitivity to noise of the algorithm increased unacceptably. However, this is a favourable situation for the use of the affine approximation, as pointed in sec. 4.3.1, that thus should be used instead.

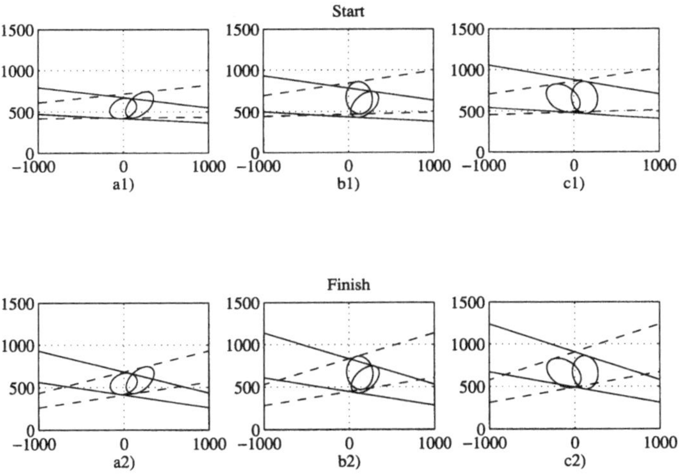

Figure 18: Initial (above) and final (below) epipolar configurations for four images of an ellipsoid. Each figure shows the superposition of adjacent images.

5 Conclusions

The recovery of the epipolar geometry between views is a key part of any algorithm to recover the 3D structure and motion compatible with the views. The structure and motion problem for curves and surfaces is more challenging since the apparent contours are viewpoint dependent and the correspondence of points between the two viewpoints in not given.

We have shown how the viewer motion can be recovered from the outlines (apparent contours) of curved surfaces by searching for epipolar tangency points both in the projective and affine cases. The results of initial experiments using these algorithms have been promising but the performance of the algorithm remains to be fully evaluated and it is still unclear whether the extraction of the motion leads to a unique solution. After computing the motion the epipolar geometry can be exploited to parameterise the apparent contours and to recover the visible surface.

We have only used apparent contours in the motion estimate. In practice one would use a combination of image features to estimate motion and could then use the apparent contours to reconstruct the surface. An important test of the usefulness of the proposed theories will be the accuracy of the reconstruction of an arbitrarily curved surface from uncalibrated viewer motion.

Acknowledgements: Roberto Cipolla acknowledges the support of the EPSRC and discussions with with Karl Åström, Peter Giblin, Stefan Rahmann and Jun Sato. Paulo R. S. Mendonça acknowledges the support of CAPES/Brazilian Ministry of Education.

References

[1] Armstrong, M., Zisserman, A. & Hartley, R. Self-calibration from image triplets. *Proc. 4th. European Conf. on Computer Vision, Cambridge (England)*, volume 1, 3–16, 1996.

[2] Åström, K., Cipolla, R. & Giblin, P.J. Generalised epipolar constraint. In *Proc. 4th . European Conf. on Computer Vision, Cambridge (England)*, volume 2, 97–108, 1996.

[3] Cham, T.J. & Cipolla, R. MDL-based curve representation using B-spline active contours. In *Proc. British Machine Vision Conference*, Edinburgh, 363–372, (September) 1996.

[4] Cipolla, R. & Blake, A. Surface shape from the deformation of apparent contours. *Int. Journal of Computer Vision*, **9(2)**, 83–112, 1992.

[5] Cipolla, R., Åström, K. & Giblin, P.J. Motion from the frontier of curved surfaces. In *Proc. IEEE 5th Int. Conf. on Computer Vision*, 269–275, Boston, (June) 1995.

[6] Cipolla, R., Fletcher, G.J. & Giblin, P.J. Following Cusps. *Int. Journal of Computer Vision*, **23(2)**, 115–129, 1997.

[7] Faugeras, O.D., Luong, Q.-T, & Maybank, S.J. Camera self-calibration: theory and experiments. In *Proc. 2nd European Conference on Computer Vision*, 321–334, Santa Margherita Ligure, Italy May 1992, Lecture Notes in Computer Science 588, Springer–Verlag.

[8] Giblin, P.J. & Weiss, R. Reconstruction of surfaces from profiles. In *Proc. 1st Int. Conf. on Computer Vision*, 136–144, London, 1987.

[9] Giblin, P.J. & Weiss, R. Epipolar fields on surfaces. In *Proc. 3rd European Conference on Computer Vision*, volume 1, pages 14–23, Stockholm, May 1994, LNCS 800, Springer–Verlag.

[10] Giblin, P.J., Pollick, F.E. & Rycroft, J.E. Recovery of an unknown axis of rotation from the profiles of a rotating surface. J. Opt. Soc. America, **11A**, 1976–1984, 1994.

[11] Koenderink, J.J. & Van Doorn, A.J. The singularities of the visual mapping. *Biological Cybernetics*, **24**, 51–59, 1976.

[12] Koenderink, J.J. & Van Doorn, A.J. The shape of smooth objects and the way contours end. *Perception*, **11**, 129–137, 1982.

[13] Koenderink, J.J. What does the occluding contour tell us about solid shape. *Perception*, **13**, 321–330, 1984.

[14] Koenderink, J.J. *Solid Shape.* MIT Press, 1990.

[15] Longuet-Higgins, H.C. A computer algorithm for reconstructing a scene from two projections. *Nature*, **293**, 133–135, 1981.

[16] Luong, Q.-T. & Faugeras, O.D. The fundamental matrix: theory, algorithms, and stability analysis. *Int. Journal of Computer Vision*, **17(1)**, 43–76, 1996.

[17] Mundy, J.L. & Zisserman, A. (eds.) *Geometric Invariance in Computer Vision*, MIT Press, 1992.

[18] Porrill, J. & Pollard, S. Curve matching and stereo calibration. *Image and Vision Computing*, **9(1)**, 45–50, 1991.

[19] Rieger, J.H. Three dimensional motion from fixed points of a deforming profile curve. *Optics Letters*, **11**, 123–125, 1986.

[20] Sato, J. & Cipolla, R. Affine Reconstruction of Curved Surfaces from Uncalibrated Views of Apparent Contours. In *Proc. IEEE 6th Int. Conf. on Computer Vision*, Bombay, 715–720, (January) 1998.

[21] Shapiro, L.S., Zisserman, A. & Brady, M. 3D Motion recovery via Affine Epipolar Geometry. *International Journal of Computer Vision*, **16**, 147–182, 1995.

[22] Vaillant, R. & Faugeras, O.D. Using extremal boundaries for 3D object modelling. *IEEE Trans. Pattern Recognition and Machine Intelligence*, **14(2)**, 157–173, 1992.

Department of Engineering, University of Cambridge, Cambridge, UK, CB2 1PZ

Department of Engineering, University of Cambridge, Cambridge, UK, CB2 1PZ

Progress in Systems and Control Theory, Vol. 25

Shape from Texture and Shading with Wavelets

M. Clerc, S. Mallat[1]

1 Introduction

The observation of a photograph or a painting demonstrates our ability to recover a three-dimensional shape from a single image. Shape cues are provided by a combination of various types of information, including shading variations, effects of perspective and texture gradients. The goal pursued here is to answer the question: given the image of a textured surface, how is it possible to retrieve the surface depth ? Recent references to the *shape from texture* problem can be found in [2, 3, 8, 10, 13, 17, 18], and the *shape from shading* problem has been studied in [7, 12, 14, 15, 16].

This work is organized in five sections. We present the image formation process, focusing on three of its aspects: the perspective projection, the relationship between scene brightness and image brightness, and lastly surface reflectance, which models the way in which a surface reflects the light it receives.

The following section introduces a stochastic model for images of textured surfaces. A texture is viewed as a stationary process, which has undergone some spatial distortion, and whose amplitude is affected by shading, giving rise in the resulting image to a non-stationary process. We model precisely the kind of non-stationarity that appears in such images. The two relevant physical terms which measure the departure from stationarity are a shading gradient and a warping gradient. We give some insight into how the actual shape of the surface may be retrieved from these two gradients in one and two dimensions.

We then focus on the estimation of the shading and warping gradient from the image, in one dimension. In a first order approximation, the warping gradient acts like a local change of scale. It finds a good representation with wavelets, because they are well localized both in space and in scale. We show that the variances of the wavelet coefficients of the image obey a transport equation in the space-scale domain. The warping gradient appears as the velocity coefficient and the shading gradient as a source term. The estimation is finally made possible thanks to an ergodicity result, and its efficiency is shown on numerical experiments.

In the last section, to solve the two-dimensional problem, we introduce a new wavelet transform, called the *warplet*. This wavelet transform is particularly well adapted to analyze the type of non-stationarity contained in

[1]Supported in part by MURI grant F49620-96-1-0028.

the images of textured surfaces, and the results in one dimension can be extended.

2 Image formation

In order to interpret an image, it is necessary to know how the image was formed in the first place. This section introduces three aspects of image formation: first the perspective projection, which is the geometrical relationship between the points in the picture and those in the actual scene. Next we analyze how the intensity of an image pixel and the brightness of the corresponding scene point are related. The third part concerns surface reflectance: it models the way points of the surface reemit the light they receive, as a function of the respective orientations of the incident light, the surface normal, and the emitted light ray.

2.1 Perspective projection

The simplest version of a camera is the pin-hole model depicted in Figure 1. The only light rays reaching the image are those coming through an ideal pin-hole O, which lies at a distance l_0 from the plane of photoreceptors. The *optical axis* is the line perpendicular to the image plane and going through O. We introduce a coordinate system centered at O, whose x_3-axis coincides with the optical axis. Figure 1 shows the orientation of the coordinate system (O, x_1, x_2, x_3). The projection of a point P' in the

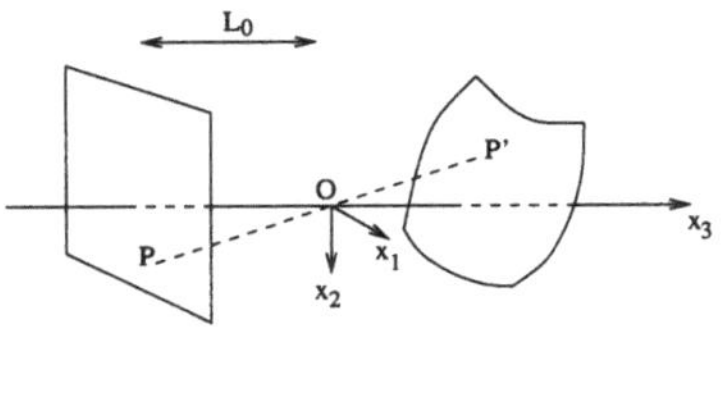

Figure 1: Ideal pin-hole camera model.

scene with coordinates $x' = (x'_1, x'_2, x'_3)$ onto a point P in the image with coordinates $x = (x_1, x_2, -l_0)$ is the well-known *perspective projection*. Since OP and OP' are collinear, the Cartesian coordinates of P and P' satisfy

$$(x_1, x_2, -l_0) = \lambda \ (x'_1, x'_2, x'_3) \ .$$

Obviously, $\lambda = -\frac{l_0}{x'_3}$, hence the image coordinates and the scene coordinates obey the relation

$$(x_1, x_2) = \left(-\frac{l_0 \ x'_1}{x'_3}, -\frac{l_0 \ x'_2}{x'_3} \right) \ ,$$

which we write as a projection from $\mathbb{R}^3$ to $\mathbb{R}^2$

$$(x_1, x_2) = p(x'_1, x'_2, x'_3) \ .$$

2.2 Image brightness

Now that we have the geometrical relationship between points in the image and in the scene, we would like to know how their respective brightnesses are related. For this we introduce two physical quantities that measure brightness. The *image irradiance* $i(x)$ is the light energy flux received by the image at position x per unit area. The scene brightness is in turn given by the *scene radiance* $l(x')$, which is the light energy flux emitted by the object surface at position x', per unit area, per unit solid angle.

At this stage, the pin-hole camera model is no longer adequate and has to be replaced by a more realistic lens model. It is clear that a unit surface patch emits light in a whole hemisphere of directions. In the pin-hole camera model, the light ray reaching a point in the image can only come from a single direction (i.e. a cone with an infinitely small solid angle). Hence the light energy flux received by a unit area of the image is actually equal to zero. If we were to replace the pin-hole by a hole with a finite diameter, the light energy flow would no longer be zero; however a point P of the scene would project onto a circle in the image, which is a drawback.

An ideal lens (Figure 2) has the property of collecting light rays from a finite solid angle at a given point, while still yielding a perspective projection. Strictly speaking, the perspective projection is only valid for points P and P' whose respective distances $-x_3$ and x'_3 from the lens satisfy the relation

$$\frac{1}{-x_3} + \frac{1}{x'_3} = \frac{1}{l_f} \ ,$$

where l_f is the focal length of the lens. Thus if the plane of photoreceptors lies at a distance $-x_3 = l_0$ from the lens, points of the surface are in focus if they are at a distance x'_3 such that

$$\frac{1}{x'_3} = \frac{1}{l_f} - \frac{1}{l_0} \ .$$

For a lens of diameter d and focal length l_f, there is a simple relationship between scene radiance l and image irradiance i

$$i(x) = l(x') \ \frac{\pi}{4} \left(\frac{d}{l_f}\right)^2 \cos^4\theta \qquad (1)$$

where θ is the angle between the optical axis and the ray connecting the object patch and the center of the lens (see Figure 2). Systems are usually

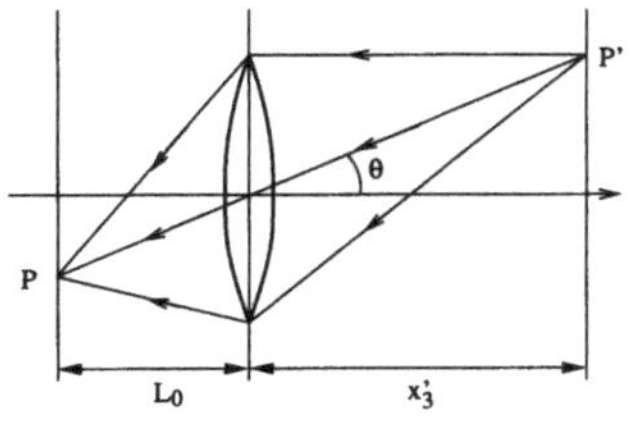

Figure 2: Lens model

calibrated so that the dependency on $\cos^4\theta$ is removed.
The proportionality result (1) indicates that by measuring the image irradiance, we have access to the scene radiance, which in turn, is the key to obtaining shape information, as we will see in the next part.

2.3 Surface reflectance

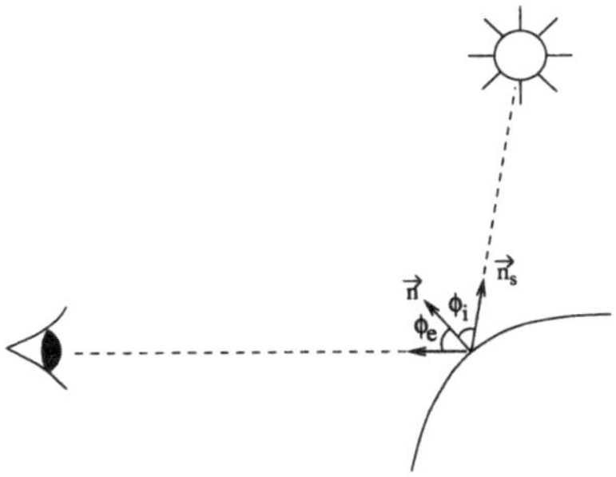

Figure 3: The *incidence* is the angle ϕ_i between the surface normal $\vec{n}$ and the illumination direction $\vec{n}_s$. The *emittance* is the angle ϕ_e between $\vec{n}$ and the viewing direction.

The radiance of a scene first depends on the amount of light falling on the surface, and also on the way the surface reflects the incident light.
The illumination is a function of the *incidence*, which is the angle ϕ_i between the surface normal $\vec{n}$ and the direction of incident light $\vec{n}_s$ (see Figure 3). We denote $e(\phi_i, x')\ \delta\omega_i$ the light energy received by the surface at a point $x' = (x'_1, x'_2, x'_3)$ from a solid angle $\delta\omega_i$ in direction ϕ_i.
The *emittance*, under which the surface patch is viewed, is the angle ϕ_e between the surface normal and the viewing direction. The surface reflectance is modelled by the *reflectance distribution* function

$$r(\phi_e, \phi_i, x') = \frac{\delta l(\phi_e, x')}{e(\phi_i, x')\delta\omega_i}\ ,$$

which determines how bright a surface patch appears under emittance ϕ_e when lit under incidence ϕ_i.
The total irradiance of the surface is

$$e_0(x') = \int e(\phi_i, x')\ \delta\omega_i\ .$$

The radiance of the surface when viewed under emittance ϕ_e is

$$l(\phi_e, x') = \int r(\phi_e, \phi_i, x')\ e(\phi_i, x')\ \delta\omega_i\ .$$

We consider the case of surfaces which reflect all the light they receive, equally in all directions: the function $r(\phi_e, \phi_i, x')$ only depends on the position x' of the surface point and is denoted $r_0(x')$. Such surfaces, called *Lambertian surfaces*, are matte surfaces with no specularity, such as snow, paper, or the surface of the moon. The surface radiance is then

$$l(x') = e_0(x')\ r_0(x').$$

For a point source of radiance e coming from a direction $\vec{n}_s$,

$$e_0(x') = e\ \vec{n}(x') \cdot \vec{n}_s\ ,$$

and therefore

$$l(x') = e\ r_0(x')\ \vec{n}(x') \cdot \vec{n}_s\ . \tag{2}$$

3 Images of textured surfaces

The sea landscape in Figure 4 conveys a definite impression of 3D depth. This impression seems to be due to the effect of perspective, even though there is no actual edge in the picture which could indicate a vanishing point. The scene depicted here is a texture (i.e. a pattern with a certain regularity), and we are looking at an example of *shape from texture.*

In this section we introduce an original model for images of textured surfaces which incorporates shading. The texture in the scene is assumed stationary, but because of the perspective projection, of the surface curvature, and of shading, the texture observed in the image is not stationary. We explain the departure from stationarity to be due to a warping and an amplitude variation. We will introduce two functions measuring each of these aspects: a warping gradient and a shading gradient. We will first introduce our model in a one-dimensional setting, and then in two dimensions. In each of these settings, we will give some insight as to how the shape of the surface can be recovered, by using shading or warping information.

Figure 4: This sea landscape conveys a definite impression of 3D depth (*shape from texture*).

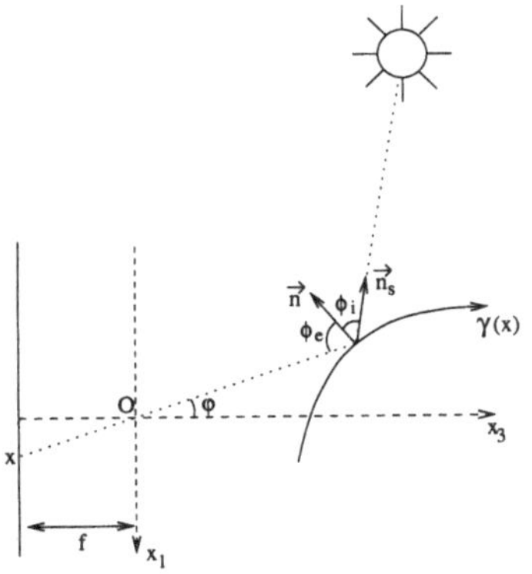

Figure 5: The arc length $\gamma(x)$ measures the length along the curve, when viewed from position x in the image.

3.1 Stochastic model in one dimension

In this section, x will represent the x_1 coordinate along the image line, and we will denote by x' the (x'_1, x'_3) coordinates along the curve (see Figure 5).

We consider the image of a textured surface, in one dimension. The scene radiance at a given point x' of a Lambertian surface is (2)

$$l(x') = e\ r_0(x')\ \vec{n}(x') \cdot \vec{n}_s\ .$$

Let $x = p(x')$ be the point in the image corresponding to x' by the perspective projection p. From (1) we know the image irradiance $i(x)$ to be proportional to the scene radiance at point x'. After a renormalization, which is common in image processing,

$$i(x) = r_0(p^{-1}(x))\ \vec{n}(p^{-1}(x)) \cdot \vec{n}_s\ . \tag{3}$$

The texture in the scene is modelled by a stochastic process R, which is wide-sense stationary. A wide-sense stationary process $R(x)$ is by definition [5] a process whose mean $\bar{R} = E\{R(x)\}$ is independent of x and whose covariance $E\left\{(R(x) - \bar{R})(R(y) - \bar{R})\right\}$ is a function of $x - y$. For short, we will write "stationary" instead of wide-sense stationary. We will assume stationary processes to be zero-mean ($\bar{R} = 0$).
Let $\gamma(x)$ be the arc-length at the surface point $p^{-1}(x) = x'$, with an arbitrarily chosen origin (see Figure 5). The texture is stationary with respect to the surface, therefore

$$r_0(p^{-1}(x)) = R(\gamma(x))\ . \tag{4}$$

Denoting

$$f(x) = \vec{n}(p^{-1}(x)) \cdot \vec{n}_s\ , \tag{5}$$

and replacing (4) and (5) in (3), the image irradiance has the following expression

$$I(x) = f(x)\ R(\gamma(x))\ . \tag{6}$$

The capital letter I points out that the irradiance is a stochastic process. It is the product of two terms, which have variations at different scales. Since the direction of the light source $\vec{n}_s$ is fixed, the variations of f occur when the surface normal varies, i.e. in the presence of curvature: they are due to the variation of the surface geometry. The variations of R are due to the pattern of the texture. We assume a *scale separation principle*, according to which R has variations at scales much smaller than the variation scale of f. Let $C_I(u, \delta u)$ (resp. $C_R(u, \delta u)$) denote the covariance of I (resp. R), centered at u:

$$C_I(u, \delta u) = E\left\{I(u - \delta u/2) I(u + \delta u/2)\right\}\ .$$

Since R is stationary, the point around which the covariance is centered is irrelevant:

$$C_R(u, \delta u) = C_R(0, \delta u)\ .$$

By replacing I with its expression (6), we have

$$C_I(u, \delta u) = f(u - \delta u/2)\ f(u + \delta u/2)\ C_R(0, \gamma(u + \delta u/2) - \gamma(u + \delta u/2))$$

Due to the scale separation principle, if δu is of the order of the variation scale of R but smaller than the variation scale of f,

$$C_I(u, \delta u) \approx f(u)^2 C_R(0, g(u)\delta u)\ .$$

If we compare the covariances of I centered at two different points u and v, we see that

$$C_I(v, \delta v) \approx \left(\frac{f(v)}{f(u)}\right)^2 C_I\left(u, \frac{g(v)}{g(u)}\delta v\right)\ . \tag{7}$$

There is thus an amplitude factor of $\left(\frac{f(v)}{f(u)}\right)^2$ and a warping (or scaling) factor of $\left(\frac{g(v)}{g(u)}\right)$ between $C_I(u, \cdot)$ and $C_I(v, \cdot)$.

When u and v are close,

$$\begin{aligned} \left(\frac{f(v)}{f(u)}\right)^2 &\approx 1 + 2(\log f)'(u)\ (v - u)\ , \\ \left(\frac{g(v)}{g(u)}\right) &\approx 1 + (\log g)'(u)\ (v - u)\ . \end{aligned}$$

The transformation beween the covariances centered at points u and v is hence locally specified by the relative variations of f and g. We call *shading gradient* the relative variation of f:

$$(\log f)'\ , \tag{8}$$

and *warping gradient* the relative variation of g:

$$(\log g)'\ . \tag{9}$$

The problem of their estimation from a single realization will be addressed in Section 4.

3.2 The importance of shading and warping gradients

The shading and warping gradients (8,9) have been shown to specify the transformation of the irradiance covariance centered between different points. This section is dedicated to the study of these two gradients, which we show to be the relevant functions in our model. First we will state the mathematical result that shows that, contrarily to the functions f and g, the shading and warping gradients are uniquely defined given an image irradiance of the form (6). Then we will investigate whether the warping and shading gradients are relevant functions from a shape recovery point of view, by pointing out which types of surfaces they may or may not determine.

3.2.1 Uniqueness

In our model, functions f and g are actually not uniquely defined, but the following proposition proved in [6] indicates that their relative variations $(\log f)'$ and $(\log g)'$ are:

Proposition 3.1 (Uniqueness of the representation) *Let $I(x)$ be of the form $I(x) = f(x)\ R(\gamma(x))$, where R is a stationary process. Under certain weak conditions on the covariance of R, there exists a stationary process $\tilde{R}$ such that $I(x) = \tilde{f}(x)\ \tilde{R}(\tilde{\gamma}(x))$ if and only if $(\log f)' = (\log \tilde{f})'$ and $(\log g)' = (\log \tilde{g})'$.*

Sketch of proof For the direct part, the idea of the proof is to write that the covariances of $f(x)\ R(\gamma(x))$ and of $\tilde{f}(x)\ \tilde{R}(\tilde{\gamma}(x))$ must be equal, which gives relations between f, $\tilde{f}$, and γ, $\tilde{\gamma}$.
For the converse, if $\tilde{f}(x) = \alpha\ f(x)$, and $\tilde{\gamma}(x) = \beta\ \gamma(x)$, it is straightforward to see that $\tilde{R}(x) = \frac{1}{\alpha} R(x/\beta)$ is a stationary process such that $I(x) = \tilde{f}(x)\ \tilde{R}(\tilde{\gamma}(x))$. □

This shows that the warping and shading functions are the relevant functions to estimate from the image irradiance of a textured surface I.

3.2.2 From the warping gradient to the shape

Recall that the warping gradient $(\log g)'$ represents the relative variations of g, which in turn is the derivative of the arc-length γ. We are going to answer two questions: supposing we know the arc-length $\gamma(x)$, is it possible to recover a curve uniquely ? And what information on the curve is lost when we know $(\log \gamma')'$ instead of γ ?
Since we are considering a perspective projection, it is better to parameterize the arc length by using φ, which is the angle between the optical axis and the light ray reaching the image (see Figure 5). There is a simple bijection between x and φ since $x(\varphi) = f\ \tan\varphi$. The arc length parameterized with φ is still denoted

$$\gamma(\varphi) = \gamma(x(\varphi)) \ .$$

Let $r(\varphi)$ denote the Euclidean distance between the optical center O and the surface point $x' = p^{-1}(x(\varphi))$. The arc-length satisfies

$$\gamma'(\varphi)^2 = r(\varphi)^2 + r'(\varphi)^2.$$

We have the following uniqueness result

Proposition 3.2 *There exists a unique curve $r(\varphi)$ with an arc length $\gamma(\varphi)$ specified up to a constant, defined for $\varphi \in [\varphi_1, \varphi_2]$, and satisfying boundary conditions*

$$\begin{cases} r(\varphi_1) = r_1 \\ r(\varphi_2) = r_2 \end{cases}$$

This brings an answer to the first question.
To answer the second question, note that if two arc-lengths γ_1 and γ_2 satisfy

$$(\log \gamma_1')' = (\log \gamma_2')'$$

then there is necessarily a linear relationship between γ_1 and γ_2. In addition to $r(\varphi_1)$ and $r(\varphi_2)$, the derivatives $r'(\varphi_1)$ and $r'(\varphi_2)$ must also be specified on the boundary in order to determine the curve $r(\varphi)$ uniquely from $(\log \gamma')'$.

3.2.3 From the shading gradient to the shape

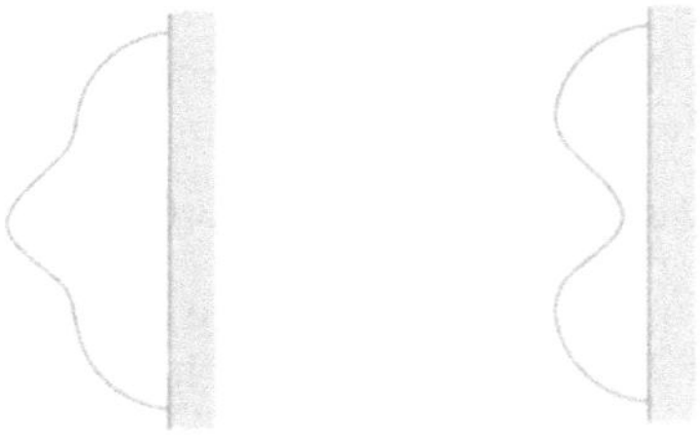

Figure 6: Two shapes yielding the same shading when the viewing direction and the lighting direction are both horizontal.

The shading term (5) is $f(x) = \cos \phi_i(p^{-1}(x))$. By extracting $\phi_i(p^{-1}(x))$ from f, it is possible to integrate the surface with a continuation method. The only difficulty is that the sign of $\phi_i(p^{-1}(x))$ is not specified, so different curves may give rise to the same shading, even if boundary conditions are given; for example in Figure 6, we do not know from the shading whether the center of the surface has a bump or a hole.
In our model, only the shading gradient, i.e. the relative variation of f, $(\log f)'$, is specified. A normalization of f is necessary; it can for instance be achieved if the surface orientation is known at a given position.

3.3 Two-dimensional model

The extension of the one-dimensional model to two dimensions is not straightforward. In one dimension, the arc length was introduced in order to give a meaning to the stationarity of a process with respect to a given curve. For the sake of simplicity, let us assume that the surface is locally developable, i.e. locally isometric to a plane. The image irradiance of a textured surface, whose texture is stationary with respect to the surface, has the following expression:
In the neighborhood of a position u in the image:

$$I(x) = f(x)\ R_u(\gamma_u(x))\ , \tag{10}$$

where

- the function f represents the shading effects (5):
$$f(x) = \vec{n}(p^{-1}(x)) \cdot \vec{n}_s \ ,$$
- the two-dimensional process R_u is stationary, and its covariance is independent of u, i.e. $E\{R_u(x)\ R_u(y)\}$ depends only on $x - y$.
- the function γ_u is a parameterization from $\mathbb{R}^2$ to $\mathbb{R}^2$ such that $\gamma_u(u) = 0$.

As in 1D, there is a scale separation principle: the texture is assumed to have variations on a far smaller scale than those of the surface geometry. Hence, if δu is of the order of the fine variation scale of R_u, the centered covariances of I and R_u obey the relation

$$C_I(u, \delta u) \approx (f(u))^2\ C_{R_u}(0, G_u \delta u) \ ,$$

where the matrix G_u is the Jacobian derivative of γ_u taken at position u. Since the covariance of R_u is assumed to be independent of u, between two neighboring points u and v in the image, we have the relation

$$C_I(v, \delta v) \approx \left(\frac{f(v)}{f(u)}\right)^2 C_I(u, G_u^{-1} G_v \delta v) \ .$$

The transformation between $C_I(u, \cdot)$ and $C_I(v, \cdot)$ is a change of amplitude by

$$\left(\frac{f(v)}{f(u)}\right)^2 \approx 1 + 2(\log f)'(u)\ (v - u) \ ,$$

and a warping by

$$\begin{aligned} G_u^{-1} G_v &\approx G_u^{-1} \left(G_u + \overrightarrow{\nabla_u G} \cdot (v-u)\right) \\ &\approx Id + G_u^{-1} \overrightarrow{\nabla_u G} \cdot (v - u) \ . \end{aligned}$$

The functions which describe this transformation are a *shading gradient* $(\log f)'(u)$ identical to the one appearing in one dimension, and a *warping gradient*

$$G_u^{-1} \overrightarrow{\nabla_u G} \ , \tag{11}$$

whose detailed expression is given in Section 5.2.

3.4 Shape recovery in two dimensions

We now study whether the parameters of our model allow an efficient recovery of 3D shape, distinguishing between shading information on the one hand and warping information on the second hand.

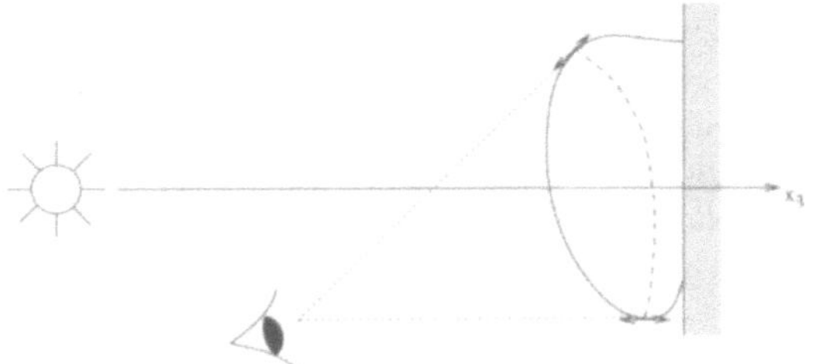

Figure 7: The dashed line on the surface depicts the *apparent contour* of the surface, i.e. points (x_1', x_2', x_3') with $(x_1', x_2') \in \partial\Omega$.

3.4.1 Shape-from-shading

The shading gradient $(\log f)'$ defines f up to a multiplicative constant. Recalling that $f(x)$ measures the cosine between the incident direction and the lighting direction, f is normalized so that its maximal value is one. Normalizations of this kind are common in image processing since some parameters, such as the lighting intensity, are unknown.
We have the following relation:

$$\vec{n}(x_1', x_2') \cdot \vec{n}_s = f(p(x')) ,$$

from which we would like to recover the surface normal $\vec{n}(x_1', x_2')$. The initial formulation of the problem was given by Horn [9] and it has since then been extensively studied [7, 12, 14, 15, 16]. The problem is more difficult than in one dimension, because the unit normal $\vec{n}$ must be specified by two angles, and we only have one scalar constraint giving its inner product against $\vec{n}_s$. The problem can however be solved by exploiting the differential nature of the surface normal. Let $\nabla x_3'$ be the 2D vector with coordinates $\left(\frac{\partial x_3'}{\partial x_1'}, \frac{\partial x_3'}{\partial x_2'}\right)$. Now, the coordinates of unit normal $\vec{n}$ in the (O, x_1, x_2, x_3) system are

$$\frac{1}{\sqrt{1+|\nabla x_3'|^2}} \left(\frac{\partial x_3'}{\partial x_1'}, \frac{\partial x_3'}{\partial x_2'}, 1\right)$$

so that if $\vec{n}_s$ has coordinates (α, β, γ),

$$f(p(x')) = \frac{\alpha\frac{\partial x_3'}{\partial x_1'} + \beta\frac{\partial x_3'}{\partial x_2'} + \gamma}{\sqrt{1+|\nabla x_3'|^2}} .$$

Note that it is always possible to change the referential so that the light source direction $\vec{n}_s$ is parallel to the x_3-axis (see Figure 7), in which case

$$f(p(x')) = \frac{1}{\sqrt{1+|\nabla x_3'|^2}} .$$

Let us consider the case of an *apparent contour* (Figure 7): Ω is the bounded connected open set representing the surface which is visible, and $\partial\Omega$ its boundary. For the sake of simplicity, let us suppose that $p(x')$ is the orthographic projection, independent of x'_3: $p(x') = (x'_1, x'_2)$.
The *shape-from-shading* problem amounts in solving for $(x'_1, x'_2) \in \Omega$

$$H(x'_1, x'_2, \nabla x'_3) = 0, \tag{12}$$

where the Hamiltonian is $H(u_1, u_2, p) = \frac{1}{\sqrt{1+|p|^2}} - f(u_1, u_2)$.
It is not reasonable to look for a differentiable solution $x'_3(x'_1, x'_2)$, because the surface normal may well not be continuous. But (12) is a Hamilton-Jacobi equation, and the appropriate class of solutions to consider is a class of generalized solutions of Hamilton-Jacobi equations called viscosity solutions, introduced by Lions in [11]. A viscosity solution of (12) can be seen as the limiting solution of the regularized problem

$$-\varepsilon \Delta x_3'^{\varepsilon} + H(x'_1, x'_2, \nabla x_3'^{\varepsilon}) = 0$$

as $\varepsilon \to 0$.

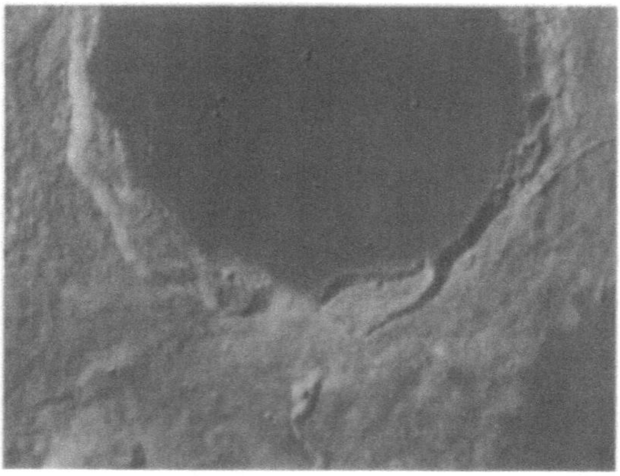

Figure 8: Lunar image acquired by the Clementine spacecraft (NASA). It is not clear visually whether the elements of relief represent cavities or bumps.

A differentiable solution of (12) is a viscosity solution, and conversely, a differentiable viscosity solution of (12) is a classical solution (i.e. (12) holds at each point).
Existence of viscosity solutions is guaranteed under regularity assumptions on H (and thus on the shading f). Viscosity solutions give rise to the same kind of indeterminacy as we noted in one dimension regarding bumps and holes. As an illustration, consider the lunar landscape in Figure 8: it is impossible to decide a priori whether the elements of relief represent cavities

or bumps.

Viscosity solutions are especially valuable from a numerical point of view, because they lead to fast and stable algorithms. Convergence results of Barles and Souganidis [1] have been applied in [15] to the *shape-from-shading* problem, to show that monotone, stable and consistent schemes coming from dynamic programming converge to the correct solution.

3.4.2 Shape-from-texture

Malik and Rosenholtz, in [13], prove a relation between the warping gradient (11) and the local surface orientation and curvature. The global surface shape can then be recovered by integrating the local shape parameters estimated by their method.
Moreover, the warping G_u can be interpreted as the derivative of the perspective map. Its knowledge at a position u in the image gives information about the surface normal at position $p^{-1}(u)$ on the surface.

4 Computing the one-dimensional gradients

We have seen (7) that the transformation between covariances of the image irradiance centered around two different positions was equivalent to a change of scale and a change of amplitude. A wavelet analysis is particularly well suited to analyze local scale changes. We observe that between the underlying stationary process R and the image irradiance I, the variance of wavelet coefficients undergoes a migration in the space-scale domain. We derive, in one dimension at first, a transport equation for the variance of wavelet coefficients of I, the velocity of which is the warping gradient $(\log g)'$. The shading gradient $(\log f)'$ appears as a source term in this equation. The main results are given with a sketch of proof, and the complete proofs may be found in [6].
In practice, given a discrete image of size N, we want to estimate sampled values of the distorsion gradient. This is made possible thanks to spatial averaging of the transport equation.

4.1 A space-scale transport PDE

We first introduce the *wavelet transform*. A wavelet is indexed by two parameters: position u and scale s. Given a mother wavelet $\psi(x)$ which is a zero-average function compactly supported inside $[-1, 1]$, a wavelet ψ_{us} is defined by

$$\psi_{us}(x) = \frac{1}{s}\psi\left(\frac{x-u}{s}\right) .$$

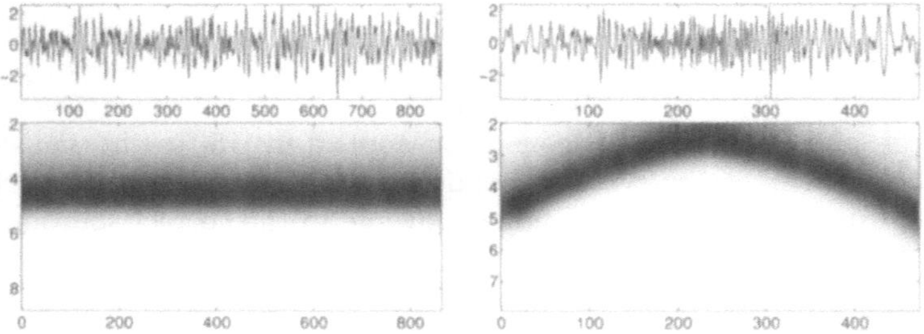

Figure 9: Top left: one realization of stationary process R; Bottom left: variance of wavelet coefficients of process R. Horizontal axis represents position u and vertical axis represents $\log s$. Dark points indicate high amplitude. Right, top and bottom: same, for non-stationary process $I(x) = R(\gamma(x))$.

It is compactly supported inside $[u - s, u + s]$.
The wavelet transform of a function $h(x)$ is a function $Wh(u, s)$ of position and scale defined as

$$Wh(u,s) = \int h(x)\ \psi^*_{us}(x)\ dx\ .$$

Because the image irradiance I is the realization of a stochastic process, so is its wavelet transform $WI(u, s)$. Its variance is denoted:

$$\sigma_{WI}(u,s) = E\left\{|WI(u,s)|^2\right\}$$

Now let R be a stationary process such that

$$I(x) = f(x)\ R(\gamma(x))\ .$$

Because R is stationary, its wavelet coefficients have a variance which is independent of position u:

Proposition 4.1 *The variance of wavelet coefficients of R,*

$$\sigma_{WR}(u,s) = E\left\{|WR(u,s)|^2\right\}$$

is independent of u, and is denoted $\sigma_{WR}(s)$.

Proof

$$\begin{aligned}
\sigma_{WR}(u,s) &= E\left\{\iint R(x)R(y)\psi_{us}(x)\psi^*_{us}(y)\ dx\ dy\right\} \\
&= \frac{1}{s^2}\iint E\left\{R(x)R(y)\right\}\psi(\frac{x-u}{s})\psi^*(\frac{y-u}{s})\ dx\ dy \\
&= \iint E\left\{R(u+sx)R(u+sy)\right\}\psi(x)\psi^*(y)\ dx\ dy
\end{aligned}$$

But since R is stationary, $E\{R(u+sx)R(u+sy)\}$ is a function of $s(x-y)$, which shows that $\sigma_{WR}(u,s)$ is independent of u. □

Figure 9(left) displays the variance of wavelet coefficients of R, which can be seen to be independent of position. The variance of wavelet coefficients of I displayed in figure 9(right), however, is not independent of position. In fact, there is a simple relationship between the two variance functions $\sigma_{WI}(u,s)$ and $\sigma_{WR}(s)$:

Proposition 4.2 (Wavelet migration) *Under certain conditions on the variance of R,*

$$\sigma_{WI}(u,s) = |f(u)|^2 \ \sigma_{WR}(s\ g(u)) \ (1+\varepsilon(s)) \ , \tag{13}$$

with $\varepsilon(s) \to 0$ when $s \to 0$.

Sketch of proof

$$\begin{aligned}
\sigma_{WI}(u,s) &= E\left\{\iint I(x)I(y)\psi_{us}(x)\psi^*_{us}(y)\ dx\ dy\right\} \\
&= \iint E\{I(u+sx)I(u+sy)\}\,\psi(x)\psi^*(y)\ dx\ dy \\
&= \iint f(u+sx)\ f(u+sy)E\{R(\gamma(u+sx))R(\gamma(u+sy))\}\,\psi(x)\psi^*(y)\ dx\ dy
\end{aligned}$$

But since R is stationary,

$$E\{R(\gamma(u+sx))R(\gamma(u+sy))\} = E\{R(\gamma(u+sx)-\gamma(u+sy))R(0)\} \ .$$

For asymptotically small s,

$$\gamma(u+sx)-\gamma(u+sy) \approx g(u)s(x-y)$$

and $f(u+sx)\ f(u+sy) \approx |f(u)|^2$, thus

$$\sigma_{WI}(u,s) \ \approx \ |f(u)|^2 \iint E\{R(g(u)sx)R(g(u)sy)\}\,\psi(x)\psi^*(y)\ dx\ dy$$

and the right-hand side is equal to

$$|f(u)|^2\ \sigma_{WR}(s\ g(u)) \ .$$

□

This result shows that between the variance of wavelet coefficients of R and the ones of I, there is a migration in the space-scale plane, as well as a change of amplitude. As a consequence, the function σ_{WI} obeys a transport PDE in the space-scale plane $(u;\log s)$:

Proposition 4.3 (Space-scale transport) *Under certain conditions on the variance of R,*

$$\frac{\partial \sigma_{WI}}{\partial u}(u,s) - \frac{g'(u)}{g(u)}\frac{\partial \sigma_{WI}}{\partial \log s}(u,s) \;=\; 2\,\frac{f'(u)}{f(u)}\,\sigma_{WI}(u,s)\,\left(1+\varepsilon(s)\right), \quad (14)$$

with $\varepsilon(s) \to 0$ *when* $s \to 0$.

Sketch of proof Formally, differentiating (13) with respect to u and s, yields

$$\frac{\partial \sigma_{WI}}{\partial u}(u,s) \;\approx\; 2\,f(u)f'(u)\,\sigma_{WR}(sg(u)) \;+\; |f(u)|^2\,sg'(u)\,\sigma'_{WR}(sg(u))$$

$$s\frac{\partial \sigma_{WI}}{\partial s}(u,s) \;\approx\; |f(u)|^2\,s\,g(u)\,\sigma'_{WR}(sg(u))$$

(14) is obtained by eliminating $\sigma'_{WR}(sg(u))$ from these two quasi-identities. □

The warping gradient $(\log g)'$ is simply the velocity of the transport equation (14), and the shading gradient $(\log f)'$ appears as a source term. In the next section, the two gradients are estimated by using equation (14) satisfied by the variance of wavelet coefficients of I.

4.2 Estimation by spatial averaging

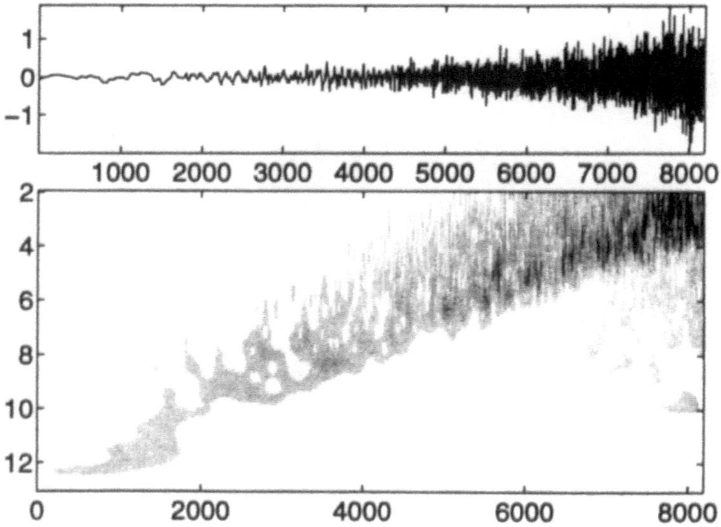

Figure 10: One realization of a process of the form $I(x) = f(x)R(\gamma(x))$, and its wavelet coefficients.

We now turn to practical considerations, showing how the shading and warping gradients may be estimated by using the transport equation (14).

In practice, an image of size N is analyzed in a subsampled wavelet basis, yielding N wavelet coefficients

$$WI(2^j k, 2^j) \ ,$$

where j and k are integers such that $\begin{cases} -\log_2(N) < j \leq 0 \\ 0 \leq k < 2^{-j} \end{cases}$.

The goal is to estimate the variance of $WI(u,s)$ and its derivatives with respect to u and $\log s$, from a realization $\left(WI(2^j k, 2^j)\right)$ of this process, of finite length N. Figure 10 shows a realization of one such process. Notice its high irregularity compared to the actual covariance, which is smooth (see as an example the covariance displayed in Figure 9(right)). The estimation of the variance of $WI(u,s)$ and its derivatives from a single realization is done with spatial averaging. The size of the averaging intervals, Δ, is much smaller than the spatial support of the image (equal to 1), but much larger than the wavelet analysis scale (equal to 2^j). Typically, $\Delta = N^{-\beta}$, with $0 < \beta < 1$ so that $\begin{cases} \Delta \to 0 \\ N\,\Delta \to \infty \end{cases}$, when $N \to \infty$.

In [6], we prove an *ergodic theorem*, which states that spatial averages (over u) of the statistics of the given realization converge to the corresponding probabilistic mean values when the sample length N goes to infinity.

The covariance and its derivatives are estimated at all points $u_0 \in [0,1]$ which are multiples of Δ, i.e. more and more densely as $N \to \infty$. For instance, the estimator for $\sigma_{WI}(u_0, 2^j)$ is

$$\widehat{\sigma_{WI}}(u_0, 2^j) = \frac{2^j}{\Delta} \sum_{|2^j k - u_0| \leq \frac{\Delta}{2}} \left|WI(2^j k, 2^j)\right|^2 \ .$$

We make two assumptions on the stationary texture R. The first is a Gaussianity assumption, and the second is a decorrelation hypothesis on the wavelet coefficients WR:

$$\left|E\left\{WR(2^j k, 2^j)\ WR(2^j k', 2^j)\right\}\right| \leq \frac{C}{|k-k'|} \ .$$

We prove that, under these two hypotheses, the estimators $\widehat{\sigma_{WI}}$ converge for fine scales, with a convergence rate equal to $\frac{1}{\sqrt{N}}$.

Let us fix p scales $2^j, 2^{j+1}, \ldots, 2^{j+p-1}$ with $2^j = N/2$ and $p > 2$. For a fixed position u_0, the set of transport equations (14) estimated at p different scales can be written as an overdetermined linear system which has to be solved for the two unknowns $(\log g)'(u_0)$ and $(\log f^2)'(u_0)$:

$$\begin{pmatrix} \widehat{\partial_{\log s}\sigma_{WI}}(u_0, 2^j) & \widehat{\sigma_{WI}}(u_0, 2^j) \\ \vdots & \vdots \\ \widehat{\partial_{\log s}\sigma_{WI}}(u_0, 2^{j+p-1}) & \widehat{\sigma_{WI}}(u_0, 2^{j+p-1}) \end{pmatrix} \begin{pmatrix} (\log g)'(u_0) \\ (\log f^2)'(u_0) \end{pmatrix} =$$

$$\begin{pmatrix} \partial_u \widehat{\sigma W_I}(u_0, 2^j) \\ \vdots \\ \partial_u \widehat{\sigma W_I}(u_0, 2^{j+p-1}) \end{pmatrix}$$

The least-squares estimator for $((\log g)'(u_0), (\log f^2)'(u_0))$, obtained by singular value decomposition, is unbiased and consistent as $N \to \infty$.

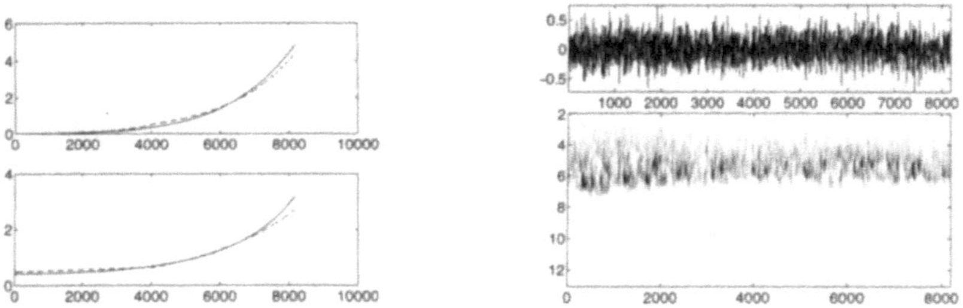

Figure 11: Left: warping g (top) and shading f (bottom). The original functions used to generate the signal presented in Figure 10 are in full line, and the functions estimated with the above algorithm are in dashed line; Right: stationary signal reconstructed by applying the inverse of the estimated warping and shading to the signal of Figure 10.

5 Warplets for two-dimensional computations

Textured images can succesfully be analyzed using wavelets, because the wavelet transform offers good localization, both in space and in (spatial) frequency [4]. We find it necessary for our purpose to introduce a new kind of wavelet transform, called the *Warplet transform*, which takes into account the kind of geometrical warping that may be found in textured images.

5.1 The Warplet transform

Let $\psi(x_1, x_2) = \rho(x_1, x_2)e^{i\xi x_1}$ be a function centered at the origin $(0, 0)$ in the space domain, and whose two-dimensional Fourier transform $\hat{\Psi}(\omega_1, \omega_2)$ is localized around $(\xi, 0)$ in the frequency domain, with $\xi \neq 0$. As in one dimension, a two-dimensional wavelet ψ_{uS} has two parameters: the position $u = (u_1, u_2)$ around which it is localized, and a warping matrix S. Such a matrix represents an orientation-preserving affine distorsion, or warping in two dimensions.

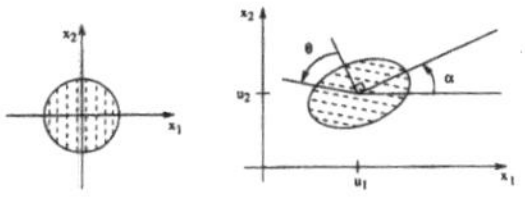

Figure 12: Top left: typical spatial support for ψ: a disk centered at $(0,0)$ with radius 1. The stripes indicate an oscillation in the x_1 direction. Top right: spatial support of ψ^θ. The direction of oscillation is rotated by θ. Bottom: spatial support of ψ_{uS}, localized around (u_1, u_2). The major (resp. minor) axis length of the ellipse is s_1 (resp. s_2).

Using the Cartan decomposition,

$$S = R_\alpha \begin{pmatrix} s_1 & 0 \\ 0 & s_2 \end{pmatrix} R_\theta$$

where R_α and R_θ are rotation matrices with angles α and θ. Since $s_1 s_2 > 0$, s_1 and s_2 are real, and have the same sign; they can be taken positive without loss of generality.
The warplet ψ_{uS} is by definition

$$\psi_{uS}(x) = \frac{1}{\det S}\, \psi\left(S^{-1} \begin{pmatrix} x_1 - u_1 \\ x_2 - u_2 \end{pmatrix}\right) . \tag{15}$$

Let ψ^θ be a rotated copy of function ψ,

$$\psi^\theta(x) = \psi\left(R_{-\theta} \begin{pmatrix} x_1 \\ x_2 \end{pmatrix}\right)$$

The function ψ^θ is still localized at $(0,0)$, but this time has a Fourier transform localized around $(\xi \, \cos\theta, \; \xi \, \sin\theta)$.
The warplet ψ_{uS} can be written with the help of function ψ_θ:

$$\psi_{uS}(x) = \frac{1}{s_1 s_2} \psi^\theta \left(\begin{pmatrix} 1/s_1 & 0 \\ 0 & 1/s_2 \end{pmatrix} R_{-\alpha} \begin{pmatrix} x_1 - u_1 \\ x_2 - u_2 \end{pmatrix}\right) . \tag{16}$$

Figure 12 displays the typical spatial supports of functions ψ, ψ^θ, and ψ_{uS}. The stripes represent the direction of oscillation (i.e. the support in the Fourier domain).

5.2 Space-warping transport PDE

The warplet transform of a function $h(x)$ is defined as the inner product of h with warplet ψ_{uS}:

$$Wh(u, S) = \iint h(x_1, x_2) \psi^*_{uS}(x_1, x_2) \; dx_1 \; dx_2 \; .$$

Recall the model (10) for the image irradiance I

$$I(x) = f(x)\ R_u(\gamma_u(x))\ .$$

The variance of the warplet coefficients of the stochastic process I is denoted

$$\sigma_{WI}(u,S) = E\left\{|WI(u,S)|^2\right\}\ .$$

The model (10) assumes that R_u is stationary, and also that $E\{R_u(x)R_u(y)\}$ is independent of u (see Section 3.3). As a consequence, $\sigma_{WR_u}(u,S)$, the variance of the warplet coefficients of R_u, is a function of the warping matrix S only. It is denoted

$$\sigma_{WR_u}(u,S) = \sigma(S).$$

Moreover, using the scale separation principle of Section 3.3, one can prove the following proposition:

Proposition 5.1 (Warplet migration) *Under some conditions on the covariance of R,*

$$\sigma_{WI}(u,S) = |f(u)|^2 \sigma_{WR}(G_u S)\ (1+\varepsilon(s_1,s_2))\ , \tag{17}$$

with $\varepsilon(s_1,s_2) \to 0$ *when* $\max(s_1,s_2) \to 0$.

Formally, differentiating (17) with respect to (u_1,u_2) and to the four elements $s_{11}, s_{12}, s_{21}, s_{22}$ of matrix S leads to a space-warping transport PDE, which is a 2D extension of the space-scale transport PDE of Proposition 4.3. Some notations must be introduced before we state the result.
The notation $\nabla_S \sigma_{WI}$ stands for the 2×2 matrix

$$\begin{pmatrix} \frac{\partial \sigma_{WI}}{\partial s_{11}} & \frac{\partial \sigma_{WI}}{\partial s_{12}} \\ \frac{\partial \sigma_{WI}}{\partial s_{21}} & \frac{\partial \sigma_{WI}}{\partial s_{22}} \end{pmatrix},$$

and the colon $':'$ denotes the inner product between two matrices:

$$M\!:\!N = \sum_{i,j} m_{ij}\ n_{ij}\ .$$

If

$$G_u = \begin{pmatrix} g_{11}(u) & g_{12}(u) \\ g_{21}(u) & g_{22}(u) \end{pmatrix},$$

the gradient of G_u is

$$\overrightarrow{\nabla_u G} = \begin{pmatrix} \overrightarrow{\nabla_u g_{11}} & \overrightarrow{\nabla_u g_{12}} \\ \overrightarrow{\nabla_u g_{21}} & \overrightarrow{\nabla_u g_{22}} \end{pmatrix}.$$

Let $\vec{A}$ denote the unknown warping gradient (11), i.e. the relative variations of G_u:

$$\vec{A} = \begin{pmatrix} \vec{a}_{11} & \vec{a}_{12} \\ \vec{a}_{21} & \vec{a}_{22} \end{pmatrix} = G_u^{-1}\ \overrightarrow{\nabla_u G}$$

The unknown shading gradient is denoted $\vec{b} = \overrightarrow{\nabla_u (\log f)^2}$.

Proposition 5.2 (Space-warping transport) *Under some conditions on the covariance of R,*

$$\overrightarrow{\nabla_u \sigma_{WI}}(u,S) - \vec{A}\, S : \nabla_S \sigma_{WI}(u,S) \;=\; \vec{b}\, \sigma_{WI}(u,S)\ (1+\varepsilon(s_1,s_2))\ , \quad (18)$$

with $\varepsilon(s_1,s_2) \to 0$ *when* $\max(s_1,s_2) \to 0$.

Note that all the ingredients of the one-dimensional transport equation (14) appear again in (18).

We briefly indicate how $\vec{A}$ and $\vec{b}$ may be recovered by using (18). Consider warplets with $\alpha = 0$; the warping matrix S now has only three parameters (s_1, s_2, θ):

$$S = \begin{pmatrix} s_1 & 0 \\ 0 & s_2 \end{pmatrix} R_\theta\ ,$$

and the warplet (16) takes a simple form

$$\psi_{uS}(x) = \frac{1}{s_1 s_2}\psi^\theta\left(\frac{x_1-u_1}{s_1}, \frac{x_2-u_2}{s_2}\right)\ .$$

After rearranging terms, (18) becomes

$$\left(s_1\frac{\partial\sigma_{WI}}{\partial s_1} \quad -\frac{s_2}{s_1}\frac{\partial\sigma_{WI}}{\partial\theta} \quad \frac{s_1}{s_2}\frac{\partial\sigma_{WI}}{\partial\theta} \quad s_2\frac{\partial\sigma_{WI}}{\partial s_2} \quad \sigma_{WI}\right)\cdot\begin{pmatrix}\vec{a}_{11}\\ \vec{a}_{12}\\ \vec{a}_{21}\\ \vec{a}_{22}\\ \vec{b}\end{pmatrix}$$

$$\approx \overrightarrow{\nabla_u \sigma_{WI}}\ . \quad (19)$$

To recover the five unknowns $(\vec{a}_{11}, \vec{a}_{12}, \vec{a}_{21}, \vec{a}_{22}, \vec{b})$, one must write (19) for a set of at least five triplets (s_1, s_2, θ). It is preferable to use even more triplets if available, yielding an overdetermined system which can be solved by singular value decomposition. The estimation of the covariance and its derivatives which appear in the equation can be done by using a spatial averaging method similar to the one described in Section 4.2.

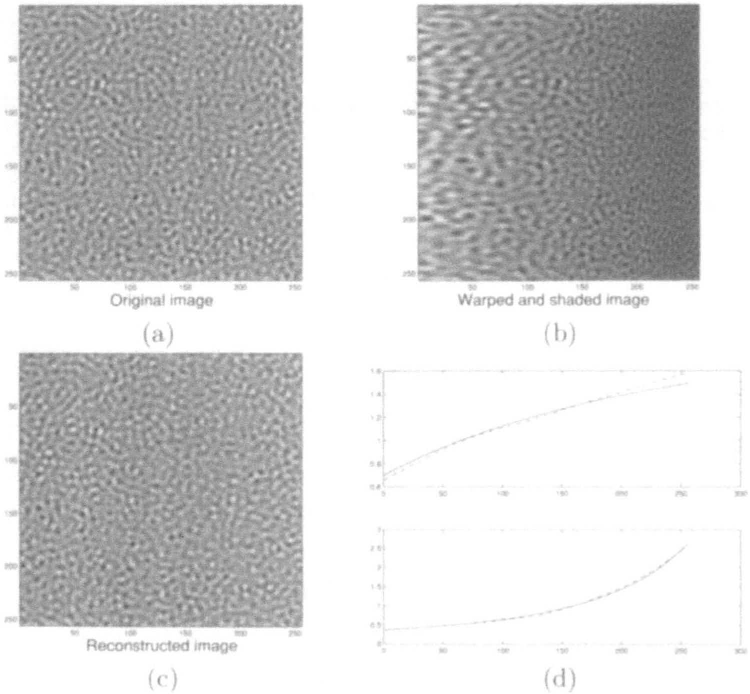

(a) A synthetic stationary texture; (b) Non-stationary texture obtained by warping and applying shading to image(a) in the horizontal direction; (c) Stationarized texture, obtained by applying the inverse of the estimated warping and shading to image(b); (d) top: the horizontal warping (full) and its estimate (dashed); bottom: the shading (full) and its estimate (dashed).

Conclusions

By taking a stochastic viewpoint on non-stationary images of textured surfaces, the double origin of the non-stationarity (a shading gradient and a warping gradient) is brought to light. The general shape-from-texture problem, which incorporates shape-from-shading, is decomposed in two stages:

- the estimation of the warping and shading gradients from the image irradiance: they specify a local rescaling and change of amplitude,
- the integration of the surface from these two gradients.

The first stage is now solved thanks to a PDE satisfied by the variance of the wavelet or warplet coefficients of the image irradiance. The second stage can be solved in a pure shape-from-shading context. Its resolution for the warping gradient component is still under study.

References

[1] Barles, G., Souganidis, P.E. (1991). Convergence of approximation schemes for fully nonlinear second order equations. *Asymptotic Analysis* **4**, pp. 271-283.

[2] Blake, A. and Marinos, C. (1990). Shape from texture: estimation, isotropy and moments. *Artificial Intelligence* **45**, pp. 323-380.

[3] Blostein, D. and Ahuja, N. (1990). Shape from texture: integrating texture-element extraction and surface estimation. *IEEE Trans. Patt. Anal. and Mach. Intell.* **11-12**, pp. 1233-1251.

[4] Bovik, A. Gopal, N., Emmoth, T. and Restrepo, A. (1992). Localized Measurement of Emergent Image Frequencies by Gabor Wavelets. *IEEE Trans. Info. Theory* **38-2**, pp. 691-712.

[5] Brockwell, P. and Davis, R. (1991). *Time Series: Theory and Methods.* Springer-Verlag.

[6] Clerc, M., Mallat, S. (1998). Shape from Texture and Shading with Wavelets, Technical Report, CERMICS, Ecole Nationale des Ponts et Chaussées.

[7] Dupuis, O. and Oliensis, J. (1992). Direct Method for Reconstructing Shape from Shading. in *Proc. IEEE Comp. Vis. Pattern Recog. Conf.*

[8] Gårding, J. (1992). Shape from Texture for Smooth Surfaces under Perspective Projection. *Journal of Mathematical Imaging and Vision* **2**, pp. 327-350.

[9] Horn, B.K.P. (1986). Robot vision. The MIT Press, McGraw-Hill Book Company.

[10] Kanatani, K. and Chou, T. (1989). Shape from Texture: General Principle. *Artificial Intelligence* **38**, pp. 1-48.

[11] Lions, P.L. (1982). *Generalized Solutions of Hamilton-Jacobi Equations.* Pitman, London.

[12] Lions, P.L., Rouy, E., Tourin, A. (1993). Shape-from-shading, viscosity solutions and edges. *Numerische Mathematik* **64**, pp. 323-353.

[13] Malik, J. and Rosenholtz, R. (1997). Computing Local Surface Orientation and Shape From Texture for Curved Surfaces. *Int. J. of Computer Vision* **23-2**, pp. 149-168.

[14] Pentland, A. (1990). Linear Shape From Shading. *Int. J. of Computer Vision* **4**, pp. 153-162.

[15] Rouy, E. and Tourin, A. (1992). A Viscosity Solutions Approach to Shape-from-Shading. *SIAM J. Numer. Anal.* **29-2**, pp. 867-884.

[16] Shah, J., Pien, H.H., and Gauch, J.M. (1996). Recovery of Surfaces with Discontinuities by Fusing Shading and Range Data Within a Variational Framework. *IEEE Trans. Image Processing* **5-8**, pp. 1243-1251.

[17] Super, B. and Bovik, A. (1995). Shape from Texture Using Local Spectral Moments. *IEEE Trans. Patt. Anal. and Mach. Intell.* **17-4**, pp. 333-343.

[18] Witkin, A. (1981) Recovering surface shape and orientation from texture. *Artificial Intelligence* **17**, pp. 17-45.

CERMICS, Ecole Nationale des Ponts et Chaussées, 77455 Marne-la-Vallée Cedex 2, France

Centre de Mathématiques Appliquées, Ecole Polytechnique, 91128 Palaiseau Cedex, France

Progress in Systems and Control Theory, Vol. 25

Anisotropic Smoothing of Posterior Probabilities [1]

Patrick C. Teo, Guillermo Sapiro, Brian A. Wandell

1 Introduction

In a large number of segmentation problems, the number of different objects or classes present in the image is known a priori. Examples are magnetic resonance images of the cortex and SAR data. A technique to introduce this prior knowledge into the segmentation process is presented and analyzed in this paper. The basic idea is to perform edge preserving anisotropic smoothing of posterior probabilities, computed via Bayes rule, followed by an independent pixelwise maximum aposterior probability (MAP) classification. In this paper, we describe the technique and develop the mathematical theory underlying it. We demonstrate that prior anisotropic smoothing of the posterior probabilities yields the MAP solution of a discrete Markov random field (MRF) with a non-interacting, analog discontinuity field. In contrast, isotropic smoothing of the posterior probabilities is equivalent to computing the MAP solution of a single, discrete MRF using continuous relaxation labeling. Combining a discontinuity field with a discrete MRF is important as it allows the disabling of clique potentials across discontinuities. Furthermore, explicit representation of the discontinuity field suggests new algorithms that incorporate properties like hysteresis and non-maximal suppression.

Following the seminal work of Perona and Malik [12], anisotropic diffusion has been used for a large number of applications. The basic idea behind anisotropic diffusion is to smooth images while preserving discontinuities (edges). This is achieved deforming the image according to a non-linear partial differential equation. A large amount of theoretical research has been devoted to these type of equations as well.

Anisotropic diffusion applied to the raw image data is well motivated only when the noise is additive and signal independent; see [4]. For example, if two objects in the scene have the same mean and differ only in variance, anisotropic diffusion of the data is not effective. In addition to this problem, anisotropic diffusion does not take into consideration the special content of the image. That is, in a large number of applications, like magnetic resonance imaging of the cortex and SAR data, it is known in advance the number of different types of objects in the scene, and directly applying

[1] This work was partially supported by ONR Grant N00014-97-1-0509, ONR Young Investigator Program Award, and NSF-LIS.

anisotropic diffusion to the raw data does not take into consideration this important information given a priori.

In [14], we proposed a possible solution to the problems described above. The proposed scheme constitutes one of the steps of a complete system for the segmentation of MRI volumes of human cortex. The technique comprises three steps; see Figure 1. First, the posterior probability of each pixel is computed from its likelihood and a homogeneous prior; i.e., a prior that reflects the relative frequency of each class (white matter, gray matter, and non-brain in the case of MRI of the cortex), but is the same across all pixels. Next, the posterior probabilities for each class are anisotropically smoothed (using a 3D-extension of the algorithm suggested by Perona and Malik [12]). Finally, each pixel is classified independently using the MAP rule. Figure 2 compares the classification of cortical white matter with and without the anisotropic smoothing step. The anisotropic smoothing produces classifications that are qualitatively smoother within regions while preserving detail along region boundaries. The intuition behind the method is straightforward. Anisotropic smoothing of the posterior probabilities results in piecewise constant posterior probabilities which, in turn, yield piecewise "constant" MAP classifications.

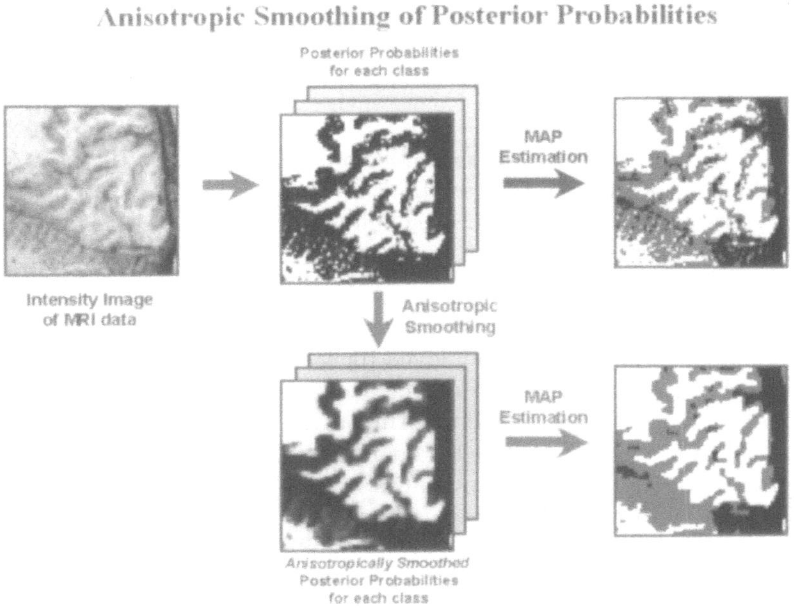

Figure 1: Schematic description of the posterior diffusion algorithm.

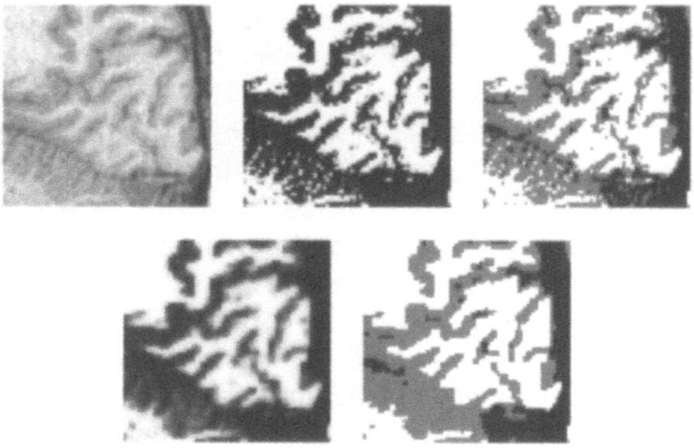

Figure 2: (Top row) Left: Intensity image of MRI data. Middle: Image of posterior probabilities corresponding to white matter class. Right: Image of corresponding MAP classification. Brighter regions in the posterior image correspond to areas with higher probability. White regions in the classification image correspond to areas classified as white matter; black regions correspond to areas classified as CSF. (Bottom row) Left: Image of white matter posterior probabilities after being anisotropically smoothed. Right: Image of MAP classification computed using smoothed posteriors.

In Figure 3 we use a toy example to further motivate the use of this technique. We compare the results of isotropic diffusion on the raw data with those obtained with anisotropic diffusion on the posterior, followed by a MAP classification.

This technique, originally developed for MRI segmentation, is quite general, and can be applied to any given (or learned) probability distribution functions, both for the priors and the likelihood. In this paper we first describe the technique in its general formulation. Then, we explore the mathematical theory underlying the technique. We demonstrate that anisotropic smoothing of the posterior probabilities yields the MAP solution of a discrete MRF with a non-interacting, analog discontinuity field. In contrast, isotropic smoothing of the posterior probabilities is equivalent to computing the MAP solution of a single, discrete MRF using continuous relaxation labeling. Combining a discontinuity field with a discrete MRF is important as it allows the disabling of clique potentials across discontinuities. Furthermore, explicit representation of the discontinuity field suggests new algorithms that incorporate hysteresis and non-maximal suppression.

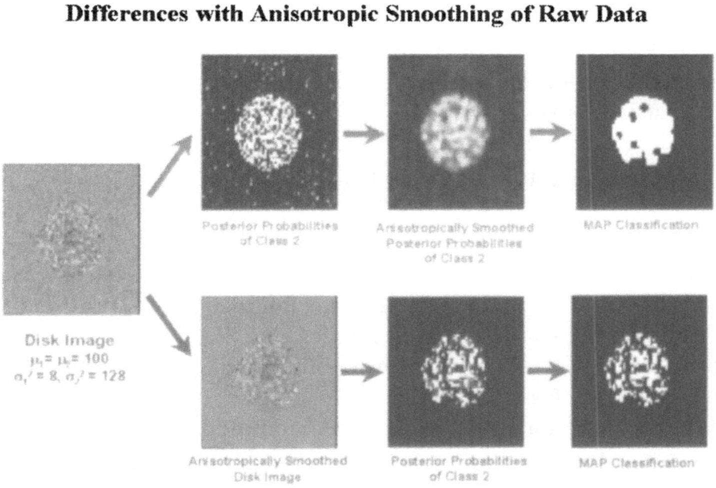

Figure 3: Toy example of the posterior diffusion algorithm. Two classes of the same average and different standard deviation are present in the image. The first row show the result of our algorithm (posterior, diffusion, MAP), while the second row shows the result of classical techniques (diffusion, posterior, MAP).

2 The general technique

Let's assume that it is given to us, in the form of a priori information, the number k of classes (different objects) in the image. In MRI of the cortex for example, these classes would be white matter, gray matter, and CSF (non-brain). In the case of SAR images, the classes can be "object" and "background." Similar classifications can be obtained for a large number of additional applications and image modalities.

In the first stage, the pixel or voxel intensities within each class are modeled as independent random variables with given or learned distributions. Thus, the likelihood $\Pr(V_i = v | C_i = c)$ of a particular pixel (or voxel in the case of 3D MRI), V_i, belonging to a certain class, C_i, is given. For example, in the case of normally distributed likelihood, we have

$$\Pr(V_i = v | C_i = c) = \frac{1}{\sqrt{2\pi}\sigma_c} \exp\left(-\frac{1}{2} \frac{(v-\mu_c)^2}{\sigma_c^2} \right). \tag{1}$$

Here, i is a spatial index ranging over all pixels or voxels in the image, and the index c stands for one of the k classes. V_i and C_i correspond to the intensity and classification of voxel i respectively. The mean and variance (μ_c and σ_c) are given, learned from examples, or adjusted by the user.

The posterior probabilities of each voxel belonging to each class are computed using Bayes' Rule:

$$\Pr(C_i = c|V_i = v) = \frac{1}{K} \Pr(V_i = v|C_i = c) \Pr(C_i = c) \tag{2}$$

where K is a normalizing constant independent of c. As in the case of the likelihood, the prior distribution, $\Pr(C_i = c)$, is not restricted, and can be arbitrarily complex. In a large number of applications, we can adopt a homogeneous prior, which implies that $\Pr(C_i = c)$ is the same over all spatial indices i. The prior probability typically reflects the relative frequency of each class.

After the posterior probabilities are computed (note that we will have now k images), the posterior images are smoothed anisotropically (in two or three dimensions), but preserving discontinuities. The anisotropic smoothing technique applied can be based on the original version proposed by Perona *et al.* [12], or any other of the extensions later proposed, e.g., [1, 4]. This step involves simulating a discretization of the following partial differential equation:

$$\frac{\partial P_c}{\partial t} = \operatorname{div}(g(||\nabla P_c||)\nabla P_c) \tag{3}$$

where $P_c = \Pr(C = c|V)$ stand for the posterior probabilities for class c, the stopping term $g(||\nabla P_c||) = \exp(-(||\nabla P_c||/\eta_c)^2)$, and η_c represents the rate of diffusion for class c. The function $g(\cdot)$ controls the local amount of diffusion such that diffusion across discontinuities in the volume is suppressed. Since we are now smoothing probabilities, to be completely formal, these evolving probabilities should be normalized each step of the iteration to add to one (vector-valued diffusions, where the vector is given by the stack of posterior images, can be applied as well).

Finally, the classifications are obtained using the maximum aposteriori probability (MAP) estimate after anisotropic diffusion. That is,

$$C_i^* = \arg\max_k \Pr^*(C_i = c|V_i = v) \tag{4}$$

where $\Pr^*(C_i = c|V_i = v)$ corresponds to the posterior following anisotropic diffusion.

Recapping, the proposed algorithm has the following steps:

1. Compute the priors and likelihood functions for each one of the classes in the images.
2. Using Bayes rule, compute the posterior for each class.
3. Apply anisotropic diffusion (combined with normalization) to the posterior images.

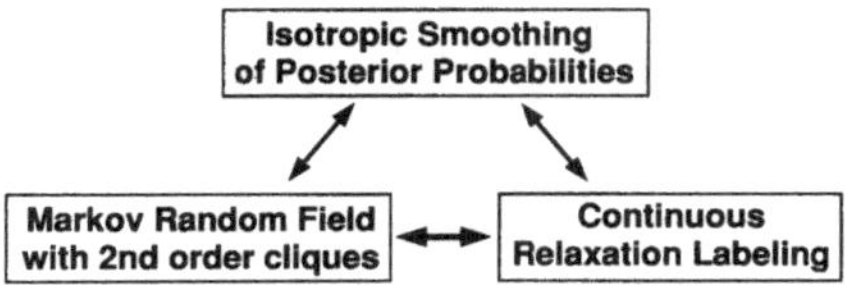

Figure 4: *Equivalence between isotropic smoothing of posterior probabilities, Markov random fields with 2nd order cliques, and continuous relaxation labeling.*

4. Use MAP to obtain the classification.

This techniques solves both problems mentioned in the introduction. That is, it can handle non-additive noise, and more important, introduces prior information about the type of images being processed. In the rest of this paper will will describe the relations of this technique with other algorithms proposed in the literature.

3 Isotropic Smoothing

In this section, we describe the relationship between maximum aposterior probability (MAP) estimation of discrete Markov random fields (MRF) and continuous relaxation labeling (CRL) [13]. This connection was originally made by Li *et. al.* [11]. We review this relationship to introduce the notation that will be used in the rest of the paper and to point out the similarities between this technique and isotropic smoothing of posterior probabilities. These relationships are depicted in Figure 4.

We specialize our notation to MRF's defined on image grids. Let $\mathcal{S} = \{1, \ldots, n\}$ be a set of sites where each $s \in \mathcal{S}$ corresponds to a single pixel in the image. For simplicity, we assume that each site can take on labels from a common set $\mathcal{L} = \{1, \ldots, k\}$. Adjacency relationships between sites are encoded by $\mathcal{N} = \{\mathcal{N}_i | i \in \mathcal{S}\}$ where $\mathcal{N}_i$ is the set of sites neighboring site i. Cliques are then defined as subsets of sites so that any pair of sites in a clique are neighbors. In this paper, we will only consider 4-neighbor adjacency for images (and 8-neighbor adjacency for volumes) and cliques of sizes no greater than two. By considering each site as a discrete random variable f_i with a probability mass function over $\mathcal{L}$, a discrete MRF $\mathbf{f}$ can be defined over the sites with a Gibbs probability distribution.

If data $d_i \in \mathbf{d}$ is observed at each site i, and is dependent only on its label f_i, then the posterior probability is itself a Gibbs distribution and by the Hammersley-Clifford theorem, also a MRF, albeit a different one [6]: $P(\mathbf{f}|\mathbf{d}) = Z^{-1} \times \exp\{-E(\mathbf{f}|\mathbf{d})\}$ where

$$E(\mathbf{f}|\mathbf{d}) = \sum_{i \in \mathcal{C}_1} V_1(f_i|d_i) + \sum_{(i,j) \in \mathcal{C}_2} V_2(f_i, f_j) \tag{5}$$

where $V_1(f_i|d_i)$ is a combination of the single site clique potential and the independent likelihood and $V_2(f_i, f_j)$ is the pairwise-site clique potential. The notation (i, j) refers to a pair of sites; thus, the sum is actually a double sum. Maximizing the posterior probability $P(\mathbf{f}|\mathbf{d})$ is equivalent to minimizing the energy $E(\mathbf{f}|\mathbf{d})$.

As a result, the MAP classification problem is one of finding the set of classes $\mathbf{f}^*$ that minimizes

$$\mathbf{f}^* = \arg\min_{\mathbf{f}\in\mathcal{L}^n} E(\mathbf{f}|\mathbf{d}). \tag{6}$$

This is a combinatorial problem since $\mathcal{L}^n$ is discrete. There are a variety of solution techniques to this problem some of which are stochastic like simulated annealing [10] while others are deterministic like ICM [2].

Continuous Relaxation Labeling. The continuous relaxation labeling approach to solving this problem was introduced by Li *et. al.* [11]. In CRL, the class (label) of each site i is represented by a vector $p_i = [p_i(f_i)|f_i \in \mathcal{L}]$ subject to the constraints: (1) $p_i(f_i) \geq 0$ for all $f_i \in \mathcal{L}$, and (2) $\sum_{f_i\in\mathcal{L}} p_i(f_i) = 1$. Within this framework, the energy $E(\mathbf{f}|\mathbf{d})$ to be minimized is rewritten as

$$\begin{aligned} E(\mathbf{p}|\mathbf{d}) &= \sum_{i\in\mathcal{C}_1}\sum_{f_i\in\mathcal{L}} V_1(f_i|d_i)p_i(f_i) \\ &\quad + \sum_{(i,j)\in\mathcal{C}_2}\sum_{(f_i,f_j)\in\mathcal{L}^2} V_2(f_i, f_j)p_i(f_i)p_j(f_j). \end{aligned}$$

Note that when $p_i(f_i)$ is restricted to $\{0, 1\}$, $E(\mathbf{p}|\mathbf{d})$ reverts to its original counterpart $E(\mathbf{f}|\mathbf{d})$. Hence, CRL embeds the actual combinatorial problem into a larger, continuous, constrained minimization problem.

The constrained minimization problem is typically solved by iterating two steps: (1) gradient computation, and (2) normalization and update. The first step decides the direction that decreases the objective function while the second updates the current estimate while ensuring compliance with the constraints. A review of the normalization techniques that have been proposed are summarized in [11]. Ignoring the need for normalization, continuous relaxation labeling is similar to traditional gradient descent: $p_i^{t+1}(f_i) \leftarrow p_i^t(f_i) - \frac{\partial E(\mathbf{p}|\mathbf{d})}{\partial p_i^t(f_i)}$ where

$$\frac{\partial E(\mathbf{p}|\mathbf{d})}{\partial p_i^t(f_i)} \doteq V_1(f_i|d_i) + 2\sum_{j:(i,j)\in\mathcal{C}_2}\sum_{f_j\in\mathcal{L}} V_2(f_i, f_j)p_j^t(f_j). \tag{7}$$

and the superscripts $t, t+1$ denote iteration numbers. The notation $j : (i, j)$ refers to a single sum over j such that (i, j) are pairs of sites belonging to a clique. Barring the different normalization techniques could be employed,

Eqn. 7 is found in the update equations of various CRL algorithms [13, 5, 9]. There are, however, two differences. First, in most CRL problems, the first term of Eqn. 7 is absent and thus proper initialization of $\mathbf{p}$ is important. We will also omit this term in the rest of the paper to emphasize the similarity with continuous relaxation labeling. Second, CRL problems typically involve maximization; thus, $V_2(f_i, f_j)$ would represent consistency as opposed to potential, and the update equation would add instead of subtract the gradient.

Isotropic Smoothing. A convenient way of visualizing the above operation is as isotropic smoothing. Since the sites represent pixels in an image, for each class f_i, $p_i(f_i)$ can be represented by an image (of posterior probabilities) such that k classes imply k such image planes. Together, these k planes form a volume of posterior probabilities. Each step of Eqn. 7 then essentially replaces the current estimate $p_i^t(f_i)$ with a weighted average of the neighboring assignment probabilities $p_j^t(f_j)$. In other words, the volume of posterior probabilities is linearly *filtered*. If the potential functions $V_2(f_i, f_j)$ favor similar labels, then the weighted average is essentially low-pass among sites with common labels and hi-pass among sites with differing labels.

This notion is best illustrated with a simple example. Consider a classification problem with three classes; i.e., $f_i \in \{1, 2, 3\}$ and the volume is made up of three planes. Define $V_2(f_i, f_j) = -1/4$ when $f_i = f_j$ and sites i and j are 4-neighbors; for example, when $f_i = 2$ and i is the middle site, $V_2(2, f_j)$ has the following values:

$$\underset{f_j=1}{\begin{bmatrix} 0 & 0 & 0 \\ 0 & 0 & 0 \\ 0 & 0 & 0 \end{bmatrix}} \quad \underset{f_j=2}{\begin{bmatrix} 0 & -1/4 & 0 \\ -1/4 & 0 & -1/4 \\ 0 & -1/4 & 0 \end{bmatrix}} \quad \underset{f_j=3}{\begin{bmatrix} 0 & 0 & 0 \\ 0 & 0 & 0 \\ 0 & 0 & 0 \end{bmatrix}}.$$

Thus, the penalty is smaller when adjacent pixels have the same class than when their classes differ. Combining these penalties with the update equation, $p_i^t(f_i)$ (with $f_i = 2$) gets replaced with a linear combination of $p_j^t(f_j)$ with weights equal to:

$$\underset{f_j=1}{\begin{bmatrix} 0 & 0 & 0 \\ 0 & 0 & 0 \\ 0 & 0 & 0 \end{bmatrix}} \quad \underset{f_j=2}{\begin{bmatrix} 0 & 1/4 & 0 \\ 1/4 & 1 & 1/4 \\ 0 & 1/4 & 0 \end{bmatrix}} \quad \underset{f_j=3}{\begin{bmatrix} 0 & 0 & 0 \\ 0 & 0 & 0 \\ 0 & 0 & 0 \end{bmatrix}}.$$

As a result, the posterior probabilities for each class is smoothed during each step of the iteration. In this example, each of the three planes of posterior probabilities is low-pass filtered separately.

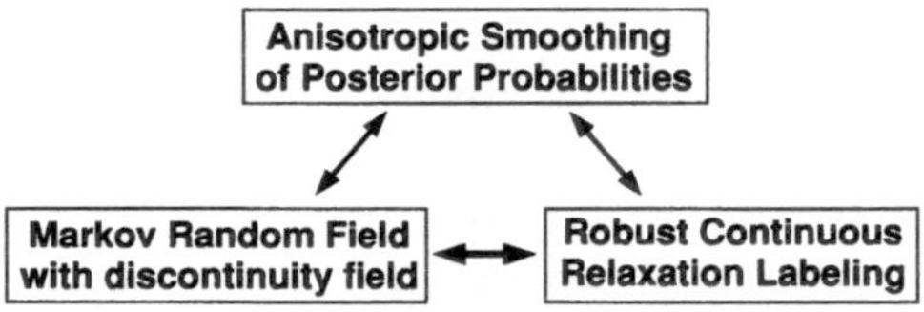

Figure 5: *Equivalence between anisotropic smoothing of posterior probabilities, Markov random fields with discontinuity fields, and robust continuous relaxation labeling.*

4 Anisotropic Smoothing

Isotropic smoothing causes significant blurring especially across region boundaries. A solution to this problem is to smooth adaptively such that smoothing is suspended across region boundaries and takes place only within region interiors. Anisotropic smoothing is often implemented by simulating nonlinear partial differential equations with the image as the initial condition [1, 12]. In this section, we show that while isotropic smoothing of posterior probabilities is the same as continuous relaxation labeling of a MRF, anisotropic smoothing of posterior probabilities is equivalent to continuous relaxation labeling of a MRF supplemented with a (hidden) analog discontinuity field. We also demonstrate that this method could also be understood as incorporating a robust consensus-taking scheme within the framework of continuous relaxation labeling. These relationships are depicted in Figure 5.

We extend the original MRF problem to include a non-interacting, analog discontinuity field on a displaced lattice. Thus, the new energy to be minimized is:

$$E(\mathbf{f}, \mathbf{l}) = \sum_{(i,j)\in\mathcal{C}_2} \left[\frac{1}{2\sigma^2} V_2(f_i, f_j) \cdot l_{i,j} + (l_{i,j} - 1 - \log l_{i,j}) \right] \tag{8}$$

where $V_1(f_i)$ has been dropped for simplicity since the discontinuity field does not interact with it. The individual sites in the discontinuity field $\mathbf{l}$ are denoted by $l_{i,j}$ which represent either the horizontal or vertical separation between sites i and j in $\mathcal{S}$. When $l_{i,j}$ is small, indicating the presence of a discontinuity, the effect of the potential $V_2(f_i, f_j)$ is suspended; meanwhile, the energy is penalized by the second term in Eqn. 8. There are a variety of penalty functions that could be derived from the robust estimation framework (see [3]). The penalty function in Eqn. 8 was derived from the Lorentzian robust estimator.

The minimization of $E(\mathbf{f}, \mathbf{l})$ is now over both $\mathbf{f}$ and $\mathbf{l}$. Since the discontinuity field is non-interacting, $\mathbf{l}$ can be minimized analytically by computing the partial derivatives of $E(\mathbf{f}, \mathbf{l})$ with respect to $l_{i,j}$ and setting that to zero. Doing so and inserting the result back into $E(\mathbf{f}, \mathbf{l})$ gives us

$$E(\mathbf{f}) = \sum_{(i,j)\in\mathcal{C}_2} \log\left[1 + \frac{1}{2\sigma^2} V_2(f_i, f_j)\right]. \tag{9}$$

Rewriting this equation in a form suitable for CRL, we get

$$E(\mathbf{p}) = \sum_{(i,j)\in\mathcal{C}_2} \log\left[1 + \frac{1}{2\sigma^2} \sum_{(f_i,f_j)\in\mathcal{L}^2} V_2(f_i, f_j) p_i(f_i) p_j(f_j)\right]. \tag{10}$$

Note that when $p_i(f_i)$ is restricted to $\{0, 1\}$, Eqn. 6 reduces to Eqn. 5.

Anisotropic Smoothing. To compute the update equation for CRL, we take the derivative of $E(\mathbf{p})$ with respect to $p_i(f_i)$:

$$\frac{\partial E(\mathbf{p})}{\partial p_i(f_i)} \doteq \sum_{j:(i,j)\in\mathcal{C}_2} w_{i,j} \left[\sum_{f_j\in\mathcal{L}} V_2(f_i, f_j) p_j(f_j)\right] \tag{11}$$

where

$$w_{i,j} = 2\sigma^2 \Big/ \left[2\sigma^2 + \sum_{(f_i,f_j)\in\mathcal{L}^2} V_2(f_i, f_j) p_i(f_i) p_j(f_j)\right]. \tag{12}$$

The term $w_{i,j}$ encodes the presence of a discontinuity. If $w_{i,j}$ is constant, then the above equation reverts to the isotropic case. Otherwise, $w_{i,j}$ either enables or disables the penalty function $V_2(f_i, f_j)$. This equation is similar to the anisotropic diffusion equation proposed by Perona and Malik [12].

However, image difference between sites i and j in Perona and Malik's equation is, in our case, replaced by a discrete version: $\sum_{f_j\in\mathcal{L}} V_2(f_i, f_j) p_j(f_j)$. The stopping term $w_{i,j}$ is also the same except that the magnitude of the image gradient is again replaced by a discrete counterpart.

Figure 6 compares between isotropic and anisotropic diffusion of the posterior, and shows the variable weight $w_{i,j}$ for one of the classes in the anisotropic case.

Robust Continuous Relaxation Labeling. Each iteration of continuous relaxation labeling can be viewed as a consensus-taking process [15]. Neighboring pixels vote on the classification of a central pixel based on their current assignment probabilities $p_j(f_j)$, and their votes are tallied using a weighted sum. The weights used are the same throughout the image; thus, pixels on one side of a region boundary may erroneously vote for pixels on the other side. Anisotropic smoothing of the posterior probabilities can be regarded as implementing a robust voting scheme since votes are tempered by $w_{i,j}$ which estimates the presence of a discontinuity. The connection between anisotropic diffusion on continuous-valued images and robust estimation was recently demonstrated by Black *et. al.* [4].

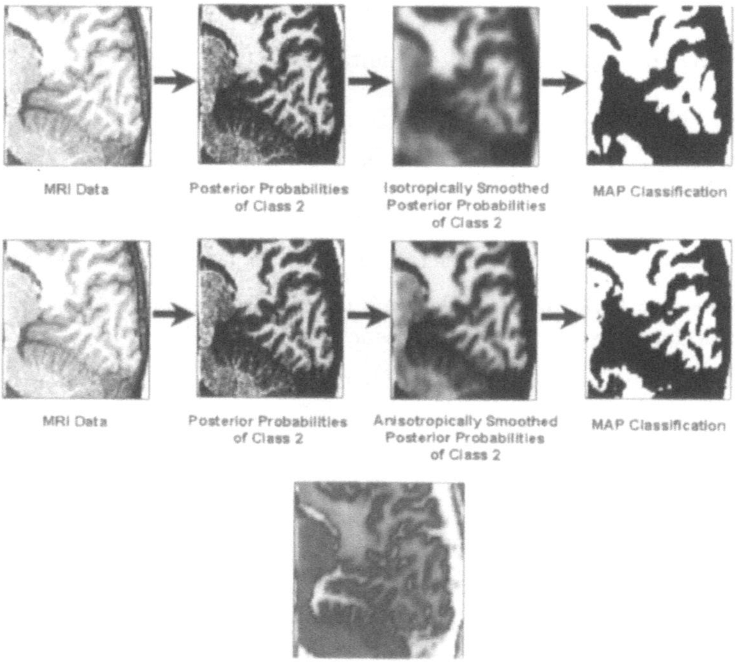

Figure 6: *Comparison between isotropic and anisotropic diffusion of the posterior, and the variable weight* $w_{i,j}$ *(last row) for one of the classes in the anisotropic case.*

5 Discussion

The anisotropic smoothing scheme was used to segment white matter from MRI data of human cortex.[2] Pixels at a given distance from the boundaries of the white matter classification were then automatically classified as gray matter. Thus, gray matter segmentation relied heavily on the white matter segmentation being accurate. Figure 7 shows comparisons between gray matter segmentations produced automatically by the proposed method and those obtained manually. More examples can be found in [14].

[2]In [8], this technique was successfully applied to SAR data as well, incorporating adaptive priors and learned distributions.

When applied to MRI, the technique being proposed bears some superficial resemblance to schemes that anisotropically smooth the raw image before classification [7]. Besides the connection between our technique and MAP estimation of Markov random fields, which is absent in schemes that smooth the image directly, there are two other important differences, since [7] suffers from the common problems of diffusion raw data detailed in the introduction. We should note again that applying anisotropic smoothing on the posterior probabilities is feasible even when the class likelihoods are described by general probability mass functions (and even multi-variate distributions!).

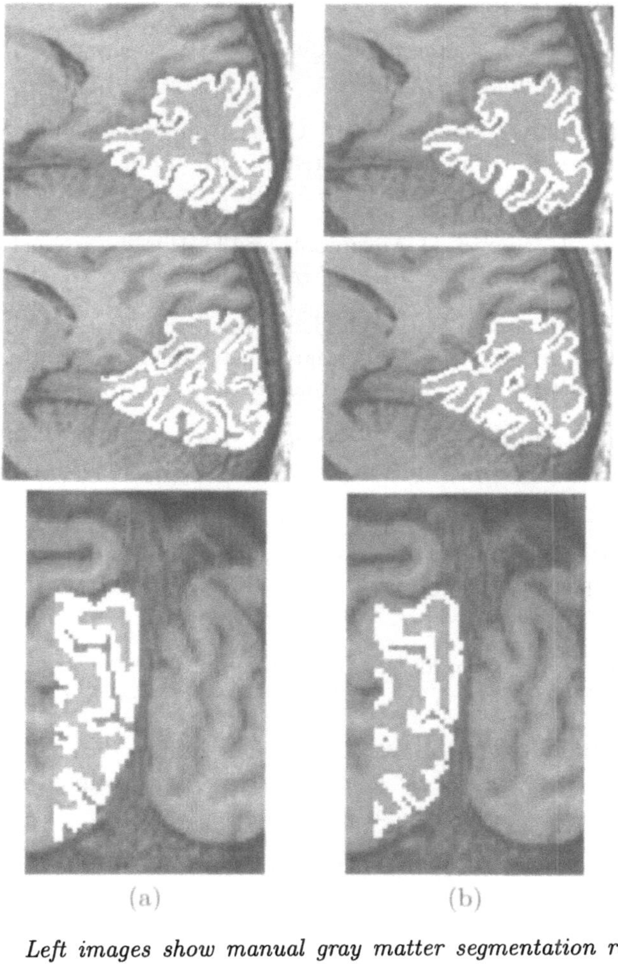

Figure 7: *Left images show manual gray matter segmentation results; right images show the automatically computed gray matter segmentation.*

The equivalence between anisotropic smoothing of posterior probabilities and MRF's with discontinuity fields also offers a solution to the problems of edge handling and missing data. These two issues can be treated in the same manner as in traditional regularization. Solving of the latter implies that MAP classification can be obtained even at locations where the pixel values are not provided.

References

[1] L. Alvarez, P. L. Lions, and J. M. Morel. Image selective smoothing and edge detection by nonlinear diffusion. *SIAM J. Numerical Analysis*, 29:845–866, 1992.

[2] J. Besag. On the statistical analysis of dirty pictures. *J. Royal Statistical Society*, 48:259–302, 1986.

[3] M. Black and A. Rangarajan. On the unification of line processes, outlier rejection, and robust statistics with applications in early vision. *Int'l J. Computer Vision*, 19(1):57–91, 1996.

[4] M. Black, G. Sapiro, D. Marimont, and D. Heeger. Robust anisotropic diffusion. *IEEE Trans. Image Processing*, March 1998.

[5] O. D. Faugeras and M. Berthod. Improving consistency and reducing ambiguity in stochastic labeling: an optimization approach. *IEEE Trans. Pattern Analysis and Machine Intelligence*, 3:412–423, 1981.

[6] S. Geman and D. Geman. Stochastic relaxation, Gibbs distributions, and the Bayesian restoration of images. *IEEE Trans. Pattern Analysis and Machine Intelligence*, 6(6):721–742, 1984.

[7] G. Gerig, O. Kubler, R. Kikinis, and F. A. Jolesz. Nonlinear anisotropic filtering of MRI data. *IEEE Trans. Medical Imaging*, 11:221–232, 1992.

[8] S. Haker, G. Sapiro, and A. Tannenbaum. Knowledge based segmentation of SAR data. *Proc. IEEE-ICIP '98*, Chicago, October 1998.

[9] R. A. Hummel and S. W. Zucker. On the foundations of relaxation labeling processes. *IEEE Trans. Pattern Analysis and Machine Intelligence*, 5(2):267–286, 1983.

[10] S. Kirkpatrick, C. D. Gellatt, and M. P. Vecchi. Optimization by simulated annealing. *Science*, 220:671–680, 1983.

[11] S. Z. Li, H. Wang, and M. Petrou. Relaxation labeling of Markov random fields. In *Int'l Conf. Pattern Recognition*, pages 488–492, 1994.

[12] P. Perona and J. Malik. Scale-space and edge detection using anisotropic diffusion. *IEEE Trans. Pattern Analysis and Machine Intelligence*, 12(7):629–639, 1990.

[13] A. Rosenfeld, R. Hummel, and S. Zucker. Scene labeling by relaxation operations. *IEEE Trans. Systems, Man, and Cybernetics*, 6(6):420–433, 1976.

[14] P. C. Teo, G. Sapiro, and B. A. Wandell. Creating connected representations of cortical gray matter for functional MRI visualization. *IEEE Trans. on Medical Imaging*, 852-863, December 1997.

[15] Y. Weiss and E. Adelson. Perceptually organized EM: a framework for motion segmentation that combines information about form and motion. In *Int'l Conf. Computer Vision and Pattern Recognition*, pages 312–326, 1996.

Computer Science, Stanford University, Stanford, CA 94305, USA;
teo@white.stanford.edu

Electrical and Computer Eng., University of Minnesota, Minneapolis, MN 55455, USA; guille@ece.umn.edu

Psychology and Neuroscience, Stanford University, Stanford, CA 94305, USA; brian@white.stanford.edu

Progress in Systems and Control Theory, Vol. 25

The Accommodation Cue in Vision

S. Soatto

1 Introduction

Any imaging system, such as the human eye, a video-camera, a CT scanner or a telescope, entails a map of the three-dimensional environment onto the two-dimensional surface of an imaging sensor. Images of any such map are characterized by a loss of information along one spatial dimension. The visual system of primates has evolved to exploit two-dimensional images to retrieve a representation of the environment that can be used to perform spatial control tasks crucial to survival.

Images must be combined with a-priori assumptions in order to retrieve a 3D representation. For instance, assuming that the visible surfaces have homogeneous reflectance properties allows associating the variations in image brightness (shading) as a cue for 3D shape [8]. Similarly, assuming that photometric patterns are statistically homogeneous allows using image "texture" to infer the local geometry of the 3D environment [11]. Shading and texture are just two examples of "pictorial cues", i.e. cues which are associated to one single still image. All such cues are intrinsically ambiguous: it is possible to construct more than one (in fact, infinitely many) 3D scenes that violate the a-priori assumptions while generating exactly the same 2D image.

"Dynamic cues", as opposed to pictorial ones, are associated with the *variation* of images over *time*. The typical a-priori assumption is that the scene is piece-wise rigid. Then the image of the same object at different times can be used to infer its 3D shape and motion. Stereo (two or more images of the same scene taken from different viewpoints) can also be considered within the context of dynamic cues: the assumption is that points seen by the left eye correspond to points of the same object on the right eye (the *correspondence* problem). Under suitable conditions, it can be proven that two different 3D rigidly moving scenes cannot produce the same sequence of 2D images, i.e. 3D structure and motion are *observable* from sequences of 2D images [5, 3, 12].

In studying dynamic and pictorial cues, ideal perspective projection is often assumed as a model of the imaging process. By "ideal" we mean that every point in space maps onto one and only one point on the imaging surface, so that the image is "in focus". Real imaging systems, however, have a finite depth of field, which means that only objects within a certain volume of space can appear in focus. We will see how this can be exploited to infer depth information. In addition to that, the control action applied in order to bring one object into focus (we call this action *accommodation*) can also be used to infer depth information. This is commonly done by auto-focusing systems in commercial video cameras.

While pictorial and dynamic cues have been studied extensively over the past few decades, there has been relatively little work on the use of accommodation in vision. The interested reader is referred to [14, 4, 10, 9] for an overview of the state of the art.

Focus and accommodation are natural complements to dynamic and pictorial cues. For instance, dynamic cues require a wide field of view and large parallax, and are well-suited to infer ego-motion, but they are ill-suited to infer the shape of small objects from far away. Depth of field, on the other hand, decreases with the aperture angle, so that accommodation is a useful cue only for small apertures.

Relation to previous work

A number of algorithms have been proposed in the literature of computational vision to estimate depth from focus and accommodation information. Sometime a distinction is drawn between algorithms that assume that a few images with different focus are given ("shape from focus"), and ones that entail a controlled search over all possible focus positions ("shape from defocus"). In our nomenclature we distinguish between the case where feedback from the actuators is or is not available for measurement ("accommodation" and "focus" cues respectively)[4].

The main assumption common to all algorithms available in the literature (overtly or covertly) is that the scene is a plane parallel to the focal plane (equifocal assumption). Not only is such an assumption severely restrictive as a model of physical surfaces, but it generates a fundamental tradeoff: image information needs to be integrated on a region that is as large as possible to defeat the effects of noise, but as small as possible to approximate the equifocal condition [14, 4, 10, 9].

Also, all papers assume that the geometric model of the camera is known (i.e. the camera is calibrated). Nothing is said with regard to what happens when there is uncertainty or errors in the calibration.

In support of the existing literature, one may argue that an arbitrary smooth scene can be approximated with equifocal surfaces to an arbitrary degree. Unfortunately this is not true, since such approximation will necessarily involve discontinuities, which the theory underlying the existing methods does not cover.

There is also a vast body of related literature in the Signal Processing community, where the problem is known as "blind deconvolution". The equifocal assumption is equivalent to assuming a shift-invariant convolution kernel, which is also common in most of the literature. The interested reader can see the special issue [1] for references.

The capability to observe scene's shape depends upon the energy distribution irradiated from the scene. To our knowledge, there are no results in the literature on the conditions that allow estimating shape.

A new analysis is therefore needed, that can address the problem of estimating an unknown shape irradiating energy according to an unknown distribution. Hopefully this analysis will result in more reliable and general algorithms to estimate both the shape of the object and the energy distribution it irradiates.

2 Basic image formation

The image of a point in space is obtained by processing its radiant energy. In natural environments – and in most artificial ones – such energy is irradiated in an incoherent fashion from sources distributed in space. Therefore, any imaging system maps *neighborhoods* (or subsets) of three-dimensional (3D) space onto points on the two-dimensional (2D) imaging surface. Owing to the additive nature of most energy transport phenomena, it is possible to write such a map as an integral over the neighborhood. To this end, call $\mathbf{X} \in \mathbb{R}^3$ the generic point in Euclidean space and $I_0(\mathbf{X})$ the radiant energy distribution[1]. The brightness (energy) $I(\mathbf{x})$ of the image at the point $\mathbf{x} \in \mathbb{R}^2$ is obtained by integrating contributions from a neighborhood $\mathcal{J}(\mathbf{x}, \mathbf{X})$ that depends both on the topology of the scene and on the geometry and optics of the camera:

$$I(\mathbf{x}) = \int_{\mathbb{R}^3} h_{\mathcal{J}}(\mathbf{x}, \mathbf{X}) I_0(\mathbf{X}) d\mathbf{X}. \tag{2.1}$$

Here $h_{\mathcal{J}}$ describes the transport of energy between $\mathbf{X} \in \mathbb{R}^3$ and $\mathbf{x} \in \mathbb{R}^2$ and the topology of the neighborhood over which the integral is performed. Since (2.1) is linear in I_0, an optical system modeled by the above integral is called *linear*. $h(\mathbf{x}, \mathbf{X})$ can be interpreted as the *point-spread function*, i.e. the image (as a function of $\mathbf{x}$) of a point-source located at $\mathbf{X}$, or as a "point-concentration map", that describes the region $\mathcal{J}$ of 3D space that contributes to the energy at the point $\mathbf{x}$. Note that under this interpretation h is a set-valued map that associates to every point $\mathbf{x}$ a region of 3D space. Also, note that physical considerations impose that I_0 and $h_{\mathcal{J}}$ (and therefore I) be non-negative.

For instance, in the context of photography, $I_0(\cdot)$ represents the luminance of the scene, $I(\cdot)$ is the (measured) photograph, and $h_{\mathcal{J}}$ is determined by the geometry and the optics of the camera, as well as by the three-dimensional shape of the scene. A "perfect" camera would have $h_{\mathcal{J}}(\mathbf{x}, \mathbf{X}) = \delta(\mathbf{X} - \rho \begin{bmatrix} \mathbf{x} & 1 \end{bmatrix}^T)$ for some $\rho = \rho(\mathbf{x}) \in \mathbb{R}$, so that the image

[1] We use the word "distribution" in the sense of functional analysis [6]: I_0 is the linear functional that, applied to the function h on a compact support, gives the real value I. In a measure-theoretic sense, I_0 plays the role of a "density". However, I_0 is not really a density, for it needs not be integrable (let alone normalized).

would reproduce exactly the two-dimensional projection of the scene onto the sensor[2]:

$$I(\mathbf{x}) = I_0(\rho \begin{bmatrix} \mathbf{x} & 1 \end{bmatrix}^T) \quad \rho \neq 0 \in \mathbb{R}. \tag{2.2}$$

We will see shortly that such an imaging device not only cannot exist, but it would indeed be rather uninteresting. It is precisely the "imperfection" of real imaging devices that can be exploited to infer interesting properties of the scene that can be useful in a variety of important applications.

Needless to say, the kernel $h_{\mathcal{J}}$ in the model (2.1) can be rather complicate. In general, it will depend upon the topology of the scene as well as upon the optical properties of the imaging system. In the following, we will assume that such contributions are independent, so that the kernel h can be factored (in the sense of convolution) into a component g that depends upon the optics, and one $\chi_{\mathcal{J}}$ that depends on the geometry of the scene:

$$h_J(\mathbf{x}, \mathbf{X}) = g(\mathbf{x}) * \chi_{\mathcal{J}}(\mathbf{x}, \mathbf{X}). \tag{2.3}$$

While g can look rather complex if one carefully models the optics, $\chi_{\mathcal{J}}$ can be represented by a simple indicator function for a locally compact set $\mathcal{J}(\mathbf{x}, \mathbf{X})$.

Consider, for instance, an optical system consisting of a round, thin lens, looking at a plane parallel to the image-plane. Then, in the integral (2.1), the coordinate X_3 orthogonal to the lens gives no contribution and if we call $\tilde{\mathbf{x}} \doteq \begin{bmatrix} X_1 & X_2 \end{bmatrix} \in \mathbb{R}^2$ we can write

$$I(\mathbf{x}) = \int_{\mathbb{R}^2} h_{\mathcal{J}}(\mathbf{x}, \tilde{\mathbf{x}}) I_0(\tilde{\mathbf{x}}) d\tilde{\mathbf{x}}. \tag{2.4}$$

In this paper we will neglect the effects of diffraction for simplicity. While such assumption seems to be realistic in commercial optical systems, we refer the interested reader to [2] for a thorough discussion on the issue.

Under the ideal assumptions of paraxial geometric optics (see [2] for details), the rays emitted from a point-source at a distance Z from the lens converge at a point at a distance Z' on the opposite side of the lens. The depths of such *conjugate* points are related by the so-called "lens equation" (see [2] for details):

$$\frac{1}{Z} + \frac{1}{Z'} = \frac{1}{f} \tag{2.5}$$

where f depends upon the material and geometric properties of the lens.

In [13], we have derived a model of the image of an equi-focal surface with a spherically symmetric imaging system. We now derive a model of image formation for non-equifocal surfaces. We first consider a planar projection system (2D to 1D) as depicted in figure 1, and then extend our considerations to a full imaging system by assuming rotational symmetry about the optical axis. The optical system described in figure 1 is characterized by f, the focal length, Z_I, the distance between the lens and the

[2]Such a camera model is called a "pin-hole": in a Cartesian reference frame centered in the pin-hole, it can be described as a map $\pi : \mathbb{R}^3 \rightarrow \mathbb{R}^2$ that associates to each point $\mathbf{X}$ in space its perspective projection $\mathbf{x} = \begin{bmatrix} \frac{X_1}{X_3} & \frac{X_2}{X_3} \end{bmatrix}$.

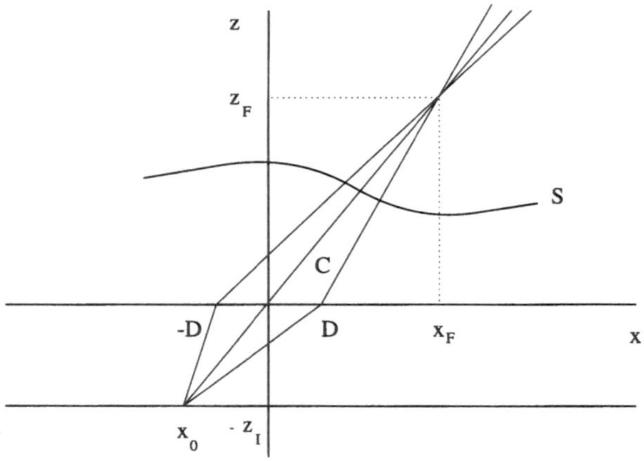

Figure 1: *The point-concentration map*

sensor and D, the lens aperture. The depth of the plane of accommodation Z_F is determined by the lens equation $\frac{1}{-Z_I} + \frac{1}{Z_F} = \frac{1}{f}$ and can be controlled by acting on Z_I or f. Therefore, we can describe the optical system using D and Z_F, which we consider as *control inputs*. Also, the point $[x_0, 0]^T$ can be uniquely described using x_F, since the two are related linearly via

$$x_F = \frac{Z_F}{Z_I} x_0. \tag{2.6}$$

We say that $[x_0, 0]^T$ is the (one and only) image point that focuses at the abscissa x_F. This allows us to concentrate our attention to the top part of figure 1, i.e. $Z > 0$. Figure 2 shows the simplified model of the imaging system.

Let $\beta_\pm$ describe the boundaries of the cone C, i.e. the lines passing through the boundaries of the aperture and the focal point $[x_F, Z_F]^T$:

$$C \doteq \{(x, Z) \mid x = \beta_\pm(Z) = \pm D + \frac{x_F \mp D}{Z_F} Z\}. \tag{2.7}$$

Occasionally we write $\beta_\pm(Z, x_F, Z_F)$ to emphasize the dependence upon the parameters x_F, Z_F. If S represents the surface of an object irradiating energy, the portion that contributes to the point focusing at x_F is the intersection between S and the cone C. If we characterize the surface S as the set $\{(x, Z) \mid Z = S(x)\}$, then the energy radiating from point $[x, Z]^T$ contributes to the point focusing at x_F if and only if

$$x \in \begin{cases} [\beta_-(Z), \beta_+(Z)] & Z < Z_F \\ [\beta_+(Z), \beta_-(Z)] & Z \geq Z_F. \end{cases} \tag{2.8}$$

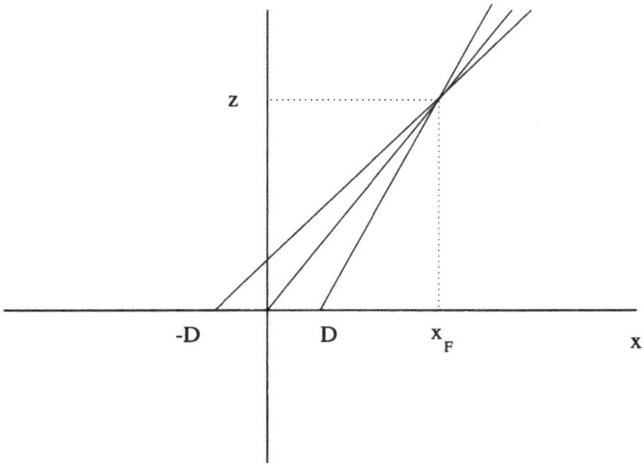

Figure 2: *A "bare-bones" imaging system*

When more than one object contributes to the energy incident x_0, we resort to superposition and consider the contribution from each object separately.

Image of a general surface

We make the assumption that physical scenes contain piecewise smooth surfaces with a finite number of discontinuities on any compact region of space. This rules out pathological cases of surfaces almost everywhere discontinuous or almost everywhere singular.

For such surfaces we can invoke Weierstrass theorem to approximate them to an arbitrary degree with piecewise linear functions. Therefore, the imaging model for any reasonable surface in a physical scene can be obtained as a limit of the composition of the imaging systems described in [13]. Therefore, we write the general model of the geometry of the imaging process as

$$I(x_F, Z_F) = \lim_{n\to\infty} \sum_{i=1}^{n} \int h(x, x_F, Z_F, S_i) I_{0_i}(x) dx \tag{2.9}$$

where each surface S_i is either a slanted plane or an occlusion. We drop the subscript $\mathcal{J}$ from h for simplicity. In order to emphasize the fact that the image depends both upon the shape of the surface S and the energy distribution I_0, we write

$$I(x_F, Z_F; S, I_0) = \int h(x, x_F, Z_F, S) I_0(x) dx + n(x_F, Z_F) \tag{2.10}$$

where the error term n comes from having neglected the effects of the optics as well as from noise in the imaging sensor.

3 Observability of shape

In the previous section we have formalized a model of the geometry of the image formation process. An image $I(x_F, Z_F)$ at the point focusing at x_F, Z_F formed from an energy I_0 distributed on a surface S is denoted by $I(x_F, Z_F; S, I_0)$ and modeled as in equation (2.10).

In this section we establish to what extent it is possible, from measurements of the image $I(x_F, Z_F; S, I_0)$ at controlled values of x_F and Z_F, to estimate the shape of the scene S, and its radiant energy I_0. In other words, we study the *observability* of shape.

In order to simplify the notation, we drop the subscript F on x_F, and substitute Z_F with u, to emphasize the fact that it is a controlled input. We first formalize the notion of *indistinguishability*.

Definition 3.1 *We say that two surfaces S_1 and S_2 are* indistinguishable *and we write $S_1 \mathcal{I} S_2$ if, for all possible energy distributions $I_0 \in \mathcal{L}_{loc}(\mathbb{R})$, we have*

$$\begin{aligned} &I(x, u; S_1, I_0) = I(x, u; S_2, I_0) \\ &\forall \ \ x \in [-B, B], \qquad \forall \ \ \{u(t)\}_{t\in[0,T]} \in \mathbb{R}^+([0,T]). \end{aligned} \tag{3.1}$$

We indicate with $\mathcal{L}_{loc}(\mathbb{R})$ the set of real-valued functions that have finite measure on any compact support (but not on necessarily on the whole line)[3].

We call $\mathcal{I}(S) = \{\tilde{S} \mid \tilde{S}\mathcal{I}S\}$ the set of surfaces indistinguishable from S.

Definition 3.2 *We say that S is* weakly observable *if*

$$\mathcal{I}(S) = \{S\}. \tag{3.2}$$

In other words, a surface S is weakly observable if, given any surface $\tilde{S}$, there exists at least a point x, a control input $u(t)$ and energy distribution I_0 for which $I(x, u(t); \tilde{S}, I_0) \neq I(x, u(t); S, I_0)$. The purpose of the next subsection is to establish that indistinguishable surfaces are equal up to a set of measure zero.

3.1 Weak observability

The following claim establishes that planar scenes are observable.

Claim 3.1 *Let $S(x) = Z_0 + \tan(\alpha)x$. Then $\mathcal{I}(S) = \{S\}$*

Proof (see [13])
An edge boundary is a special case of a planar surface, $S(x) = Z_0 + \alpha x$, where $\alpha = 0$ and the energy distributions I_0 are such that $I_0(x) = 0 \quad x < 0$.

[3] In particular, we allow "delta" in the class of admissible energy distributions to simulate an ideal point-source.

Naturally the left half plane $x < 0$ contains no information and therefore we can say nothing about observability there. However, for the right half plane $x \geq 0$ we can conclude that depth is observable:

Claim 3.2 *Let $S(x) = Z_0$ for all $x \geq 0$ and $I_0(x) = 0$ for all $x < 0$. Then*

$$\mathcal{I}(S) = \{\tilde{S} \mid S(x) = \tilde{S}(x) \; \forall \; x \geq 0\}. \tag{3.3}$$

Consider now the image of an occluding boundary as modeled in [13]. The surface discontinuity can be described locally by the depth of the occluding surface Z_1 and that of the background Z_2. Therefore, we describe the surface as S_{Z_1,Z_2}. The image of the occluding boundary can be obtained from [13] as

$$I(x, u, Z_1, Z_2) = \int h_1(x, x, u, Z_1) I_1(x) dx + \int h_2(x, x, u, Z_1, Z_2) I_2(x) dx \tag{3.4}$$

The shape of an occluding boundary, characterized by Z_1 and Z_2, is observable.

Claim 3.3 *Let S_{Z_1,Z_2} describe an occluding boundary:*

$$\mathcal{I}(S_{Z_1,Z_2}) = \{S_{Z_1,Z_2}\} \tag{3.5}$$

Let $x_c = x_c(u, Z_1)$ be defined as the abscissa of the intersection of the cone delimited by $\beta_\pm$ with the boundary $[0, Z_1]^T$. Then for $x > x_c$ it follows that the background plane gives no contribution, for $h_2(x, x) = 0 \;\; \forall \, x > x_c$. Therefore, the analysis is identical to the planar case. Similarly, for $x < -x_c$, the occluding plane gives no contribution, and the analysis also follows immediately. Therefore, in this section we only need to prove observability for $x \in [-x_c, x_c]$.

Proof

Assume that for all $x \in [-x_c, x_c], u(t) > 0, I_1$ and I_2 we have

$$I(x, u, Z_1, Z_2) = I(x, u, \tilde{Z}_1, \tilde{Z}_2) \tag{3.6}$$

while $\tilde{Z}_1 \neq Z_1$ and $\tilde{Z}_2 \neq Z_2$. We write the above equation as

$$\int \Big(h_1(x, y, u) - \tilde{h}_1(x, y, u)\Big) I_1 dy + \int \Big(h_2(x, y, u) - \tilde{h}_2(x, y, u)\Big) I_2 dy = 0 \tag{3.7}$$

where we have defined $\tilde{h}_1 = h_1(x, y, u, \tilde{Z}_1)$ and $\tilde{h}_2 = h_2(x, y, u, \tilde{Z}_1, \tilde{Z}_2)$. Since the equality holds for all I_1 and I_2, the only way in which the sum of the two integrals is zero is if each integral is zero. But the first integral implies that

$$\tilde{Z}_1 = Z_1 \tag{3.8}$$

following section 3.1. As for the second integral, choose $I_2(x) = \delta(x - x)$, $u(t) = t$ *and compute* $\frac{\partial h_2}{\partial t} = \delta(x, x_l) + \delta(x, x_r)$, *where we have*

$$x_l = \begin{cases} \beta_+ & t < Z_1 \\ \gamma & Z_1 \leq t < Z_2 \\ \beta_- & t \geq Z_2. \end{cases} \tag{3.9}$$

Therefore, we must have $\gamma(x, t, Z_1, Z_2) = \gamma(x, t, Z_1, \tilde{Z}_2)$ *for all* t *and* x, *which is*

$$\frac{Z_1 - Z_2}{Z_1 - t} x = Z_1 - \tilde{Z}_2 Z_1 - tx \tag{3.10}$$

and from the fact that we can choose x *and* t *arbitrarily we conclude that*

$$\tilde{Z}_2 = Z_2 \tag{3.11}$$

and therefore $S_{\tilde{Z}_1, \tilde{Z}_2} = S_{Z_1, Z_2}$, *which proves the claim.*

If we neglect the portion of surface that is occluded, we can approximate any physical surface to an arbitrary precision using piecewise linear functions and discontinuities up to a set of measure zero. For this class of surfaces we can therefore conclude observability:

Claim 3.4 *For any physical surface* S *(i.e. piecewise smooth with countable discontinuities), we have* $\mathcal{I}(S) = \{S\}$.

4 Strong observability

The definition of weak observability given in the previous section requires that, for every surface S, there exists at least one energy distribution I_0 that allows distinguishing it from any other surface $\tilde{S}$. In particular, we have used delta-distributions to prove that any two surfaces can be distinguished, and therefore shape is "weakly" observable[4].

Besides the fact that delta-distributions are a mathematical idealization and cannot be physically realized, it is most often the case that we cannot choose the distribution I_0 at will. Rather, I_0 is a property of the scene being viewed, over which we have no control. In this section we will develop a theory of observability of shape that depends upon the particular energy distribution. Let us define as $\mathcal{I}(S|I_0)$ as the set of surfaces that cannot be distinguished from S *given* the energy distribution I_0:

$$\mathcal{I}(S|I_0) = \{\tilde{S} \mid \int h(x, u, y; \tilde{S}) I_0(y) dy = \int h(x, u, y; S) I_0(y) dy \quad \forall\ x, u\} \tag{4.1}$$

[4]We remind the reader that we use the word "surface" even though we are restricting our considerations to one-dimensional curves. Generalizations of the arguments to two-dimensional surfaces embedded in Euclidean space require only minor modifications.

Clearly, not all energy distributions will allow distinguishing two surfaces. For instance, the distribution $I_0 = 0$ does not allow distinguishing any surface. A "white-noise image", thought of as the limit of differences of a Brownian motion, would instead allow to distinguish any two surfaces; however, it is obviously not physically realizable. In the next subsection we characterize the power of a distribution to distinguish different shapes.

4.1 Sufficient excitation and resolution

Definition 4.1 *We say that the distribution I_0 is* sufficiently exciting *for S if*

$$\mathcal{I}(S|I_0) = \mathcal{I}(S). \tag{4.2}$$

Using the terminology above, we say that a shape S is observable if there exists at least one sufficiently exciting distribution I_0. A white-noise image is sufficiently exciting for any surface S. The following claim gives a necessary characterization of sufficiently exciting distributions.

Claim 4.1 *The distribution $\{I_0(y)\}$ is sufficiently exciting in the interval $y \in [y_0,\ y_1]$ (for any surface S) only if*

$$dI_0(y) \neq 0 \qquad \forall\ y \in [y_0,\ y_1]. \tag{4.3}$$

Proof (see [13])

The above are necessary conditions for a distribution to be sufficiently exciting. On the other hand, there exists at least one "universally exciting" distribution (a white noise image), although it has unbounded variation. The next proposition links the degree of "complexity" of the distribution with the "resolution" at which two surfaces can be distinguished.

Claim 4.2 *Let I_0 be a band-limited distribution, and $\{\theta_i(\cdot)\}$ be a complete orthonormal set of bases of $\mathcal{L}^2$, so that there exists N such that*

$$I_0(y) = \sum_{i=0}^{N} \alpha_i \theta_i(y). \tag{4.4}$$

Then two surfaces S and $\tilde{S}$ can can only be distinguished up to the resolution determined by N; that is, if we write

$$h(x,u,y;S) = \sum_{j=0}^{\infty} \beta_j(x,u;S)\theta_j(y) \tag{4.5}$$

then

$$S\mathcal{I}\tilde{S} \Leftrightarrow \beta_j(x,u;S) = \beta_j(x,u;\tilde{S}) \quad \forall x,u,\ j = 0 \ldots N. \tag{4.6}$$

Proof

$$\begin{aligned} I(x,u;S) &= \sum_{i,j=0}^{N} \beta_j(x,u;S)\alpha_i \int \theta_j(y)\theta_i(y)dy + \\ & + \int \sum_{i,j=N+1}^{\infty} \beta_j(x,u;S)\alpha_i\theta_i(y)\theta_j(y)dy \end{aligned} \quad (4.7)$$

since $\alpha_i = 0 \ \forall i > N$, *the second term vanishes. From the orthonormality of the basis we obtain that*

$$I(x,u;S) = \sum_{i=0}^{N} \beta_i(x,u;S)\alpha_i \quad (4.8)$$

from which the claim follows.

4.2 Joint observability

In this section we consider the possibility that two different surfaces which support two different energy distribution, give raise to the same observations. To this end we say that the pair $(\tilde{S}, \tilde{I}_0)$ is indistinguishable from the pair (S, I_0), and we write $(\tilde{S}, \tilde{I}_0)\mathcal{I}(S, I_0)$ if

$$\int h(x,y,u;\tilde{S})\tilde{I}_0(y)dy = \int h(x,y,u;S)I_0(y)dy + \text{const} \quad (4.9)$$

We only require equality up to a constant, since the overall brightness can be rescaled or normalized.

We define, as usual, the set of pairs $(\tilde{S}, \tilde{I}_0)$ that are indistinguishable from (S, I_0) as

$$\mathcal{I}(S, I_0) \doteq \{(\tilde{S}, \tilde{I}_0) \mid (\tilde{S}, \tilde{I}_0)\mathcal{I}(S, I_0)\} \quad (4.10)$$

Claim 4.3 *On all points* x *where* $S(x)$ *is differentiable, the set* $\mathcal{I}(S, I_0)$ *is given by the pair* $(\tilde{S}, \tilde{I}_0)$ *that satisfies*

$$\frac{I_0(x_l)S}{-u+(x+1)S'} - \frac{I_0(x_r)S}{u+(x-1)S'} = \frac{\tilde{I}_0(\tilde{x}_l)\tilde{S}}{-u+(x+1)\tilde{S}'} - \frac{\tilde{I}_0(\tilde{x}_r)\tilde{S}}{u+(x-1)\tilde{S}'} \quad (4.11)$$

where $x_l, x_r, \tilde{x}_l, \tilde{x}_r$ *are defined by*

$$\begin{cases} x_l = -1 + \frac{x+1}{u} S(x_l) \\ x_r = 1 + \frac{x-1}{u} S(x_l) \\ \tilde{x}_l = -1 + \frac{x+1}{u} \tilde{S}(\tilde{x}_l) \\ \tilde{x}_r = 1 + \frac{x-1}{u} \tilde{S}(\tilde{x}_l) \end{cases} \quad (4.12)$$

Proof: *Call* $I(x,u;S,I_0) \doteq \int h(x,y,u;S)I_0(y)dy$. *First note that* $I(x,u;S,I_0)$ *is continuous almost everywhere as a function of* x *and* u. *The points of discontinuity are those* (x,u) *such that* $x_l(x,u) = x_r(x,u)$ *and such that* $I_0(x)$ *is discontinuous. Therefore, since we require equality up to a constant, we have that* $I(x,u;S,I_0) = I(x,u;\tilde{S},\tilde{I}_0) + \text{const}$ *if an only if the partial derivative with respect to* x *of both sides of the equation are equal for all values of* x *and* u. *Taking the derivative in the sense of distributions, we have*

$$\begin{aligned}\frac{\partial I}{\partial x}(x,u;S,I_0) &= \int \left(\delta(y-x_l)\frac{\partial x_l}{\partial x} - \delta(y-x_r)\frac{\partial x_r}{\partial x}\right) I_0(y)dy = \\ & \qquad I_0(x_l)\frac{\partial x_l}{\partial x} - I_0(x_r)\frac{\partial x_r}{\partial x})I_0(y) \end{aligned} \tag{4.13}$$

In order to compute the partials of x_l *and* x_r, *we can differentiate the equations that define them*

$$\frac{\partial}{\partial x}(x_l + 1 - \frac{x+1}{u}S(x_l)) = 0 \tag{4.14}$$

and similarly for x_r. *From the above we obtain that*

$$\frac{\partial}{\partial x_l} = \frac{S(x_l)}{-u + (x+1)S'(x_l)} \tag{4.15}$$

from which the claim follows.

5 Estimation

While section 3 establishes that it is "generically" possible to observe the shape of a surface, the proofs are not constructive, and therefore they do not suggest how such an estimation can be carried out. It is the goal of this section to introduce algorithms to perform the estimation. While in [13] we have derived algorithms in closed-form for the case of equi-focal surfaces, here we are interested in extensions to the general case.

Let the imaging process be modeled by

$$I(x,u;S) = \int h(x,y,u;S)I_0(y)dy. \tag{5.1}$$

In a digital imaging system x takes values on a discrete and bounded set (the pixel grid); also, $u(t)$ takes values on a discrete bounded set, due to temporal sampling, so that we measure I at the points $(x_i, u_j; S)$ with $i = 1 \dots N$ and $j = 1 \dots M$. Typically also I takes only a discrete set of values, usually quite large (for instance 2^{16}); we will therefore assume that I takes values on a bounded subset of the positive real line. We also assume that S can be described using a finite number of parameters $s_1, \dots, s_p$.

Although, due to physical considerations, energy distributions can be considered to be bounded positive functions, $I_0 \in \mathcal{L}^\infty$, we will at first restrict our attention to a smaller (and more structured) class by assuming further that $I_0 \in \mathcal{L}^2$. Our assumptions will then be that we measure values

$$I(x_i, u_j; s_1, \dots s_p) = \int h(x_i, y, u_j; s_1, \dots, s_p) I_0(y) dy \quad \forall i = 1 \dots N,\ j = 1 \dots M. \tag{5.2}$$

If we evaluate the above equation at all i and j, and stack the result into a column vector of dimension NM, we can write

$$I(s) = \int h(y, s) I_0(y) dy \tag{5.3}$$

where h is also a column vector of dimension NM. The map h defines a linear operator

$$\begin{aligned} L_h : \mathcal{L}^2(\mathbb{R}) &\longrightarrow \mathbb{R}^{NM} \\ I_0 &\mapsto \int h(y, s) I_0(y) dy. \end{aligned} \tag{5.4}$$

Since both $\mathcal{L}^2$ and $\mathbb{R}^{NM}$ are Hilbert spaces, we can define an adjoint operator L_h^*

$$\begin{aligned} L_h^* : \mathbb{R}^{NM} &\longrightarrow \mathcal{L}^2(\mathbb{R}) \\ I &\mapsto \int h^T(y, \cdot) I dy \end{aligned} \tag{5.5}$$

which satisfies

$$< L_h I_0, I >_{\mathbb{R}^{NM}} = < I_0, L_h^* I >_{\mathcal{L}^2(\mathbb{R})} \tag{5.6}$$

for all I_0, I. We also define the *Grammian* operator $\mathcal{G}_h = L_h^* L_h$.

With the new notation we can characterize the imaging process as

$$I(s) = L_h(s) I_0. \tag{5.7}$$

Now, we can apply the operator L_h to both sides to obtain $L_h^* I(s) = \mathcal{G}_h(s) I_0$. Let us suppose for a moment that the Grammian is *"invertible"*, i.e. that there exists an operator $\mathcal{G}_h^{-1}$ that satisfies

$$\int \mathcal{G}_h^{-1} \mathcal{G}_h I_0 dy = I_0 \quad \forall I_0 \tag{5.8}$$

then we could "reconstruct" I_0 as

$$I_0 = \mathcal{G}_h^{-1} L_h^* I(s) \tag{5.9}$$

and substitute the results into equation (5.7) to obtain

$$L_h^\perp(s) I(s) = 0 \tag{5.10}$$

where we have defined

$$L_h^{\perp}(s) \doteq Id(NM) - L_h \mathcal{G}_h^{-1} L_h^* \tag{5.11}$$

and $Id(NM)$ is the identity matrix of dimension $NM \times NM$. This constraint would depend solely on the shape of the scene, represented by s, and not by its radiant energy I_0, so that we could try to estimate s by solving the following optimization problem

$$\hat{s} = \arg\min_s \|L_h^{\perp}(s) I(s)\|. \tag{5.12}$$

However, in general the inverse Grammian will not exist, for that would imply that

$$\int \mathcal{G}_h^{-1}(t,\tau) h^T(t) h(v) d\tau = \delta(t-v) \tag{5.13}$$

which clearly does not have a solution on linear operators in $\mathcal{L}^2(\mathbb{R})$. Intuitively, this means that from measurements on I, which carry a finite amount of information, we cannot reconstruct I_0, which belongs to an infinite-dimensional space. So the best we can do is to reconstruct a finite-dimensional approximation of I_0.

Since I_0 lives on a Hilbert space, we can use Gram-Schmidt orthogonalization approximate the restriction of I_0 to a compact domain as a linear combination of "basis functions" ϕ_i:

$$I_0 = \sum_{i=1}^{\infty} \phi_i(y) k_i \tag{5.14}$$

where $k_i \doteq < I_0, \phi_i >_{\mathcal{L}^2}$ are the Fourier coefficients. In general we can approximate I_0 by truncating the above sum after, say, i_f terms. This corresponds to approximating I_0 with its orthogonal projection onto the finite-dimensional linear subspace of $\mathcal{L}^2$ spanned by the first i_f basis functions ϕ_i, $i = 1 \ldots i_f$. If I_0 is band-limited, Shannon's sampling theorem guarantees that we commit no errors in such an approximation.

We write the first i_f terms of the above sum in matrix notation as

$$I_0(y) = \Phi(y) K \tag{5.15}$$

so that the imaging process can be modeled as

$$I(s) = L_{h\phi}(s) K I_0 \tag{5.16}$$

where the operator $L_{h\phi} : \mathbb{R}^{i_f} \longrightarrow \mathbb{R}^{NM}$ defined by $I_0(y) \mapsto \int h(y,s)\Phi(y)dy$ is now finite-dimensional. Provided that $i_f > NM$, we can now safely ask for the Grammian $\mathcal{G}_{h\phi}$ to be invertible. If p is the dimension of the unknown parameter space, it follows that s is observable if the operator

$$L_{h\phi}^{\perp} \doteq Id(NM) - \int \frac{hh^T}{\int h^T \Phi \Phi^T h dw} dy \tag{5.17}$$

has rank p. In this case we can look for the surface parameters s that solve the following optimization problem

$$s = \arg\min_{s} \|L_h^{\perp}(s)I\|_2^2. \tag{5.18}$$

Before doing that, however, there is an issue that needs to be cleared. We need to guarantee that the solutions to the reduced problem (5.18) are in one-to-one correspondence to the solutions obtained by simultaneously estimating s and the unknown distribution I_0 from (2.1). A theorem of Golub and Pereyra comes handy at this point. It guarantees that the orthogonal projector does not alter the structure of the solution of the system (2.1). See [7] for details on this issue.

Remark 5.0.1 *The estimation methods described in the previous section do not enforce the fact that energy distributions are non-negative. Therefore, in the presence of noise, it is possible that the method thus described will reconstruct an I_0 which is negative on some region of its domain. It is necessary to develop methods that guarantee the non-negativity of the estimated distributions.*

6 Conclusions

Accommodation provides an unequivocal cue to reconstruct shape from images obtained with an imaging device of varying geometry. It provides a complement to pictorial and motion cues.

In practice, the sensitivity of accommodation depends upon the depth of field. For commercial cameras, which are constructed so as to have a depth of field as large as possible (so that pictures are "sharp"), accommodation is very sensitive to noise, since the variations in the image at different focus settings are negligible. In applications, accommodation can be employed where the depth of field is limited, as for instance in microscopy.

Significant work needs to be performed both at the level of analysis, and at the level of algorithms, for instance to embody the constraint that the distributions involved are non-negative.

References

[1] Various authors. *IEEE Transactions on Signal Processing, special issue on signal processing for advanced communications.* 1997.

[2] M. Born and E. Wolf. *Principle of optics.* Pergamon Press, 1980.

[3] W. Dayawansa, B. Ghosh, C. Martin, and X. Wang. A necessary and sufficient condition for the perspective observability problem. *Systems and Control Letters*, 25(3):159–166, 1994.

[4] J. Ens and P. Lawrence. An investigation of methods for determining depth from focus. *IEEE Trans. Pattern Anal. Mach. Intell.*, 15:97–108, 1993.

[5] O. Faugeras. *Three dimensional vision, a geometric viewpoint.* MIT Press, 1993.

[6] F. Friedlander. *Introduction to the theory of distributions.* Cambridge University Press, 1982.

[7] G. Golub and V. Pereyra. The differentiation of pseudo-inverses and nonlinear least-squares problems whose variables separate. *SIAM J. Numer. Anal.*, 10(2):413–532, 1973.

[8] B. Horn. *Robot vision.* MIT press, 1986.

[9] J. Marshall, C. Burbeck, and D. Ariely. Occlusion edge blur: a cue to relative visual depth. *Intl. J. Opt. Soc. Am. A*, 13:681–688, 1996.

[10] A. Pentland. A new sense for depth of field. *IEEE Trans. Pattern Anal. Mach. Intell.*, 9:523–531, 1987.

[11] R. Rosenholtz and J. Malik. A differential method for computing local shape-from-texture for planar and curved surfaces. UCB-CSD 93-775, Computer Science Division, University of California at Berkeley, 1993.

[12] S. Soatto. 3-d structure from visual motion: modeling, representation and observability. *Automatica*, 33:1287–1312, 1997.

[13] S. Soatto. Observability of shape from focus. In *Proc. of the IEEE Conf. on Decision and Control.*, December 1998, to appear.

[14] M. Subbarao and G. Surya. Depth from defocus: a spatial domain approach. *Intl. J. of Computer Vision*, 13:271–294, 1994.

Department of Electrical Engineering, Washington University, One Brookings dr., St. Louis - MO 63130, USA, and Dipartimento di Matematica ed Informatica, Università di Udine, Udine 33100 - Italy.
Email soatto@ee.wustl.edu.

Progress in Systems and Control Theory, Vol. 25

Hybrid Control in Automotive Applications

A. Balluchi, M.D. Di Benedetto,
C. Pinello, and A. Sangiovanni–Vincentelli [1]

1 Introduction

Hybrid systems have captured the attention of the research community because of their intrinsic power and of the challenging mathematical problems they pose. We believe that much needs to be done to understand how to use effectively this mathematical formalism for important applications. We have embarked on an ambitious project aiming at the complete redesign of an automotive engine control unit for the next generation automobiles equipped with complex devices such as drive-by-wire and electronic valve control. The design problem has been formulated in an innovative way and makes extensive use of hybrid system technology. We present first the overall methodology and its basic components. Then we focus on the highest levels of abstraction that involve the formulation of the control problem and its solution. A hybrid model which describes the torque generation mechanism and the powertrain dynamics is developed. Then, two particular control sub-problems (cut-off and fast positive force tracking) are formulated as hybrid optimal control problems, whose solutions are obtained by relaxing the problems to the continuous domain and mapping their solutions back into the hybrid domain. A formal analysis as well as experimental results demonstrate the properties and the quality of the control laws.

1.1 The Framework

Hybrid systems have been the subject of intensive study in the past few years with particular emphasis placed on a unified representation of the models in terms of rigorous mathematical foundations (see [12], [19], [15], [14], [8], [9]). The class of hybrid control problems is extremely broad since it obviously contains continuous control problems as well as discrete event control problems as special cases. For this reason, it is very difficult to devise a general yet effective strategy to solve such problems. In our opinion, it is important to address significant domains of application of hybrid control to develop further understanding of the implications of the model on the control algorithms.

In this paper, we focus on a domain of application that is of great industrial interest: automotive engine control. In this domain, the ever increasing computational power of micro–controllers has made it possible

[1]Research sponsored in part by a CNR grant

to extend the performance and the functionality of electronic sub–systems controlling the motion of the car. This opportunity has exposed the need for control algorithms with guaranteed properties that can reduce substantially emission and gas consumption while maintaining the performance of the car [18, 5, 12]. We argue that most of the engine control problems are hybrid since the behavior of four–stroke gasoline engines is determined by the combination of a discrete event sub-system modeling the torque generation process, and a continuous time sub-system modeling the powertrain behavior. It is then the plant to be controlled that is hybrid, no matter what the control strategy is. This is an important point to make since some interesting general approaches to hybrid control assume that the plant to be controlled has a continuous dynamics and hence they are not directly applicable here [13, 10, 7].

The design problem addressed with the proposed approach can be described as follows. From a set of specifications given by a car manufacturer to the embedded system provider:

- find a set of algorithms that control the engine to match the specifications;
- implement the algorithms on a mixed mechanical-electrical architecture consisting of programmable components, such as microprocessors and DSP's, application specific integrated circuits, and physical devices such as sensors and actuators.

In current practice, the implementation architecture of the electronic component is fixed a priori (micro-controllers with fixed-point arithmetic units). The specifications and the control algorithms are so much tuned to this architecture that it becomes very difficult to verify in an effective way the behavior of the system. In addition, the control algorithms are based on simple, intuitive and often open loop rules. Our team embarked on an ambitious project whose goal has been to redefine the entire design process according to these principles:

- the specifications are given at a high level of abstraction so that they are independent of implementation decisions;
- the control algorithms are designed so that formal correctness and optimality properties are ensured by a rigorous mathematical framework;
- the abstract representation of the design is mapped to an architecture using estimation techniques that guide the selection process;
- the implementation architecture of the electronic component is selected among a set of candidates (e.g. a 32-bit micro-controller with Floating Point Unit, a 32-bit micro-processor with a DSP co-processor,

a multiprocessor architecture with an 8-bit I/O processor and a 16-bit micro-controller with a DSP unit) in such a way that cost, reliability and time to market are optimized.

The goal of the presented design process is to shorten the design cycle (now of the order of 5 years) by a very sizable amount (by 3 years and more!) while reducing the cost of the electronic system and increasing its quality and functionality. The conscious reuse of previously designed blocks and the capability of taking decisions at early stages of the design process are key points to this end. Our approach has been tuned and tested on industrial problems and is the result of a long standing working relationship with Magneti-Marelli, a large European automotive electronic subsystem provider.

1.2 Design methodology overview

The design process, shown in figure 1, is described as a set of successive refinement steps from a level of abstraction as high as possible all the way down to the details needed for the final implementation. The process has been conceived to be applicable in almost all embedded design problems; however, to be more concrete, we will use the Engine Control Sub-system design to illustrate the basic principles. In our approach, we identify five main levels of abstraction:

System level. At this level, the specifications are captured and analyzed. The supplier and the customer (in the case of the engine control unit, a car manufacturer) in general agree upon the specifications. The system "car" is viewed as the system converting the driver's commands conveyed by a variety of input devices such as clutch, gas and brake pedals, gear stick, cruise control settings, into actual motion and other outputs of interest to the driver or to some regulatory body, (governments or car manufacturers associations) such as emission levels, noise, fuel consumption. Constraints are placed on these outputs and objective functions are set to determine the quality of the car as specified by market analyses performed by the car manufacturer. According to the market segment targets, a car in the high-performance range will be required to respond to a fast push of the accelerator pedal within a very short time, with little regard to the fuel consumption needed to achieve this goal, while an economy car will be required to respond as well as possible to the command maintaining the fuel consumption below a certain threshold. Since some of the key mechanical components of the car, such as the engine and the chassis, are in general not part of the embedded system design, they can be considered as given. Thus, the goal is to design the engine control sub-system so that the car with that particular engine and chassis performs as specified. Unfortunately, most often than not, the specifications are determined according to an informal process that leads to misunderstanding and late product delivery. To support precise specifications, a simulation environment together with an engineering spreadsheet tool is highly desirable and is being developed in collaboration with Cadence Design Systems, Inc.

Function level. Once the specifications are determined, the design process inside the supplier company begins. Because of the complexity of the problem, a decomposition of the system into interacting simpler subsystems, called functions, is clearly a key step towards a good quality design. The decomposition is useful if it leads to a design process that can be carried out as independently as possible for each component. The decomposition process has to "spread" constraints and objectives among the components so that the composition of the behaviors of the components, made feasible and possibly optimal with their own constraints and cost functions, is guaranteed to meet the constraints and the objectives of the overall controlled system. Since in general it is difficult to decompose the system into independent parts, the determination of the local objectives and constraints has to be the result of a careful trade-off between the desire of maintaining optimality at the global level and the one of easing the design task for each of the components considered in isolation. This decomposition has to be based on the understanding of the physical process of interest. For the engine control case, the variables to be controlled are the air-fuel mix, the injection timing and the spark timing while the objective and constraints are stated in terms of emissions and motion of the car. Motion is the result of the application of torque to the driveline. The engine, via the combustion process, produces the torque. Emissions are the result of the chemical process in the catalytic converter that is fed the exhaust gases of the combustion process. The combustion process in turns is controlled by the quantities introduced above. The air-fuel mix is the result of two processes each determining the air and fuel quantity to mix. From this simple analysis, it appears that motion generation and emission generation are the results of a cascade of chemical-electrical-mechanical processes and that each can be controlled somewhat independently once the local constraints and objectives are set. In Section 3, we will present this example with more details.

Operation level. Once the decomposition of the problem has been performed, each sub-problem has to be solved with its constraints and objectives. Each sub-problem solution is expressed in terms of basic building blocks, called operations, for re-use. Each operation is given a classification in terms of its nature, e.g., measurement, actuation, and control operations. At this level, we start deciding "how to solve" the problem while at the system and function level we were dealing with "what is the problem" to solve. If, indeed, one or more of these sub-problems are not solvable, we need to revisit the previous design step to adapt the local objectives and constraints or to change the decomposition itself. Note that we try to close the design loop as early as possible to avoid major re-design steps at a later stage that can yield so much damage to schedule and profitability.

Architecture level. The operations are still defined abstractly and not directly related to physical devices. At the architecture level, we determine the actual interconnection of components (architecture) that implement one or more operations. This step consists of matching abstract operations with a collection of available or "to-be-designed" devices trying to optimize cost, size, reliability, time-to-market. Components can be classified as electrical, chemical, or mechanical devices. Some examples of components are the pressure sensor, the intake manifold, and the microprocessor. Note that the microprocessor or any other software programmable component is very flexible since it can be used to implement a large class of operations. At this level, the actual performance of the system can be more precisely estimated. If the estimation indicates that the objectives and constraints are not met, the choice of operations has to change accordingly. Once more the engineering change loop is closed as early as possible.

Component level. This step involves the actual design of the components of the architecture determined in the previous step if they are not available in the library or if their use requires some degree of customization as in the case of electronic programmable components, where software must be developed to customize the processor for the particular operation under consideration.

The operation and architecture design process is tightly linked. Decisions taken on what operations to implement have an obvious impact on the architecture and vice-versa. Hence it is very difficult to de-couple these two steps. Indeed, the mapping process from operations to architecture is critical to obtain a good design. In our methodology, mapping is performed continuously so that cost and performance estimation can guide the design decisions.

Note that operation and architecture abstractions are shown in parallel to stress that the two design processes have to be tightly linked. It is in this phase that we address the key problem of software design. Software is the by-product of mapping a set of operations to the programmable components of the electronic architecture. In summary, the overall methodology is hierarchical and consists of five main levels and of the mapping steps that take from one level to the next. Each step is characterized by analysis and optimization that can be carried out with the appropriate tools. The goal is to make the design process rigorous and as well defined as possible and to re-use as much as possible any of the design components for the present application as well as for future applications.

1.3 Paper organization

The focus of this paper is on the various aspects of the mathematical models used to represent the engine and the power-train and of the design of the control algorithms. A companion paper describes the main issues related to algorithm implementation and architecture selection [11]. In particular, the paper is organized as follows. In Section 2, the way we construct the reference model for the engine control problem is described. In Section 3, the torque–generation and powertrain models are presented as hybrid systems. In Section 4, a control sub-problem related to the region of operation of the engine when the driver pushes quickly the accelerator pedal, is formulated and solved in the hybrid system domain. The strategy is to first find an optimal solution to the continuous time relaxed control problem and then to map it back in the hybrid domain. In Section 5, another sub-problem is considered related to the region of operation of the engine when the driver releases quickly the accelarator pedal. For cars equipped with drive-by-wire subsystems, this problem can be formulated as a minor variation of the previously considered problem. However, a substantial difference exists if the throttle angle is not available as a control variable. In this case, we devised a strategy, following the same general ideas, that yields a solution demonstrably close to the upper bounding optimal solution to the relaxed problem. Some concluding remarks are offered in Section 6.

2 System behavior specifications: constructing the reference model

A car manufacturer gives the system specifications in terms of how the vehicle should react to the inputs given by the driver. In this paper, we restrict our attention to the specifications related to force requests, i.e. we do not deal with requirements on trajectory control of the vehicle. The assumption is that the driver requests force by using the gas, brake and clutch pedals, and a manual gear shift. In general, the capture of the specifications as interpreted by the supplier is often obscured by other considerations involving the implementation of a system, thus making it difficult to re-use part of the solution as well as to verify whether the specifications are met. In our approach, the specifications are captured using a hybrid model as shown in figure 1. The states of the top-level Finite State Machine (FSM) correspond to different regions of operations of the engine. The transitions are determined by the action of the driver on the input signals or by engine conditions. Each region of operation is characterized by a set of constraints related to driving performance, such as comfort and safety, or gas and noise emission, and a cost function that identifies the desired behavior of the controlled system. The controlled system is represented by a model that includes ordinary differential equations as well as discrete components. The goal of the controller is to act on the inputs to the plant (the throttle plate angle, the injection pulse duration and the spark advance angle) so that it behaves according to the specifications summarized in the FSM. Figure 1 shows the FSM that specifies the behavior of the vehicle. The initial state is the Stop state that corresponds to the engine being off. From this state, the driver causes the transition to the Startup state, turning the ignition Key to the startup value. From the Startup state the FSM enters the Idle state, if the gas pedal signal is 0, otherwise enters the RPM Tracking state. The desired behavior of the controlled system in the Idle state, for example, is the following: the force F_G acting on the vehicle has to be zero and the crankshaft revolution speed n must be regulated to a reference value, e.g. 500 rpm, with an excursion of up to 20 rpm while minimizing fuel consumption. Moreover, the transient response with respect to sudden loads of 25 Nm must have a 'settling time' of 2 s and an excursion less than 40 rpm. Last, the air-fuel ratio must remain within (10/00 of the stoichiometric desired level). In the RPM Tracking state, the engine is supposed to adjust its RPM to match a prescribed relationship with the gas pedal signal. When the transmission is engaged, the Force Tracking state is entered. In this state, a particular torque profile, that depends on the gas pedal position, the gear and the RPM, has to be realized with constraints on the settle time and with cost function related to fuel consumption and drive comfort.

The Fast Negative Force Transient (FNFT) state is entered if the gas pedal is suddenly released. In this state, a step reference for the requested force is considered and a different control law is used. This control law should minimize the transient time subject to very tight drive comfort constraints. However, if no drive-by-wire control is available, a feasible solution may not exist. In this case, the optimization problem is reformulated so that the driving comfort constraint becomes the cost function. When the gas pedal is pushed again or the transient is elapsed, the transition to the Force Tracking (FT) state is entered. If the gas pedal is completely released and the minimum RPM is approached, the transition to the Idle with transmission on state is enabled. In this state, the RPM is kept constant at a prescribed value until the gas pedal position or the RPM triggers the transition to the Force tracking state. If the gas pedal is pushed quickly, then the transition to the Fast Positive Force Transient (FPFT) state is enabled. This state is the mirror image of the FNFT state. Here the torque profile to be followed is a positive step and the control law minimizes the transient time subject to fuel consumption and comfort constraints. Finally, when a cruise control is available, a Speed Tracking (ST) state is present. If the driver turns the Key to the stop value or if the engine turns off, the Stop state is reached. Although this scheme is very simple, it has quite an impact in clarifying the objectives of the design and how to decompose the problem into well defined sub-problems. In addition, this way of specifying the problem has an impact on the final implementation, reducing the amount of complexity and redundancy in the code. The requirements captured by the diagram can be considered as the specifications for the control problem. In addition, the structure of the FSM is the same independently of the car manufacturer or of the car range. Only the parameters of the cost and constraint functions change, thus favoring maximum re-use.

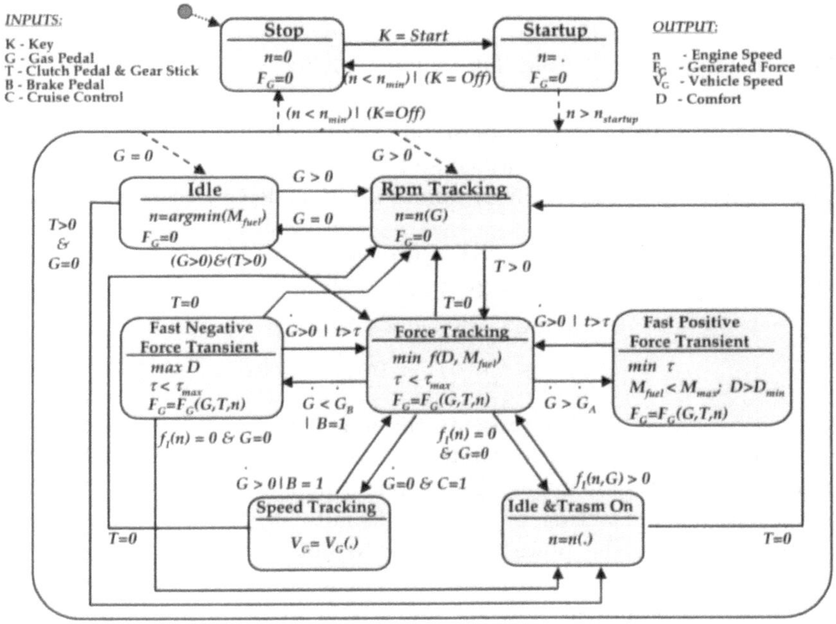

Figure 1: FSM of the regions of operation describing system specifications.

3 Plant model

In this section, the hybrid model proposed in [2][3] for vehicles with 4–stroke N–cylinder gasoline engine is illustrated. The hybrid model $\mathcal{M}_{1cyl}$ of a vehicle with single cylinder engine is reported in Figure 2. This model is composed of three parts:

1. a Finite State Machine[2](FSM) describing the behavior of the pistons;

[2]A deterministic Finite State Machine is a six-tuple $M = \{I, Y, \mathcal{S}, s_0, \lambda, \gamma\}$ where I is the (finite) set of inputs, Y is the (finite) set of outputs, $\mathcal{S}$ is the finite set of states, s_0 is the initial state, $\lambda : \mathcal{S} \times I \to \mathcal{S}$ is the next state function, $\gamma : \mathcal{S} \times I \to Y$ is the output function.

2. a Discrete Event system[3](DE) modeling torque generation (dashed boxes and lines);

3. a Continuous Time (CT) system modeling the powertrain and air dynamics (solid boxes and lines).

The combination of the FSM and the DE system describes the torque generation mechanism for a single cylinder, whose behavior is affected by the CT powertrain and air evolution. The hybrid model for a N-cylinder vehicle is obtained composing N FSM/DE cylinder models and the CT model.

Powertrain Model. The powertrain is described by a continuous time system model developed, identified and validated at Magneti Marelli Engine Control Division. Powertrain dynamics are modeled by the linear system

$$\dot{\zeta} = A\zeta + bu - b_0 \tag{3.1}$$

$$\dot{\phi}_c = [0,\ 1,\ 0]\,\zeta \tag{3.2}$$

$$a = c\zeta - d_0 \tag{3.3}$$

$$j = (cA)\,\zeta + (cb)\,u - (cb_0) \tag{3.4}$$

Where $\zeta = [\alpha_e, \omega_c, \omega_p]^T$ represents the axle torsion angle, the crankshaft revolution speed, the wheel revolution speed, and the 4–th variable ϕ_c represents the crankshaft angle. The input signal u is the torque produced by the engine and acting on the crankshaft. Vector b_0 models the resistant actions on the powertrain, due to both internal friction and external forces. Dynamics (3.1) is asymptotically stable and is characterized by a real dominant pole λ_1, and a pair of conjugate complex poles $\lambda \pm j\mu$. Vehicle acceleration (3.3) and jerk (3.4) ($j = \frac{da}{dt}$) are the powertrain outputs of interest for drivability and comfort. Assuming vehicle speed equal to wheel speed and denoting by R_w the wheel radius, we have $a = R_w\dot{\omega}_p = R_w[0,0,1](A\zeta + bu - b_0)$, and since the third entry in b is 0, $a = R_w[0,0,1]A\zeta - R_w[0,0,1]b_0$, i.e. $c = R_w[0,0,1]A$, $d_0 = R_w[0,0,1]b_0$. Model parameters A, b, c, b_0 and d_0 depend on the transmission gear, and it is assumed that, during a cut–off operation, the driver does not act on the gear.

Cylinder's behavior. The behavior of each cylinder in the engine is abstractly represented by the Finite State Machine shown in Figure 2. FSM state $\mathcal{S}$ assumes values in the set $\{H, I, C, E\}$ as follows.

- *Exhaust run* represented by state H. The piston goes up from the bottom to the top of its run (bottom and top dead centers resp.), expelling combustion exhaust gases.

[3]A DE system is intended in the sense of [11].

- *Intake run* represented by state I. During its down–run the piston loads the air–fuel mix.
- *Compression run* represented by state C. During its up movement the piston compresses the loaded mix.
- *Expansion run* represented by state E. The compressed mix combustion, generated by a spark signal, produces a sudden pressure increase which pushes the piston downwards.

The transitions occur when the piston reaches the bottom or top dead center. The guard conditions enabling the transitions in the FSM are written in terms of the piston position expressed by $\tilde{\phi}$: the absolute value of the crank angle with respect to the upper dead center position, related to the crankshaft angle ϕ_c as depicted in Figure 2, where ϕ_{co} corresponds to the angle the crank is mechanically mounted on the shaft.

Torque generation. The timing of the spark can be set at each cycle in order to modulate the generated torque (see [2] [4]). Moreover this process is characterized by the delays between the times in which fuel injection and spark advance are set and the time in which such decisions have an effect. Control signals are then subject to a transport process which can be represented by a DE system active at every FSM transition. Such DE system, reported in Figure 2 with dashed boxes, increments its time counter k by one at each transition. Its inputs are: the mass of air-fuel mix $q(k) \in \mathbb{R}^+$ loaded during the intake phase; a parameter $\gamma(k) \in [\gamma_{\min}, \gamma_{\max}]$ which represents the ratio between the mass of fuel injected and the mass of air loaded, normalized with respect to its "optimal" value in reference to burning efficiency; the modulation factor $r(k) \in [r_{min}, 1]$ due to not optimal spark timing. The DE system output is the torque $u(k)$.

At the FSM transition $E \to H$ the DE system reads its inputs $q(k)$ and $\gamma(k)$, and stores in its state $z \in \mathbb{R}$ the maximum amount of torque achievable during the next expansion phase, obtained by the mix-to-torque gain G. Such value is corrected at the $I \to C$ transition applying the modulation factor $r(k)$, due to the chosen spark advance. The DE output $u(k)$ is always zero except at the $C \to E$ transition when it is set to the value stored in z. Input $u(t)$ to the CT powertrain dynamics is obtained from $u(k)$ by a zero order hold block, i.e. between two transitions of the FSM occurring at times t_k and t_{k+1}, it holds $u(t) = u(k)$ for $t \in [t_k, t_{k+1})$.

Air dynamics. The model of the quantity of air entering the cylinder during the intake run is obtained from the air flow balance equation of the manifold. Manifold dynamics is a continuous time process controlled by the throttle valve that changes the effective section of the intake manifold. The air mass loaded during an intake run from $t = t_k$, is subject to manifold pressure dynamics.

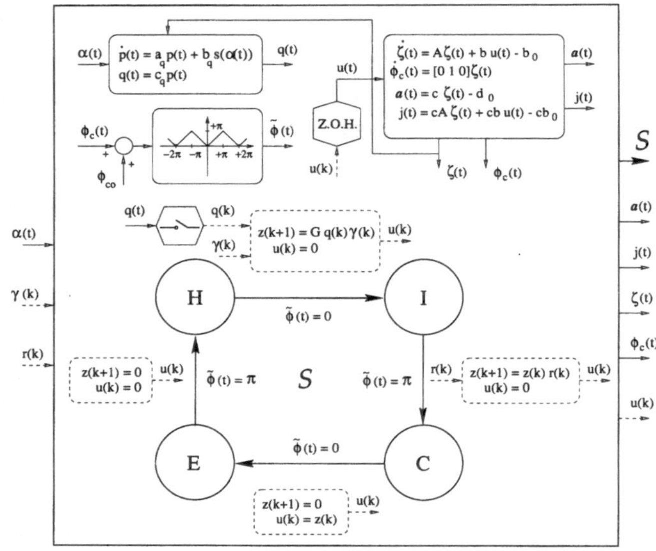

Figure 2: Hybrid model for a vehicle with 4–stroke single cylinder engine.

Let p and α denote the manifold pressure and the throttle angle, respectively. Pressure dynamics is modeled as:

$$\begin{aligned} \dot{p} &= a_q(\omega_c, p)\; p + b_q(p)s(\alpha) && (3.5)\\ q &= c_q(\omega_c, p)\; p && (3.6)\end{aligned}$$

where $s(\alpha)$ is the so called equivalent throttle area, given in terms of throttle angle α, and

$$\begin{aligned} a_q(\omega_c, p) &= -\frac{V_d}{4\pi V_c}\eta_v(\omega_c, p)\;\omega_c, && (3.7)\\ b_q(p) &= \frac{p_{atm}\sqrt{R_g T_{atm}}}{V_c}\beta\left(\frac{p}{p_{atm}}\right), && (3.8)\\ c_q(\omega_c, p) &= \frac{V_d}{4R_g T_{atm}}\eta_v(\omega_c, p) && (3.9)\end{aligned}$$

where V_d, V_c are respectively the displacement and the displacement plus manifold volume, p_{atm}, T_{atm} are the ambient pressure and air temperature, R_g is the ideal gas constant. The volumetric efficiency in cylinder intakes is denoted by $\eta_v(\omega_c, p)$. Finally,

$$\beta\left(\frac{p}{p_{atm}}\right) = \begin{cases} \sqrt{\frac{2c_v/c_p}{c_p/c_v-1}\left[\left(\frac{p}{p_{atm}}\right)^{\frac{2}{c_p/c_v}} - \left(\frac{p}{p_{atm}}\right)^{\frac{c_p/c_v+1}{c_p/c_v}}\right]} \\ \qquad \text{if } \frac{p}{p_{atm}} \geq \left(\frac{2}{c_p/c_v+1}\right)^{\frac{c_p/c_v}{c_p/c_v-1}} \\ \sqrt{c_p/c_v}\left(\frac{2}{c_p/c_v+1}\right)^{\frac{c_p/c_v+1}{2(c_p/c_v-1)}} \\ \qquad \text{if } \frac{p}{p_{atm}} < \left(\frac{2}{c_p/c_v+1}\right)^{\frac{c_p/c_v}{c_p/c_v-1}} \end{cases}$$

While in the past throttle valve was directly connected to the gas pedal, in the future most cars will be equipped with drive–by-wire where the throttle valve is controlled by the engine control unit to achieve better performance. In this paper, we consider the case of cars equipped with drive-by-wire and for the FNFT region of operations, we also consider the case of traditional cars.

Engine hybrid model. An engine is characterized by the number of cylinders[4]. The pistons are connected to the crankshaft, so that the phases of their behavior are related to each other. The overall model of torque generation for a N–cylinder engine is then the combination of N FSMs as in Figure 2 and of N DE systems representing the behavior of each piston. The hybrid model of the complete engine is obtained by adding to torque generation model the powertrain CT dynamics (3.1), (3.2) and air CT dynamics (3.5) which are shared among all cylinders.

In this paper we focus on the most relevant case of a 4-cylinder engine. Its model, referred to in the rest of the paper as $\mathcal{M}_{4cyl}$, has input signals

1. throttle angle α, a scalar continuous time signal in the class $\mathcal{A}_{4cyl}$ of functions $\mathbb{R}^+ \to [0, \frac{\pi}{2}]$;
2. parameters $\boldsymbol{\gamma} = [\gamma_1, \gamma_2, \gamma_3, \gamma_4]^T$, a four dimensional discrete time signal in the class $\mathcal{G}_{4cyl}$ of functions $\mathbb{N} \to (\gamma_{\min}, \gamma_{\max})^4$, with components synchronized with the corresponding DE model;
3. spark advance modulation factors $\mathbf{r} = [r_1, r_2, r_3, r_4]^T$, a four dimensional discrete time signal in the class $\mathcal{R}_{4cyl}$ of functions $\mathbb{N} \to (r_{\min}, 1)^4$ with components synchronized with the corresponding DE model;

In a 4-cylinder engine, no FSM is in the same state at any time and all transitions are synchronous, i.e., offset angles ϕ_{c0_i} are either 0 or π. Without loss of generality, we assume that in the model $\mathcal{M}_{4cyl}$ the crankshaft offsets ϕ_{co_i} are $\phi_{c0_1} = \phi_{c0_3} = \pi, \phi_{c0_2} = \phi_{c0_4} = 0$.

[4]Most cars have four cylinders, but there are engines that have a different number of cylinders. For example, Formula 1 racing cars can have 8, 10 or 12 cylinders. Fiat Coupè 2000 Turbo has 5.

4 Fast Positive Force Transient design

4.1 The Fast Force Transient (FFT) optimal control problem

In the Fast (both Positive and Negative) Force Transient (FFT) regions of operation the objective is to steer the system from a given point, characterized by torque delivered to the crankshaft $u(0) = u_0$, to a new point with torque value u_R in minimum time. Since we consider drive-by-wire subsystems, available controls are: air mass that depends on the throttle angle, fuel injection and spark advance. The control action has to be designed so that comfort is maintained. "Peak–to–peak" acceleration and jerk have been experimentally identified as the most important factors in passenger comfort. We formulated the control problem as: *steer the powertrain elastic state, keeping the jerk within the bounds* $[0, j_{\max}]$, *to a point such that the application of the new requested value of transmitted torque produces an oscillating acceleration evolution* $\tilde{a}(t)$ *bounded above by a threshold of perception* $\tilde{a}_{th} > 0$. The oscillating acceleration evolution is computed using natural mode decomposition. Introduce

$$\begin{bmatrix} x' \\ x \end{bmatrix} = N_{\zeta x} \left(\zeta + A^{-1} b \, u_R - A^{-1} b_0 \right) , \tag{4.10}$$

with $x' \in \mathbb{R}$, $x \in \mathbb{R}^2$ and $N_{\zeta x} \in \mathbb{R}^{3\times 3}$ obtained from the eigenvectors of A, and input transformation $v = u - u_R$. Dynamics (3.1) becomes

$$\begin{bmatrix} \dot{x}' \\ \dot{x} \end{bmatrix} = \begin{bmatrix} \lambda_1 & 0 \\ 0 & A_x \end{bmatrix} \begin{bmatrix} x' \\ x \end{bmatrix} + \begin{bmatrix} b_{x'} \\ b_x \end{bmatrix} v , \tag{4.11}$$

with $A_x = \begin{bmatrix} \lambda & -\mu \\ \mu & \lambda \end{bmatrix}$, and (3.3) is rewritten as $a = -cA^{-1}(bu_R - b_0) + [c_{x'} \; c_x] \begin{bmatrix} x' \\ x \end{bmatrix}$ with $[c_{x'} \; c_x] = c \, N_{\zeta x}^{-1}$. Under constant control v in (4.11), the oscillating component of the acceleration can then be expressed as

$$\begin{aligned} \dot{x} &= A_x x + b_x v , & (4.12) \\ \tilde{a} &= c_x x . & (4.13) \end{aligned}$$

Let $\hat{\rho} = \tilde{a}_{th} \| c_x^T \|^{-1}$. Consider the manifold

$$\mathcal{C}_x = \left\{ \begin{bmatrix} x' \\ x \end{bmatrix} \in \mathbb{R}^3 \,|\, x' \in \mathbb{R}, \; x = \hat{\rho} \begin{bmatrix} \cos(\theta) \\ \sin(\theta) \end{bmatrix} \text{ with } \theta \in [0, \pi) \right\} \tag{4.14}$$

Note that the norm of $x(t)$ in (4.11) decreases over time when $v = 0$. Hence, if at some time $\bar{t}$, x has been driven to $\mathcal{C}_x$, control $v(t) = 0$, which corresponds to torque $u(t) = u_R$, keeps the trajectory inside $\mathcal{C}_x$ with acceleration

$\tilde{a}(t)$ bounded above by the threshold value $\tilde{a}_{th}$, for all $t \geq \bar{t}$. The FFT control problem can then be formulated as the optimal control problem of steering the state x to manifold $\mathcal{C}_x$ in minimum time with $j \in [0, j_{\max}]$. Let $\mathcal{C}_\zeta$ be the image of $\mathcal{C}_x$ in the ζ state space

$$\mathcal{C}_\zeta = \left\{ \zeta \in \mathbb{R}^3 | \zeta = N_{\zeta x}^{-1} y - A^{-1}(b\, u_R - b_0) \text{ with } y \in \mathcal{C}_x \right\} . \tag{4.15}$$

Problem 4.1.1 *Given the engine hybrid model $\mathcal{M}_{4cyl}$, for any ζ_0 not inside the region delimited by $\mathcal{C}_\zeta$ as in (4.15),*

$$\min_{\substack{\alpha \in \mathcal{A}_{4cyl} \\ \gamma \in \mathcal{G}_{4cyl} \\ \mathbf{r} \in \mathcal{R}_{4cyl}}} \int_0^T dt \tag{4.16}$$

$$\textit{subject to:} \quad \left\{ \begin{array}{l} \text{Dynamics of Hybrid Model } \mathcal{M}_{4cyl} \text{ with} \\ \textit{FSM initial state: } \mathcal{S}_1 = H, \mathcal{S}_2 = I, \mathcal{S}_3 = C, \mathcal{S}_4 = E, \\ z_1(0) = 0, \;\; z_2(0) = z_3(0) = z_4(0) = G q_0, \\ \zeta(0) = \zeta_0, \;\; \phi_c(0) = 0, \;\; p(0) = p_0 \\ \zeta(T) \in \mathcal{C}_\zeta \\ 0 \leq j(t) \leq j_{\max} \textit{ for all } t \in [0, T] \end{array} \right. \tag{4.17}$$

with $(\zeta_0, 0)^T$ the powertrain dynamics state value at the initial time and p_0, $q_0 = c(\omega_c(0), p_0) p_0$ the corresponding manifold pressure and mass of air.

The above choice for the FSM initial state is not a limitation since any other admissible initial state can be obtained by a reordering of the cylinders.

4.2 FFT functional level design

In this Section the design of the functional level in the Fast Force Transient region is presented. At the functional level the plant is abstractly described relaxing the hybrid model $\mathcal{M}_{4cyl}$ to the continuous–time domain. Hence, the desired behavior is expressed by an optimal control problem which represents a relaxation of the original hybrid Problem 4.1.1. In Section 4.3, where the operation level design is described, relaxed problem solutions are mapped back to the hybrid domain yielding a sub–optimal solution to Problem 4.1.1. The relaxed problem is concerned with comfort requirements, for minimum time optimal trajectories of system (3.1) to manifold (4.15), assuming no constraint on torque signal. The solution is easily obtained by rewriting dynamics (3.1) with input j in place of u. From (3.1) and (3.4),

$$\dot{\zeta} = (I - b(cb)^{-1}c) A\, \zeta + (cb)^{-1} b\, j - (I - b(cb)^{-1}c) b_0 \, , \tag{4.18}$$

$$u = -(cb)^{-1}(cA)\zeta + (cb)^{-1} j + (cb)^{-1}(cb_0) \, . \tag{4.19}$$

System (4.18)-(4.19) corresponds to the inverse of system (3.1)-(3.4). For $a = c\zeta - d_0 = 0$, hence for $j = 0$ (4.18) represents the zero-dynamics of system (3.1)-(3.3), which has a pole in the origin and two poles equal to the zeros of $c(sI - A)^{-1}b$. Hence, it is characterized by one pole in the origin and two poles equal to the zeros of the system $c(sI - A)^{-1}b$. Since, c is proportional to the third row of A and the third entry in b is 0, $s = 0$ is also a zero of $c(sI - A)^{-1}b$. Hence, dynamics (4.18) has two poles in the origin and a real pole η. A solution to the relaxed control problem is easily derived in the transformed space $\xi = N_{\zeta\xi}(\zeta + A^{-1}b\ u_R - A^{-1}b_0)$ of the natural modes,

$$\dot{\xi} = A_\xi \xi + b_\xi j \quad \text{with} \quad A_\xi = \begin{bmatrix} 0 & 1 & 0 \\ 0 & 0 & 0 \\ 0 & 0 & \eta \end{bmatrix} \tag{4.20}$$

where the target manifold is expressed as

$$\mathcal{C}_\xi = \{\xi \in \mathbb{R}^3 | \xi = N_{x\xi} x \text{ with } x \in \mathcal{C}_x\} \tag{4.21}$$

with $N_{x\xi} = N_{\zeta\xi} N_{\zeta x}^{-1}$. Let $\mathcal{J}$ denote the class of signals $j : \mathbb{R} \to [0, j_{\max}]$. The functional level relaxed control problem in the FFT region is:

Problem 4.2.1 *For any ξ_0 not inside the region delimited by $\mathcal{C}_\xi$ as in (4.21),*

$$\min_{j \in \mathcal{J}} \int_0^T dt \tag{4.22}$$

$$\textit{subject to:} \quad \begin{cases} \dot{\xi} = A_\xi\ \xi + b_\xi\ j \\ \xi(0) = \xi_0 \\ \xi(T) \in \mathcal{C}_\xi \end{cases} \tag{4.23}$$

The optimal solutions to Problem 4.2.1 are characterized as follows.

Proposition 4.2.1 *Let $\hat{j}(t)$, with $t \in [0, T]$, be an optimal solution to Problem 4.2.1 and let $\hat{\xi}(t)$ be the corresponding minimum time trajectory to manifold $\mathcal{C}_\xi$ as in (4.21). Then, $\hat{j}(t) \in \{0, j_{\max}\}$ for all $t \in [0, T]$ and*

$$\hat{j}(T) = \begin{cases} 0 & \text{if } b_\xi^T Q \xi(T) > 0 \\ j_{\max} & \text{if } b_\xi^T Q \xi(T) < 0 \end{cases}, \tag{4.24}$$

where $Q = N_2^T N_2 + N_3^T N_3$, with N_2, N_3 the second and third row of $N_{\xi x} = N_{x\xi}^{-1}$, respectively. Finally, there exist at most two times $t_1, t_2 \in [0, T]$ where a switching of $\hat{j}(t)$ takes place. These points are the solution of

$$[b_\xi^1 + b_\xi^2 (T - t_i),\ b_\xi^2,\ b_\xi^3 e^{\eta(T - t_i)}]\ Q\xi(T) = 0,\ i = 1, 2\ . \tag{4.25}$$

The proof of the above proposition is obtained applying Pontryagin's Maximum Principle. The minimum time control to a given $\bar{\xi} = \xi(T) \in \mathcal{C}_\xi$

reachable in time T with jerk $j(\cdot) \in \mathcal{J}$, is derived as follows

$$\hat{\jmath}(t) = \begin{cases} \hat{\jmath}(T) \ \ \forall t \in [0,T] & \text{if (4.25) has no solution in } (0,T) \\ \begin{cases} j_{\max} - \hat{\jmath}(T) & \forall t \in [0,t_1) \\ \hat{\jmath}(T) & \forall t \in [t_1,T] \end{cases} & \text{if (4.25) has a solution } t_1 \in (0,T) \\ \begin{cases} \hat{\jmath}(T) & \forall t \in [0,t_1) \\ j_{\max} - \hat{\jmath}(T) & \forall t \in [t_1,t_2) \\ \hat{\jmath}(T) & \forall t \in [t_2,T] \end{cases} & \text{if (4.25) has solutions } t_1, t_2 \in (0,T) \end{cases} \tag{4.26}$$

with $\hat{\jmath}(T)$ as in (4.24). From any $\bar{\xi} \in \mathcal{C}_\xi$ reachable with $j(\cdot) \in \mathcal{J}$, the minimum time trajectory is obtained integrating backwards dynamics (4.20) with jerk signal as in (4.26). Let $\mathrm{Cont}(\mathcal{C}_\xi, T)$ denote the set of points controllable to $\mathcal{C}_\xi$, with $j(\cdot) \in \mathcal{J}$, in time lower than or equal to T. The backwards integration defines a partition $S_\xi^{j_{\max}}$, S_ξ^0 of set $\mathrm{Cont}(\mathcal{C}_\xi, T)$, such that for any $\xi \in \mathrm{Cont}(\mathcal{C}_\xi, T)$, the minimum time jerk control to $\mathcal{C}_\xi$ is expressed as

$$\hat{\jmath}(\xi) \ = \ \begin{cases} 0 & \text{if } \xi \in S_\xi^0 \\ j_{\max} & \text{if } \xi \in S_\xi^{j_{\max}} \end{cases} \ . \tag{4.27}$$

From (4.19) and (4.27), the minimum time control to $\mathcal{C}_\zeta$ in the physical state space, in terms of torque on the crankshaft, is

$$\hat{u}(\zeta) = \begin{cases} (cb)^{-1}(-(cA)\zeta + (cb_0)), & \text{if } \zeta \in S_\zeta^0 \\ (cb)^{-1}(-(cA)\zeta + (cb_0) + j_{\max}), & \text{if } \zeta \in S_\zeta^{j_{\max}} \end{cases} \tag{4.28}$$

where $S_\zeta^{j_{\max}}$, S_ζ^0 are obtained mapping $S_\xi^{j_{\max}}$, S_ξ^0 to the ζ space.

4.3 FFT operation level design

The minimum time torque signal (4.28) is clearly not feasible for the hybrid model $\mathcal{M}_{4cyl}$ in Section 3. We derive in this Section an implementable approximation of torque (4.28), consistent with the hybrid model $\mathcal{M}_{4cyl}$, in terms of: throttle angle α, mix composition factor γ, and spark modulation factor $\mathbf{r}$.

The main difficulties are:

(a) available torque is limited to the set of values $\{0\} \cup G[r_{\min}\gamma_{\min}q, \gamma_{\max}q]$, where the air mass q is subject to manifold pressure dynamics;

(b) torque generation has to be synchronized with the powertrain dynamics;

(c) there is a delay between the time at which control signal γ_i (composition factor) and r_i (spark advance modulation) are set and the time at which the corresponding torque is generated.

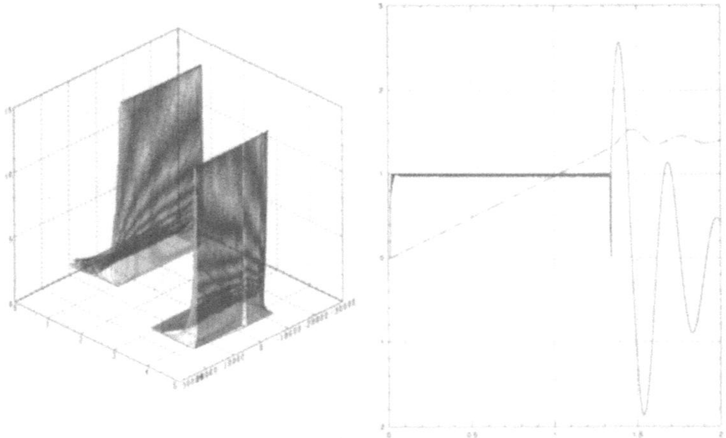

Figure 3: Switching surfaces in the minimum time control to $\mathcal{C}_x$ in cylindric coordinates (left). Acceleration and jerk profile in the hybrid closed loop system (right).

To simplify discussion, set $\zeta(k) \triangleq \zeta(t_k)$ for all k. As discussed in Section 3, the torque $u(k)$ provided by the i–th cylinder at time t_k depends on

1. the mass of air $q(k-2)$ loaded and the composition factor $\gamma_i(k-2)$ of the mix at time t_{k-2} (corresponding to an $H \to I$ transition of the i–th cylinder's FSM);
2. the spark modulation factor $r_i(k-1)$ applied at time t_{k-1} (corresponding to an $I \to C$ transition of the i–th cylinder's FSM).

Such torque feeds the powertrain dynamics in the interval $[t_k, t_{k+1})$, steering the state from $\zeta(k)$ to $\zeta(k+1)$. In order to properly set feedback signals γ_i and r_i, respectively at times t_{k-2} and t_{k-1}, a prediction of $\zeta(k)$ is needed at both instants. Assuming powertrain state available to measurement, one and two–step predictions are obtained by a forward integration of (3.1).

The torque generation process can be viewed as a MIMO system composed of two interconnected systems. The air mass evolution $q(t)$ is subject to manifold pressure dynamics (3.5), which is controlled by throttle angle α and depends on the crankshaft revolution speed ω_c. The torque to the crankshaft is provided by a two–step–delay discrete event system with inputs $\gamma_i(k)$ and $r_i(k)$. The coupling between the discrete event system and pressure dynamics is given by the driveline dynamics (3.1), through ω_c. While the impact on the discrete event system of q is not negligible, the driveline impact on air dynamics is weak enough to allow decentralization.

4.3.1 Air operation design

The reference evolution $\hat{q}(t)$ for air mass is obtained considering a rigid model for the driveline and solving the optimal control problem stated in the previous section assuming that $r_i = \gamma_i = 1$. Let τ_{drv} denote the driveline transmission ratio and let M_v denote the mass of the vehicle. In a rigid model of the driveline, the force acting on the vehicle is given by $\frac{u}{R\ \tau_{drv}}$ and $j = \frac{da}{dt} = (M_v\ R\ \tau_{drv})^{-1}\frac{du}{dt}$. A bound on the vehicle jerk limits the derivative of the torque. The reference signal $\hat{q}(t)$ is chosen as the one that yields the reference value u_R in minimum time for the rigid model, i.e.

$$\hat{q}(t) = \begin{cases} q(0) + j_{\max}\frac{M_v\ R\ \tau_{drv}}{G}t & \text{if } t < \frac{G(q_R - q(0))}{j_{\max} M_v R \tau_{drv}} \\ q_R & \text{if } t \geq \frac{G(q_R - q(0))}{j_{\max} M_v R \tau_{drv}} \end{cases} \tag{4.29}$$

where $q_R = u_R/G$. A decentralized scheme in terms of the throttle angle for perfect tracking of (4.29) is devised using Variable Structure Control (VSC) technique (see [17]). Rewrite dynamics (3.5) as

$$\dot{p} = a_q(\omega_c, p)\, p + w \tag{4.30}$$

Consider $\hat{q}(t)$ as in (4.29) and introduce the function $\sigma(t) = q(t) - \hat{q}(t)$. By definition of $\hat{q}(t)$, $\hat{q}(t) = q(0)$ i.e. $\sigma(0) = 0$. The *equivalent control* is such signal w_{eq} which ensures $\sigma(t) = 0$ if $\sigma(0) = 0$. Imposing $\dot{\sigma} = 0$, from (3.6) and (4.30), the equivalent control is obtained as follows

$$w_{eq} = -a_q(\omega_c, p)\, p + \left(c_q(\omega_c, p) + \frac{\partial c_q}{\partial p}(\omega_c, p)\ p\right)^{-1}\left(\dot{\hat{q}} - \frac{\partial c_q}{\partial \omega_c}(\omega_c, p)\ p\ \dot{\omega}_c\right).$$

In practice, model (3.5), (3.6) is not perfectly known, so that one should modify control w_{eq} to cope with modeling uncertainties. Further, while $\dot{\hat{q}}$ is accessible, $\dot{\omega}_c$ is not available. However, driveline weak coupling allows to simplify the control by using a decentralized scheme (see [20],[16]). Interaction is compensated by a robust control with respect to nonlinearities and coupling (see e.g. [6]). Given a value $\bar{\eta}_v$, $\bar{\omega}_c$ of volumetric efficiency and crankshaft speed, introduce $a_{q1} = -\frac{V_d}{4\pi V_c}\bar{\eta}_v\bar{\omega}_c$, $c_{q1} = \frac{V_d}{4R_g T_{atm}}\bar{\eta}_v$, and $\tilde{a}_q(\omega_c, p) = a_q(\omega_c, p) - a_{q1}$, $\tilde{c}_q(\omega_c, p) = c_q(\omega_c, p) - c_{q1}$. If $\tilde{a} = \tilde{c} = 0$, the VSC

$$w_{vsc} = -a_{q1}\ p + \frac{\dot{\hat{q}}}{c_{q1}} - \frac{W}{c_{q1}}\ \text{sign}\ (\sigma) \tag{4.31}$$

with $W > 0$, guarantees a sliding regime along the manifold $\sigma(p, \hat{q}) = 0$, during which perfect tracking of reference signal $\hat{q}(t)$ is achieved.

Proposition 4.3.1 *Assume that* $p, \omega_c, \dot{\omega}_c, \dot{\hat{q}}$ *satisfy* $p \in [p_1, p_{atm}]$, $\omega_c \in [\omega_1, \omega_2]$, $|\dot{\omega}_c| \leq \dot{\omega}_{\max}$, $|\dot{\hat{q}}| \leq \dot{\hat{q}}_{\max}$. *Choose a* $\bar{\eta}_v$, $\bar{\omega}$ *and let* $\tilde{C}, \tilde{C}_p, \tilde{C}_{\omega_c}, \tilde{A}_q$

be such that $\tilde{C} \geq \max_{\omega_c,p} \left|\frac{\tilde{c}_q}{c_{q1}}\right|$, $\tilde{C}_p \geq \max_{\omega_c,p} \left|\frac{\partial \tilde{c}_q}{\partial p}\right|$, $\tilde{C}_{\omega_c} \geq \max_{\omega_c,p} \left|\frac{\partial \tilde{c}_q}{\partial \omega_c}\right|$, $\tilde{A}_q \geq \max_{\omega_c,p} |\tilde{a}_q|$. *Given* $\epsilon > 0$, *if* $\tilde{C} + \tilde{C}_p \frac{p_{atm}}{c_{q1}} < 1$ *the VSC (4.31) with*

$$W = c_{q1}\tilde{A}_q p_{atm} + \frac{\left(\tilde{C} + \tilde{C}_p \frac{p_{atm}}{c_{q1}}\right)\dot{\hat{q}}_{\max} + \left(\tilde{C}_{\omega_c} p_{atm}\right)\dot{\omega}_{\max} + \epsilon}{1 - \tilde{C} - \tilde{C}_p \frac{p_{atm}}{c_{q1}}} \tag{4.32}$$

guarantees a sliding motion on $\sigma = 0$, *during which perfect tracking of* $\hat{q}(t)$ *as in (4.29) is achieved, robustly w.r.t. to driveline evolution.*

A feedback in terms of throttle angle α is obtained from (4.31) by inverting the nonlinear input map $w = b_q(p)s(\alpha)$. In order to reduce the undesired chattering phenomena, typical in VSC, the sign function is replaced by $\frac{\sigma}{\delta+|\sigma|}$, where the smoothing parameter $\delta > 0$ is properly tuned so to maintain satisfactory tracking. The decentralized feedback is

$$\alpha = s^{-1}\left(\frac{1}{b_q(p)}\left(-a_{q1}\, p + \frac{\dot{\hat{q}}}{c_{q1}} - \frac{V}{c_{q1}}\frac{\sigma}{\delta + |\sigma|}\right)\right) \tag{4.33}$$

with $\sigma(t) = c_q(\omega_c, p)\, p - \hat{q}(t)$.

4.3.2 Motion generation operation design

A feedback in terms of mix composition factor $\gamma(k)$ and spark modulation factor $\mathbf{r}(k)$ which follows law (4.28) is presented in the sequel. Assuming that cranckshaft revolution speed does not change significantly between two successive FSM transitions in $\mathcal{M}_{4cyl}$, i.e. $t_{k+1} - t_k \approx \tau_k = \pi/\omega_c(t_k)$, since $u(t) = u(k)$ for $t \in [t_k, t_{k+1})$, powertrain dynamics (3.1),(3.3) is discretized as follows

$$\zeta(k+1) = \hat{A}\zeta(k) + \hat{b}u(k) \tag{4.34}$$

$$a(k) = c\zeta(k) \tag{4.35}$$

where $\hat{A} = e^{A\tau_k}$, $\hat{b} = (\hat{A}-I)A^{-1}b$. The jerk mean value in the time intervall $[t_k, t_{k+1}]$ can be expressed as

$$\frac{\int_{t_k}^{t_{k+1}} j(t)dt}{t_{k+1} - t_k} = \frac{a(t_{k+1}) - a(t_k)}{t_{k+1} - t_k} \approx \frac{a(k+1) - a(k)}{\tau_k} = \frac{c(\hat{A} - I)}{\tau_k}\zeta(k) + \frac{c\hat{b}}{\tau_k}u(k).$$

Hence, the torque law

$$\hat{u}_d(k) = (c\hat{b})^{-1}c(\hat{A} - I)\zeta(k) + (c\hat{b})^{-1}\tau_k\, \hat{j}(\zeta(k))\,, \tag{4.36}$$

where $\hat{j}(\zeta)$ is chosen according to (4.27), that is

$$\hat{j}(\zeta) = \begin{cases} 0 & \text{if } \zeta \in S_\zeta^0 \\ j_{\max} & \text{if } \zeta \in S_\zeta^{j_{\max}} \end{cases} \tag{4.37}$$

produces a jerk with mean value equal to $\hat{\jmath}(\zeta)$. The actual jerk profile in the evolution of the hybrid model $\mathcal{M}_{4cyl}$ exhibits a ripple on the average value due to the fact that, being $u(t)$ piecewise–constant, between two samples the natural modes of dynamics (3.1) evolve. Due to this ripple, jerk signal exceeds interval $[0, \jmath_{\max}]$, and a more conservative feedback than (4.36) has to be devised.

Proposition 4.3.2 *Define*

$$\begin{aligned}
\hat{u}_c(\zeta) &= (cb)^{-1}(\hat{\jmath}(\zeta) - cA\zeta) && (4.38)\\
\hat{u}_n(\zeta) &= (cA\hat{b} + cb)^{-1}(\hat{\jmath}(\zeta) - cA\hat{A}\zeta) && (4.39)\\
\hat{u}_m(\zeta) &= (c(I+\tau_k A)b)^{-1}(\jmath_{\max} - c(I+\tau_k A)A\zeta) && (4.40)
\end{aligned}$$

with $\hat{\jmath}(\zeta)$ as in (4.37). Let $\jmath_c(t)$,$\jmath_n(t)$,$\jmath_m(t)$ and $\jmath_d(t)$ denote the jerk profiles corresponding to torques (4.38), (4.39), (4.40) and (4.36) respectively. Assume that, under the feedback laws $\hat{u}_c$,$\hat{u}_n$,$\hat{u}_m$ and $\hat{u}_d$, $\frac{d^2\jmath(t)}{dt^2} \leq 0$ for all $t \in [t_k, t_{k+1}]$. The following control

$$\hat{u}(k) = \begin{cases} \max(\hat{u}_c(\zeta), \hat{u}_n(\zeta)) & \text{if } \zeta(k) \in S^0_\zeta \\ \begin{cases} \hat{u}_c(\zeta) & \text{if } \jmath_d(t_k) > \jmath_{\max} \wedge \frac{d\jmath_c}{dt}(t_k) \leq 0 \\ \hat{u}_n(\zeta) & \text{if } \jmath_d(t_{k+1}) > \jmath_{\max} \wedge \frac{d\jmath_n}{dt}(t_{k+1}) \geq 0 \\ \hat{u}_m(\zeta) & \text{otherwise} \end{cases} & \text{if } \zeta(k) \in S^{\jmath_{\max}}_\zeta \end{cases} \tag{4.41}$$

produces a jerk profile $\jmath(t)$ such that $0 \leq \jmath(t) \leq \jmath_{\max}$.

A feedback control in terms of the mix composition factor $\gamma(k)$ of the cylinder which enters the I state, and the spark advance modulation $r(k)$ of the cylinder which enters the C state, which produces the torque $\hat{u}(k)$ as in (4.41), is reported in Figure 4. As discussed at the beginning of the section, control $\gamma(k)$ affects torque $u(k+2)$, while control $r(k)$ affects torque $u(k+1)$. The former depends also on the mass of air $q(k)$, while the latter depends on $q(k-1)$. Equation (4.41) gives the desired torque at time t_{k+1} and t_{k+2} in terms of future states $\zeta(k+1)$ and $\zeta(k+2)$ respectively. Figure 3 shows an evolution of jerk and acceleration in the hybrid system $\mathcal{M}_{4cyl}$ under the proposed feedback.

5 Fast Negative Force Transient design

5.1 The FNFT optimal control problem

For drive–by–wire cars, where the throttle valve is actuated by an electrical motor, the mass of air q is a regulated variable and the designer has full control on torque generation. For these systems the FNFT problem can be formulated as the FPFT problem where $u_R = 0$. The corresponding minimum time control problem is solved using the same arguments and algorithms described for the FPFT problem. However, for traditional systems, where the throttle valve is directly connected to the gas pedal and there is no possibility of acting on the mass of air loaded by the cylinders, a different approach has to be used. When the air quantity is not a controlled variable, the FNFT problem is typically referred to as the cut–off control problem, where the goal is to shut off fuel injection minimizing passengers' discomfort due to powertrain oscillations. In a cut–off operation, the hybrid engine model $\mathcal{M}_{4cyl}$ is fed by a decreasing sequence $q_a(k)$ of air intakes, which converges to the steady-state air quantity with pedal released, q_a^o. This results in an upper bound for the available torque. We identify as the starting point of the cut–off control horizon the time $t_0 = t_{k_0}$ at which $q_a(k_0) = q_a(k_0 - 1) = q_a(k_0 - 2) = q_a^o$, i.e. the time at which all the loaded cylinders' potential torques are at the steady–state value. To simplify notation, we set $t_0 = 0$ and $k_0 = 0$, and we use u for v since $u_R = 0$.

$\zeta_k = \zeta(t_k), \quad q_k = q(t_k)$ (current driviline state and mass of air)
$\tau_k = \pi/[0\ 1\ 0]\,\zeta_k, \quad \hat{A} = e^{A\,\tau_k}, \quad \hat{b} = (\hat{A} - I)A^{-1}b$ (current driviline discrete model)

$\zeta_{k+1} = \hat{A}\zeta(k) + \hat{b}\,z_2^c(k)$ (future driviline state)

if $\| N_{\zeta x}(\zeta_{k+1} + A^{-1}bu_R)\| > \hat{\rho}$ then (exit condition: state ζ in $\mathcal{C}_\zeta$)
$\hat{u}_{k+1} = \hat{u}(k+1)$ (according to (4.41) evaluated at ζ_{k+1})
else
$\hat{u}_{k+1} = u_R$ (target torque u_R applied)
endif

$r = \hat{u}_{k+1}/z_1^c(k)$
if $r < r_{\min}$ then $r = r_{\min}$ elseif $r > 1$ then $r = r_{\min}$ endif
$r(k) = r$ (spark advance control)

$\zeta_{k+2} = \hat{A}\zeta_{k+1} + \hat{b}\,r_k\,z_1^c(k)$ (next driviline state)
if $\zeta_{k+2} \in S_\zeta^0$ then
$\hat{j} = 0$
endif

if $\| N_{\zeta x}(\zeta_{k+2} + A^{-1}bu_R)\| > \hat{\rho}$ then (exit condition: state ζ in $\mathcal{C}_\zeta$)
$\hat{u}_{k+2} = \hat{u}(k+2)$ (according to (4.41) evaluated at ζ_{k+2})
else
$\hat{u}_{k+2} = u_R$ (target torque u_R applied)
endif

$\gamma = \hat{u}_{k+2}/(Gq_k)$
if $\gamma < \gamma_{\min}$ then $\gamma = \gamma_{\min}$ elseif $\gamma > \gamma_{\max}$ then $\gamma = \gamma_{\max}$ endif
$\gamma(k) = \gamma$ (mix composition control)

$z_1^c(k+1) = G\,q_k\,\gamma_k$ (next potential torque)
$z_2^c(k+1) = z_1^c(k)\,r$ (next predicted torque)

Figure 4: Discrete event algorithm for mix composition factor $\gamma(k)$ and spark advance modulation $r(k)$.

The objective of cut–off control is to minimize the peak of the acceleration $\tilde{a}(t)$ until it is below the threshold of acceleration perception $\tilde{a}_{th}$. Remind that, once the powertrain state ζ is inside the region delimited by $\mathcal{C}_\zeta$ as in (4.15) (with $u_R = 0$), fuel injection can be shut off with vehicle oscillations below threshold. Hence, the cut–off control can be formulated as the optimal control problem of steering state ζ to $\mathcal{C}_\zeta$ minimizing the acceleration peaks, acting on the mix composition factor γ and the spark modulation factor $\mathbf{r}$ inputs.

Problem 5.1.1 *Given the engine hybrid model $\mathcal{M}_{4cyl}$, for any initial state ζ_0 not inside the region delimited by $\mathcal{C}_\zeta$ as in (4.15) (with $u_R = 0$),*

$$\min_{\substack{\gamma \in \mathcal{G}_{4cyl} \\ \mathbf{r} \in \mathcal{R}_{4cyl}}} \sup_{0 \le t \le T} |\tilde{a}(t)|$$

$$\text{subject to:} \quad \begin{cases} \text{Dynamics of Hybrid Model } \mathcal{M}_{4cyl} \text{ with} \\ p(t) = p_0, \quad \text{for all } t \ge 0, \\ \textit{FSM initial state: } \mathcal{S}_1 = H, \mathcal{S}_2 = I, \mathcal{S}_3 = C, \mathcal{S}_4 = E, \\ z_1(0) = 0, \ z_2(0) = z_3(0) = z_4(0) = G q_a^o, \\ \zeta(0) = \zeta_0, \phi_c(0) = 0, \\ \zeta(T) \in \mathcal{C}_\zeta, \end{cases}$$

where $\tilde{a}(\cdot)$ is as in (4.13), p_0 is the steady-state manifold pressure with gas pedal released, i.e. $q_a^o = c_q(\omega_c(0), p_0)p_0$ and the final time $T < \infty$ is unspecified.

Note that, the above choice for the FSM initial state does not represent a limitation since any other admissible initial state can be obtained by a reordering of the cylinders.

5.2 FNFT functional level design

The main difficulties in Problem 5.1.1 are that the plant to be controlled is hybrid and that the input signals γ and $\mathbf{r}$ are bounded. The design at the functional level is developed relaxing the hybrid problem into the continuous–time domain, maintaining the bound on the value the torque signal u can take. In the operation level design the solution will be mapped back to the hybrid domain obtaining feedback laws in terms of the plant inputs γ and $\mathbf{r}$. Consider the transformed state space introduced in (4.10) associated with powertrain natural mode decomposition. Let

$$B_{\hat{\rho}} = \{x \in \mathbb{R}^2 : \|x\| \le \hat{\rho}, \quad \hat{\rho} = \tilde{a}_{th} \|c_x^T\|^{-1}\} \tag{5.42}$$

be the projection of the volume delimited by $\mathcal{C}_x$ as in (4.14) on the x sub-space. Since $u_R = 0$, we use u for v. The relaxed problem is as follows.

Problem 5.2.1 *Let*

$$\mathcal{U} = \{u : [0, +\infty) \to \mathbb{R} \mid u(t) \text{ measurable and } 0 \le u(t) \le M, \forall t \ge 0\} \tag{5.43}$$

with $M = G\gamma_{\max}q_a^o$. *For any initial state* $x_0 \notin B_{\hat\rho}$ *as in (5.42),*

$$\min_{u \in \mathcal{U}} \sup_{0 \le t \le T} |\tilde a(t)|$$

$$\text{subject to:} \quad \begin{cases} \dot x(t) = A_x x(t) + b_x u(t) \\ x(0) = x_0 \\ x(T) \in \partial B_{\hat\rho} \end{cases}$$

where $\tilde a(\cdot)$ *is as in (4.13) and the final time* $T < \infty$ *is unspecified.*

Note that, for some initial conditions a solution to Problem 5.2.1 may not exist. In the sequel existence is discussed and a solution is given. Let $R(\theta) \in \mathbb{R}^2$ be the counterclockwise rotation matrix of θ radians in $\mathbb{R}^2$ and, for any $x \in \mathbb{R}^2$, let $x_\perp$ denote $R(\frac{\pi}{2})x$. Let $x_M = -A_x^{-1}b_x M$ be the equilibrium point of (4.12) with $u = M$ and let $v = -\ \overrightarrow{vers}\ (x_M)_\perp$, where $\overrightarrow{vers}\ (x) \triangleq \frac{x}{\|x\|}$. In our previous paper [2], it was shown that there exists $u \in \mathcal{U}$, which steers some initial states x_0 of (4.12) to the origin along a straight line (see Figure 5). However, under this control the time to reach $B_{\hat\rho}$ becomes unbounded as x_0 approaches the line $v^T x = 0$. We propose a modified control law that eliminates this problem by giving up (slightly) optimality for particular trajectories. However, optimality can be approached infinitely close as shown in the following theorem.

Theorem 5.2.1 *Consider an initial state* $x_0 \notin B_{\hat\rho}$. *Let* $\mathcal{U}' \subset \mathcal{U}$ *be the class of all input signals* $u' \in \mathcal{U}$ *which steer the state* x *of system (4.12) from* x_0 *to a point on* $\partial B_{\hat\rho}$. *Let* $\psi(u', x_0, t)$, *with* $t \in [0, T]$, *denote the trajectory from* x_0 *to* $x_T = \psi(u', x_0, T) \in \partial B_{\hat\rho}$ *corresponding to some* $u' \in \mathcal{U}'$, *where* T *depends on* u' *and* x_0. *Let*

$$\tilde A(x_0) = \inf_{u' \in \mathcal{U}'} \sup_{0 \le t \le T} |c_x \psi(u', x_0, t)| = \inf_{u' \in \mathcal{U}'} \sup_{0 \le t \le T} |\tilde a(t)|\Big|_{u = u'} .$$

For any $x \notin B_{\hat\rho}$, *consider the feedback control*

$$\hat u = \begin{cases} \begin{cases} 0 & \text{if } v^T x \ge 0 \\ M & \text{if } v^T x < 0 \end{cases} & \text{if } x \notin \mathcal{D}_M \\ \\ \begin{cases} -\frac{x^T R(-\frac{\pi}{2}) A_x x}{x^T R(-\frac{\pi}{2}) b} & \text{if } (R(\theta)v)^T x \le 0 \\ 0 & \text{if } (R(\theta)v)^T x > 0 \end{cases} & \text{if } x \in \mathcal{D}_M \end{cases} . \tag{5.44}$$

If $\|x_M\| \ge \hat\rho$ *then*

A1. *if* $(c_x x_M)(c_x b_x) > 0$ *control (5.44) with* θ *set to any value in the interval* $(0, \sin^{-1}(\frac{-\lambda}{\sqrt{\lambda^2+\mu^2}}))$, *steers in finite time state* x_0 *to* $\partial B_{\hat\rho}$ *and is optimal for Problem 5.2.1, i.e.* $\tilde A(x_0) = \sup_{0 \le t \le T} |\tilde a(t)|\big|_{u = \hat u}$.

A2. *if $(c_x x_M)(c_x b_x) \leq 0$ there is a set of initial conditions $\mathcal{X}_0 \subset \mathbb{R}^2/B_{\hat{\rho}}$ such that for any $x_0 \in \mathcal{X}_0$ no optimal solution in finite time exists. However, for any $\epsilon > 0$ there exists $\theta > 0$ such that control (5.44) steers any x_0 to $\partial B_{\hat{\rho}}$ in finite time with, $\sup_{0\leq t\leq T} |\tilde{a}(t)|\big|_{u=\hat{u}} - \tilde{A}(x_0) < \epsilon$ if $x_0 \in \mathcal{X}_0$, and $\sup_{0\leq t\leq T} |\tilde{a}(t)|\big|_{u=\hat{u}} = \tilde{A}(x_0)$ if $x_0 \notin \mathcal{X}_0$.*

else, if $\|x_M\| < \hat{\rho}$,

B. *control (5.44) with $\theta = 0$, is optimal for Problem 5.2.1.*

It is interesting to note that for all models of existing cars available to us, $(c_x x_M)(c_x b_x) > 0$ and, hence, (5.44) was actually optimal. Figure 5 shows the closed loop phase space under control (5.44) with $(c_x x_M)(c_x b_x) > 0$ and $\theta \simeq 0$. The proof of Theorem 5.2.1 is based on the analysis of the reachable sets for system (4.12) with control $u \in \mathcal{U}$ (see [3],[4]).

5.3 FNFT operation level design

Torque signal (5.44) is clearly not feasible for the hybrid model $\mathcal{M}_{4cyl}$, introduced in Section 3. Following the approach illustrated in Section 4.3, feedback laws for the mix composition $\boldsymbol{\gamma}$ and the spark modulation $\mathbf{r}$, such that the produced torque approximates signal (5.44) are devised. Torque generation mechanism constraints described in Section 4.3, i.e. (*a*) bounds, (*b*) synchronization and (*c*) delay, have to be taken into account.

Our solution is to over–constrain the optimal control problem limiting our attention to bang–bang torque that can take the two extreme values 0 and M only. This corresponds to setting $r_i = 1$, and to using as feedback input γ_i which assumes values 0 (fuel not injected) and $\gamma_{\max}$ (maximum amount of fuel injected). Synchronization of γ_i, and hence of the generated torque, with the powertrain dynamics is the main difficulty. To compensate the torque generation 2-step delay, a prediction $\tilde{x}_2(k)$ of $x(t_{k+2})$ is obtained from $x(k)$ by forward integration of (4.12). The injection control is defined according to a switching curve $\sigma(x) = 0$, with $\sigma : \mathbb{R}^2 \to \mathbb{R}$ as follows

$$\hat{\gamma}_i(k) = \begin{cases} 0 & \text{if } \tilde{x}_2(k) \in B_{\hat{\rho}} \\ \begin{cases} 0 & \text{if } \sigma(\tilde{x}_2(k)) \geq 0 \\ 1 & \text{if } \sigma(\tilde{x}_2(k)) < 0 \end{cases} & \text{if } \tilde{x}_2(k) \notin B_{\hat{\rho}} \,. \end{cases} \tag{5.45}$$

In the sequel we assume that the prediction $\tilde{x}_2(k)$ is correct, i.e. $\tilde{x}_2(k) = x(t_{k+2})$. The bang-bang part of the control law for the continuous case is characterized by the switching line

$$\sigma(x) = v^T x. \tag{5.46}$$

It is natural to propose this bang-bang control for the hybrid case. However, convergence of this bang–bang control law is not obvious. The properties of (5.45) depend on the position of x_M with respect to $B_{\hat{\rho}}$. If $\|x_M\| \leq \hat{\rho}$, control law (5.45), with $\sigma(x)$ as in (5.46), has been proven to yield trajectories that converge to $B_{\hat{\rho}}$ if the engine speed is larger than an appropriate bound (see [4]). If $\|x_M\| > \hat{\rho}$, we have shown that this control law may not converge and $\sigma(x)$ needs to be modified to guarantee convergence. Since for all model parameters seen in commercial cars $\|x_M\| > \hat{\rho}$, we discuss only this case.

5.3.1 Convergence analysis and performances

Since $\|x(t)\|$ monotonically converges to zero when torque $u = 0$ is applied, we need to worry only when the generated torque under feedback (5.45) continues to be $u = M$, while the continuous torque feedback (5.44) calls for $u = 0$. The set of points such that a torque $u = M$ applied for time Δ causes an increase in norm is $\bar{\mathcal{N}}_\Delta = \{x \in \mathbb{R}^2 \mid \|x\| < \|e^{A_x\Delta}(x - x_M) + x_M\|\}$. Since $\lim_{\Delta\to 0} \bar{\mathcal{N}}_\Delta \neq \bar{\mathcal{N}}_\Delta\big|_{\Delta=0}$, $\bar{\mathcal{N}}_\Delta$ is extended by continuity as follows

$$\mathcal{N}_\Delta = \begin{cases} \bar{\mathcal{N}}_\Delta & \text{if } \Delta > 0, \\ \{x \in \mathbb{R}^2 \mid x^T A_x(x - x_M) \geq 0\} & \text{if } \Delta = 0. \end{cases} \tag{5.47}$$

Define the sequence $\Delta_k \stackrel{\Delta}{=} t_{k+1} - t_k$. If $\Delta_k = \frac{\pi}{\mu}$ the generated torque under feedback (5.45) switches to 0 when it should switch to M and vice versa and the resulting performance is poor. We restrict engine speed to $\omega_c(t) > 2\mu$ for all $t > 0$, i.e. $\Delta_k < \frac{\pi}{2\mu}$ for all $k > 0$. For all Δ' such that $\partial B_{\hat{\rho}} \cap \partial\mathcal{N}_{\Delta'} = \emptyset$, $B_{\hat{\rho}} \subset \mathcal{N}_{\Delta'}$. Thus trajectories with $\omega_c(t) \leq \frac{\pi}{\Delta'}$ for some t, may "jump" $B_{\hat{\rho}}$ under $u(k) = M$. Hence, to prevent this undesirable behavior, the switching line is defined considering the intersections between $\partial\mathcal{N}_\Delta$ and $\partial B_{\hat{\rho}}$.

Lemma 5.3.1 *There exists $\Delta_\mathcal{N} > 0$ such that for all $\Delta \in [0, \Delta_\mathcal{N})$,*

$$\partial\mathcal{N}_\Delta \cap \partial B_{\hat{\rho}} = \{x_c^{(1)}, x_c^{(2)}\} \textit{ with } x_c^{(1)} \neq x_c^{(2)}, \;\; x_M(x_c^{(2)} - x_c^{(1)}) > 0. \tag{5.48}$$

Let $\Delta_\mathcal{N}^0 = \min\{\frac{\pi}{2\mu}, \Delta_\mathcal{N}\}$. Assume that the crankshaft velocity satisfy $\omega_c(t) > \frac{\pi}{\Delta}$ for all $t > 0$, for some $\Delta < \Delta_\mathcal{N}^0$, so that $\partial B_{\hat{\rho}} \cap \partial\mathcal{N}_\Delta \neq \emptyset$.

Consider point $x_c^{(2)}$ as in (5.48) and introduce the arc $\psi_c^{(2)}$ from x_D to $x_B = \psi(M, x_D, \pi/\mu)$ such that $v^T x_D = v^T x_B = 0$ and $x_c^{(2)} \in \psi_c^{(2)}$, i.e.,

$$\psi_c^{(2)} = \left\{x \in \mathbb{R}^2 | x = \psi(M, x_D, \tau) \text{ with } \tau \in [0, \pi/\mu]\right\}. \tag{5.49}$$

The switching function (5.46) is modified to include the arc $\psi_c^{(2)}$:

$$\sigma(x) = \begin{cases} (v^T b_x/\|b_x\|)^{-1} v^T x, & \text{if } ({b_{x\perp}}^T x \leq {b_{x\perp}}^T x_B) \vee \\ & \quad ({b_{x\perp}}^T x \geq {b_{x\perp}}^T x_D) \\ (b_x/\|b_x\|)^T (x - x_b), & \text{if } {b_{x\perp}}^T x_B < {b_{x\perp}}^T x < {b_{x\perp}}^T x_D \\ \text{with } x_b \text{ s.t. } ({b_{x\perp}}^T(x - x_b) = 0) \;\wedge (\; x_b \in \psi_c^{(2)}) & \end{cases} \tag{5.50}$$

Note that, since $x_c^{(2)}$ depends on Δ, $\sigma(x)$ is chosen conservatively based on the lower bound for the crankshaft speed ω_c. Let $\varphi(\hat{\gamma}_i, x_0, t)$ denote a generic trajectory in the reduced state space x obtained under control (5.45) with $x(0) = x_0$.

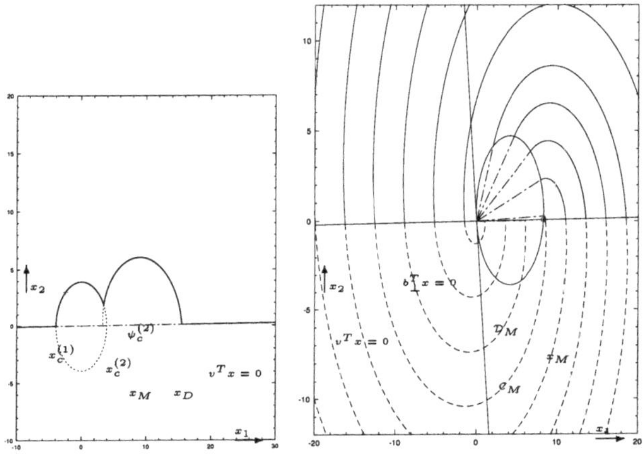

Figure 5: Optimal trajectories for the relaxed problem. If $x \notin \mathcal{D}_M$, u is equal to either M (solid line) or 0 (dashed line). Otherwise $0 < u < M$ (dash-dot line). Hybrid switching line for $\|x_M\| > \rho$.

Proposition 5.3.1 *Assume $\omega_c(t) > \omega_1$ for all $t \geq 0$ (with $\omega_1 \geq \frac{\pi}{\Delta_{\mathcal{N}}^0}$) and consider all trajectories $\varphi(\hat{\gamma}_i, x_0, t)$ under control $\hat{\gamma}_i(k)$ as in (5.45) with $\sigma(x)$ as in (5.50) with $\Delta = \frac{\pi}{\omega_1}$. There exists engine speed $\omega_{\min}$ such that, if $\omega_1 > \omega_{\min}$, for all initial conditions x_0 trajectories $\varphi(\hat{\gamma}_i, x_0, t)$ converge to $B_{\hat{\rho}}$ in finite time.*

The proof reported in [4] is constructive and gives the value $\omega_{\min}$. An optimal solution to Problem 5.1.1 is unknown, however bounds on the attainable cost can be obtained comparing the relaxed problem optimal solution cost to the cost attainable by the proposed hybrid control, thus proving the quality of the approach.

Given $x_0 \notin B_{\hat{\rho}}$, let $\psi(\hat{u}, x_0, t)$ denote the closed loop trajectory in the reduced state space under relaxed optimal torque control $\hat{u}(t)$ as in (5.44) from x_0 to $\partial B_{\hat{\rho}}$. Let $\bar{T}$ be such that $\psi(\hat{u}, x_0, \bar{T}) \in \partial B_{\hat{\rho}}$ and let $\bar{x} = \psi(\hat{u}, x_0, \bar{t})$, with $\bar{t} \in [0, \bar{T}]$, be such that $|c_x \bar{x}| = \max_{t \in [0,\bar{T}]} |c_x \psi(\hat{u}, x_0, t)|$. Let $\hat{T}$ denote the time needed to steer under control $\hat{\gamma}_i(k)$ as in (5.45) the initial state x_0 to $\partial B_{\hat{\rho}}$, i.e. $\varphi(\hat{\gamma}_i, x_0, \hat{T}) \in \partial B_{\hat{\rho}}$. Let $\hat{x} = \varphi(\hat{\gamma}_i, x_0, \hat{t})$, with $\hat{t} \in [0, \hat{T}]$, be such that $|c_x \hat{x}| = \max_{t \in [0,\hat{T}]} |c_x \varphi(\hat{\gamma}_i, x_0, t)|$.

Proposition 5.3.2 *(See [4].) Assume $\omega_c(t) > \omega_{\min} = \frac{\pi}{\Delta_{\max}}$ $\forall t \geq 0$, where $\omega_{\min}$ is given by Proposition 5.3.1. If $(c_x x_M)(c_x b_x) > 0$, there exists $\rho_w > 0$ such that, given x_0 with $\|x_0\| > \rho_w$, the following holds:*

$$0 \leq |c_x \hat{x}| - |c_x \bar{x}| \leq w_a^0(-\rho^0 v_\perp, \Delta_{\max}) \qquad \textit{if } v^T x_0 < 0$$

$$0 \leq |c_x \hat{x}| - |c_x \bar{x}| \leq w_a^M(\rho^M, v_\perp, \Delta_{\max}) \qquad \textit{if } v^T x_0 \geq 0$$

where $\rho^0 = \rho_w e^{\frac{\pi\lambda}{\mu}} - \|x_M\| \left(1 + e^{\frac{\pi\lambda}{\mu}}\right)$, $\rho^M = \rho_w e^{\frac{\pi\lambda}{\mu}}$,

$$\begin{aligned} w_a^0(x,\Delta) &= |c_x h_\perp| (\|\psi(0, x_a^0, \mu\angle(x_a^0, h_\perp))\| - \|\psi(0, x, \mu\angle(-v, h))\|) \\ w_a^M(x,\Delta) &= |c_x h_\perp| \left(\|\psi(M, x_a^M, \mu\alpha^M)\| - \|\psi(M, x, \mu\angle(-v, h))\|\right) \end{aligned}$$

where $h = \overrightarrow{vers}\left((cA_x)^T\right)$, $x_a^0 = \psi(M, x, \Delta)$, $x_a^M = \psi(0, x, \Delta)$ *and* $\alpha^M = \angle((x_a^M - x_M), -h_\perp)$.

If x_0 belongs to either $\mathcal{R}_3 = \left\{x \in \mathbb{R}^2 \,|\, v^T x < 0,\ (c_x x_M) h^T (x - x_M) < 0\right\}$ *or* $\mathcal{R}_4 = \left\{x \in \mathbb{R}^2 \,|\, v^T x > 0,\ (c_x x_M) h^T x > 0\right\}$ *the hybrid control (5.45) is optimal for Problem 5.1.1.*

5.4 Experimental results

The proposed cut–off control strategy has been implemented and tested at Magneti-Marelli Engine Control Division on a commercial car, a 16 valve 1400 cc engine car. The engine control electronics is a 4LV Magneti Marelli on board computer based on a 25MHz 32–bit Althair Motorola microprocessor with fixed point arithmetic unit. The experiment was carried out driving the car in the test ring and measuring the important parameters and variables that determine the performance of the control strategy.
The convergence analysis for the nominal values of the parameters, ensures convergence for engine speeds greater than $\omega_{\min} \simeq 396$ rpm. To prevent engine from stopping, cut-off strategies are usually not applied for ω_c less than 1000 rpm. The proposed control requires knowledge of the powertrain state. The problem of devising a robust observation scheme for the powertrain dynamics in any operating condition is out of the scope of this paper and is a topic currently under investigation. In our experiments, a discrete-time Luenberger observer has been shown to perform satisfactorily. In Figures 6 the performance achieved by the proposed cut-off strategy are compared with the performance of a currently implemented open-loop strategy. On the left, the evolutions in the $\hat{x}$ sub–space of the observer are reported along with the switching curve $\sigma(x) = 0$, as it is approximated in the implementation, and the target set $B_{\hat{\rho}}$. On the right, the resulting evolutions of the oscillating component $\tilde{a}(t)$ of the acceleration are shown. With the proposed control strategy the state is steered to $B_{\hat{\rho}}$ with no encirclement and $\tilde{a}(t)$ monotonically decreases to a_{th}. As expected from the

theoretical results, once injection is set to zero permanently $\tilde{a}(t)$ remains bounded within the perception threshold. In the open-loop strategy an encirclement of $B_{\hat{\rho}}$ produces a peak of the acceleration.

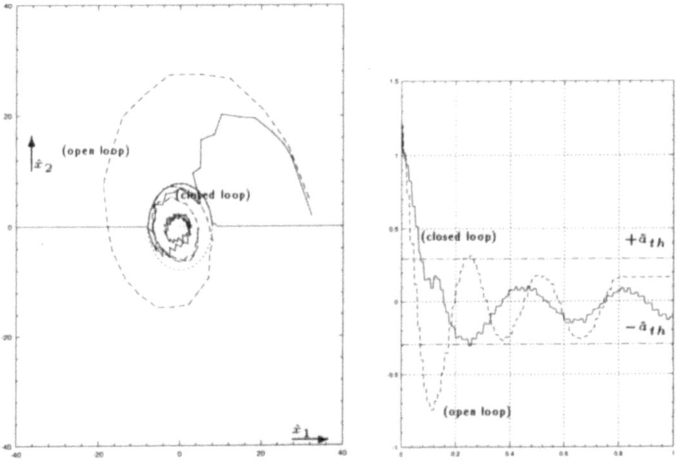

Figure 6: Observer sub-space $\hat{x}$ evolution with target set $B_{\hat{\rho}}$ and switching line $\sigma(x) = 0$ (on the left) and oscillating acceleration (in m/second^2) $\tilde{a}(t)$ signal profiles (on the right), in a controlled cut–off (solid line) and an uncontrolled cut-off (dashed line).

6 Conclusions

We have presented a top-down methodology for the design of embedded controllers for automotive applications, that encompasses all levels of design abstractions from specifications to final implementation. In this paper, we focussed on the highest levels of abstraction, i.e. at the specification level and at the control algorithm design level. At these levels we made use of an abstract model of the plant (engine and drive-line) that hides details unnecessary for the synthesis of the control algorithms. The abstract model represents the torque generation process as a finite-state machine and the drive-line as a third order system of ordinary differential equations. Hence, the control problems are cast in the hybrid system frame. Two important sub-problems have been considered corresponding to two specific regions of operations of the engine: Fast Positive and Fast Negative Force Tracking. For both problems, we derived a quasi-optimal control law by solving a relaxed problem in the continuous domain and then mapping back the solution in the hybrid space. Because of the particular aspects of our model, we have been able to bound rigorously the deterioration in performance of the control law with respect to the super-optimal strategy derived for the relaxed problem. This scheme appears to be extensible to all regions of operation of the engine thus providing a general paradigm for the entire engine control problem. We expect to conclude the design of the complete engine controller by the end of 1998. The successive step of mapping the algorithms into a computing architecture is carried out using a design methodology that is the basis of the Polis system developed at the University of California at Berkeley [1] and incorporated in the Felix VCC environment by Cadence Design Systems, Inc.

Acknowledgments: This research has been partially sponsored by PARADES, a Cadence, Magneti-Marelli and SGS-Thomson GEIE, and by CNR. We wish to acknowledge the support of the management of Magneti Marelli, and, in particular of Drs. Rossi, who pointed the cut-off problem out as a challenge, Pecchini, Mortara and Romeo. ISI provided the X-Math environment to carry out the simulation and the development of the control law. Dr. Alberto Ferrari developed two powerful simulation environments, both including X-Math. One is based on Ptolemy, a University of California simulation and design tool for heterogeneous systems, the other on BoNES, a commercial simulation tool from the Alta group of Cadence; these environments have been essential to develop, tune and implement the control strategies. Dr. Luca Benvenuti offered insightful comments on the results of the paper.

References

[1] F. Balarin, M. Chiodo, P. Giusto, H. Hsieh, A. Jurecska, L. Lavagno, C. Passerone, A. Sangiovanni-Vincentelli, E. Sentovich, K. Suzuki, and B. Tabbara, *Hardware-software co-design of embedded systems, the polis approach*, VLSI, Computer Architecture and Digital Signal Processing, Kluwer Academic Publishers, Boston, Dordrecht, London, 1997.

[2] A. Balluchi, M. Di Benedetto, C. Pinello, C. Rossi, and A. Sangiovanni-Vincentelli, *Cut-off in engine control: a hybrid system approach*, 36th CDC (San Diego, CA), 1997, pp. 4720–4725.

[3] A. Balluchi, M. Di Benedetto, C. Pinello, C. Rossi, and A. Sangiovanni-Vincentelli, *Hybrid control for automotive engine management: The cut-off case.*, Hybrid Systems: Computation and Control (T. A. Henzinger and S. Sastry, eds.), Lecture Notes in Computer Science, vol. 1386, Springer–Verlag, London, U.K., 1998, pp. 13–32.

[4] A. Balluchi, M. Di Benedetto, C. Pinello, C. Rossi, and A. Sangiovanni-Vincentelli, *Hybrid control in automotive applications: the cut-off control.*, Tech. Report No. M98/34, UCB ERL, June 1998.

[5] K. Butts, I. Kolmanovsky, N. Sivashankar, and J. Sun, *Hybrid systems in automotive control applications*, Control using logic–based switching (A. S. Morse, ed.), Lecture notes in control and information sciences, vol. 222, Springer–Verlag, London, U.K., 1997, pp. 173–189.

[6] Y.H. Chen, G. Leitmann, and Z.K. Xiong, *Robust control design for interconnected systems with time–varying uncertainties*, Inter. J. Contr. **54**,(1991), 1119–1142.

[7] X. Ge, W. Kohn, A. Nerode, and J. B. Remmel, *Hybrid systems: Chattering approximation to relaxed controls*, Hybrid Systems III (R. Alur, T. Henzinger, and E. D. Sontag, eds.), Lecture Notes in Computer Science, vol. 1066, Springer–Verlag, London, U.K., 1996, pp. 76–100.

[8] C. Horn and P. J. Ramadge, *Robustness issues for hybrid systems*, 34th CDC (New Orleans, LA), 1995, pp. 1467–1472.

[9] L. Hou, A. Michel, and H. Ye, *Stability analysis of switched systems*, 35th CDC (Kobe, Japan), 1996, pp. 1208–1212.

[10] W. Kohn, A. Nerode, and J. B. Remmel, *Hybrid systems as Finsler manifolds: Finite state control as approximation to connections*, Hybrid Systems II (P. Antsaklis, W. Kohn, A. Nerode, and S. Sastry, eds.), Lecture Notes in Computer Science, vol. 999, Springer–Verlag, London, U.K., 1995, pp. 294–321.

[11] E. Lee and A. Sangiovanni-Vincentelli, *Comparing models of computation*, Proc. International Conference on CAD (Santa Clara, CA), 1996, pp. 234–241.

[12] A. S. Morse (ed.), *Control using logic–based switching*, Lecture notes in control and information sciences, vol. 222, Springer–Verlag, London, U.K., 1997.

[13] A. Nerode and W. Kohn, *Models for hybrid systems: Automata, topologies, controllability, observability*, Hybrid Systems (R. L. Grossman, A. Nerode, A. P. Ravn, and Hans Rishel, eds.), Lecture Notes in Computer Science, vol. 736, Springer–Verlag, London, U.K., 1993, pp. 317–356.

[14] T. Niinomi, B. H. Krogh, and J. E. R. Cury, *Synthesis of supervisory controllers for hybrid systems based on approximating automata*, 34th CDC (New Orleans, LA), 1995, pp. 1461–1466.

[15] S. Pettersson and B. Lennartson, *Stability and robustness for hybrid systems*, 35th CDC (Kobe, Japan), 1996, pp. 1202–1207.

[16] H. H. Rosenbrock, *Computer-aided control system design*, Academic Press, London, U.K., 1974.

[17] V.I. Utkin, *Variable structure systems with sliding modes: a survey*, IEEE Trans. on Aut. Contr. **22** (1977), 212–222.

[18] L. Y. Wang, A. Beydoun, J. Cook, J. Sun, and I. Kolmanovsky, *Optimal hybrid control with applications to automotive powertrain systems*, Control using logic–based switching (A. S. Morse, ed.), Lecture notes in control and information sciences, vol. 222, Springer–Verlag, London, U.K., 1997, pp. 190–200.

[19] Hui Ye, A. N. Michel, and L. Hou, *Stability theory for hybrid dynamical systems*, 34th CDC (New Orleans, LA), 1995, pp. 2679–2684.

[20] L. F. Yeung and G. F. Bryant, *New dominance concepts for multivariable control systems design*, Inter. J. Contr. **55** (1992), no. 4, 969–988.

PARADES G.E.I.E., Via San Pantaleo, 66, 00186 Roma, Italy

Dipartimento di Ingegneria Elettrica, Università dell'Aquila, Poggio di Roio, 67040 L'Aquila, Italy

PARADES G.E.I.E., Via San Pantaleo, 66, 00186 Roma, Italy

Department of Electrical Engineering and Computer Science, University of California at Berkeley, CA 94720, USA; PARADES G.E.I.E., Via San Pantaleo, 66, 00186 Roma, Italy

Progress in Systems and Control Theory, Vol. 25

Control Synthesis for Discrete Event Systems

J.G. Thistle[1]

1 Introduction

Among the control challenges posed by modern digital technology are problems of subsystem coordination in large-scale, complex systems. These are common in communications, transportation, manufacturing and other fields. In many instances, the essence of such coordination problems can be captured by "lumping" the trajectories of low-level subsystems into sequences of discrete, instantaneous events, thus abstracting away from any continuous dynamics. The resulting *discrete event systems* (DES) may be modelled by means of automata, formal languages, formal logic, etc.

Supervisory control theory [1] aims to provide a comprehensive approach to the development of control logic for large-scale DES. Framed in terms of formal languages (i.e., sets of event sequences), it explains the decomposition of complex control and coordination problems through modular synthesis and decentralized control architecture. Notions of *controllable* and *observable* languages determine the existence of solutions to control problems, while properties of *hierarchical consistency* and *nonconflicting languages* determine the efficacy of hierarchical and modular decomposition.

Supervisory control's central theme of problem decomposition has been treated thoroughly in other surveys [2, 3]. The present article will therefore focus on some of the computational aspects of the "monolithic" subproblems that such decomposition yields. In so doing, it will emphasize links with related work on formal synthesis in computer science, and with computer science approaches to hybrid control system synthesis.

In particular, the paper will consider algorithms for the synthesis of supervisory controls under complete observations, in the case where all pertinent formal languages are represented as finite automata (finite state machines equipped with *acceptance conditions*). Supervisory control synthesis can then be viewed as the problem of controlling an automaton to the satisfaction of a specification based on its acceptance condition, and suitable control laws may be computed by calculating fixpoints of monotone operators on state subsets. A quantifier-style fixpoint notation called the μ-calculus provides an elegant means of representing such fixpoints. Because this fixpoint calculus is a well-established tool for the formal specification and verification of computer systems, considerable research effort has been

[1]The author's research was supported in part by Research Grant number 155594 of the Natural Sciences and Engineering Research Council of Canada

directed toward the efficient evaluation of μ-calculus formulas over large finite state machines.

The next section of the paper gives an overview of discrete event control synthesis in the case of complete observations, introducing the μ-calculus and illustrating its use in control. The treatment is deliberately informal and sketchy: details may be found in the many references. Section 3 discusses related work on formal synthesis and verification of computer systems, and applications to control synthesis for hybrid systems with simple continuous dynamics (namely, timed discrete-event systems).

2 Monolithic Control Synthesis

2.1 Supervisory Control Theory

Supervisory control was framed by Ramadge and Wonham [1] in a setting of *formal languages*. A formal language is a set of strings of symbols, which in the context of supervisory control represents a set of sequences of discrete events. Standard formal languages contain only finite strings, but there also exists a well-developed theory of infinite-string languages, or ω*-languages*.

The abstract nature of the formal language framework facilitates the separation of representational and computational questions from more fundamentally control-theoretic issues. Indeed, the main results of supervisory control identify basic notions of *controllability* and *observability* – which underpin the existence of solutions to "monolithic" control synthesis problems – and properties of *nonconflicting* languages and *hierarchical consistency*, which underlie effective modular decomposition of complex control synthesis. These results are independent of the choice of concrete language representation used to perform computations. See [2, 3] for a survey.

Monolithic (or nonmodular) control synthesis amounts, in the case of complete observations, to the computation of the *supremal controllable sublanguage* [4] of a given specification language – or the supremal ω-*controllable sublanguage*, if infinite event strings are taken into account [5, 6]. This and other calculations have been studied using various representations of the formal languages in question, many having special structure that may simplify the computation – some examples are so-called product systems, hierarchical state machines, vector DES and exchange systems (see [3]). However, a simple representation that highlights universal issues is that of finite (but otherwise unstructured) automata, or state machines.

2.2 The (ω)-Regular Case

Indeed, if all of the pertinent languages can be represented in terms of finite automata (i.e., if they are all *regular* languages, or ω-*regular* languages in the infinite-string case), then the computation of the supremal

(ω)-controllable sublanguage can itself be viewed as a control problem: given a plant represented as a finite automaton equipped with a control mechanism, compute the set of initial states from which the automaton can be controlled to satisfy a certain specification. In the case where only finite event strings are considered (regular languages) [1], the specification typically requires that the controlled automaton always be able potentially to reach a designated subset of so-called "marked" states. Thus it might be required that the automaton of figure 1 be controlled in such a way that the state subset $\{-1, 1\}$ can always potentially be reached. The infinite-string case (ω-regular languages) entails more complicated specifications involving conditions on the state subset that the controlled automaton enters infinitely often. It turns out to be sufficient to consider boolean combinations of conditions of the following form: state subset R is visited infinitely often; and state subset I is visited almost always (or co-finitely often). Thus, in the case of figure 1, it might be required that the automaton infinitely often visit $\{-4, 4\}$, and that the it remain almost always in state subset $\{-4, -3, -2, -1, 1, 2, 3, 4\}$ – this would hold if it were simply to cycle forever through states 1,2,3 and 4 or through states -1,-2,-3 and -4.

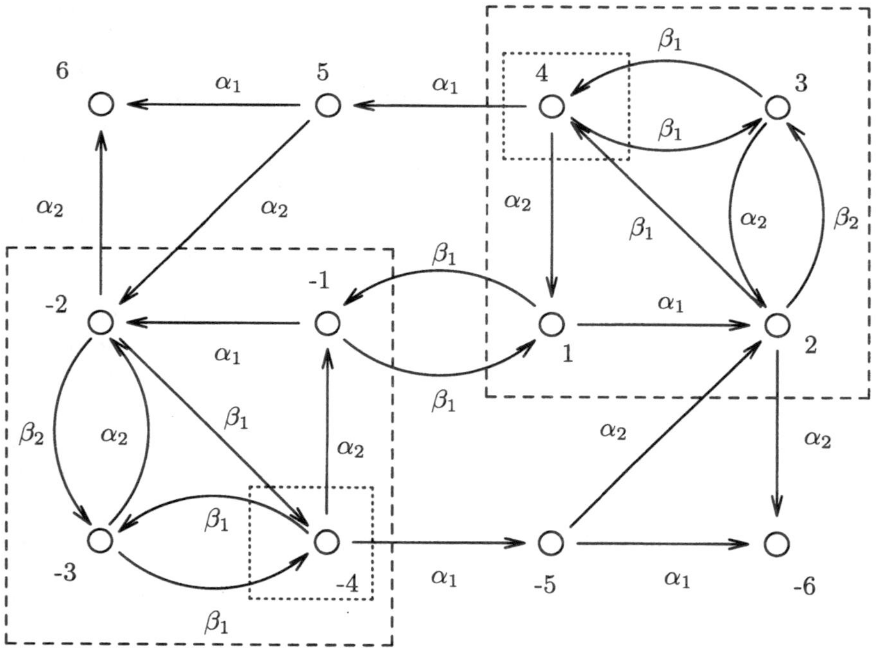

Figure 1: Plant automaton.

The details of the control mechanism are not critical to the discussion that follows. Adopt for the sake of simplicity the basic model that has the event symbols partitioned into *controllable* and *uncontrollable* symbols: state transitions labelled with controllable symbols can be *disabled*, or prevented, by the controller, whereas those labelled with uncontrollable symbols cannot. In figure 1, for example, it might be that transitions labelled with α's can be disabled while those labelled with β's cannot.

2.3 Inverse Dynamics and Fixpoint Calculus

The state subset from which the automaton can be controlled to satisfy its specification may be represented as a fixpoint of a monotone operator on automaton state subsets; such a characterization leads immediately to an algorithm for the computation of the desired subset. Rather than explaining the computations in full detail we shall consider some simple examples.

Let the state set of the automaton under control be X. Define the *inverse dynamics operator* $\theta^{\mathcal{A}} : \mathcal{P}(X) \longrightarrow \mathcal{P}(X)$ as that which takes any state subset $X' \subseteq X$ to the set of states from which the automaton can be controlled to enter X' in a single transition.

The fixpoint characterization is greatly simplified by the adoption of a quantifier-style notation. Let X_1 be a variable ranging over state subsets and let $\phi(X_1)$ be an expression that denotes a state subset. Suppose that $\phi(X_1)$ is monotone in X_1, meaning that the larger the subset X_1 the larger the subset $\phi(X_1)$. Then

$$\mu X_1.\ \phi(X_1)$$

denotes the *least fixpoint* of $\phi(\cdot)$ – that is, the smallest state subset $X' \subseteq X$ such that $X' = \phi(X')$. Similarly,

$$\nu X_1.\ \phi(X_1)$$

denotes the *greatest fixpoint* of $\phi(\cdot)$. (The existence of such fixpoints is guaranteed by monotonicity.)

This so-called "μ-calculus" notation is widely used for formal specification and verification of computer systems [7]. When coupled with the inverse dynamics operator, the fixpoint calculus allows the succinct denotation of a variety of state subsets that are significant for the control of the automaton.

For example, the "control-reachability subset," or the set of all states from which the automaton can be controlled eventually to enter a given state subset $R \subseteq X$, is denoted by

$$\mu X_1.\ \theta^{\mathcal{A}}(R \cup X_1)$$

Indeed, this least fixpoint is computed as the limit of the nondecreasing sequence

$$R_0 = \emptyset$$
$$R_{i+1} = \theta^{\mathcal{A}}(R \cup R_i), \quad i \geq 0$$

By induction, R_i is the set of states from which the automaton can be controlled to enter R in at most i transitions; thus the limit is indeed the set of states from which it can be controlled eventually to enter R.

The "supremal control-invariant subset" of a given state subset I is given by

$$\nu X_1.\ [\theta^{\mathcal{A}}(X_1) \cap I]$$

Indeed, by elementary fixpoint theory this greatest fixpoint is the largest state subset $X_1 \subseteq X$ such that

$$X_1 \subseteq \theta^{\mathcal{A}}(X_1) \cap I$$

and this condition holds if and only if X_1 is control-invariant and contained in I.

2.4 The Finite-String Problem

The above examples do not fully illustrate the usefulness of the fixpoint quantifier notation, which comes into its own when quantifiers are nested. Consider therefore the basic finite-string synthesis problem [4], which essentially entails computing the set of states from which an automaton can be controlled so that a given state subset $M \subseteq X$ (the set of *marked* states) can always potentially be reached. For this, let $\mathcal{A}(\not\rightarrow X')$ denote the automaton obtained from an automaton $\mathcal{A}$ by altering the control mechanism in such a way that state transitions leading to the state subset $X' \subseteq X$ can always be disabled, irrespective of other control actions. Then the least fixpoint

$$\mu X_1.\ \theta^{\mathcal{A}(\not\rightarrow X')}(X' \cap (M \cup X_1))$$

represents the set of states from which the automaton $\mathcal{A}(\not\rightarrow X')$ can be forced eventually to enter $X' \cap M$ by way of a path that is contained within the subset X' (cf. the example of the control-reachability subset). Thus, if

$$X' \subseteq \mu X_1.\ \theta^{\mathcal{A}(\not\rightarrow X')}(X' \cap (M \cup X_1))$$

then whenever $\mathcal{A}(\not\rightarrow X')$ is in X', it can be controlled to remain within X' and eventually to reach M. But this means that whenever the automaton $\mathcal{A}$ is in state subset X', it too can be controlled to remain within X' (because any transitions that can be disabled in $\mathcal{A}(\not\rightarrow X')$ but not in $\mathcal{A}$ simply lead to X'), and it can be so controlled in such a way that it can always (potentially) reach $X' \cap M$. It follows that the greatest fixpoint

$$\nu X_0.\ \mu X_1.\ \theta^{\mathcal{A}(\not\rightarrow X_0)}(X_0 \cap (M \cup X_1))$$

– being (by fixpoint theory) the largest state subset that satisfies the above inclusion – is the set of all states from which $\mathcal{A}$ can be controlled so that it can always potentially reach the state subset M [8].

2.5 The Infinite-String Case: Büchi and Rabin Conditions

Consider now a specification that requires that a state subset $R \subseteq X$ be visited infinitely often – termed a *Büchi* acceptance condition in the theory of automata on infinite strings [9]. The set of states from which the automaton can be controlled to satisfy this condition is given by

$$\nu X_1.\ \mu X_2.\ \theta^{\mathcal{A}}((R \cap X_1) \cup X_2)$$

Note that $\mu X_2.\ \theta^{\mathcal{A}}((R \cap X_1) \cup X_2)$ is just the reachability subset of $R \cap X_1$. Thus the greatest fixpoint denoted by the above expression is precisely the largest set $X_1 \subseteq X$ from which the automaton can be controlled to enter $R \cap X_1$. By extension, this X_1 is therefore the largest subset from which the automaton can be controlled to enter R infinitely often.

The more general *Rabin* and *Streett* acceptance conditions constitute boolean combinations of Büchi conditions. As a simple example consider the logical complement of a Büchi condition: if $I = X \setminus R$ then the negation of the above Büchi acceptance criterion is that the automaton visit $I \subseteq X$ "almost always." The set of states from which this can be ensured through control is denoted by

$$\mu X_1.\ \nu X_2.\ [\theta^{\mathcal{A}}(X_1) \cup [\theta^{\mathcal{A}}(X_2) \cap I]]$$

The state set X being finite, this least fixpoint can be computed as the limit of the sequence

$$\begin{aligned} I_0 &= \emptyset \\ I_{i+1} &= \nu X_2.\ [\theta^{\mathcal{A}}(I_i) \cup [\theta^{\mathcal{A}}(X_2) \cap I]], \quad i \geq 0 \end{aligned}$$

By induction, I_i is the state subset from which the automaton can be controlled to visit $X \setminus I (= R)$ at most $i-1$ times; indeed, I_1 is precisely the supremal control-invariant subset of I. Thus the least fixpoint is precisely the set of states from which the automaton can be controlled to visit $X \setminus I$ only finitely often.

If it is required that the automaton visit a subset $R \subseteq X$ infinitely often *and* some other subset $I \subseteq X$ almost always, then the set of initial states from which it can be suitably controlled is denoted by the following fixpoint expression, which generalizes the previous two examples:

$$\mu X_1.\ \nu X_2.\ \mu X_3.\ [\theta^{\mathcal{A}}(X_1) \cup [\theta^{\mathcal{A}}(X_1 \cup (X_2 \cap R) \cup X_3) \cap I]]$$

Indeed, a so-called Rabin acceptance condition is a disjunction of conditions such as the above. Results of formal language and automata theory show that we can restrict attention without loss of generality to Rabin acceptance conditions, in the sense that any infinite-string formal language

(ω-language) that is accepted by a finite automaton is accepted by a deterministic finite automaton equipped with a Rabin acceptance condition. A detailed description of the fixpoint corresponding to a full Rabin condition is, however, beyond the scope of this paper. That given in [10] involves inductions on the number of disjuncts in the acceptance condition, and on the structure of the automaton. It combines the use of the inverse dynamics operator and the fixpoint calculus with operators that alter the structure of the automaton in question, somewhat in the manner of the fixpoint characterization of the previous section.

3 Bibliography

3.1 Formal Synthesis, Games and Tree Automata

The fixpoints of the above formulas can all be computed by means of a straightforward iteration of the appropriate state subset operators. As originally formulated by Ramadge and Wonham, the finite-string computation of the supremal controllable sublanguage was in fact defined as the calculation of the greatest fixpoint of a monotone operator on formal languages [4], but the present automata-theoretic algorithm illuminates more clearly the ties between control synthesis and related, well-known problems in the synthesis and verification of computer systems, particularly in the infinite-string case.

Indeed, computation of the state subset from which a finite automaton can be controlled to the satisfaction of a Rabin acceptance condition is formally equivalent to a number of problems in the literature of computer science and engineering. *Church's problem* [11] entails the synthesis of a finite automaton whose input-output relation that satisfies a specification represented by a Rabin automaton on infinite strings [11]. Formulated in the late fifties as a problem in circuit synthesis, it has more recently been newly proposed as a formal paradigm for the synthesis of reactive computer systems [12]. Church's problem was first solved via a reformulation as the problem of computing winning positions in *ω-regular infinite games* [13, 14], and later by a reduction to the equivalent *emptiness problem for automata on infinite trees*. The latter problem is at the root of many decision procedures for modal logics used in the specification and verification of computer systems [15, 16, 9]. These problems are NP-complete; the algorithms are polynomial in the number of automaton states, but exponential in the number of disjuncts in the Rabin acceptance condition.

The control-theoretic version was solved in [10], by means of a synthesis of the methods of [15] and [16]. See [17, 18] for extensions, and [8] for the application of the μ-calculus to the finite-string problem. The special case of a Büchi acceptance condition was treated in [19], though without the use of the μ-calculus; similar results on *stabilization* of discrete-event systems were reported in [20]. The infinite-string formulation of supervisory control was initiated by Ramadge [5].

3.2 Fixpoint Computation

Because of the μ-calculus' applications to formal specification and verification, considerable effort has been devoted to the efficient evaluation of fixpoints over large state machines. Simple expressions with no nesting of fixpoint quantifiers can be computed in time linear in the number of states [21] by the method of the examples of the previous section: to compute the least (resp., greatest) fixpoint with respect to a state-subset variable, the variable is set to the empty set (resp., the complete state set), the value of the quantified expression is computed, and the variable is reset to the resulting value; this process is repeated, generating a nondecreasing (resp., nonincreasing) sequence of state subsets that converges finitely to the desired fixpoint.

By nesting such iterations, μ-calculus formulas with nested quantifiers may be evaluated. This yields an algorithm that runs in about n^d steps, where n is the number of states in the state machine and d is the nesting depth of fixpoint quantifiers. It has been shown that by carefully exploiting monotonicity, this can essentially be reduced to about $n^{d/2}$ steps [22].

In practice, n may be very large because of combinatorial explosion in the construction of models or specifications. For some applications, a particular data structure has proven useful in reducing computation time: *binary decision diagrams* (BDD) represent boolean functions as directed acyclic graphs [23]. The BDD of figure 2 encodes the formula $(x_1 \vee x_2) \wedge (x_3 \vee x_4)$. The truth value of the formula is evaluated by tracing a path from the root to a leaf according to the truth values of the respective variables.

Experience suggests that appropriate ordering of the variables in a BDD (according to heuristic rules) can allow rapid evaluation of complex boolean functions. The various components of finite automata (transition functions, marked subsets, etc.) can be coded as boolean functions and subsequently represented by BDDs; this approach is claimed to allow practical evaluation of μ-calculus formulas over machines with 10^{20} states [24, 25].

3.3 Timed DES and Other Hybrid Systems

The methods of control synthesis outlined in the previous section have also been applied to real-time discrete event systems. A *timed automaton* is one that is equipped with a number of real-valued clocks [26]. State transitions may be disabled by conditions on clock values, and when state transitions occur, they may reset certain clocks. Such models thus couple the discrete-event dynamics of an automaton to the continuous evolution of their real-valued clocks. Alur and Dill [26] identify a class of disabling conditions that admit a finite-state congruence for the continuous dynamics, and thus

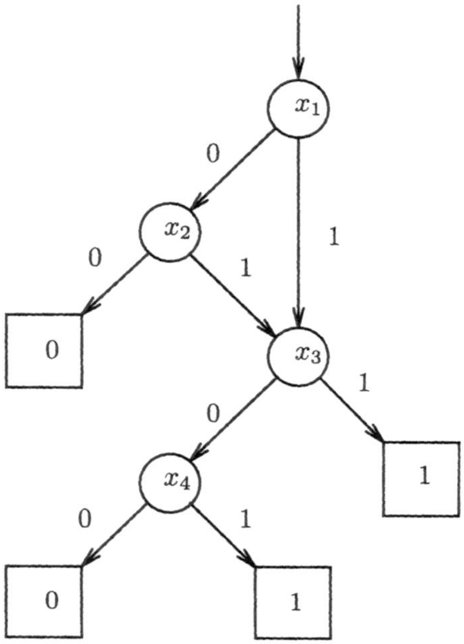

Figure 2: Binary decision diagram for evaluation of $(x_1 \vee x_2) \wedge (x_3 \vee x_4)$.

allow for these simple hybrid systems to be modelled by finite-state quotient systems. Indeed, this approach essentially summarizes clock evolution into a series of discrete events that correspond to passage from one coset of the congruence to another. In [27] this method is used to convert supervisory control problems in real-valued time to equivalent problems with discrete event timing [28], and then solve them using methods of [4, 29, 10].

A more direct application to real-time control synthesis is that of [30, 31], where the continuous evolution of time is accommodated by appropriately generalizing the inverse dynamics operator. In this case the fixpoint computations involve sequences of regions in Euclidean space. It has been proposed that the complexity of these regions might be limited by restricting them to polyhedra through appropriate structural assumptions [30] and by approximating them by their convex hulls [32]. Indeed, there is an extensive literature on the use of approximations in fixpoint computations for real-time systems [33]. Such techniques have also been applied to control synthesis problems for more general classes of hybrid systems, through the approximation of hybrid models with discrete event systems [34].

4 Conclusion

The μ-calculus fixpoint notation provides an elegant conceptual and computational paradigm for control synthesis – especially for pure discrete event systems – and links control synthesis with a large body of research on the verification and synthesis of computer systems.

References

[1] P. J. Ramadge and W. M. Wonham. Supervisory control of a class of discrete event processes. *SIAM J. Control and Optimization*, 25(1):206–230, 1987.

[2] P. J. Ramadge and W. M. Wonham. The control of discrete event systems. *Proceedings of the IEEE*, 77(1):81–98, January 1989.

[3] John G. Thistle. Supervisory control of discrete-event systems. *Mathematical and Computer Modelling*, 23:25–53, 1996.

[4] W. M. Wonham and P. J. Ramadge. On the supremal controllable sublanguage of a given language. *SIAM J. Control and Optimization*, 25(3):637–659, 1987.

[5] Peter J. G. Ramadge. Some tractable supervisory control problems for discrete-event systems modeled by Büchi automata. *IEEE Trans. Automatic Control*, 34(1):10–19, January 1989.

[6] J. G. Thistle and W. M. Wonham. Supervision of infinite behaviour of discrete-event systems. *SIAM J. Control and Optimization*, 32(4):1098–1113, July 1994.

[7] Dexter Kozen. Results on the propositional μ-calculus. *Theoretical Computer Science*, 27:333–354, 1983.

[8] J. G. Thistle and R. P. Malhamé. Multiple marked languages and noninterference in modular supervisory control. In *35th Annual Allerton Conference*, pages 523–532, 1997.

[9] Wolfgang Thomas. Automata on infinite objects. In Jan van Leeuwen, editor, *Handbook of Theoretical Computer Science, vol. B: Formal Models and Semantics*, pages 134–191. Elsevier, The MIT Press, 1990.

[10] J. G. Thistle and W. M. Wonham. Control of infinite behaviour of finite automata. *SIAM J. Control and Optimization*, 32(4):1075–1097, July 1994.

[11] Alonzo Church. Logic, arithmetic and automata. In *Proceedings of the International Congress of Mathematicians, 15-22 August, 1962*, pages 23–35, Djursholm, Sweden, 1963. Institut Mittag-Leffler.

[12] Amir Pnueli and Roni Rosner. On the synthesis of an asynchronous reactive module. In *Automata, Languages and Programming,* 16th International Colloquium, Stresa, Italy, July 1989, Proceedings (Lecture Notes in Computer Science no. 372), pages 652–671. Association for Computing Machinery, Springer-Verlag, January 1989.

[13] J. Richard Büchi and Lawrence H. Landweber. Solving sequential conditions by finite-state strategies. *Transactions of the American Mathematical Society*, 138:295–311, 1969.

[14] David L. Dill. *Trace Theory for Automatic Hierarchical Verification of Speed-Independent Circuits.* ACM Distinguished Dissertations. MIT Press, 1989.

[15] Michael O. Rabin. *Automata on Infinite Objects and Church's Problem.* Conference Board of the Mathematical Sciences Regional Conference Series in Mathematics No. 13. American Mathematical Society, Providence, Rhode Island, 1972. Lectures from the CBMS Regional Conference held at Morehouse College, Atlanta, Georgia, September 8 – 12, 1969.

[16] E. Allen Emerson and Charanjit S. Jutla. The complexity of tree automata and logics of programs (extended abstract). In *29th Annual Symposium on Foundations of Computer Science*, pages 328 – 337, 1988.

[17] J. G. Thistle. On control of systems modelled as deterministic Rabin automata. *Discrete Event Dynamic Systems: Theory and Applications*, 5(4):357–381, September 1995.

[18] J. G. Thistle and R. P. Malhamé. Control of ω-automata under state fairness assumptions. *Systems and Control Letters*, 33:265–274, 1998.

[19] S. Young, D. Spanjol, and V. K. Garg. Control of discrete event systems modeled with deterministic Büchi automata. In *Proceedings of 1992 American Control Conference*, pages 2809–2813, 1992.

[20] Cüynet M. Özveren and Alan S. Willsky. Output stabilizability of discrete-event dynamic systems. *IEEE Trans. Automatic Control*, 36(8):925–935, August 1991.

[21] André Arnold and Paul Crubille. A linear algorithm to solve fixed-point equations on transition systems. *Information Processing Letters*, 29:57–66, September 1988.

[22] David E. Long, Anca Browne, Edmund M. Clarke, Somesh Jha, and Wilfredo R. Marrero. An improved algorithm for the evaluation of fixpoint expressions. In David L. Dill, editor, *Computer aided verification : 6th international conference, CAV '94, Stanford, California, USA, June 21-23, 1994: proceedings, Lecture notes in computer science, No. 818*, pages 338–350. Springer-Verlag, 1994.

[23] R.E. Bryant. Graph-based algorithms for boolean function manipulation. *IEEE Transactions on Computers*, C-35:677–691, 1986.

[24] E. Clarke, O. Grumberg, and D. Long. Verification tools for finite-state concurrent systems. In *A Decade of Concurrency: Reflections and Perspectives. REX School/Symposium, Noordwijkerhout, The Netherlands, 1-4 June 1994*, pages 124–175, Berlin, 1994. Springer-Verlag.

[25] S. Balemi, G. J. Hoffmann, P. Gyugyi, H. Wong-Toi, and G. F. Franklin. Supervisory control of a rapid thermal multiprocessor. *IEEE Trans. on Automatic Control*, 38(7):1040–1059, July 1993.

[26] Rajeev Alur and David Dill. Automata for modeling real-time systems. In *Proc. Int'l Colloquium on Automata, Languages and Programming*, pages 322–335, 1990.

[27] Howard Wong-Toi and Gérard Hoffmann. The control of dense real-time discrete-event systems. Unpublished manuscript, 1992.

[28] B. A. Brandin and W. M. Wonham. Supervisory control of timed discrete-event systems. *IEEE Transactions on Automatic Control*, 39(2):329–342, February 1994.

[29] John Graham Thistle. *Control of Infinite Behaviour of Discrete-Event Systems.* PhD thesis, University of Toronto, Toronto, Canada, January 1991. Available as Systems Control Group Report No. 9012, Systems Control Group, Dept. of Electl. Engrg., Univ. of Toronto, January 1991.

[30] O. Maler, A. Pnueli, and J. Sifakis. On the synthesis of discrete controllers for timed systems (an extended abstract). In Ernst W. Mayr, editor, *STACS 95, 12th Annual Symposium on Theoretical Aspects of Computer Science, Lecture Notes in Computer Science, vol. 900*, pages 229–242. Springer-Verlag, 1995.

[31] E. Asarin, O. Maler, and A. Pnueli. Symbolic controller synthesis for discrete and timed systems. In P. Antsaklis, W. Kohn, A. Nerode, and S. Sastry, editors, *Hybrid Systems II* (Lecture Notes in Computer Science no. 999), pages 1–20, New York, 1995. Springer-Verlag.

[32] Nicolas Halbwachs, Pascal Raymond, and Yann-Eric Proy. Verification of linear hybrid systems by means of convex approximations. In *International Symposium on Static Analysis, SAS'94, Lecture Notes in Computer Science 864*, pages 223–237. Springer-Verlag, 1994.

[33] David L. Dill and Howard Wong-Toi. Verification of real-time systems by successive over and under approximation. In Pierre Wolper, editor, *Computer Aided Verification, 7th International Conference, Liege, Belgium, July, 3-5, 1995, Proceedings. Lecture Notes in Computer Science, Vol. 939*, pages 409–422. Springer-Verlag, 1995.

[34] Tohsihiko Niinomi, Bruce Krogh, and José E.R. Cury. Synthesis of supervisory controllers for hybrid systems based on approximating automata. In *Proceedings of the 34th Conference on Decision and Control*, pages 1461–1466. IEEE, 1995.

Department of Electrical and Computer Engineering, École Polytechnique de Montréal, C.P. 6079, succ. Centre-ville, Montreal, Quebec, Canada H3C 3A7